The graph of $y = c + a \sin b(x - d)$ or $y = c + a \cos b(x - d)$, where $b > 0$, has amplitude $|a|$, period $2\pi/b$, a vertical translation c units up if $c > 0$ or $|c|$ units down if $c < 0$, and a phase shift d units to the right if $d > 0$ or $|d|$ units to the left if $d < 0$. The graph of $y = a \tan bx$ or $y = a \cot bx$ has period π/b, where $b > 0$.

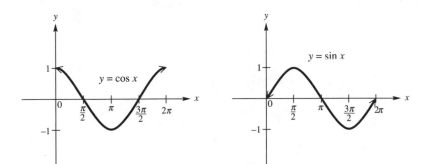

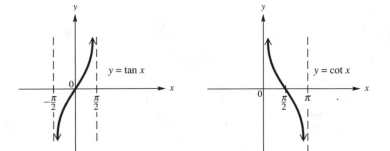

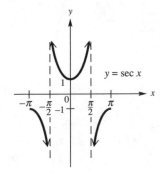

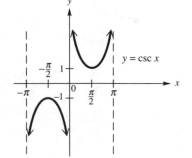

Trigonometry
SIXTH EDITION ▼▼▼

Sixth Edition

Trigonometry

MARGARET L. LIAL ▼▼▼
American River College

E. JOHN HORNSBY, JR. ▼▼▼
University of New Orleans

DAVID I. SCHNEIDER ▼▼▼
University of Maryland

▲ **ADDISON-WESLEY**

An imprint of Addison Wesley Longman, Inc.

Reading, Massachusetts • Menlo Park, California • New York • Harlow, England
Don Mills, Ontario • Sydney • Mexico City • Madrid • Amsterdam

Sponsoring Editor: Anne Kelly
Developmental Editor: Lynn Mooney
Project Editor: Lisa A. De Mol
Design Administrator: Jess Schaal
Text Design: Lesiak/Crampton Design Inc.: Lucy Lesiak
Cover Design: Lesiak/Crampton Design Inc.: Lucy Lesiak
Cover Illustration: Precision Graphics
Production Administrator: Randee Wire
Compositor: Interactive Composition Corporation
Printer and Binder: R.R. Donnelley & Sons
Cover Printer: Phoenix Color Corporation

Trigonometry, Sixth Edition
Copyright © 1997 by Addison-Wesley Educational Publishers Inc.

Library of Congress Cataloging-in-Publication Data
Lial, Margaret L.
    Trigonometry / Margaret L. Lial, E. John Hornsby, Jr., David I. Schneider.—6th ed.
       p.   cm.
    Includes index.
    ISBN 0–673–99553–4—0–673–97659–9 (annotated instructor's edition)
    1. Trigonometry.   I. Hornsby, E. John.   II. Schneider, David I.,
1942–   .   III. Title.
QA531.L5   1996
516.24—dc20   96–656
                                                        CIP

Reprinted with corrections, May 1998.

5 6 7 8 9 10 11 12-DOC-02 01 00 99 98

Contents

Preface

Trigonometry, **Sixth Edition,** is written for students in a traditional college trigonometry course. We assume students have had at least one course in algebra. Geometry is a desirable prerequisite, but many students reach trigonometry with little or no background in geometry. Because of this, we explain the necessary ideas from geometry as needed. Although this book is intended for a traditional course, we have acknowledged the growing interest in using graphing calculators to augment and deepen the concepts typically presented in trigonometry.

Changes in Content ▼▼▼

We have given less emphasis to showing specific calculator keystrokes, to allow for more variation in calculators. We assume that all students will be using at least a scientific calculator, and that many will use graphing calculators.

New Features ▼▼▼

Several new features have been incorporated in this edition. The design has been developed to enhance the pedagogical features and increase their accessibility.

- Each chapter opens with a genuine application of the material to be presented. Corresponding examples and exercises, identified with a special icon, are located throughout the chapter.
- We have made an effort to point out the many connections between mathematical topics in this course and those studied earlier, as well as connections between mathematics and the "real world." Optional Connections boxes presenting such topics are included in many sections throughout the book. Most of them include thought-provoking questions for writing or class discussion. A few topics, such as the sum and product identities, are now included in a Connections box, rather than in a complete section. In addition, we have included a feature in many exercise sets called Discovering Connections. These groups of exercises tie together different topics and highlight the relationships among various concepts and skills.

3

Radian Measure and the Circular Functions

3.1 Radian Measure

3.2 Applications of Radian Measure

3.3 Circular Functions of Real Numbers

3.4 Linear and Angular Velocity

Finding sources of energy has been an important concern since the beginning of civilization. During the past 100 years, people have relied on fossil fuels for a large portion of their energy requirements. Fossils fuels are finite and limited. The heavy use of fossil fuels has caused irreversible damage to our environment and may be accelerating a greenhouse effect. Nuclear energy as an alternative has a potential for providing almost unlimited amounts of energy. Unfortunately, it creates health risks and dangerous nuclear wastes. Currently there is no completely safe disposal method for nuclear wastes. As a result, no new nuclear power plants have been ordered in the United States since 1978.

Over the past twenty-five years the production of solar energy has evolved from a mere kilowatt of electricity to hundreds of megawatts. Solar energy has many advantages over traditional energy sources in that it does not pollute and has the potential of being an unlimited, cheap source of energy. Its use and production is not limited to a small number of countries but is readily available throughout the United States and the world. The North American Southwest has some of the brightest sunlight in the world with a potential to provide up to 2500 kilowatt-hours per square meter.

In the design of solar power plants, engineers need to position solar panels perpendicular to the sun's rays so that maximum energy can be collected. Understanding the movement and position of the sun at any time and

Source: Winter, C., R. Sizmann, and Vant-Hunt (Editors), *Solar Power Plants,* Springer-Verlag, 1991.

Chapter Openers present a genuine application of the material to be presented.

Titled Examples include detailed, step-by-step solutions and descriptive side comments. Examples relating to the Chapter Openers are marked with a special symbol.

▶ **EXAMPLE 4**
Finding the angle of elevation of the sun

3.3 CIRCULAR FUNCTIONS OF REAL NUMBERS   **127**

Find the value of s in the interval $[0, \pi/2]$ that has $\cos s = .96854556$.

The value of s can be found with a calculator set for radian mode. Recall from Section 2.3 how we found an angle measure given a trigonometric function value of the angle. The same procedure is repeated here with the calculator set to radian mode to find that

$$\cos .25147856 = .96854556,$$

and $0 < .25147856 < \pi/2$, so $s = .25147856$.

Find the exact value of s in the interval $[0, \pi/2]$ for which $\sin s = \sqrt{2}/2$.

Sketch a triangle in quadrant I and use the definition of $\sin s$ to label the sides as shown in Figure 11. To relate it to the definition of the trigonometric function $\sin \theta$, multiply the lengths of each side by 2. We recognize this as a right triangle with the two acute angles of 45°. To find s, convert 45° to radians, to get $s = \pi/4$. ▶

The next example answers the first question posed in the application presented at the beginning of the chapter.

 Knowing the position of the sun in the sky is essential for solar-power plants. Solar panels need to be positioned perpendicular to the sun's rays for maximum efficiency. The angle of elevation θ of the sun in the sky at any latitude L can be calculated using the formula

$$\sin \theta = \cos D \cos L \cos \omega + \sin D \sin L$$

where $\theta = 0$ corresponds to sunrise and $\theta = \pi/2$ occurs if the sun is directly overhead. ω is the number of radians that the Earth has rotated through since noon when $\omega = 0$. D is the declination of the sun which varies because the Earth is tilted on its axis. (*Source:* Winter, C., R. Sizmann, and Vant-Hunt (Editors), *Solar Power Plants,* Springer-Verlag, 1991.)

Sacramento, California, has a latitude of $L = 38.5°$ or .6720 radians. Find the angle of elevation θ of the sun at 3 P.M. on February 29, 2000, where at that time, $D \approx -.1425$ and $\omega \approx .7854$.

Use the formula for $\sin \theta$.

$$\sin \theta = \cos D \cos L \cos \omega + \sin D \sin L$$
$$= \cos(-.1425) \cos(.6720) \cos(.7854) + \sin(-.1425) \sin(.6720)$$
$$\approx .4593$$

Thus, $\theta \approx .4773$ radians or 27.3°. ▶

X

Boxes highlight words, definitions, rules, and procedures.

Connections Boxes point out the many connections between mathematics and the "real world" or other mathematical concepts.

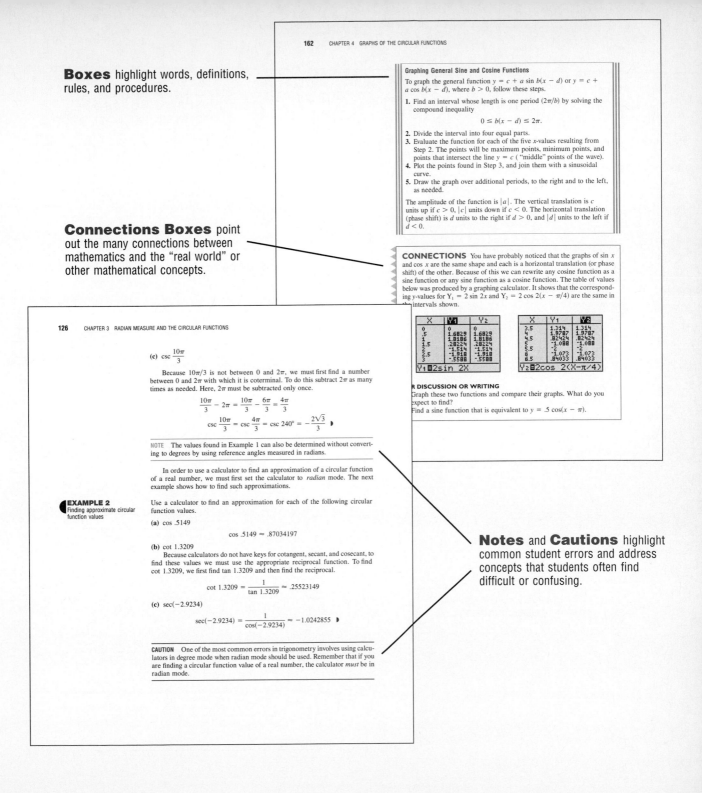

Graphing General Sine and Cosine Functions

To graph the general function $y = c + a \sin b(x - d)$ or $y = c + a \cos b(x - d)$, where $b > 0$, follow these steps.

1. Find an interval whose length is one period $(2\pi/b)$ by solving the compound inequality

$$0 \le b(x - d) \le 2\pi.$$

2. Divide the interval into four equal parts.
3. Evaluate the function for each of the five x-values resulting from Step 2. The points will be maximum points, minimum points, and points that intersect the line $y = c$ ("middle" points of the wave).
4. Plot the points found in Step 3, and join them with a sinusoidal curve.
5. Draw the graph over additional periods, to the right and to the left, as needed.

The amplitude of the function is $|a|$. The vertical translation is c units up if $c > 0$, $|c|$ units down if $c < 0$. The horizontal translation (phase shift) is d units to the right if $d > 0$, and $|d|$ units to the left if $d < 0$.

CONNECTIONS You have probably noticed that the graphs of $\sin x$ and $\cos x$ are the same shape and each is a horizontal translation (or phase shift) of the other. Because of this we can rewrite any cosine function as a sine function or any sine function as a cosine function. The table of values below was produced by a graphing calculator. It shows that the corresponding y-values for $Y_1 = 2 \sin 2x$ and $Y_2 = 2 \cos 2(x - \pi/4)$ are the same in the intervals shown.

X	Y1	Y2
0	0	0
.5	1.6829	1.6829
1	1.8186	1.8186
1.5	.28224	.28224
2	-1.514	-1.514
2.5	-1.918	-1.918
3	-.5588	-.5588

Y1∎2sin 2X

X	Y1	Y2
3.5	1.314	1.314
4	1.9787	1.9787
4.5	.82424	.82424
5	-1.088	-1.088
5.5	-2	-2
6	-1.073	-1.073
6.5	.84033	.84033

Y2∎2cos 2(X-π/4)

R DISCUSSION OR WRITING

Graph these two functions and compare their graphs. What do you expect to find?
Find a sine function that is equivalent to $y = .5 \cos(x - \pi)$.

(c) $\csc \dfrac{10\pi}{3}$

Because $10\pi/3$ is not between 0 and 2π, we must first find a number between 0 and 2π with which it is coterminal. To do this subtract 2π as many times as needed. Here, 2π must be subtracted only once.

$$\frac{10\pi}{3} - 2\pi = \frac{10\pi}{3} - \frac{6\pi}{3} = \frac{4\pi}{3}$$

$$\csc \frac{10\pi}{3} = \csc \frac{4\pi}{3} = \csc 240° = -\frac{2\sqrt{3}}{3} \quad \blacktriangleright$$

NOTE The values found in Example 1 can also be determined without converting to degrees by using reference angles measured in radians.

In order to use a calculator to find an approximation of a circular function of a real number, we must first set the calculator to *radian* mode. The next example shows how to find such approximations.

◀EXAMPLE 2
Finding approximate circular function values

Use a calculator to find an approximation for each of the following circular function values.

(a) $\cos .5149$

$$\cos .5149 \approx .87034197$$

(b) $\cot 1.3209$

Because calculators do not have keys for cotangent, secant, and cosecant, to find these values we must use the appropriate reciprocal function. To find $\cot 1.3209$, we first find $\tan 1.3209$ and then find the reciprocal.

$$\cot 1.3209 = \frac{1}{\tan 1.3209} \approx .25523149$$

(c) $\sec(-2.9234)$

$$\sec(-2.9234) = \frac{1}{\cos(-2.9234)} \approx -1.0242855 \quad \blacktriangleright$$

CAUTION One of the most common errors in trigonometry involves using calculators in degree mode when radian mode should be used. Remember that if you are finding a circular function value of a real number, the calculator *must* be in radian mode.

Notes and **Cautions** highlight common student errors and address concepts that students often find difficult or confusing.

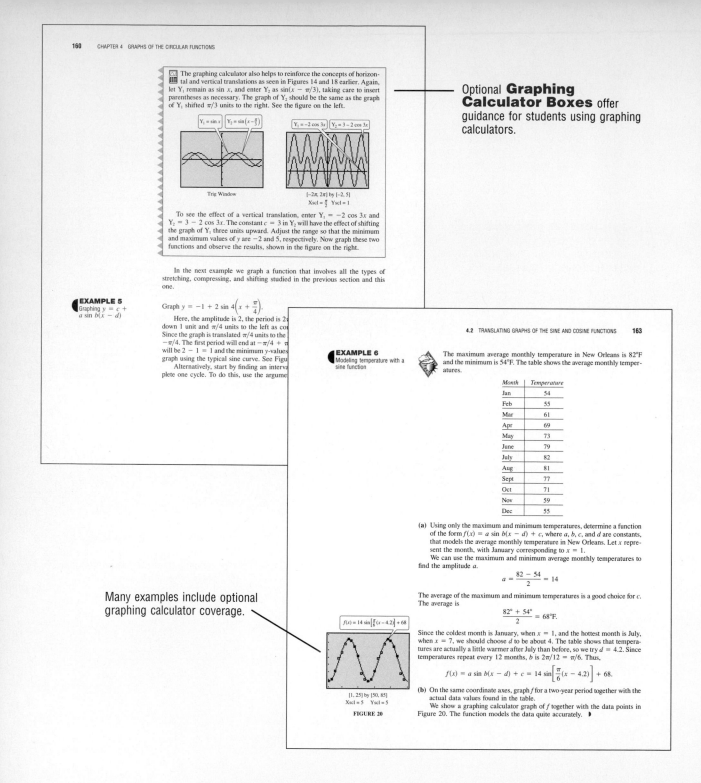

Optional **Graphing Calculator Boxes** offer guidance for students using graphing calculators.

Many examples include optional graphing calculator coverage.

---

160 CHAPTER 4 GRAPHS OF THE CIRCULAR FUNCTIONS

The graphing calculator also helps to reinforce the concepts of horizontal and vertical translations as seen in Figures 14 and 18 earlier. Again, let Y_1 remain as sin x, and enter Y_2 as $\sin(x - \pi/3)$, taking care to insert parentheses as necessary. The graph of Y_2 should be the same as the graph of Y_1 shifted $\pi/3$ units to the right. See the figure on the left.

$Y_1 = \sin x$ $Y_2 = \sin\left(x - \frac{\pi}{3}\right)$ $Y_1 = -2 \cos 3x$ $Y_2 = 3 - 2 \cos 3x$

Trig Window

$[-2\pi, 2\pi]$ by $[-2, 5]$
$X\text{scl} = \frac{\pi}{2}$ $Y\text{scl} = 1$

To see the effect of a vertical translation, enter $Y_1 = -2 \cos 3x$ and $Y_2 = 3 - 2 \cos 3x$. The constant $c = 3$ in Y_2 will have the effect of shifting the graph of Y_1 three units upward. Adjust the range so that the minimum and maximum values of y are -2 and 5, respectively. Now graph these two functions and observe the results, shown in the figure on the right.

In the next example we graph a function that involves all the types of stretching, compressing, and shifting studied in the previous section and this one.

EXAMPLE 5
Graphing $y = c + a \sin b(x - d)$

Graph $y = -1 + 2 \sin 4\left(x + \frac{\pi}{4}\right)$.

Here, the amplitude is 2, the period is 2[...] down 1 unit and $\pi/4$ units to the left as co[...] Since the graph is translated $\pi/4$ units to the [...] $-\pi/4$. The first period will end at $-\pi/4 + \pi$[...] will be $2 - 1 = 1$ and the minimum y-values[...] graph using the typical sine curve. See Figu[...]

Alternatively, start by finding an interva[...] plete one cycle. To do this, use the argume[...]

4.2 TRANSLATING GRAPHS OF THE SINE AND COSINE FUNCTIONS   163

EXAMPLE 6
Modeling temperature with a sine function

The maximum average monthly temperature in New Orleans is 82°F and the minimum is 54°F. The table shows the average monthly temperatures.

Month	Temperature
Jan	54
Feb	55
Mar	61
Apr	69
May	73
June	79
July	82
Aug	81
Sept	77
Oct	71
Nov	59
Dec	55

(a) Using only the maximum and minimum temperatures, determine a function of the form $f(x) = a \sin b(x - d) + c$, where $a, b, c,$ and d are constants, that models the average monthly temperature in New Orleans. Let x represent the month, with January corresponding to $x = 1$.

We can use the maximum and minimum average monthly temperatures to find the amplitude a.

$$a = \frac{82 - 54}{2} = 14$$

The average of the maximum and minimum temperatures is a good choice for c. The average is

$$\frac{82° + 54°}{2} = 68°F.$$

Since the coldest month is January, when $x = 1$, and the hottest month is July, when $x = 7$, we should choose d to be about 4. The table shows that temperatures are actually a little warmer after July than before, so we try $d = 4.2$. Since temperatures repeat every 12 months, b is $2\pi/12 = \pi/6$. Thus,

$$f(x) = a \sin b(x - d) + c = 14 \sin \left[\frac{\pi}{6}(x - 4.2)\right] + 68.$$

(b) On the same coordinate axes, graph f for a two-year period together with the actual data values found in the table.

We show a graphing calculator graph of f together with the data points in Figure 20. The function models the data quite accurately. ▶

$f(x) = 14 \sin \left[\frac{\pi}{6}(x - 4.2)\right] + 68$

$[1, 25]$ by $[50, 85]$
$X\text{scl} = 5$ $Y\text{scl} = 5$

FIGURE 20

Exercises corresponding to the Chapter Openers are marked with a special symbol.

Writing and **Conceptual Exercises** are included to aid students in applying the concepts presented.

Many exercises and examples are based on **Real Data,** and many require reading graphs and charts.

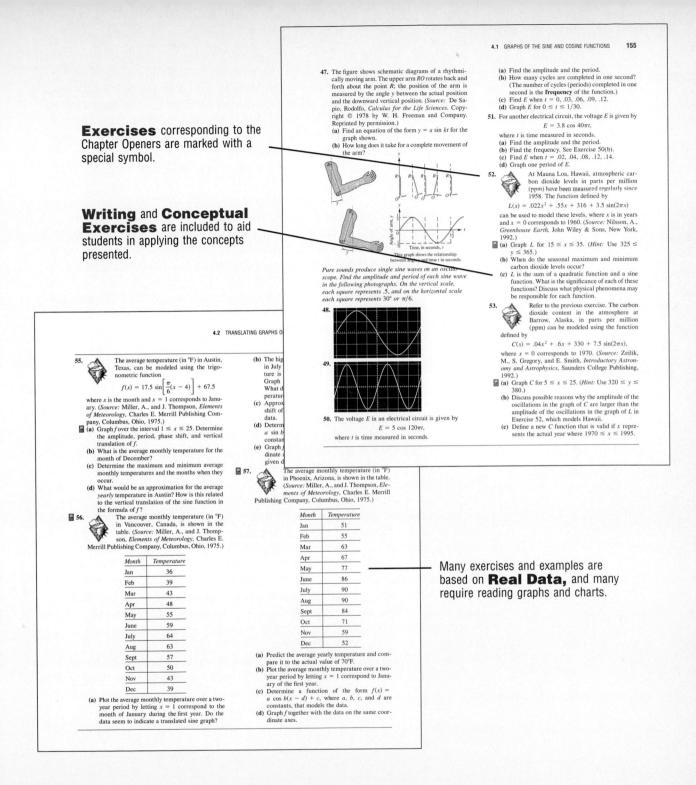

47. The figure shows schematic diagrams of a rhythmically moving arm. The upper arm RO rotates back and forth about the point R; the position of the arm is measured by the angle y between the actual position and the downward vertical position. (*Source:* De Sapio, Rodolfo, *Calculus for the Life Sciences.* Copyright © 1978 by W. H. Freeman and Company. Reprinted by permission.)
  (a) Find an equation of the form $y = a \sin kt$ for the graph shown.
  (b) How long does it take for a complete movement of the arm?

Pure sounds produce single sine waves on an oscilloscope. Find the amplitude and period of each sine wave in the following photographs. On the vertical scale, each square represents .5, and on the horizontal scale each square represents 30° or π/6.

48.

49.

50. The voltage E in an electrical circuit is given by
$$E = 5 \cos 120\pi t,$$
where t is time measured in seconds.

(a) Find the amplitude and the period.
(b) How many cycles are completed in one second? (The number of cycles (periods) completed in one second is the **frequency** of the function.)
(c) Find E when $t = 0, .03, .06, .09, .12$.
(d) Graph E for $0 \leq t \leq 1/30$.

51. For another electrical circuit, the voltage E is given by
$$E = 3.8 \cos 40\pi t,$$
where t is time measured in seconds.

(a) Find the amplitude and the period.
(b) Find the frequency. See Exercise 50(b).
(c) Find E when $t = .02, .04, .08, .12, .14$.
(d) Graph one period of E.

52. At Mauna Loa, Hawaii, atmospheric carbon dioxide levels in parts per million (ppm) have been measured regularly since 1958. The function defined by
$$L(x) = .022x^2 + .55x + 316 + 3.5 \sin(2\pi x)$$
can be used to model these levels, where x is in years and $x = 0$ corresponds to 1960. (*Source:* Nilsson, A., *Greenhouse Earth*, John Wiley & Sons, New York, 1992.)

(a) Graph L for $15 \leq x \leq 35$. (*Hint:* Use $325 \leq y \leq 365$.)
(b) When do the seasonal maximum and minimum carbon dioxide levels occur?
(c) L is the sum of a quadratic function and a sine function. What is the significance of each of these functions? Discuss what physical phenomena may be responsible for each function.

53. Refer to the previous exercise. The carbon dioxide content in the atmosphere at Barrow, Alaska, in parts per million (ppm) can be modeled using the function defined by
$$C(x) = .04x^2 + .6x + 330 + 7.5 \sin(2\pi x),$$
where $x = 0$ corresponds to 1970. (*Source:* Zeilik, M., S. Gregory, and E. Smith, *Introductory Astronomy and Astrophysics*, Saunders College Publishing, 1992.)

(a) Graph C for $5 \leq x \leq 25$. (*Hint:* Use $320 \leq y \leq 380$.)
(b) Discuss possible reasons why the amplitude of the oscillations in the graph of C are larger than the amplitude of the oscillations in the graph of L in Exercise 52, which models Hawaii.
(c) Define a new C function that is valid if x represents the actual year where $1970 \leq x \leq 1995$.

4.2 TRANSLATING GRAPHS OF

55. The average temperature (in °F) in Austin, Texas, can be modeled using the trigonometric function
$$f(x) = 17.5 \sin\left[\frac{\pi}{6}(x - 4)\right] + 67.5$$
where x is the month and $x = 1$ corresponds to January. (*Source:* Miller, A., and J. Thompson, *Elements of Meteorology*, Charles E. Merrill Publishing Company, Columbus, Ohio, 1975.)

(a) Graph f over the interval $1 \leq x \leq 25$. Determine the amplitude, period, phase shift, and vertical translation of f.
(b) What is the average highest temperature for the month of December?
(c) Determine the maximum and minimum average monthly temperatures and the months when they occur.
(d) What would be an approximation for the average *yearly* temperature in Austin? How is this related to the vertical translation of the sine function in the formula of f?

56. The average monthly temperature (in °F) in Vancouver, Canada, is shown in the table. (*Source:* Miller, A., and J. Thompson, *Elements of Meteorology*, Charles E. Merrill Publishing Company, Columbus, Ohio, 1975.)

Month	Temperature
Jan	36
Feb	39
Mar	43
Apr	48
May	55
June	59
July	64
Aug	63
Sept	57
Oct	50
Nov	43
Dec	39

(a) Plot the average monthly temperature over a two-year period by letting $x = 1$ correspond to the month of January during the first year. Do the data seem to indicate a translated sine graph?

(b) The high in July is Graph What peratu
(c) Approx shift of data.
(d) Determ $a \sin b$ constan
(e) Graph dinate given d

57. The average monthly temperature (in °F) in Phoenix, Arizona, is shown in the table. (*Source:* Miller, A., and J. Thompson, *Elements of Meteorology*, Charles E. Merrill Publishing Company, Columbus, Ohio, 1975.)

Month	Temperature
Jan	51
Feb	55
Mar	63
Apr	67
May	77
June	86
July	90
Aug	90
Sept	84
Oct	71
Nov	59
Dec	52

(a) Predict the average yearly temperature and compare it to the actual value of 70°F.
(b) Plot the average monthly temperature over a two-year period by letting $x = 1$ correspond to January of the first year.
(c) Determine a function of the form $f(x) = a \cos b(x - d) + c$, where a, b, c, and d are constants, that models the data.
(d) Graph f together with the data on the same coordinate axes.

- Graphing calculator comments and screens are given throughout the book as appropriate. These are identified with an icon, so that an instructor may choose whether or not to use them. We know that many students own graphing calculators and may need guidance for using them, even if they are not a required part of the course.
- Many examples and exercises are based on real data, and many require reading charts and graphs.
- The exercise sets have been completely rewritten and contain many new exercises, including more conceptual and writing exercises (which are marked by symbols in the instructor's edition), as well as optional graphing calculator exercises. Those exercises that require applying the topics in a section to ideas beyond the examples are marked as challenging in the instructor's edition.
- Cautions and notes are included to highlight common student errors and misconceptions. Some of these address concepts that students often find difficult or confusing.

Supplements ▼▼▼

For the Instructor

Annotated Instructor's Edition   With this volume, instructors have immediate access to the answers to every exercise in the text, excluding proofs and writing exercises. In a special section at the end of the book, each answer is printed next to or below the corresponding text exercise. In addition, challenging exercises, which will require most students to stretch beyond the concepts discussed in the text, are marked with the symbol ▲. The conceptual (◉) and writing (⬚) exercises are also marked in this edition so instructors may assign these problems at their discretion. (Graphing calculator exercises will be marked by ▦ in both the student's and instructor's editions.)

Instructor's Resource Manual   Included here are four versions of a chapter test for each chapter; additional test items for each chapter; and two forms of a final examination. Answers to all tests and additional exercises also are provided. Answers to most of the textbook exercises are included as well.

Instructor's Solution Manual   This manual includes complete, worked-out solutions to every even exercise in the textbook (excluding most writing exercises).

Test Generator/Editor for Mathematics with QuizMaster is a computerized test generator that lets instructors select test questions by objective or section or use a ready-made test for each chapter. The software is algorithm driven so that regenerated number values maintain problem types and provide a large number of test items in both multiple-choice and open-response formats for one or more test forms. The **Editor** lets instructors modify existing questions or create their own including graphics and accurate math symbols. Tests created with the **Test Generator** can be used with **QuizMaster,** which records student scores as they take tests on a single computer or network, and prints reports for students, classes, or courses. CLAST and TASP versions of this package are also available. (IBM, DOS/Windows, and Macintosh)

For the Student

Student's Solution Manual   Complete, worked-out solutions are given for odd-numbered exercises and chapter review exercises in a volume available for purchase by students. In addition, a practice chapter test, with answers, is provided for each chapter. All-new cumulative review exercises with worked-out solutions are also included.

Videotapes   A new videotape series has been developed to accompany *Trigonometry,* Sixth Edition. In a separate lesson for each section of the book, the series covers all objectives, topics, and problem-solving techniques within the text.

Interactive Mathematics Tutorial Software with Management System   is an innovative software package that is objective-based, self-paced, and algorithm driven to provide unlimited opportunity for review and practice. Tutorial lessons provide examples, progress-check questions, and access to an on-line glossary. Practice problems include hints for the first incorrect responses, solutions, textbook page references, and on-line tools to aid in computation and understanding. Quick Reviews for each section focus on major concepts. The optional **Management System** records student scores on disk and lets instructors print diagnostic reports for individual students or classes. Student versions, which include record-keeping and practice tests, may be purchased by students for home use.

Acknowledgments ▼▼▼

We are grateful to the many users of the fifth edition and to our reviewers for their insightful comments and suggestions. It is because they take the time to write thoughtful reviews that our textbooks continue to meet the needs of students and their instructors.

Reviewers

William A. Armstrong, Phoenix College

Marilyn Barrick, North Central Texas College

Carole A. Bauer, Triton College

Karin Beaty, Midlands Technical College

Carolyn J. Case, Vincennes University

Denise Brown, Collin County Community College

Marte Carter, Central Missouri State

Julane B. Crabtree, Johnson County Community College

Martha Diehl, Central Missouri State

Paul A. Dirks, Miami-Dade Community College

Richard Dreessen, Langston University

Irma T. Holm, Long Beach City College

Donald R. Hunt, Central Florida Community College

Ken Hurley, Polk Community College

Raja Khoury, Houston Community College

Jaclyn LeFebvre, Illinois Central College

John C. Matovsky, Louisiana Technical University

Judy S. McInerney, Sandhills Community College

Sandy Morris, College of DuPage

Marnie Pearson, Foothill College

Mary Beth Pederson, Illinois Central College

Janice Roy, Montcalm Community College

Janet S. Schachtner, San Jacinto College

Jerry A. Schuitman, Delta College

Cynthia Floyd Sikes, Georgia Southern University

Debbye Stapleton, Georgia Southern University

Glynna Strait, Odessa College

Mark Swetnam, Ashland Community College

Mahbobeh Vezvaei, Kent State University

Dr. Lee Witt, Davenport College of Business

Accuracy Checkers

Norma F. James

Matthew T. Lazar, University of California

Paul O'Heron, Broome Community College

Mary Beth Pederson, Illinois Central College

We are thankful for the assistance given by Gary Rockswold, Mankato State University, who researched and provided the chapter opener applications and exercises. As always, we are grateful to Paul Eldersveld, College of DuPage, for an outstanding job of coordinating the print supplements. We also thank Kitty Pellissier, who checked the answers to all the exercises in her usual careful and thorough manner, and Paul Van Erden, American River College, who created an accurate, complete index. As always, Ed Moura and Anne Kelly were there to lend support and guidance. Special thanks go out to Lisa De Mol and Dee Netzel, who coordinated the production of an extremely complex project. It is only through the cooperation of these and many other individuals that we are able to produce texts that successfully serve both instructors and students.

Margaret L. Lial
E. John Hornsby, Jr.
David I. Schneider

An Introduction to Scientific and Graphing Calculators

In the past, some of the most brilliant minds in mathematics and science spent long, laborious hours calculating values for logarithmic and trigonometric tables. These tables were essential to solve equations in real applications. Because they could not predict what values would be needed, the tables were sometimes incomplete. During the second half of the twentieth century, computers and sophisticated calculators appeared. These computing devices are able to evaluate mathematical expressions and generate tables in a fraction of a second. As a result, the study of mathematics is changing dramatically.

Although computers and calculators have made a profound difference, they have *not* replaced mathematical thought. Calculators cannot decide whether to add or subtract two numbers in order to solve a problem—only you can do that. Once you have made this decision, calculators can efficiently determine the solution to the problem. In addition, graphing calculators also provide important graphical and numerical support to the validity of a mathematical solution. They are capable of exposing errors in logic and pointing to patterns. These patterns can lead to conjectures and theorems about mathematics. The human mind is capable of mathematical insight and decision making, but is not particularly proficient at performing long arithmetic calculations. On the other hand, calculators are incapable of possessing mathematical insight, but are excellent at performing arithmetic and other routine computations. In this way, calculators complement the human mind.

If this is your first experience with a scientific or graphing calculator, the numerous keys and strange symbols that appear on the keyboard may be intimidating. Like any learning experience, take it a step at a time. You do not have to understand every key before you begin using your calculator. Some keys may not even be needed in this course. The following explanations and suggestions are intended to give you a brief overview of scientific and graphing calculators.

It is not intended to be complete or specific toward any particular type of calculator. You may find that some things are different on your calculator. *Remember, a calculator comes with an owner's manual.* This manual is essential in learning how to use your calculator.

Scientific Calculators ▼▼▼

Two basic parts of any calculator are the keyboard and the display. The keyboard is used to input data—the display is used to output data. Without correct input, the displayed output is meaningless. Most scientific calculators do not display the entire arithmetic expression that is entered, but only display the most recent number inputted or outputted. If an arithmetic expression is entered incorrectly, it is not possible to edit it. The entire expression must be entered again.

Order of Operation

The order in which expressions are entered into a calculator is essential to obtaining correct answers. Operations on a calculator can usually be divided into two basic types: unary and binary. Unary operations require that only one number be entered. Examples of unary operations are $\sqrt{x}$, x^2, $x!$, $\sqrt[3]{x}$, and $\log x$. When entering a unary operation on a scientific calculator, the number is usually entered first, followed by the unary operation. For example, to find the square root of 4, press the key $\boxed{4}$, followed by the square root key. Binary operations require that two numbers are entered. Examples of binary operations are $+$, $-$, $\times$, $\div$, and x^y. When evaluating a binary operation on a scientific calculator, the operation symbol is usually entered between the numbers. Thus, to add the two numbers 4 and 5, enter $\boxed{4}$ $\boxed{+}$ $\boxed{5}$ $\boxed{=}$. However, on some calculators, such as those made by Hewlett Packard, it is necessary to use *Reverse Polish Notation* (RPN). In RPN the operation is entered last, after the operands. One advantage of RPN is that parentheses are usually not necessary.

Every calculator has a set of built-in precedence rules that can be found in the owner's manual. For example, suppose that the expression $3 + 4 \times 2 =$ is entered, from left to right, into a scientific calculator. The output will usually be 11, and not 14. This is because multiplication is performed before addition in the absence of parentheses. Parentheses can always be used to override existing precedence rules. *When in doubt, use parentheses.* Try evaluating $\frac{24}{4-2}$. It should be entered as $24 \div (4 - 2) =$ in order to obtain the correct answer of 12. This is because division has precedence over subtraction.

Scientific Notation

Numbers that are either large or small in absolute value are often displayed using scientific notation. The numeric expression 2.46 E12 refers to the large number 2.46×10^{12}, while the expression 2.46 E$-$12 refers to the small positive number 2.46×10^{-12}. Try multiplying one billion times ten million. Observe the output on your calculator.

Precision and Accuracy

Precision refers to the number of digits a calculator will display. When $\frac{1}{3}$ is evaluated, a calculator may display 0.333333333. This answer is approximate. The displayed precision of most calculators is between 8 and 12 digits. Accuracy is different from precision. It refers to the number of correct digits that an answer contains, compared to the true value. If a scale is misread as 129.6 pounds, when the actual answer is 145.8 pounds, then the number 129.6 has four digits of precision, but only one digit of accuracy. Many times when using a calculator to solve a real application, it will display ten digits of precision, but only a few digits will be accurate or meaningful. For example, suppose you drive 100 miles on 3 gallons of gas. A calculator would say that your mileage is $100 \div 3 \approx 33.33333333$. The precision of this answer is ten digits. The accuracy is probably not ten digits unless both the mileage and amount of gasoline were measured in an exceedingly accurate manner. It would be more reasonable or accurate to say that the mileage is about 33 miles per gallon, rather than 33.33333333 miles per gallon.

Second and Inverse Keys

Because the size of the keyboard is limited, there is often a 2nd or INV key. This key can be used to access additional features. These additional features are usually labeled above the key in a different color.

Graphing Calculators ▼▼▼

Graphing calculators provide several features beyond those found on scientific calculators. The bottom rows of keys on a graphing calculator are often similar to those found on scientific calculators. Graphing calculators have additional keys that can be used to create graphs, make tables, analyze data, and change settings. One of the major differences between graphing and scientific calculators is that a graphing calculator has a larger viewing screen with graphing capabilities.

Editing Input

The screen of a graphing calculator can display several lines of text at a time. This feature allows the user to view both previous and current expressions. If an incorrect expression is entered, a brief error message is displayed. It can be viewed and corrected by using various editing keys—much like a word-processing program. You do not need to enter the entire expression again. Many graphing calculators can also recall past expressions for editing or updating.

Order of Operation

Arithmetic expressions on graphing calculators are usually entered as they are written in mathematical equations. As a result, unary operations like $\sqrt{x}$, $\sqrt[3]{x}$, and $\log x$ are entered first, followed by the number. Unary operations like x^2 and $x!$ are entered after the number. Binary operations are entered in a manner similar to most scientific calculators. The order of operation on graphing calculators is also important. For example, try evaluating the expression $\sqrt{2} \times 8$. If this expression is entered as it is written, without any parentheses, a graphing calculator may display 11.3137085 and not 4. This is because a square root is performed before multiplication. To prevent this error, use parentheses around 2×8.

Calculator Screen

If you look closely at the screen of a graphing calculator, you will notice that the screen is composed of many tiny rectangles or points called pixels. The calculator can darken these rectangles so that output can be displayed. Many graphing calculator screens are approximately 96 pixels across and 64 pixels high. Computer screens are usually 640 by 480 pixels or more. For this reason, you will notice that the resolution on a graphing calculator screen is not as clear as on most computer terminals. With a graphing calculator, a straight line will not always appear to be exactly straight and a circle will not be precisely circular. Because of the screen's low resolution, graphs generated by graphing calculators may require mathematical understanding to interpret them correctly.

Viewing Window

The viewing window for a graphing calculator is similar to the viewfinder in a camera. A camera cannot take a picture of an entire view in a single picture. The camera must be centered on some object and can only photograph a subset of the available scenery. A person may want to photograph a close-up of a face or a person standing in front of a mountain. A camera with a zoom lens can capture different views of the same scene by zooming in and out. Graphing calculators have similar capabilities. The xy-coordinate plane is infinite. The calculator screen can show only a finite, rectangular region in the xy-coordinate plane. This rectangular region must be specified before a graph can be drawn. This is done by setting minimum and maximum values for both the x- and y-axes. Determining an appropriate viewing window is often one of the most difficult things to do. Many times it will take a few attempts before a satisfactory window size is found. Like many cameras, the graphing calculator can also zoom in and out. Zooming in shows more detail in a small region of a graph, whereas zooming out gives a better overall picture of the graph.

Graphing and the Free-moving Cursor

Once a viewing window has been determined, an equation in the form of $y = f(x)$ can be graphed. A simple example of this form is $y = 3x$. Four or more equations of this type can be graphed at once in the same viewing window. A graphing calculator has a free-moving cursor. By using the arrow keys, a small cross-hair can be made to move about on the screen. Its x- and y- coordinates are usually displayed on the screen. The cursor can be used to approximate the locations of features on the graph, such as x-intercepts and points of intersection. Using the trace key, the free-moving cursor can also be made to trace over the graph, displaying the corresponding x- and y- coordinates located on the graph.

Tables

Some graphing calculators have the ability to display tables. For example, if $y = x^2$, then a vertical table like the following can be generated automatically. This is an efficient way to evaluate an equation at selected values of x.

X	Y
0	0
1	1
2	4
3	9
4	16
5	25
6	36

Programming

Graphing calculators can be programmed, much like computers. Complex problems can be solved with the aid of programs. In this course, it will not be necessary for you to program your calculator. However, the capability is there, if you choose to use it.

Additional Features

Graphing calculators have additional features too numerous to list completely. They may be able to generate sequences, find maximums and minimums on graphs, do arithmetic with complex numbers, solve systems of equations using matrices, and analyze data with statistics. The most advanced calculators are capable of performing *symbolic manipulation.* Using these calculators, one can factor $x^2 - 1$ into $(x - 1)(x + 1)$ and simplify $\dfrac{x^2 y^3}{x^{-2} y}$ to $x^4 y^2$ automatically. If the solution to a problem is π, symbolic manipulation routines will display π rather than 3.141592654.

Final Comments ▼▼▼

Mathematicians from the past would have been amazed by today's calculators. Calculators are powerful computing devices that can perform difficult computational tasks. The solutions to many important equations in mathematics cannot be determined by hand. However, the solutions to these equations often can be approximated using a calculator. Calculators also provide the capability to ask questions like "What if . . . ?" more easily. Values in algebraic expressions can be altered and conjectures tested quickly.

At the heart of all mathematics is deductive thought and proof. No robot or artificial intelligence program has been effective at this task. Only the human mind is capable of this. Calculators are an important tool in mathematics. Like any tool, they must be used *appropriately* in order to enhance our ability to understand mathematics. Mathematical insight may often be the quickest and easiest way to solve a problem; a calculator may neither be needed nor appropriate. By using mathematical concepts, you can decide when to use or not to use a calculator.

1

The Trigonometric Functions

Trigonometry has been used for millennia to solve problems related to astronomy, surveying, and construction. Prior to the fifteenth century, astronomy had the greatest influence on the development of trigonometry. The Greek astronomer Hipparchus is usually given credit for first studying the trigonometric properties of angles. For centuries astronomers wanted to determine how far it was to the stars. It was not until 1838 that the astronomer Friedrich Bessel determined the distance to a star called 61 Cygni. He used a *parallax* method that relied on the measurement of very small angles. This measurement confirmed that the heliocentric model of Copernicus was correct and gave scientists a better understanding about the size and structure of the universe.

You observe parallax when you ride in an automobile and see a nearby object apparently moving backward with respect to more distant objects. The same is true for some stars that are relatively close to Earth. As Earth revolves around the sun, the observed angle θ of some nearby stars changes due to parallax as shown in the figure below.

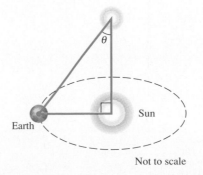

Not to scale

The table lists the size of the angle θ in seconds for five stars. (One second is equal to 1/3600th of a degree.)

Star	θ
alpha Centauri	.763
Barnard's Star	.546
Sirius	.377
61 Cygni	.292
Procyon	.287

Since stars are very distant objects, the parallax of a star is small and always less than one second. How can we use θ to find the distance to these stars? In order to solve problems like estimating distances to stars, determining the feasibility of a total solar eclipse, or approximating the depth of a crater on the moon, we must understand angles, triangles, and trigonometric functions. These concepts are introduced in this chapter.*

1.1 Basic Concepts ▼▼▼

Many ideas in trigonometry are best explained with a graph of a plane. Each point in the plane corresponds to an **ordered pair,** two numbers written inside parentheses, such as $(-2, 4)$. Graphs are set up with two axes, one for each number in an ordered pair. The horizontal axis is called the **x-axis,** and the vertical axis is the **y-axis.** The two axes intersect at a point called the **origin.** To locate the point that corresponds to the ordered pair $(-2, 4)$, start at the origin, and move 2 units left and 4 units up. The point $(-2, 4)$ and other sample points are shown in Figure 1.

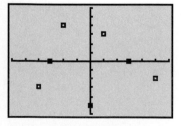

[−6, 6] by [−6, 6]
Xscl = 1   Yscl = 1

The points plotted in Figure 1 are shown here on a typical graphing calculator screen. The minimum and maximum x-values are −6 and 6, while those of the y-values are also −6 and 6. For both axes, the distance between tick marks (scale) is 1. We designate these "window dimensions" as shown above.

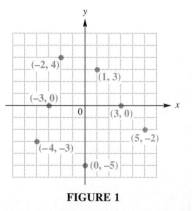

FIGURE 1

*Sources: Freebury, H. A., *A History of Mathematics,* MacMillan Company, New York, 1968. Zeilik, M., S. Gregory, and E. Smith, *Introductory Astronomy and Astrophysics,* Saunders College Publishers, 1992.

The axes divide the plane into four regions called **quadrants.** The quadrants are numbered in a counterclockwise direction, as shown in Figure 2. The points on the axes themselves belong to none of the quadrants. Figure 2 also shows that in quadrant I both the x-coordinate and the y-coordinate are positive; in quadrant II the value of x is negative while y is positive, and so on.

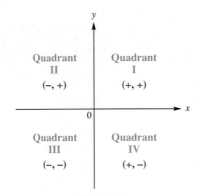

FIGURE 2

The distance between any two points on a plane can be found by using a formula derived from the **Pythagorean theorem.**

Pythagorean Theorem

If the **legs** (the two shorter sides) of a right triangle have lengths a and b, respectively, and if the length of the **hypotenuse** (the longest side, opposite the 90° angle) is c, then

$$a^2 + b^2 = c^2.$$

A proof of the Pythagorean theorem is outlined in Exercise 93.

EXAMPLE 1
Finding a distance from Earth to the center of the sun

FIGURE 3

The maximum distance from the surface of Earth to a point on the sun is 92,955,600 miles. See Figure 3. The diameter of the sun is about 864,930 miles.* Find the smallest distance from Earth's surface to the center of the sun.

From the given diameter, we know that the radius is $864,930/2 = 432,465$ miles. As Figure 3 shows, we are given the lengths of the hypotenuse and one leg of a right triangle, so we use the Pythagorean theorem.

$$c^2 = a^2 + b^2$$
$$92,955,600^2 = 432,465^2 + b^2$$
$$b = 92,954,594 \quad \blacktriangleright$$

*The Universal Almanac 1993, John W. Wright, General Editor, Andrews and McMeel, p. 51.

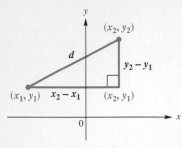

FIGURE 4

To find the distance between the two points (x_1, y_1) and (x_2, y_2), start by drawing the line segment connecting the points, as shown in Figure 4. Complete a right triangle by drawing a line through (x_1, y_1) parallel to the x-axis and a line through (x_2, y_2) parallel to the y-axis. The ordered pair at the right angle of this triangle is (x_2, y_1).

The horizontal side of the right triangle in Figure 4 has length $x_2 - x_1$, while the vertical side has length $y_2 - y_1$. If d represents the distance between the two original points, then by the Pythagorean theorem,

$$d^2 = (x_2 - x_1)^2 + (y_2 - y_1)^2.$$

Upon solving for d, we obtain the *distance formula*.

Distance Formula

The distance between the points (x_1, y_1) and (x_2, y_2) is given by the **distance formula,**

$$d = \sqrt{(x_2 - x_1)^2 + (y_2 - y_1)^2}.$$

EXAMPLE 2
Using the distance formula

Use the distance formula to find the distance, d, between each of the following pairs of points.

(a) $(2, 6)$ and $(5, 10)$

Either point can be used as (x_1, y_1). If we choose $(2, 6)$ as (x_1, y_1) and $(5, 10)$ as (x_2, y_2), then $x_1 = 2$, $y_1 = 6$, $x_2 = 5$, and $y_2 = 10$.

$$
\begin{aligned}
d &= \sqrt{(x_2 - x_1)^2 + (y_2 - y_1)^2} \\
&= \sqrt{(5 - 2)^2 + (10 - 6)^2} \qquad x_1 = 2,\, y_1 = 6,\, x_2 = 5,\, y_2 = 10 \\
&= \sqrt{3^2 + 4^2} \\
&= \sqrt{9 + 16} \\
&= \sqrt{25} \\
&= 5
\end{aligned}
$$

(b) $(-7, 2)$ and $(3, -8)$

$$
\begin{aligned}
d &= \sqrt{[3 - (-7)]^2 + (-8 - 2)^2} \\
&= \sqrt{10^2 + (-10)^2} \\
&= \sqrt{100 + 100} \\
&= \sqrt{200} \\
&= 10\sqrt{2}
\end{aligned}
$$

Here $\sqrt{200}$ was simplified as $\sqrt{200} = \sqrt{100} \cdot \sqrt{2} = 10\sqrt{2}$. ▶

CONNECTIONS The *midpoint formula* is used to find the coordinates of the midpoint of a line segment. (Recall that the midpoint of a line segment is equidistant from the endpoints of the segment.) To develop the midpoint formula, let (x_1, y_1) and (x_2, y_2) be any two distinct points in a plane. (Although the figure shows $x_1 < x_2$, no particular order is required.) Assume that the two points are not on a horizontal or vertical line. Let (x, y) be the midpoint of the segment connecting (x_1, y_1) and (x_2, y_2). Draw vertical lines from each of the three points to the x-axis, as shown in the figure.

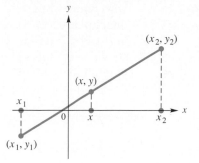

Since (x, y) is the midpoint of the line segment connecting (x_1, y_1) and (x_2, y_2), the distance between x and x_1 equals the distance between x and x_2, so that

$$x_2 - x = x - x_1$$
$$x_2 + x_1 = 2x$$
$$x = \frac{x_1 + x_2}{2}.$$

Thus, the x-coordinate of the midpoint is the average of the x-coordinates of the endpoints of the segment. In a similar way, the y-coordinate of the midpoint is

$$y = \frac{y_1 + y_2}{2},$$

the average of the y-coordinates of the endpoints of the segment.

FOR DISCUSSION OR WRITING
1. Find the midpoint of the segment with endpoints $(8, -4)$ and $(-9, 6)$.
2. A line segment has an endpoint at $(2, -8)$ and midpoint $(-1, -3)$. Find the other endpoint of the segment.
3. Verify the formula given above for the y-coordinate of the midpoint.
4. In the discussion, one assumed that the x-value of the midpoint is halfway between x_1 and x_2 on the x-axis. Why is this true?

INTERVAL NOTATION It is often necessary to specify sets of numbers defined by inequalities. One way of doing this is by using **set-builder notation,** such as $\{x \mid x < 5\}$ (read "the set of all x such that x is less than 5"). Another type of notation, called **interval notation,** can be used for writing intervals. For example, using this notation, the interval $\{x \mid x < 5\}$ is written as $(-\infty, 5)$. The symbol $-\infty$ does not indicate a number; it is used to show that the interval includes all real numbers less than 5. The parenthesis indicates that 5 is not included. Since there is no smallest number, and 5 is not included, the interval $(-\infty, 5)$ is an example of an **open interval.** Intervals that include the endpoints, as in the example $\{x \mid 0 \le x \le 5\}$, are **closed intervals.** Closed intervals are indicated with square brackets. The interval $\{x \mid 0 \le x \le 5\}$ is written as $[0, 5]$. An interval like $(2, 5]$, that is open on one end and closed on the other, is a **half-open interval.** Examples of other sets written in interval notation are shown in the following chart. In these intervals, assume that $a < b$. Note that a parenthesis is always used with the symbols $-\infty$ or ∞.

Type of Interval	Set	Interval Notation	Graph
Open interval	$\{x \mid x > a\}$	(a, ∞)	
	$\{x \mid a < x < b\}$	(a, b)	
	$\{x \mid x < b\}$	$(-\infty, b)$	
Half-open interval	$\{x \mid x \ge a\}$	$[a, \infty)$	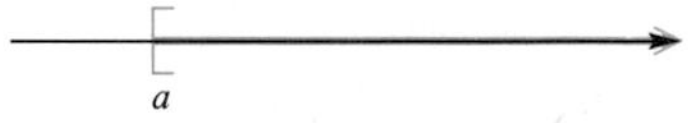
	$\{x \mid a < x \le b\}$	$(a, b]$	
	$\{x \mid a \le x < b\}$	$[a, b)$	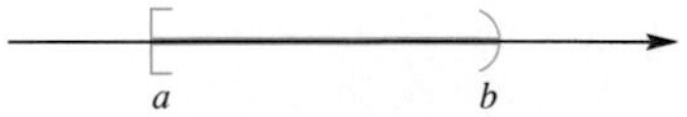
	$\{x \mid x \le b\}$	$(-\infty, b]$	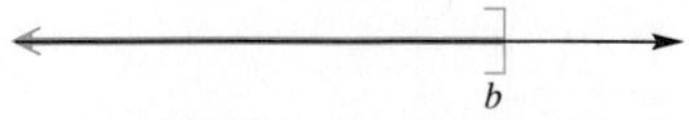
Closed interval	$\{x \mid a \le x \le b\}$	$[a, b]$	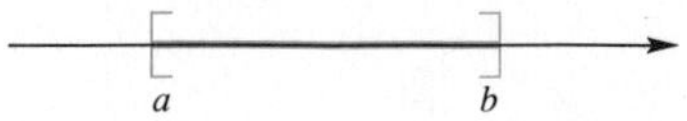

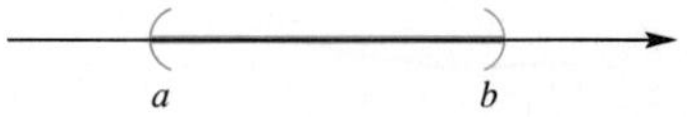

It is also customary to use $(-\infty, \infty)$ to represent the set of all real numbers.

RELATIONS AND FUNCTIONS   A **relation** is defined as a set of ordered pairs. Many relations have a rule or formula showing the connection between the two components of the ordered pairs. For example, the formula

$$y = -5x + 6$$

shows that a value of y can be found from a given value x by multiplying the value of x by -5 and then adding 6. According to this formula, if $x = 2$, then $y = -5 \cdot 2 + 6 = -4$, so that $(2, -4)$ belongs to the relation. In the relation $y = -5x + 6$, the value of y depends on the value of x, so that y is the **dependent variable** and x is the **independent variable.**

NOTE   A relation is a set of points, often defined by an equation such as $y = -5x + 6$. While precise language would require that we say "the relation defined by the equation $y = -5x + 6$," we will often use the less cumbersome language "the relation $y = -5x + 6$."

Most of the relations in trigonometry are also *functions.*

> **Function**
>
> A relation is a **function** if each value of the independent variable leads to exactly one value of the dependent variable. This means that each value of x must produce exactly one value of y.

It is customary for x to be considered the independent variable and y the dependent variable, and we shall follow that convention.

For example, $y = -5x + 6$ defines a function. For any one value of x that might be chosen, $y = -5x + 6$ gives exactly one value of y. In contrast, $y^2 = x$ defines a relation that is not a function. If we choose the value $x = 16$, then $y^2 = x$ becomes $y^2 = 16$, from which $y = 4$ or $y = -4$. The one x-value, 16, leads to two y-values, 4 and -4, so that $y^2 = x$ does not define a function.

Functions are often named with letters such as f, g, or h. For example, the function $y = -5x + 6$ can be written as

$$f(x) = -5x + 6,$$

where $f(x)$, read "f of x," replaces y. For the function $f(x) = -5x + 6$, if $x = 3$ then $f(x) = f(3) = -5 \cdot 3 + 6 = -15 + 6 = -9$, or

$$f(3) = -9,$$

indicating that the ordered pair $(3, -9)$ belongs to function f. Also,

$$f(-7) = -5(-7) + 6 = 41,$$

so $(-7, 41)$ belongs to f.

Recall that $|a|$ represents the absolute value of a. By definition, $|a| = a$ if $a \geq 0$ and $|a| = -a$ if $a < 0$. Thus $|4| = 4$ and $|-4| = 4$.

◀EXAMPLE 3
Using function notation

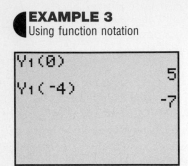

Function notation is available on some graphing calculators. By defining Y_1 as $-X^2 + |X - 5|$, the function values at 0 and -4 are 5 and -7, respectively. Compare with Example 3(a) and (b).

Let $f(x) = -x^2 + |x - 5|$. Find each of the following.

(a) $f(0)$

Use $f(x)$ and replace x with 0.

$$f(0) = -0^2 + |0 - 5| = -0 + |-5| = 5$$

(b) $f(-4) = -(-4)^2 + |-4 - 5| = -16 + |-9| = -16 + 9 = -7$

(c) $f(a) = -a^2 + |a - 5|$ Each x was replaced with a.

(d) Why does f define a function?

For each value of x, there is exactly one value of $f(x)$; therefore f defines a function.   ▶

The set of all possible values that can be used as a replacement for the independent variable in a relation is called the **domain** of the relation. The set of all possible values for the dependent variable is the **range** of the relation. By observing the graph of a relation, it is often easy to determine the domain and the range. For example, in Figure 5 the domain is the set of real numbers between -6 and 6, inclusive, and the range is the set of real numbers between -2 and 2, inclusive. Using interval notation, these sets are written $[-6, 6]$ and $[-2, 2]$, respectively.

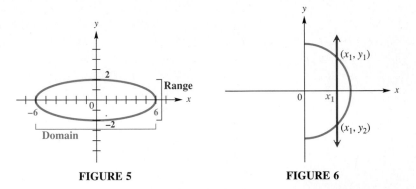

FIGURE 5   **FIGURE 6**

For a relation to be a function, each value of x in the domain of the function must lead to exactly one value of y. Figure 6 shows the graph of a relation. A point x_1 has been chosen on the x-axis. A vertical line drawn through x_1 intersects the graph in more than one point. Since the x-value x_1 leads to more than one value of y, this graph is not the graph of a function. This example suggests the *vertical line test* for a function.

Vertical Line Test

If any vertical line intersects the graph of a relation in more than one point, then the graph is not the graph of a function.

EXAMPLE 4
Identifying domains, ranges, and functions

Find the domain and range for the following relations. Identify any functions.

(a) $y = x^2$

Here x, the independent variable, can take on any value, so the domain is the set of all real numbers, $(-\infty, \infty)$. Since the dependent variable y equals the square of x, and since a square is never negative, the range is the set of all nonnegative numbers, $[0, \infty)$.

Each value of x leads to exactly one value of y, so $y = x^2$ defines a function. The graph of $y = x^2$ in Figure 7 shows that it satisfies the conditions of the vertical line test.

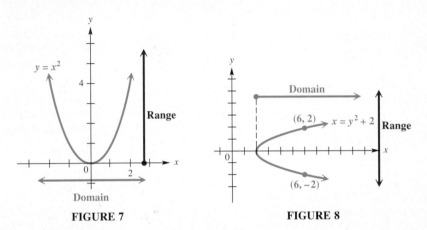

FIGURE 7   **FIGURE 8**

(b) $3x + 2y = 6$

In this relation x and y can take on any value at all. Both the domain and range are $(-\infty, \infty)$. For any value of x that might be chosen, the equation $3x + 2y = 6$ would lead to exactly one value of y. Therefore, $3x + 2y = 6$ defines a function.

(c) $x = y^2 + 2$ (Figure 8)

For any value of y, the square of y is nonnegative; that is, $y^2 \geq 0$. Since $x = y^2 + 2$, this means that $x \geq 0 + 2 = 2$, making the domain of the relation $[2, \infty)$. Any real number may be squared, so the range is the set of all real numbers, $(-\infty, \infty)$. To decide whether the relation is a function, choose a sample value of x greater than 2 from the domain. Choosing 6 for x gives

$$6 = y^2 + 2$$
$$4 = y^2$$
$$y = 2 \quad \text{or} \quad y = -2$$

Since one x-value, 6, leads to two y-values, 2 and -2, the relation $x = y^2 + 2$ does not define a function. As can be seen in Figure 8, the graph does not satisfy the requirements of the vertical line test.

(d) $y = \sqrt{1 - x}$

The domain is found from the requirement that the quantity under the radical, $1 - x$, must be greater than or equal to 0 for y to be a real number.

$$1 - x \geq 0$$
$$x \leq 1 \qquad \text{Add } x \text{ to both sides.}$$

The domain is $(-\infty, 1]$. To determine the range, note that the given radical is nonnegative, so the range is $[0, \infty)$.

Since each value of x in the domain leads to a single value of y, this relation is a function. ▶

◀ **EXAMPLE 5**
Finding the domain of a function from its rule

Find the domain for each of the following functions.

(a) $y = \dfrac{1}{x - 2}$

Since division by 0 is undefined, x cannot equal 2. (This value of x would make the denominator become $2 - 2$, or 0.) Any other value of x is acceptable, so the domain is all values of x other than 2, $(-\infty, 2) \cup (2, \infty)$.

(b) $y = \dfrac{8 + x}{(2x - 3)(4x - 1)}$

This denominator is 0 if either

$$2x - 3 = 0 \qquad \text{or} \qquad 4x - 1 = 0.$$

Solve each of these equations.

$$
\begin{array}{ll}
2x - 3 = 0 & 4x - 1 = 0 \\
2x = 3 & 4x = 1 \\
x = \dfrac{3}{2} & x = \dfrac{1}{4}
\end{array}
$$

The domain here includes all real numbers x such that $x \neq 3/2$ and $x \neq 1/4$. Using interval notation, this is written $(-\infty, 1/4) \cup (1/4, 3/2) \cup (3/2, \infty)$.

(c) $y = \dfrac{1}{\sqrt{x^2 + 16}}$

For the expression to be defined, the denominator cannot be zero, and the radicand, $x^2 + 16$, must be positive. To find the domain we must solve $x^2 + 16 > 0$.

$$x^2 + 16 > 0$$
$$x^2 > -16 \qquad \text{Subtract 16.}$$

Since this last inequality is true for all real numbers, the domain is $(-\infty, \infty)$. ▶

X	Y1
-1	-.3333
0	-.5
1	-1
2	ERROR
3	1
4	.5
5	.33333

Y₁ = 1/(X-2)

Many modern graphing calculators are capable of generating tables of ordered pairs for functions. Here Y_1 is defined as shown in Example 5(a). Notice that an ERROR message is returned for $X = 2$, since this leads to 0 in the denominator.

NOTE   In later work we will encounter trigonometric functions that are defined in such a way that restrictions are necessary, since their denominators cannot equal zero. Example 5 illustrates this idea with algebraic functions.

1.1 Exercises ▼▼▼▼▼▼▼▼▼▼▼▼▼▼▼▼▼▼▼▼▼▼▼▼▼▼▼▼▼▼▼▼▼▼▼▼

Graph the points on a coordinate system and identify the quadrant or axis for each point.

1. $(3, 2)$

2. $(-7, 6)$

3. $(-7, -4)$

4. $(8, -5)$

5. $(0, 5)$

6. $(-8, 0)$

7. $(4.5, 7)$

8. $(-7.5, 8)$

Give the quadrant in which each point lies.

9. $(-5, \pi)$

10. $(\pi, -3)$

11. $(-\sqrt{2}, -2\sqrt{2})$

12. $\left(1 + \sqrt{3}, \dfrac{\pi}{2}\right)$

13. Suppose the point (a, b) lies in the first quadrant. Describe how you would move from the point (a, b) to the point $(a, -b)$.

14. Suppose the point (a, b) lies in the first quadrant. Describe how you would move from the point (a, b) to the point $(-a, b)$.

15. If $xy = 1$ is graphed, in which quadrants will the points of the graph lie?

16. If (a, b) represents a point that lies in quadrant II, in which quadrant will each point lie?
   (a) $(-a, b)$ **(b)** $(-a, -b)$ **(c)** $(a, -b)$

Use the distance formula to find the distance between the following pairs of points. See Example 2.

17. $(2, -1)$ and $(-3, -4)$

18. $(-5, 2)$ and $(3, -7)$

19. $(-1, 0)$ and $(-4, -5)$

20. $(-2, -3)$ and $(-6, 4)$

21. $(\sqrt{2}, -\sqrt{5})$ and $(3\sqrt{2}, 4\sqrt{5})$

22. $(5\sqrt{7}, -\sqrt{3})$ and $(-\sqrt{7}, 8\sqrt{3})$

23. Determine the distance between $(5, -6)$ and the x-axis.

24. Determine the distance between $(5, -6)$ and the y-axis.

25. The accompanying graphing calculator screen shows two ways to find the distance between two points in the plane. What are the two points?

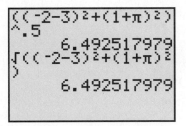

26. Suppose the point (a, b) lies in the first quadrant and is 6 units from the origin. In which quadrant is the point $(-a, -b)$ located and how far is it from the origin?

27. State the Pythagorean theorem in your own words.

28. Suppose the point (a, b) lies in the first quadrant and is 4 units from the origin. Explain why $a < 4$ and $b < 4$.

A triple of positive integers (a, b, c) is called a Pythagorean triple if it satisfies the equation of the Pythagorean theorem, $a^2 + b^2 = c^2$. Determine whether each of the following is a Pythagorean triple.

29. $(9, 12, 15)$

30. $(6, 8, 10)$

31. $(5, 10, 15)$

32. $(7, 24, 25)$

33. Show by an example that the following statement is true: If, for the positive integers a, b, and c, $a^2 + b^2 = c^2$, then it is not necessarily true that $a + b = c$.

34. Show that $(3, 4, 5)$ is a Pythagorean triple. Then, show that $(3k, 4k, 5k)$ is a Pythagorean triple for $k = 2$, $k = 3$, and $k = 4$. What general conclusion seems likely from this observation? (Although this is not a proof, the conclusion is indeed true, and can be proved using algebra.)

The converse of the Pythagorean theorem says that if a triangle has sides of lengths a, b, and c, where c is the longest side, and $a^2 + b^2 = c^2$, then the triangle is a right triangle. Use this result and the distance formula to decide if the following points are the vertices of right triangles.

GRAPH + FIND DISTANCE, than Use P.T.

35. $(-2, 5), (1, 5), (1, 9)$ **36.** $(-9, -2), (-1, -2), (-9, 11)$

37. $(\sqrt{3}, 2\sqrt{3} + 3), (\sqrt{3} + 4, -\sqrt{3} + 3), (2\sqrt{3}, 2\sqrt{3} + 4)$

38. $(4 - \sqrt{3}, -2\sqrt{3}), (2 - \sqrt{3}, -\sqrt{3}), (3 - \sqrt{3}, -2\sqrt{3})$

Find all values of x or y such that the distance between the given points is as indicated.

39. $(x, 7)$ and $(2, 3)$ is 5   **40.** $(5, y)$ and $(8, -1)$ is 5

41. $(3, y)$ and $(-2, 9)$ is 12   **42.** $(x, 11)$ and $(5, -4)$ is 17

43. Use the distance formula to write an equation for all points that are 5 units from $(0, 0)$. Sketch a graph showing these points.

44. Write an equation for all points 3 units from $(-5, 6)$. Sketch a graph showing these points.

The following exercises require the Pythagorean theorem. See Example 1. (Source: Trigonometry with Calculators by Lawrence S. Levy. Reprinted by permission of the author.)

45. A 1000-ft section of railroad track expands 6 inches because the day is very hot. This causes end C (see the figure) to break off and move to position B, forming right triangle ABC. Find BC. (The surprising answer to this simple problem explains why railroad tracks, bridges, and similar structures must be designed to allow for expansion.)

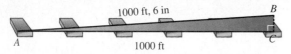

1000 ft, 6 in

A   1000 ft   B, C

46. Clothing manufacturers sometimes cut their material "on the bias" (that is, at 45° to the direction the threads run) to give it more elasticity. A tie maker wants to cut twenty 8-in strips of silk on the bias from

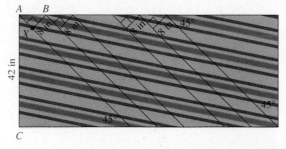

a rectangular piece of material that costs $10 per (linear) yd of material 42 in wide (see the figure). Find the total cost of the material. *Note:* This unappealing combination of units—inches, yards, and dollars—is typical of many practical problems, not just in the clothing industry. (*Hint:* First find length AB, using isosceles triangle ABX.)

47. The height (h) of the Great Pyramid of Egypt is 144 m. The apothem (a in the figure) measures 184.7 m. Assuming the base is a square, find the length l of a side of the base.

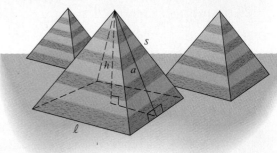

48. Use the result of Exercise 47 to find the length of the edge of the pyramid labeled s in the figure.

49. The derivation of the midpoint formula makes the assumption that the two points are not on a horizontal or vertical line. Explain why the midpoint formula is still valid in these two cases.

Use the midpoint formula given in the Connections box to find the midpoint of the line segment joining the two points.

50. (1, 3) and (7, 5)

51. (4, −3) and (−1, 2)

52. (8, 0) and (−6, 5)

53. (π, 3.5) and (4 − π, −5.5)

54. A line segment has an endpoint at (3, 2) and midpoint (5, 3). Find the other endpoint.

55. A line segment has an endpoint at (−4, 1) and midpoint (5, 6). Find the other endpoint.

56. A line segment has an endpoint at (a, b) and midpoint (0, 0). Find the other endpoint.

Write each of the following sets using interval notation.

57. $\{x \mid x > 6\}$

58. $\{y \mid y > 9\}$

59. $\{p \mid p \geq 10\}$

60. $\{x \mid -3 < x < 7\}$

61. $\{y \mid 8 \leq y \leq 13\}$

62.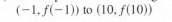

63. ←——+——+——+——)——+——→
　　　　−1　0　1　2　3

64. ←—+——+——(——+——+——]——+—→
　　　　−1　0　1　2　3　4　5

65. Explain why the set $\{y \mid |y| \leq 1\}$ is the same as $[-1, 1]$.

66. Explain why the set $\{y \mid |y| \geq 1\}$ is the same as $(-\infty, -1] \cup [1, \infty)$.

Let $f(x) = -2x^2 + 4x + 6$. Find each of the following. See Example 3.

67. $f(0)$

68. $f(-2)$

69. $f(-1)$

70. $f(-m)$

71. $f(1 + a)$

72. $f(2 - p)$

73. The accompanying table was generated by a graphing calculator for a linear function $Y_1 = f(x)$. Use the table to answer the following questions.
　(a) What is $f(2)$?
　(b) If $f(x) = -2.4$, what is the value of x?
　(c) At what point does the graph of $Y_1 = f(x)$ cross the y-axis?
　(d) At what point does the graph of $Y_1 = f(x)$ cross the x-axis?

74. Let $f(x)$ be the function in the accompanying graph. Find each of the following.
　(a) $f(0)$
　(b) $f(6)$
　(c) a negative number a for which $f(a) = 0$
　(d) three positive values of x for which $f(x) = 10$
　(e) the distance between $(8, f(8))$ and $(10, f(10))$
　(f) the midpoint of the line segment from $(-1, f(-1))$ to $(10, f(10))$

X	Y1
0	3.6
1	2.4
2	1.2
3	0
4	-1.2
5	-2.4
6	-3.6

X=0

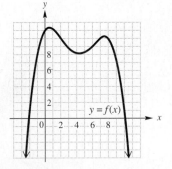

Find the domain and range. Identify any equations that define functions. See Example 4.

75. $y = 4x - 3$

76. $2x + 5y = 10$

77. $y = x^2 + 4$

78. $y = 2x^2 - 5$

79. $y = -2(x - 3)^2 + 4$

80. $y = 3(x + 1)^2 - 5$

81. $x = y^2$

82. $y = \sqrt{4 + x}$

83. $y = \sqrt{x^2 + 1}$

84. $y = \sqrt{1 - x^2}$

85.

86.

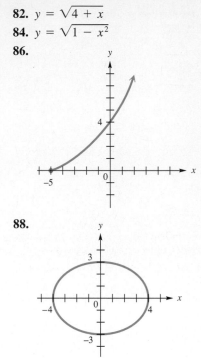

87.

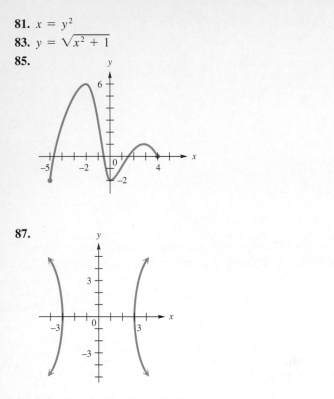

88.

Find the domain. See Example 5.

89. $y = \dfrac{1}{x}$

90. $y = \dfrac{-2}{x + 1}$

91. $y = \dfrac{-1}{\sqrt{x^2 + 25}}$

92. $y = \dfrac{3 + x}{\sqrt{2x - 5}}$

▼▼▼▼▼▼▼▼▼▼▼▼▼▼ **DISCOVERING CONNECTIONS** (Exercises 93–97) ▼▼▼▼▼▼▼▼▼▼▼▼▼▼

The figure shown is a square made up of four right triangles and a smaller square. By using the method of equal areas, the Pythagorean theorem may be proved. Work Exercises 93–97 in order, filling in the blanks with the missing information.

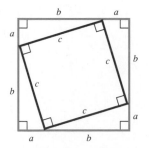

93. The length of a side of the large square is _____ , so its area is (_____)² or _____ .

94. The area of the large square may also be found by obtaining the sum of the areas of the four right triangles and the smaller square. The area of each right triangle is _____ , so the sum of the areas of the four right triangles is _____ . The area of the smaller square is _____ .

95. The sum of the areas of the four right triangles and the smaller square is _____ .

96. Since the areas in Exercises 93 and 95 represent the area of the same figure, the expressions there must be equal. Setting them equal to each other we obtain _____ = _____ .

97. Subtract $2ab$ from each side of the equation in Exercise 96 to obtain the desired result _____ = _____ .

1.2 Angles ▼▼▼

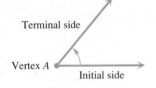

Line *AB*

Segment *AB*

Ray *AB*

FIGURE 9

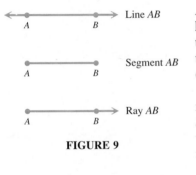

Terminal side

Vertex *A*

Initial side

FIGURE 10

A line may be drawn through the two distinct points *A* and *B*. This line is called **line *AB*.** The portion of the line between *A* and *B*, including points *A* and *B* themselves, is **segment *AB*.** The portion of line *AB* that starts at *A* and continues through *B*, and on past *B*, is called **ray *AB*.** Point *A* is the endpoint of the ray. (See Figure 9.)

An **angle** is formed by rotating a ray around its endpoint. The ray in its initial position is called the **initial side** of the angle, while the ray in its location after the rotation is the **terminal side** of the angle. The endpoint of the ray is the **vertex** of the angle. Figure 10 shows the initial and terminal sides of an angle with vertex *A*.

If the rotation of the terminal side is counterclockwise, the angle is **positive.** If the rotation is clockwise, the angle is **negative.** Figure 11 shows two angles, one positive and one negative.

An angle can be named by using the name of its vertex. For example, the angle on the right in Figure 11 can be called angle *C*. Alternatively, an angle can be named using three letters, with the vertex letter in the middle. Thus, the angle on the right also could be named angle *ACB* or angle *BCA*.

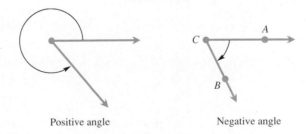

Positive angle

Negative angle

FIGURE 11

There are two systems in common use for measuring the size of angles. The most common unit of measure is the **degree.** (The other common unit of measure is called the *radian,* which is discussed in Chapter 3.) Degree measure was

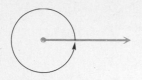

A complete rotation of a ray gives an angle whose measure is 360°.

FIGURE 12

developed by the Babylonians, 4000 years ago. To use degree measure, we assign 360 degrees to a complete rotation of a ray. In Figure 12, notice that the terminal side of the angle corresponds to its initial side when it makes a complete rotation.

One degree, written 1°, represents 1/360 of a rotation. Therefore, 90° represents $90/360 = 1/4$ of a complete rotation, and 180° represents $180/360 = 1/2$ of a complete rotation. Angles of measure 1°, 90°, and 180° are shown in Figure 13.

1° angle   90°   180°

FIGURE 13

Special angles are named as shown in the following chart.

Types of Angles		
Name	*Angle Measure*	*Example*
Acute angle	Between 0° and 90°	60° 82°
Right angle	Exactly 90°	90°
Obtuse angle	Between 90° and 180°	97° 138°
Straight angle	Exactly 180°	180°

If the sum of the measures of two angles is 90°, the angles are called **complementary.** Two angles with measures whose sum is 180° are **supplementary.**

EXAMPLE 1
Finding measures of
complementary and
supplementary angles

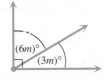

FIGURE 14

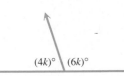

FIGURE 15

Find the measure of each angle in Figures 14 and 15.

(a) In Figure 14, since the two angles form a right angle (as indicated by the symbol),

$$6m + 3m = 90$$
$$9m = 90$$
$$m = 10.$$

The two angles have measures of $6 \cdot 10° = 60°$ and $3 \cdot 10° = 30°$.

(b) The angles in Figure 15 are supplementary, so

$$4k + 6k = 180$$
$$10k = 180$$
$$k = 18.$$

These angle measures are $4(18°) = 72°$ and $6(18°) = 108°$. ▶

Angles can be measured with an instrument called a **protractor.** Figure 16 shows a protractor measuring an angle of 35°.

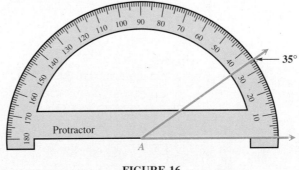

FIGURE 16

NOTE   Much of the study of trigonometry involves finding angle measures. We will see later in this book that trigonometry does not rely on using protractors to find these measures. For example, if we know the lengths of three sides of a triangle, we can use the Law of Cosines (Section 7.3) to find the angle measures mathematically, and with more precision than we could get from using a protractor.

Do not confuse an angle with its measure. Angle A of Figure 16 is a rotation; the measure of the rotation is 35°. This measure is often expressed by saying that $m(\text{angle } A)$ is 35°, where $m(\text{angle } A)$ is read "the measure of angle A." It saves a lot of work, however, to abbreviate $m(\text{angle } A) = 35°$ as simply angle $A = 35°$.

Traditionally, portions of a degree have been measured with minutes and seconds. One **minute,** written $1'$, is 1/60 of a degree.

$$1' = \frac{1}{60}° \qquad \text{or} \qquad 60' = 1°$$

One **second,** $1''$, is $1/60$ of a minute.

$$1'' = \frac{1}{60}' = \frac{1}{3600}^{\circ} \qquad \text{or} \qquad 60'' = 1'$$

The measure $12°\ 42'\ 38''$ represents 12 degrees, 42 minutes, 38 seconds.

The next example shows how to perform calculations with degrees, minutes, and seconds.

◖EXAMPLE 2
Calculating with degrees, minutes, and seconds

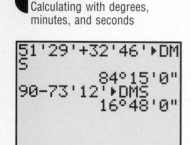

The calculations explained in Example 2 can be done with a graphing calculator capable of working with degrees, minutes, and seconds.

Perform each calculation.

(a) $51°\ 29' + 32°\ 46'$

Add the degrees and the minutes separately.

$$\begin{array}{r} 51°\ 29' \\ +\ 32°\ 46' \\ \hline 83°\ 75' \end{array}$$

Since $75' = 60' + 15' = 1°\ 15'$, the sum is written

$$\begin{array}{r} 83° \\ +\ \ 1°\ 15' \\ \hline 84°\ 15'. \end{array}$$

(b) $90° - 73°\ 12'$

Write $90°$ as $89°\ 60'$. Then

$$\begin{array}{r} 89°\ 60' \\ -\ 73°\ 12' \\ \hline 16°\ 48'. \end{array} \ ◗$$

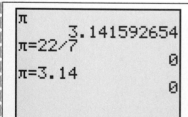

A graphing calculator will express the first few digits of the irrational number π, as shown in the first line of the display. If a statement is *false*, such as those in the second and third displays, the logic feature returns a zero. If a statement is *true*, a one is returned.

The second and third lines display false statements because 22/7 and 3.14 are rational *approximations* of π; they are not equal to π.

▦ Many graphing calculators have the capability to add and subtract angles given in degrees, minutes, and seconds. They are also able to convert from degrees, minutes, and seconds to decimal degrees (see below) and vice versa. Read your owner's manual for details on these capabilities. Also find out whether your calculator has a function to convert between degree measure and radian measure.

The real number π is used extensively to express angle measures in trigonometry. Learn where the π key is located on your calculator keyboard. Be aware that when *exact* values involving π are required, such as $\pi/3$ and $\pi/4$, decimal approximations given by the calculator are not acceptable.

Because calculators are an integral part of our world today, it is now common to measure angles in **decimal degrees.** For example, 12.4238° represents

$$12.4238° = 12\frac{4238}{10{,}000}°.$$

The next example shows how to change between decimal degrees and degrees, minutes, and seconds.

◀ EXAMPLE 3
Converting between decimal degrees and degrees, minutes, seconds

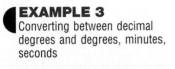
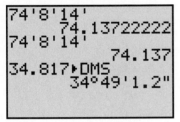

The conversions in Example 3 can be done on some graphing calculators. The second displayed result was obtained by setting the calculator to show only three places after the decimal point.

(a) Convert 74° 8′ 14″ to decimal degrees. Round to the nearest thousandth of a degree.

Since $1' = \frac{1}{60}°$ and $1'' = \frac{1}{3600}°$,

$$74° \, 8' \, 14'' = 74° + \frac{8}{60}° + \frac{14}{3600}°$$
$$\approx 74° + .1333° + .0039°$$
$$= 74.137° \text{ (rounded)}.$$

(b) Convert 34.817° to degrees, minutes, and seconds.

$$\begin{aligned}
34.817° &= 34° + .817° \\
&= 34° + (.817)(60') &\quad 1\text{ degree} = 60\text{ minutes} \\
&= 34° + 49.02' \\
&= 34° + 49' + .02' \\
&= 34° + 49' + (.02)(60'') &\quad 1\text{ minute} = 60\text{ seconds} \\
&= 34° + 49' + 1'' \text{ (rounded)} \\
&= 34° \, 49' \, 1''
\end{aligned}$$

An angle is in **standard position** if its vertex is at the origin and its initial side is along the positive x-axis. The two angles in Figure 17 are in standard position. An angle in standard position is said to lie in the quadrant in which its terminal side lies. For example, an acute angle is in quadrant I and an obtuse angle is in quadrant II. Angles in standard position having their terminal sides along the x-axis or y-axis, such as angles with measures 90°, 180°, 270°, and so on, are called **quadrantal angles.**

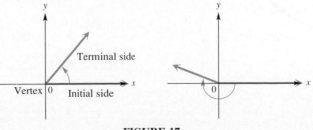

FIGURE 17

A complete rotation of a ray results in an angle of measure 360°. But there is no reason why the rotation need stop at 360°. By continuing the rotation, angles of measure larger than 360° can be produced. The angles in Figure 18(a) have measures 60° and 420°. These two angles have the same initial side and the same terminal side, but different amounts of rotation. Angles that have the same initial side and the same terminal side are called **coterminal angles.** As shown in Figure 18(b), angles with measures 110° and 830° are coterminal. ▶

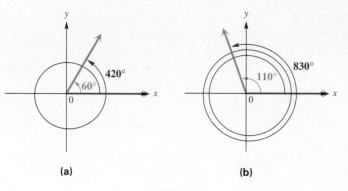

(a)   (b)

FIGURE 18

◀ EXAMPLE 4
Finding measures of coterminal angles

Find the angles of smallest possible positive measure coterminal with the following angles.

(a) 908°

Add or subtract 360° as many times as needed to get an angle with measure greater than 0° but less than 360°. Since 908° − 2 · 360° = 908° − 720° = 188°, an angle of 188° is coterminal with an angle of 908°. See Figure 19.

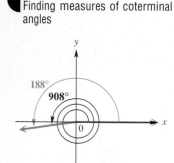

FIGURE 19

(b) −75°

Use a rotation of 360° + (−75°) = 285°. See Figure 20. ▶

Sometimes it is necessary to find an expression that will generate all angles coterminal with a given angle. For example, suppose we wish to do this for a 60° angle.

Since any angle coterminal with 60° can be obtained by adding an appropriate integer multiple of 360° to 60°, we can let n represent any integer, and the expression

$$60° + n \cdot 360°$$

will represent all such coterminal angles.

The table below shows a few possibilities.

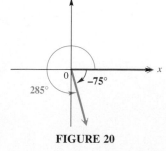

FIGURE 20

Value of n	Angle Coterminal with 60°
2	60° + 2 · 360° = **780°**
1	60° + 1 · 360° = **420°**
0	60° + 0 · 360° = **60°** (the angle itself)
−1	60° + (−1) · 360° = **−300°**

EXAMPLE 5
Analyzing the revolutions of a phonograph record

A phonograph record makes 45 revolutions per minute. Through how many degrees will a point on the edge of the record move in 2 seconds?

The record revolves 45 times per minute or $45/60 = 3/4$ time per second (since there are 60 seconds in a minute). In 2 seconds, the record will revolve $2 \cdot (3/4) = 3/2$ times. Each revolution is $360°$, so a point on the edge will revolve $(3/2) \cdot 360° = 540°$ in 2 seconds. ▶

1.2 Exercises ▼▼▼▼▼▼▼▼▼▼▼▼▼▼▼▼▼▼▼▼▼▼▼▼▼▼▼▼▼▼▼▼▼▼▼

1. Explain the difference between a segment and a ray.
2. What part of a complete revolution is an angle of 45°?
3. What angle is its own complement?
4. What angle is its own supplement?
5. Does a merry-go-round turn in a clockwise or counterclockwise direction?
6. Does the shadow of a sundial move in a clockwise or counterclockwise direction?

Find the measure of each angle in Exercises 7–12. See Example 1.

7. $(7x)°$ $(11x)°$

8. $(2y)°$ $(4y)°$

9. $(5k + 5)°$ $(3k + 5)°$

10. Supplementary angles with measures $10m + 7$ and $7m + 3$ degrees
11. Supplementary angles with measures $6x - 4$ and $8x - 12$ degrees
12. Complementary angles with measures $9z + 6$ and $3z$ degrees
13. If an angle measures x degrees, how can we represent its complement?
14. If an angle measures x degrees, how can we represent its supplement?
15. If a positive angle has measure x between $0°$ and $60°$, how can we represent the first negative angle coterminal with it?
16. If a negative angle has measure x between $0°$ and $-60°$, how can we represent the first positive angle coterminal with it?

Perform each calculation. See Example 2.

17. $62° 18' + 21° 41'$
18. $75° 15' + 83° 32'$
19. $71° 58' + 47° 29'$
20. $90° - 73° 48'$
21. $90° - 51° 28'$
22. $180° - 124° 51'$
23. $90° - 72° 58' 11''$
24. $90° - 36° 18' 47''$

Convert each angle measure to decimal degrees. Use a calculator, and round to the nearest thousandth of a degree. See Example 3.

25. $20° 54'$
26. $38° 42'$
27. $91° 35' 54''$
28. $34° 51' 35''$
29. $274° 18' 59''$
30. $165° 51' 9''$

Convert each angle measure to degrees, minutes, and seconds. Use a calculator as necessary. See Example 3.

31. 31.4296°   **32.** 59.0854°   **33.** 89.9004°

34. 102.3771°   **35.** 178.5994°   **36.** 122.6853°

37. Read about the degree symbol (°) in the manual for your graphing calculator. How is it used?

38. Show that 1.21 hours is the same as 1 hour, 12 minutes, and 36 seconds. Discuss the similarity between converting hours, minutes, and seconds to decimal hours and converting degrees, minutes, and seconds to decimal degrees.

Find the angles of smallest positive measure coterminal with the following angles. See Example 4.

39. −40°   **40.** −98°   **41.** −125°   **42.** −203°

43. 539°   **44.** 699°   **45.** 850°   **46.** 1000°

Give an expression that generates all angles coterminal with the given angle. Let n represent any integer.

47. 30°   **48.** 45°   **49.** 60°   **50.** 90°   **51.** 135°   **52.** 270°   **53.** −90°   **54.** −135°

55. Explain why the answers to Exercises 52 and 53 give the same set of angles.

56. Which two of the following are not coterminal with $r°$?
   (a) $360° + r°$ **(b)** $r° − 360°$
   (c) $360° − r°$ **(d)** $r° + 180°$

Consider the function $Y_1 = 360((X/360) − \text{int}(X/360))$ specified on a graphing calculator. (Note: The value of $\text{int}(x)$ is the largest integer less than or equal to x. With some calculators, int is found in the MATH menu.) The screen here shows that for X = 908 and X = −75, the function returns the smallest possible positive measure coterminal with the angle. See Example 4.

```
Y₁(908)
              188
Y₁(-75)
              285
```

57. Rework Exercise 39 with a graphing calculator.

58. Rework Exercise 40 with a graphing calculator.

Sketch the angle in standard position. Draw an arrow representing the correct amount of rotation. Find the measure of two other angles, one positive and one negative, that are coterminal with the given angle. Give the quadrant of each angle.

59. 75°   **60.** 89°   **61.** 122°   **62.** 174°   **63.** 234°   **64.** 250°

65. 300°   **66.** 512°   **67.** 624°   **68.** −52°   **69.** −61°   **70.** −159°

Locate the following points in a coordinate system. Draw a ray from the origin through the given point. Indicate with an arrow the angle in standard position having smallest positive measure. Then find the distance r from the origin to the point, using the distance formula of Section 1.1.

71. $(-3, -3)$ **72.** $(-5, 2)$ **73.** $(-3, -5)$

74. $(\sqrt{3}, 1)$ **75.** $(-2, 2\sqrt{3})$ **76.** $(4\sqrt{3}, -4)$

77. A windmill makes 90 revolutions per minute. How many revolutions does it make per second?

78. A turntable makes 45 revolutions per minute. How many revolutions does it make per second?

Solve each problem. See Example 5.

79. A tire is rotating 600 times per minute. Through how many degrees does a point on the edge of the tire move in 1/2 second?

80. An airplane propeller rotates 1000 times per minute. Find the number of degrees that a point on the edge of the propeller will rotate in 1 second.

81. A pulley rotates through 75° in one minute. How many rotations does the pulley make in an hour?

82. One student in a surveying class measures an angle as 74.25°, while another student measures the same angle as 74° 20′. Find the difference between these measurements, both to the nearest minute and to the nearest hundredth of a degree.

83. 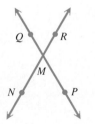 Due to Earth's rotation, celestial objects like the moon and the stars appear to move across the sky, rising in the east and setting in the west. As a result, if a telescope on Earth remains stationary while viewing a celestial object, the object will slowly move outside of the viewing field of the telescope. For this reason a motor is often attached to telescopes so that the telescope rotates at the same rate as Earth. Determine how long it should take the motor to turn the telescope through an angle of 1 minute in a direction perpendicular to Earth's axis.

1.3 Angle Relationships and Similar Triangles ▼▼▼

FIGURE 21

In this section we look at some geometric properties that will be used in the study of trigonometry.

In Figure 21, the sides of angle *NMP* have been extended to form another angle, *RMQ*. The pair of angles *NMP* and *RMQ* are called **vertical angles.** Another pair of vertical angles, *NMQ* and *PMR*, are formed at the same time. Vertical angles have the following important property.

> **Vertical Angles**
>
> Vertical angles have equal measures.

Parallel lines are lines that lie in the same plane and do not intersect. Figure 22 shows parallel lines *m* and *n*. When a line *q* intersects two parallel lines, *q* is called a **transversal.** In Figure 22, the transversal intersecting the

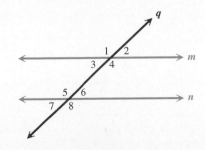

FIGURE 22

parallel lines forms eight angles, indicated by numbers. It is shown in geometry that angles 1 through 8 in Figure 22 possess some special properties regarding their degree measures. The following chart gives their names with respect to each other, and rules regarding their measures.

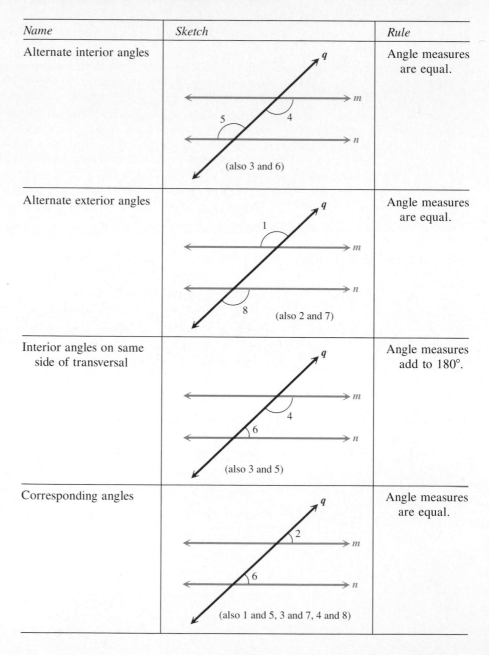

Name	*Sketch*	*Rule*
Alternate interior angles	(also 3 and 6)	Angle measures are equal.
Alternate exterior angles	(also 2 and 7)	Angle measures are equal.
Interior angles on same side of transversal	(also 3 and 5)	Angle measures add to 180°.
Corresponding angles	(also 1 and 5, 3 and 7, 4 and 8)	Angle measures are equal.

EXAMPLE 1
Finding angle measures

Current models of graphing calculators feature numerical solve capability. The equation in Example 1 is solved in this screen. Consult your owner's manual to find the syntax required by your particular model.

Find the measure of each marked angle, given that lines m and n are parallel. (See Figure 23.)

The marked angles are alternate exterior angles, which are equal. This gives

$$3x + 2 = 5x - 40$$
$$42 = 2x$$
$$21 = x.$$

One angle has a measure of $3 \cdot 21 + 2 = 65$ degrees, and the other has a measure of $5x - 40 = 5 \cdot 21 - 40 = 65$ degrees. ▶

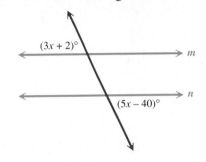

FIGURE 23

TRIANGLES An important property of triangles that was first proved by the Greek geometers deals with the sum of the measures of the angles of any triangle.

Angle Sum of a Triangle

The sum of the measures of the angles of any triangle is $180°$.

While it is not an actual proof, a rather convincing argument for the truth of this statement can be given using any size triangle cut from a piece of paper. Tear each corner from the triangle, as suggested in Figure 24(a). You should be able to rearrange the pieces so that the three angles form a straight angle, as shown in Figure 24(b).

Suppose that two of the angles of a triangle are $48°$ and $61°$. Find the measure of the third angle, x, by using the fact that all three angle measures add to $180°$.

$$48° + 61° + x = 180°$$
$$109° + x = 180°$$
$$x = 71°$$

The third angle of the triangle measures $71°$.

Triangles are classified according to angles and sides as shown in the chart on the following page.

(a)

(b)

FIGURE 24

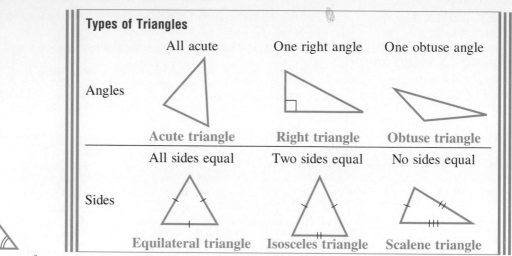

Types of Triangles

	All acute	One right angle	One obtuse angle
Angles	Acute triangle	Right triangle	Obtuse triangle
	All sides equal	Two sides equal	No sides equal
Sides	Equilateral triangle	Isosceles triangle	Scalene triangle

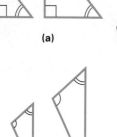

(a)

(b)

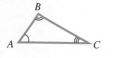

(c)

FIGURE 25

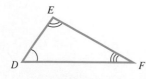

FIGURE 26

Many key ideas of trigonometry depend on **similar triangles,** which are triangles of exactly the same shape but not necessarily the same size. Figure 25 shows three pairs of similar triangles.

The two triangles in the third pair have not only the same shape but also the same size. Triangles that are both the same size and the same shape are called **congruent triangles.** If two triangles are congruent, then it is possible to pick one of them up and place it on top of the other so that they coincide. If two triangles are congruent, then they must be similar. However, two similar triangles need not be congruent.

The triangle supports for a child's swing are congruent (and thus similar) triangles, machine-produced with exactly the same dimensions each time. These supports are just one example of similar triangles. The supports of a long bridge, all the same shape but decreasing in size toward the center of the bridge, are examples of similar (but not congruent) triangles.

Suppose a correspondence between two triangles *ABC* and *DEF* is set up as shown in Figure 26.

Angle *A* corresponds to angle *D*.	Side *AB* corresponds to side *DE*.
Angle *B* corresponds to angle *E*.	Side *BC* corresponds to side *EF*.
Angle *C* corresponds to angle *F*.	Side *AC* corresponds to side *DF*.

For triangle *ABC* to be similar to triangle *DEF*, the following conditions must hold.

Conditions for Similar Triangles

1. Corresponding angles must have the same measure.
2. Corresponding sides must be proportional. (That is, their ratios must be equal.)

These conditions are used in the following examples.

EXAMPLE 2
Finding angle measures in similar triangles

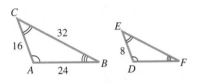

FIGURE 27

In Figure 27, triangles *ABC* and *NMP* are similar. Find the measures of angles *B* and *C*.

Since the triangles are similar, corresponding angles have the same measure. Since *C* corresponds to *P* and *P* measures 104°, angle *C* also measures 104°. Since angles *B* and *M* correspond, *B* measures 31°. ▶

NOTE The small arcs found at the angles in Figures 25–28 are used to denote the corresponding angles in the triangle. This symbolism will be used when appropriate in this book. We will also use △ to denote "triangle."

EXAMPLE 3
Finding side lengths in similar triangles

FIGURE 28

Given that △*ABC* and △*DFE* in Figure 28 are similar, find the lengths of the unknown sides of △*DFE*.

As mentioned before, similar triangles have corresponding sides in proportion. Use this fact to find the unknown side lengths in △*DFE*. Side *DF* of △*DFE* corresponds to side *AB* of △*ABC*, and sides *DE* and *AC* correspond. This leads to the proportion

$$\frac{8}{16} = \frac{DF}{24}.$$

Recall the *cross-multiplication* property of proportions.

$$\text{If} \quad \frac{a}{b} = \frac{c}{d}, \quad \text{then} \quad ad = bc.$$

Use cross-multiplication to solve the equation for *DF*.

$$\frac{8}{16} = \frac{DF}{24}$$

$$8 \cdot 24 = 16 \cdot DF \qquad \text{Cross-multiply.}$$

$$192 = 16 \cdot DF$$

$$12 = DF$$

Side *DF* has a length of 12.

Side *EF* corresponds to *CB*. This leads to another proportion:

$$\frac{8}{16} = \frac{EF}{32}.$$

Cross-multiplication gives

$$8 \cdot 32 = 16 \cdot EF$$

$$16 = EF.$$

Side *EF* has a length of 16. ▶

Applied problems can sometimes be solved using properties of similar triangles.

EXAMPLE 4
Finding the height of a flagpole

The people at the Arcade Fire Station need to measure the height of the station flagpole. They notice that at the instant when the shadow of the station is 18 feet long, the shadow of the flagpole is 99 feet long. The station is 10 feet high. Find the height of the flagpole.

Figure 29 shows the information given in the problem. The two triangles shown there are similar, so that corresponding sides are in proportion, with

$$\frac{MN}{10} = \frac{99}{18}$$

or

$$\frac{MN}{10} = \frac{11}{2}$$

$$2 \cdot MN = 110$$

$$MN = 55.$$

The flagpole is 55 feet high. ▶

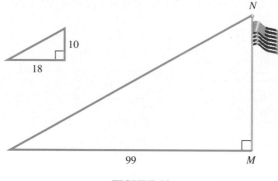

FIGURE 29

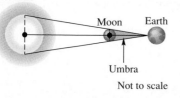

FIGURE 30

EXAMPLE 5
Determining when a solar eclipse can occur

The sun has a diameter of about 865,000 miles with a maximum distance from Earth's surface of about 94,500,000 miles. The moon has a smaller diameter of 2,159 miles. For a total solar eclipse to occur, the moon must pass between Earth and the sun. The moon must also be close enough to Earth for the moon's umbra (shadow) to reach the surface of Earth. See Figure 30.*

(a) Calculate the maximum distance that the moon can be from Earth and still have a total solar eclipse occur.

_____

* Karttunen, H., P. Kröger, H. Oja, M. Putannen, and K. Donners (editors), *Fundamental Astronomy,* Springer-Verlag, 1994.
The Guinness Book of Records 1995.

Let D_s be the Earth-sun distance, d_s the diameter of the sun, D_m the Earth-moon distance, and d_m the diameter of the moon. Then, by similar triangles

$$\frac{D_s}{D_m} = \frac{d_s}{d_m}$$

$$D_m = \frac{D_s d_m}{d_s} = \frac{94,500,000 \times 2,159}{865,000} \approx 236,000 \text{ miles.}$$

(b) During this century, the closest approach of the moon to Earth's surface was 217,000 miles and the farthest was 249,000 miles. Can a total solar eclipse occur every time that the moon is between Earth and the sun? Explain. No. The moon must be less than 236,000 miles away from Earth and sometimes it is farther than this. ▶

1.3 Exercises ▼▼▼▼▼▼▼▼▼▼▼▼▼▼▼▼▼▼▼▼▼▼▼▼▼▼▼▼▼▼▼▼▼▼▼▼▼▼

1. A geometry book states, "When two straight lines intersect, the opposite angles are equal." What term do we use for "opposite angles"?

2. Consider Figure 22. If the measure of one of the angles is known, can the measures of the remaining seven angles be determined?

Use the properties of angle measures given in this section to find the measure of each marked angle. In Exercises 11–14, m and n are parallel. See Example 1.

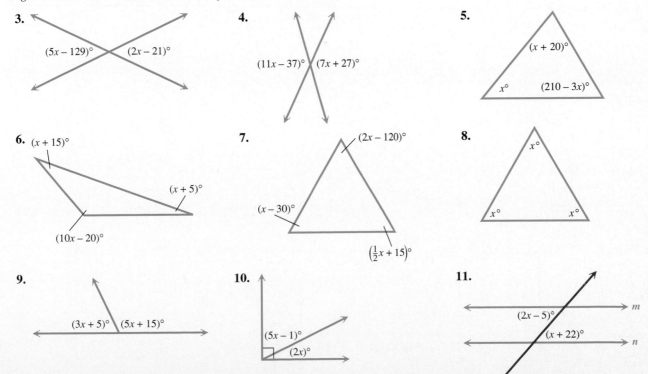

3. $(5x - 129)°$ $(2x - 21)°$

4. $(11x - 37)°$ $(7x + 27)°$

5. $(x + 20)°$ $x°$ $(210 - 3x)°$

6. $(x + 15)°$ $(x + 5)°$ $(10x - 20)°$

7. $(2x - 120)°$ $(x - 30)°$ $\left(\frac{1}{2}x + 15\right)°$

8. $x°$ $x°$ $x°$

9. $(3x + 5)°$ $(5x + 15)°$

10. $(5x - 1)°$ $(2x)°$

11. $(2x - 5)°$ m $(x + 22)°$ n

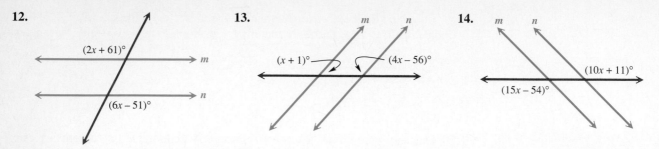

12.

13.

14.

The measures of two angles of a triangle are given. Find the measure of the third angle.

15. 37°, 52°

16. 29°, 104°

17. 147° 12′, 30° 19′

18. 136° 50′, 41° 38′

19. 74.2°, 80.4°

20. 29.6°, 49.7°

21. Can a triangle have two angles of measures 85° and 100°? Explain.

22. Can a triangle have two obtuse angles? Explain.

23. Use the given figure to find the measures of the numbered angles, given that lines *m* and *n* are parallel.

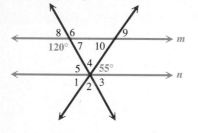

24. Find the measures of the marked angles, given that *x* + *y* = 40. (*Hint:* You must solve a system of equations.)

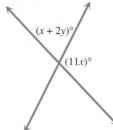

Classify each triangle in Exercises 25–36 as either acute, right, *or* obtuse. *Also classify each as either* equilateral, isosceles, *or* scalene.

25.

26.

27.

28.

29.

30.

31.

32.

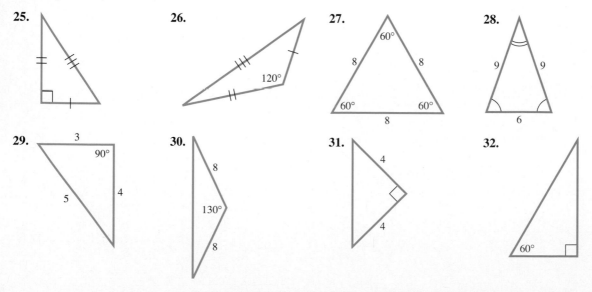

33. **34.** **35.** **36.**

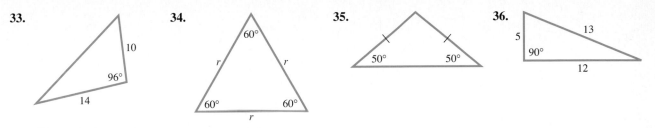

37. Write a definition of *isosceles right triangle*.

38. Explain why the sum of the lengths of any two sides of a triangle must be greater than the length of the third side.

39. Must all equilateral triangles be similar? Explain.

40. In the classic 1939 movie *The Wizard of Oz,* the scarecrow, upon getting a brain, says the following: "The sum of the square roots of any two sides of an isosceles triangle is equal to the square root of the remaining side." Give an example to show that his statement is incorrect.

Name the corresponding angles and the corresponding sides for each of the following pairs of similar triangles.

41. **42.**

43. (*EA* is parallel to *CD*.) **44.** (*HK* is parallel to *EF*.)

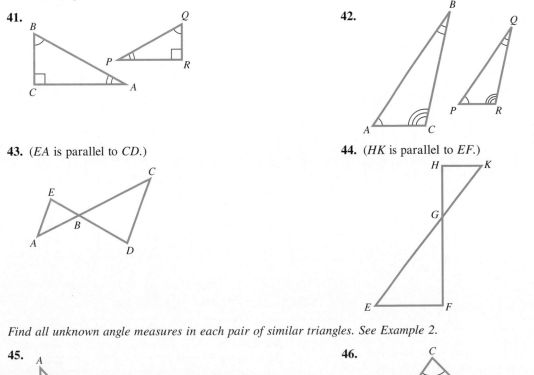

Find all unknown angle measures in each pair of similar triangles. See Example 2.

45. **46.**

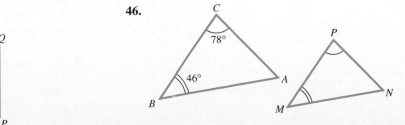

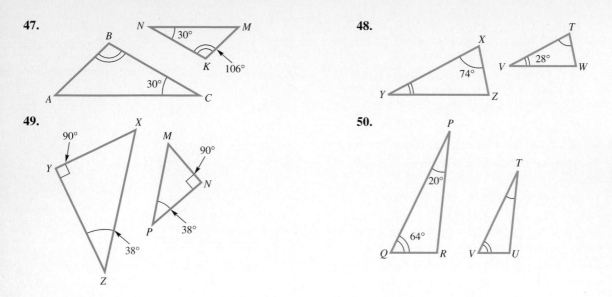

47.

48.

49.

50.

Find the unknown side lengths in each pair of similar triangles. See Example 3.

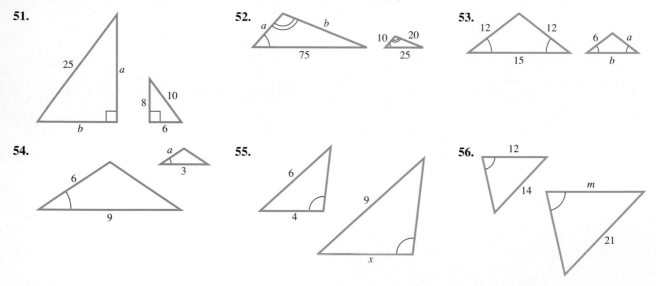

51.

52.

53.

54.

55.

56.

Solve the following problems. See Example 4.

57. A tree casts a shadow 45 m long. At the same time, the shadow cast by a vertical 2-m stick is 3 m long. Find the height of the tree.

58. A forest fire lookout tower casts a shadow 180 ft long at the same time that the shadow of a 9-ft truck is 15 ft long. Find the height of the tower.

59. On a photograph of a triangular piece of land, the lengths of the three sides are 4 cm, 5 cm, and 7 cm, respectively. The shortest side of the actual piece of land is 400 m long. Find the lengths of the other two sides.

60. The Santa Cruz lighthouse is 14 m tall and casts a shadow 28 m long at 7 P.M. At the same time, the shadow of the lighthouse keeper is 3.5 m long. How tall is she?

61. A house is 15 ft tall. Its shadow is 40 ft long at the same time the shadow of a nearby building is 300 ft long. Find the height of the building.

62. By drawing lines on a map, a triangle can be formed by the cities of Phoenix, Tucson, and Yuma. On the map, the distance between Phoenix and Tucson is 8 cm, the distance between Phoenix and Yuma is 12 cm, and the distance between Tucson and Yuma is 17 cm. The actual straight-line distance from Phoenix to Yuma is 230 km. Find the distances between the other pairs of cities.

In each diagram, there are two similar triangles. Find the unknown measurement in each. (Hint: In the sketch for Exercise 63, the side of length 100 in the small triangle corresponds to a side of length 100 + 120 = 220 in the larger triangle.)

63.

64.

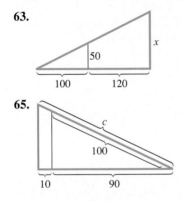

65.

66.

Work each of the following problems.

67. Two quadrilaterals (four-sided figures) are similar. The lengths of the three shortest sides of the first quadrilateral are 18 cm, 24 cm, and 32 cm. The lengths of the two longest sides of the second quadrilateral are 48 cm and 60 cm. Find the unknown lengths of the sides of these two figures.

68. Assume that Lincoln was 6 1/3 ft tall and his head 3/4 ft long. Knowing that the carved head of Lincoln at Mount Rushmore is 60 ft tall, find out how tall his entire body would be if it were carved into the mountain.

In each figure, two similar triangles are present. Find the value of each variable in the figures.

69.

70.

71. Refer to Example 5. The sun's distance from the surface of Mars is approximately 142,000,000 miles. One of Mars' two moons, Phobos, has a maximum diameter of 17.4 miles. (*Source:* Zeilik, M., S. Gregory, and E. Smith, *Introductory Astronomy and Astrophysics,* Saunders College Publishers, 1992.)

(a) Calculate the maximum distance that the moon Phobos can be from Mars in order for a total eclipse of the sun to occur.

(b) Phobos is approximately 5800 miles from Mars. Is it possible for Phobos to cause a total eclipse on the planet Mars?

72. Refer to Exercise 71. The sun's distance from the surface of Jupiter is approximately 484,000,000 miles. One of Jupiter's moons, Ganymede, is the largest moon in the solar system with a diameter of 3270 miles.

(a) Calculate the maximum distance that the moon Ganymede can be from Jupiter in order for a total eclipse of the sun to occur on Jupiter.

(b) Ganymede is approximately 665,000 miles from Jupiter. Is it possible for Ganymede to cause a total eclipse on the planet Jupiter?

73. Estimate the height of an actual flagpole. During the early morning or late afternoon of a sunny day, hold a ruler perpendicular to the ground and measure the length of the ruler's shadow. Then measure the length of the flagpole's shadow. (You can either use your ruler or you can measure your foot and walk off steps along the shadow.) Use the two measurements to obtain the height of the flagpole.

1.4 Definitions of the Trigonometric Functions ▼▼▼

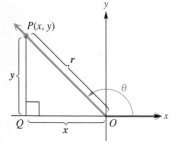

FIGURE 31

The study of trigonometry covers the six trigonometric functions defined in this section. Most sections in the remainder of this book involve at least one of these functions. To define these six basic functions, start with an angle θ (the Greek letter *theta**) in standard position. Choose any point P having coordinates (x, y) on the terminal side of angle θ. (The point P must not be the vertex of the angle.) See Figure 31.

A perpendicular from P to the x-axis at point Q determines a triangle having vertices at O, P, and Q. (More will be said about such triangles in Section 1.5.) The distance r from $P(x, y)$ to the origin, $(0, 0)$, can be found from the distance formula.

$$r = \sqrt{(x - 0)^2 + (y - 0)^2}$$
$$r = \sqrt{x^2 + y^2}$$

Notice that $r > 0$, since distance is never negative.

The six trigonometric functions of angle θ are called **sine, cosine, tangent, cotangent, secant,** and **cosecant.** In the following definitions, we use the customary abbreviations for the names of these functions.

---

* Greek letters are often used to name angles. A list of Greek letters appears inside the back cover of this book.

Trigonometric Functions

Let (x, y) be a point other than the origin on the terminal side of an angle θ in standard position. The distance from the point to the origin is $r = \sqrt{x^2 + y^2}$. The six trigonometric functions of θ are:

$$\sin \theta = \frac{y}{r} \qquad\qquad \csc \theta = \frac{r}{y} \quad (y \neq 0)$$

$$\cos \theta = \frac{x}{r} \qquad\qquad \sec \theta = \frac{r}{x} \quad (x \neq 0)$$

$$\tan \theta = \frac{y}{x} \quad (x \neq 0) \qquad \cot \theta = \frac{x}{y} \quad (y \neq 0).$$

NOTE Although Figure 31 shows a second quadrant angle, these definitions apply to any angle θ. Because of the restrictions on the denominators in the definitions of tangent, cotangent, secant, and cosecant, some angles will have undefined function values. This will be discussed in more detail later.

◀**EXAMPLE 1**
Finding the function values of an angle

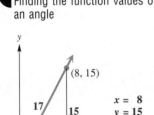

FIGURE 32

The terminal side of an angle α in standard position goes through the point $(8, 15)$. Find the values of the six trigonometric functions of angle α.

Figure 32 shows angle α and the triangle formed by dropping a perpendicular from the point $(8, 15)$ to the x-axis. The point $(8, 15)$ is 8 units to the right of the y-axis and 15 units above the x-axis, so that $x = 8$ and $y = 15$.

Since $r = \sqrt{x^2 + y^2}$,

$$r = \sqrt{8^2 + 15^2}$$
$$= \sqrt{64 + 225}$$
$$= \sqrt{289}$$
$$= 17.$$

The values of the six trigonometric functions of angle α can now be found with the definitions given above.

$$\sin \alpha = \frac{y}{r} = \frac{15}{17} \qquad \csc \alpha = \frac{r}{y} = \frac{17}{15}$$

$$\cos \alpha = \frac{x}{r} = \frac{8}{17} \qquad \sec \alpha = \frac{r}{x} = \frac{17}{8}$$

$$\tan \alpha = \frac{y}{x} = \frac{15}{8} \qquad \cot \alpha = \frac{x}{y} = \frac{8}{15} \quad ▶$$

◖EXAMPLE 2
Finding the function values of
an angle

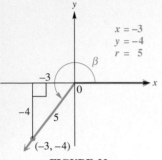

FIGURE 33

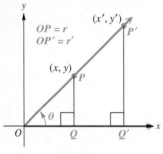

FIGURE 34

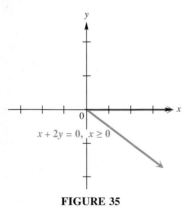

FIGURE 35

The terminal side of angle β in standard position goes through $(-3, -4)$. Find the values of the six trigonometric functions of β.

As shown in Figure 33, $x = -3$ and $y = -4$. The value of r is

$$r = \sqrt{(-3)^2 + (-4)^2}$$
$$r = \sqrt{25}$$
$$r = 5.$$

(Remember that $r > 0$.) Then by the definitions of the trigonometric functions,

$$\sin \beta = \frac{-4}{5} = -\frac{4}{5} \qquad \csc \beta = \frac{5}{-4} = -\frac{5}{4}$$

$$\cos \beta = \frac{-3}{5} = -\frac{3}{5} \qquad \sec \beta = \frac{5}{-3} = -\frac{5}{3}$$

$$\tan \beta = \frac{-4}{-3} = \frac{4}{3} \qquad \cot \beta = \frac{-3}{-4} = \frac{3}{4}. \quad ◗$$

The six trigonometric functions can be found from *any* point on the terminal side of the angle other than the origin. To see why any point may be used, refer to Figure 34, which shows an angle θ and two distinct points on its terminal side. Point P has coordinates (x, y) and point P' (read "P-prime") has coordinates (x', y'). Let r be the length of the hypotenuse of triangle OPQ, and let r' be the length of the hypotenuse of triangle $OP'Q'$. Since corresponding sides of similar triangles are in proportion,

$$\frac{y}{r} = \frac{y'}{r'},$$

so that $\sin \theta = y/r$ is the same no matter which point is used to find it. A similar result holds for the other five functions.

We can also find the trigonometric function values of an angle if we know the equation of the line coinciding with the terminal ray. Recall from algebra that the graph of the equation

$$Ax + By = 0$$

is a line that passes through the origin. If we restrict x to have only nonpositive or only nonnegative values, we obtain as the graph a ray with endpoint at the origin. For example, the graph of $x + 2y = 0$, $x \geq 0$, is shown in Figure 35. A ray such as the one described above can serve as the terminal side of an angle in standard position. By finding a point on the ray, the trigonometric function values of the angle can be found.

EXAMPLE 3
Finding the function values of an angle

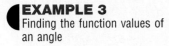

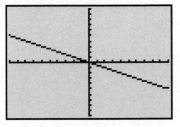

[–10, 10] by [–10, 10]

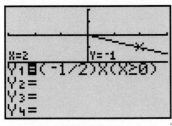

[–3, 3] by [–2, 2]

The line $x + 2y = 0$ is graphed in a standard viewing window in the first screen above. It is done by entering the equation as $Y_1 = (-1/2)x$ or $Y_1 = -.5x$.

The second screen is in split-screen mode. With the restriction $x \geq 0$ we are able to simulate the angle θ in standard position as shown in Figure 36.

Find the six trigonometric function values of the angle θ in standard position, if the terminal side of θ is defined by $x + 2y = 0$, $x \geq 0$.

The angle is shown in Figure 36. We can use *any* point except $(0, 0)$ on the terminal side of θ to find the trigonometric function values, so if we let $x = 2$, we can find the corresponding value of y.

$$x + 2y = 0, \ x \geq 0$$
$$2 + 2y = 0 \qquad \text{Arbitrarily choose } x = 2.$$
$$2y = -2$$
$$y = -1$$

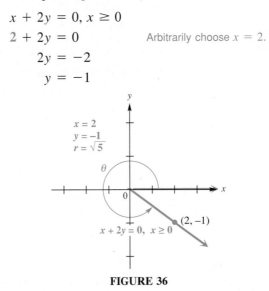

FIGURE 36

The point $(2, -1)$ lies on the terminal side, and the corresponding value of r is $r = \sqrt{2^2 + (-1)^2} = \sqrt{5}$. Now use the definitions of the trigonometric functions.

$$\sin \theta = \frac{y}{r} = \frac{-1}{\sqrt{5}} = \frac{-1}{\sqrt{5}} \cdot \frac{\sqrt{5}}{\sqrt{5}} = -\frac{\sqrt{5}}{5} \qquad \csc \theta = \frac{r}{y} = \frac{\sqrt{5}}{-1} = -\sqrt{5}$$

$$\cos \theta = \frac{x}{r} = \frac{2}{\sqrt{5}} = \frac{2}{\sqrt{5}} \cdot \frac{\sqrt{5}}{\sqrt{5}} = \frac{2\sqrt{5}}{5} \qquad \sec \theta = \frac{r}{x} = \frac{\sqrt{5}}{2}$$

$$\tan \theta = \frac{y}{x} = \frac{-1}{2} = -\frac{1}{2} \qquad \cot \theta = \frac{x}{y} = \frac{2}{-1} = -2 \ \blacktriangleright$$

Recall that when the equation of a line is written in the form $y = mx + b$, the coefficient of x is the slope of the line. In Example 3, $x + 2y = 0$ can be written as $y = (-1/2)x$, so the slope is $-1/2$. Notice that $\tan \theta = -1/2$. In general, it is true that $m = \tan \theta$. See the Connections box in Section 1.5 for a proof of this.

NOTE   The trigonometric function values we found in Examples 1–3 are *exact*. If we were to use a calculator to approximate these values, the decimal results would not be acceptable if exact values were required.

CONNECTIONS A convenient way to see the three basic trigonometric ratios geometrically is shown in Figure 37 below for θ in quadrants I and II. The circle, which has a radius of 1, is called a *unit circle*. We will see the unit circle again in Chapter 3. By memorizing this figure and the segments that represent the sine, cosine, and tangent functions, one can quickly recall the properties of the trigonometric functions. Horizontal line segments to the left of the origin and vertical line segments below the x-axis represent negative values. Note that the tangent line that contains the tangent segment must be tangent to the circle at $(1, 0)$, no matter which quadrant θ lies in.

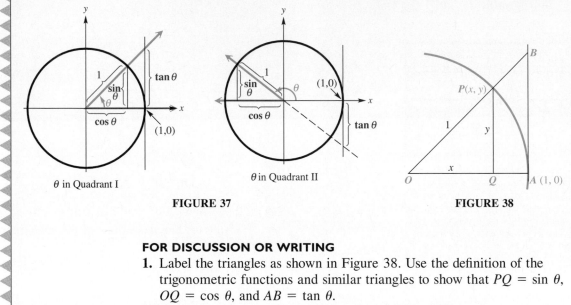

FIGURE 37   **FIGURE 38**

FOR DISCUSSION OR WRITING

1. Label the triangles as shown in Figure 38. Use the definition of the trigonometric functions and similar triangles to show that $PQ = \sin\theta$, $OQ = \cos\theta$, and $AB = \tan\theta$.
2. Sketch similar figures for θ in quadrants III and IV.

If the terminal side of an angle in standard position lies along the y-axis, any point on this terminal side has x-coordinate 0. Similarly, an angle with terminal side on the x-axis has y-coordinate 0 for any point on the terminal side. Since the values of x and y appear in the denominators of some of the trigonometric functions, and since a fraction is undefined if its denominator is 0, some of the trigonometric function values of quadrantal angles (i.e., those with terminal side on an axis) will be undefined.

EXAMPLE 4
Finding trigonometric function values of a quadrantal angle

Find the values of the six trigonometric functions for the following angles.

(a) an angle of 90°

First, select any point on the terminal side of a 90° angle. Let us select the point $(0, 1)$, as shown in Figure 39(a). Here $x = 0$ and $y = 1$. Verify that $r = 1$. Then, by the definition of the trigonometric functions,

$$\sin 90° = \frac{1}{1} = 1 \qquad \csc 90° = \frac{1}{1} = 1$$

$$\cos 90° = \frac{0}{1} = 0 \qquad \sec 90° = \frac{1}{0} \text{ (undefined)}$$

$$\tan 90° = \frac{1}{0} \text{ (undefined)} \qquad \cot 90° = \frac{0}{1} = 0.$$

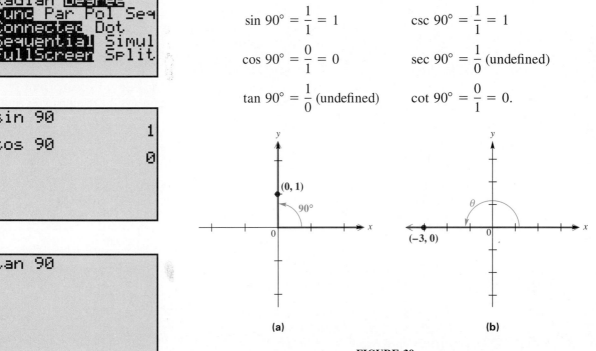

(a) (b)

FIGURE 39

(b) an angle in standard position with terminal side through $(-3, 0)$

Figure 39(b) shows the angle. Here, $x = -3$, $y = 0$, and $r = 3$, so the trigonometric functions have the following values.

$$\sin \theta = \frac{0}{3} = 0 \qquad \csc \theta = \frac{3}{0} \text{ (undefined)}$$

$$\cos \theta = \frac{-3}{3} = -1 \qquad \sec \theta = \frac{3}{-3} = -1$$

$$\tan \theta = \frac{0}{-3} = 0 \qquad \cot \theta = \frac{-3}{0} \text{ (undefined)} \blacktriangleright$$

When the calculator is set to degree mode, it returns the correct values for sin 90° and cos 90°. Notice that it returns an ERROR message for tan 90°, since 90° is not in the domain of the tangent function. Compare these results to those found in Example 4.

The conditions under which the trigonometric function values of quadrantal angles are undefined are summarized here.

Undefined Function Values

If the terminal side of a quadrantal angle lies along the *y*-axis, the tangent and secant functions are undefined. If it lies along the *x*-axis, the cotangent and cosecant functions are undefined.

Since the most commonly used quadrantal angles are 0°, 90°, 180°, 270°, and 360°, the values of the functions of these angles are summarized in the following table. This table is for reference only; you should be able to reproduce it quickly.

Quadrantal Angles

θ	$\sin \theta$	$\cos \theta$	$\tan \theta$	$\cot \theta$	$\sec \theta$	$\csc \theta$
0°	0	1	0	Undefined	1	Undefined
90°	1	0	Undefined	0	Undefined	1
180°	0	−1	0	Undefined	−1	Undefined
270°	−1	0	Undefined	0	Undefined	−1
360°	0	1	0	Undefined	1	Undefined

The values given in this table can also be found with a calculator that has trigonometric function keys. First, make sure the calculator is set for *degree mode*.

CAUTION   One of the most common errors involving calculators in trigonometry occurs when the calculator is set for *radian measure,* rather than *degree measure.* (Radian measure of angles is studied in Chapter 3.) For this reason, be sure that you know how to set your calculator to *degree mode*.

There are no calculator keys for finding the function values of cotangent, secant, or cosecant. The next section shows how to find these function values with a calculator.

1.4 Exercises ▼▼▼▼▼▼▼▼▼▼▼▼▼▼▼▼▼▼▼▼▼▼▼▼▼▼▼▼▼▼▼▼▼▼▼▼▼

In Exercises 1–4, sketch an angle θ in standard position such that θ has the smallest possible positive measure, and the given point is on the terminal side of θ.

1. $(-3, 4)$ **2.** $(-4, -3)$ **3.** $(5, -12)$ **4.** $(-12, -5)$

Find the values of the six trigonometric functions for the angles in standard position having the following points on their terminal sides. Rationalize denominators when applicable. Use a calculator in Exercises 11 and 12. See Examples 1, 2, and 4.

5. $(-3, 4)$ **6.** $(-4, -3)$ **7.** $(0, 2)$ **8.** $(-4, 0)$

9. $(1, \sqrt{3})$ **10.** $(-2\sqrt{3}, -2)$ **11.** $(8.7691, -3.2473)$ **12.** $(-5.1021, 7.6132)$

13. For any nonquadrantal angle θ, sin θ and csc θ will have the same sign. Explain why this is so.

14. If cot θ is undefined, what is the value of tan θ?

15. How is the value of r interpreted geometrically in the definitions of the sine, cosine, secant, and cosecant functions?

16. If the terminal side of an angle β is in quadrant III, what is the sign of each of the trigonometric function values of β?

Suppose that the point (x, y) is in the indicated quadrant. Decide whether the given ratio is positive or negative. (Hint: It may be helpful to draw a sketch.)

17. II, $\dfrac{y}{r}$ **18.** II, $\dfrac{x}{r}$ **19.** III, $\dfrac{y}{r}$ **20.** III, $\dfrac{x}{r}$

21. IV, $\dfrac{x}{r}$ **22.** IV, $\dfrac{y}{r}$ **23.** IV, $\dfrac{y}{x}$ **24.** IV, $\dfrac{x}{y}$

In Exercises 25–30, an equation with a restriction on x is given. This is an equation of the terminal side of an angle θ in standard position. Sketch the smallest positive such angle θ, and find the values of the six trigonometric functions of θ. See Example 3.

25. $2x + y = 0, x \geq 0$ **26.** $3x + 5y = 0, x \geq 0$ **27.** $-4x + 7y = 0, x \leq 0$

28. $-6x - y = 0, x \leq 0$ **29.** $-5x - 3y = 0, x \leq 0$ **30.** $6x - 5y = 0, x \geq 0$

31. Rework Example 3 using a different value for x. Find the corresponding y-value, and then show that the six trigonometric function values you obtain are the same as the ones obtained in Example 3.

32. Rework Example 3 using the values of x and y for which the point (x, y) is on the circle of radius one having center at the origin, and then show that the six trigonometric function values you obtain are the same as those obtained in Example 3.

Use the trigonometric function values of quadrantal angles given in this section to evaluate each of the following. An expression such as $\cot^2 90°$ means $(\cot 90°)^2$ which is equal to $0^2 = 0$.

33. $\cos 90° + 3 \sin 270°$ **34.** $\tan 0° - 6 \sin 90°$

35. $3 \sec 180° - 5 \tan 360°$ **36.** $4 \csc 270° + 3 \cos 180°$

37. $\tan 360° + 4 \sin 180° + 5 \cos^2 180°$ **38.** $2 \sec 0° + 4 \cot^2 90° + \cos 360°$

39. $\sin^2 180° + \cos^2 180°$ **40.** $\sin^2 360° + \cos^2 360°$

41. $\sec^2 180° - 3 \sin^2 360° + 2 \cos 180°$ **42.** $5 \sin^2 90° + 2 \cos^2 270° - 7 \tan^2 360°$

If n is an integer, n · 180° represents an integer multiple of 180°, and (2n + 1) · 90° represents an odd integer multiple of 90°. Decide whether each of the following is equal to 0, 1, −1, or is undefined.

43. $\sin[n \cdot 180°]$ **44.** $\cos[(2n + 1) \cdot 90°]$ **45.** $\tan[(2n + 1) \cdot 90°]$ **46.** $\tan[n \cdot 180°]$

---

47. The angles 15° and 75° are complementary. With your calculator determine sin 15° and cos 75°. Make a conjecture about the sines and cosines of complementary angles and test your hypothesis with other pairs of complementary angles. (*Note:* This relationship will be discussed in detail in Section 2.1.)

48. The angles 25° and 65° are complementary. With your calculator determine tan 25° and cot 65°. Make a conjecture about the tangents and cotangents of complementary angles and test your hypothesis with other pairs of complementary angles. (*Note:* This relationship will be discussed in detail in Section 2.1.)

49. With your calculator determine sin 10° and sin (−10°). Make a conjecture about the sines of an angle and its negative and test your hypothesis with other angles. Also, use a geometry argument with the definition of sin θ to justify your hypothesis. (*Note:* This relationship will be discussed in detail in Section 5.1.)

50. With your calculator determine cos 20° and cos(−20°). Make a conjecture about the cosines of an angle and its negative and test your hypothesis with other angles. Also, use a geometry argument with the definition of cos θ to justify your hypothesis. (*Note:* This relationship will be discussed in detail in Section 5.1.)

Figure 37 suggests that the coordinates of the intersection of the unit circle and the terminal side of an angle x is the point (cos x, sin x). Use this fact for the following exercises.

51. Define the cosine function in terms of the *x*-coordinate of a point on the unit circle.

52. Define the sine function in terms of the *y*-coordinate of a point on the unit circle.

In Exercises 53–58, place your graphing calculator in parametric and degree modes. Set the window and functions as shown here, and graph. A circle of radius 1 will appear on the screen. Use the trace feature to move a short distance around the circle. In the accompanying graphing screen the point on the circle corresponds to an angle of T = 25°, cos 25° is .90630779, and sin 25° is .42261826.

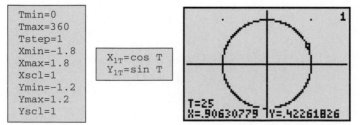

```
Tmin=0
Tmax=360
Tstep=1
Xmin=-1.8
Xmax=1.8
Xscl=1
Ymin=-1.2
Ymax=1.2
Yscl=1
```

```
X₁T=cos T
Y₁T=sin T
```

53. Use the right- and left-arrow keys to move to the point corresponding to 20°. What are cos 20° and sin 20°?

In Exercises 54–56 assume 0° ≤ T ≤ 90°.

54. For what angle *T* is cos *T* ≈ .766?

55. For what angle *T* is sin *T* ≈ .574?

56. For what angle *T* does cos *T* = sin *T*?

57. As T increases from $0°$ to $90°$ does the cosine increase or decrease? How about the sine?

58. As T increases from $90°$ to $180°$ does the cosine increase or decrease? How about the sine?

59. 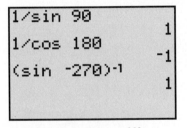 Suppose that a star has parallax θ with respect to Earth and the sun. Let the coordinates of Earth be (x, y), the star be $(0, 0)$, and the sun be $(x, 0)$. See the figure. We would like to find an equation for x, the distance between the sun and the star.

(a) Write an equation involving a trigonometric function that relates x, y, and θ.

(b) Solve your equation for x.

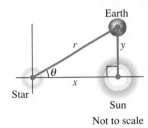

Not to scale

1.5 Using the Definitions of the Trigonometric Functions ▼▼▼

```
1/sin 90
                    1
1/cos 180
                   -1
(sin -270)⁻¹
                    1
```

This screen shows how csc 90°, sec 180° and csc (–270°) can be found, using the appropriate reciprocal identities. Compare these results with the ones found in the chart of quadrantal angle function values in Section 1.4.

Be sure not to use the *inverse trigonometric function* keys to find the reciprocal function values. Consult your owner's manual if further information is needed.

In this section several useful results are derived from the definitions of the trigonometric functions given in the previous section. First, recall the definition of a reciprocal: the **reciprocal** of the nonzero number x is $1/x$. For example, the reciprocal of 2 is $1/2$, and the reciprocal of $8/11$ is $11/8$. There is no reciprocal for 0. Scientific calculators have a reciprocal key, usually labeled $\boxed{1/x}$ or $\boxed{x^{-1}}$. Using this key gives the reciprocal of any nonzero number entered in the display.

THE RECIPROCAL IDENTITIES   The definitions of the trigonometric functions in the previous section were written so that functions on the same line are reciprocals of each other. Since $\sin \theta = y/r$ and $\csc \theta = r/y$,

$$\sin \theta = \frac{1}{\csc \theta} \quad \text{and} \quad \csc \theta = \frac{1}{\sin \theta}.$$

Also, $\cos \theta$ and $\sec \theta$ are reciprocals, as are $\tan \theta$ and $\cot \theta$. In summary, we have the **reciprocal identities** that hold for any angle θ that does not lead to a zero denominator.

Reciprocal Identities

$$\sin \theta = \frac{1}{\csc \theta} \qquad \csc \theta = \frac{1}{\sin \theta}$$

$$\cos \theta = \frac{1}{\sec \theta} \qquad \sec \theta = \frac{1}{\cos \theta}$$

$$\tan \theta = \frac{1}{\cot \theta} \qquad \cot \theta = \frac{1}{\tan \theta}$$

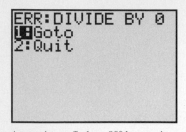

Attempting to find sec 90° by entering 1/cos 90° leads to an ERROR message, indicating that division by 0 is not allowed. Compare to the chart of quadrantal angle function values in Section 1.4, where we indicate that sec 90° is undefined.

◀EXAMPLE 1
Using the reciprocal identities

Identities are equations that are true for all meaningful values of the variable. For example, both $(x + y)^2 = x^2 + 2xy + y^2$ and $2(x + 3) = 2x + 6$ are identities. Identities are studied in more detail in Chapter 5.

NOTE   When studying identities, be aware that various forms exist. For example,

$$\sin \theta = \frac{1}{\csc \theta}$$

can also be written

$$\csc \theta = \frac{1}{\sin \theta} \quad \text{and} \quad (\sin \theta)(\csc \theta) = 1.$$

You should become familiar with all forms of these identities.

Find each function value.

(a) $\cos \theta$, if $\sec \theta = \dfrac{5}{3}$

Since $\cos \theta$ is the reciprocal of $\sec \theta$,

$$\cos \theta = \frac{1}{\sec \theta} = \frac{1}{5/3} = \frac{3}{5}.$$

(b) $\sin \theta$, if $\csc \theta = -\dfrac{\sqrt{12}}{2}$

$$\sin \theta = \frac{1}{-\sqrt{12}/2}$$

$$= \frac{-2}{\sqrt{12}}$$

$$= \frac{-2}{2\sqrt{3}} \qquad \sqrt{12} = \sqrt{4 \cdot 3} = 2\sqrt{3}$$

$$= \frac{-1}{\sqrt{3}}$$

$$= \frac{-\sqrt{3}}{3} \qquad \text{Multiply by } \frac{\sqrt{3}}{\sqrt{3}} \text{ to rationalize the denominator. ▶}$$

In the definition of the trigonometric functions, r is the distance from the origin to the point (x, y). Distance is never negative, so $r > 0$. If we choose a point (x, y) in quadrant I, then both x and y will be positive. Since $r > 0$, all six of the fractions used in the definitions of the trigonometric functions will be positive, so that the values of all six functions will be positive in quadrant I.

A point (x, y) in quadrant II has $x < 0$ and $y > 0$. This makes the values of sine and cosecant positive for quadrant II angles, while the other four functions take on negative values. Similar results can be obtained for the other quadrants, as summarized on the following page.

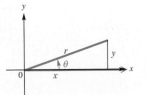

cos 14879
        -.4848096202
sin 14879
        .8746197071

This screen was obtained with the calculator in degree mode. How can we use it to justify that an angle of 14,879° is a second quadrant angle?

Signs of Function Values

θ in quadrant	$\sin \theta$	$\cos \theta$	$\tan \theta$	$\cot \theta$	$\sec \theta$	$\csc \theta$
I	+	+	+	+	+	+
II	+	−	−	−	−	+
III	−	−	+	+	−	−
IV	−	+	−	−	+	−

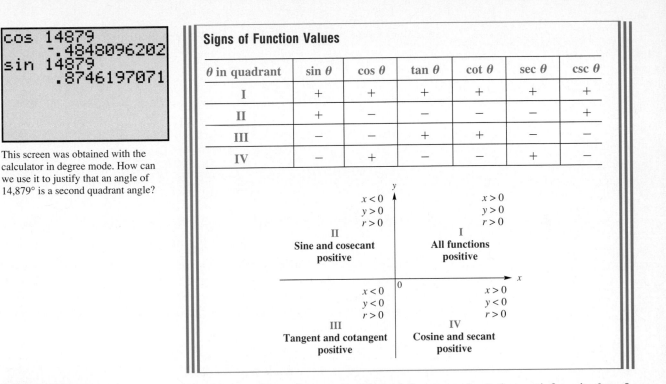

$x < 0$, $y > 0$, $r > 0$ — **II** — Sine and cosecant positive

$x > 0$, $y > 0$, $r > 0$ — **I** — All functions positive

$x < 0$, $y < 0$, $r > 0$ — **III** — Tangent and cotangent positive

$x > 0$, $y < 0$, $r > 0$ — **IV** — Cosine and secant positive

◀**EXAMPLE 2**
Identifying the quadrant of an angle

Identify the quadrant (or quadrants) of any angle θ that satisfies $\sin \theta > 0$, $\tan \theta < 0$.

Since $\sin \theta > 0$ in quadrants I and II, while $\tan \theta < 0$ in quadrants II and IV, both conditions are met only in quadrant II. ▶

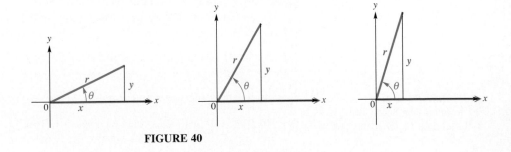

FIGURE 40

Figure 40 shows an angle θ as it increases in measure from near 0° toward 90°. In each case, the value of r is the same. As the measure of the angle increases, y increases but never exceeds r, so that $y \le r$. Dividing both sides by the positive number r gives

$$y \le r$$

$$\frac{y}{r} \le 1.$$

In a similar way, angles in the fourth quadrant suggest that

$$-1 \leq \frac{y}{r},$$

so

$$-1 \leq \frac{y}{r} \leq 1.$$

Since $y/r = \sin \theta$,

$$-1 \leq \sin \theta \leq 1$$

for any angle θ. In the same way,

$$-1 \leq \cos \theta \leq 1.$$

The tangent of an angle is defined as y/x. It is possible that $x < y$, that $x = y$, or that $x > y$. For this reason y/x can take on any value at all, so $\tan \theta$ can be any real number, as can $\cot \theta$.

The functions $\sec \theta$ and $\csc \theta$ are reciprocals of the functions $\cos \theta$ and $\sin \theta$, respectively, making

$$\sec \theta \leq -1 \text{ or } \sec \theta \geq 1,$$
$$\csc \theta \leq -1 \text{ or } \csc \theta \geq 1.$$

In summary, the ranges of the trigonometric functions are as follows.

Ranges of Trigonometric Functions

For any angle θ for which the indicated functions exist:

1. $-1 \leq \sin \theta \leq 1$ and $-1 \leq \cos \theta \leq 1$;
2. $\tan \theta$ and $\cot \theta$ may be equal to any real number;
3. $\sec \theta \leq -1$ or $\sec \theta \geq 1$ and $\csc \theta \leq -1$ or $\csc \theta \geq 1$.

(Notice that $\sec \theta$ and $\csc \theta$ are *never* between -1 and 1.)

EXAMPLE 3
Deciding whether a trigonometric function value is in the range

Decide whether the following statements are *possible* or *impossible*.

(a) $\sin \theta = \sqrt{8}$
For any value of θ, $-1 \leq \sin \theta \leq 1$. Since $\sqrt{8} > 1$, there is no value of θ with $\sin \theta = \sqrt{8}$.

(b) $\tan \theta = 110.47$
Tangent can take on any value. Thus, $\tan \theta = 110.47$ is possible.

(c) $\sec \theta = .6$
Since $\sec \theta \leq -1$ or $\sec \theta \geq 1$, the statement $\sec \theta = .6$ is impossible. ▶

The six trigonometric functions are defined in terms of x, y, and r, where the Pythagorean theorem shows that $r^2 = x^2 + y^2$ and $r > 0$. With these relationships, knowing the value of only one function and the quadrant in which the angle lies makes it possible to find the values of all six of the trigonometric functions. This procedure is shown in the next example.

EXAMPLE 4
Finding all function values given one value and the quadrant

Suppose that angle α is in quadrant II and $\sin \alpha = 2/3$. Find the values of the other five functions.

We can choose any point on the terminal side of angle α. For simplicity, since $\sin \alpha = y/r$, we choose the point with $r = 3$. Then

$$\frac{y}{r} = \frac{2}{3}.$$

If $r = 3$, then y will be 2. To find x, use the result $x^2 + y^2 = r^2$.

$$x^2 + y^2 = r^2$$
$$x^2 + 2^2 = 3^2$$
$$x^2 + 4 = 9$$
$$x^2 = 5$$
$$x = \sqrt{5} \quad \text{or} \quad x = -\sqrt{5}$$

Since α is in quadrant II, x must be negative, as shown in Figure 41, so $x = -\sqrt{5}$. This puts the point $(-\sqrt{5}, 2)$ on the terminal side of α.

Now that the values of x, y, and r are known, the values of the remaining trigonometric functions can be found.

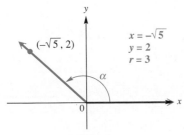

$x = -\sqrt{5}$
$y = 2$
$r = 3$

FIGURE 41

$$\cos \alpha = \frac{x}{r} = \frac{-\sqrt{5}}{3}$$

$$\tan \alpha = \frac{y}{x} = \frac{2}{-\sqrt{5}} = \frac{-2\sqrt{5}}{5}$$

$$\cot \alpha = \frac{x}{y} = \frac{-\sqrt{5}}{2}$$

$$\sec \alpha = \frac{r}{x} = \frac{3}{-\sqrt{5}} = \frac{-3\sqrt{5}}{5}$$

$$\csc \alpha = \frac{r}{y} = \frac{3}{2} \quad \blacktriangleright$$

THE PYTHAGOREAN IDENTITIES   We can derive three very useful new identities from the relationship $x^2 + y^2 = r^2$. Dividing both sides by r^2 gives

$$\frac{x^2}{r^2} + \frac{y^2}{r^2} = \frac{r^2}{r^2},$$

or

$$\left(\frac{x}{r}\right)^2 + \left(\frac{y}{r}\right)^2 = 1.$$

Since $\sin \theta = y/r$ and $\cos \theta = x/r$, this result becomes

$$(\sin \theta)^2 + (\cos \theta)^2 = 1,$$

or, as it is usually written,

$$\sin^2 \theta + \cos^2 \theta = 1.$$

Starting with $x^2 + y^2 = r^2$ and dividing through by x^2 gives

$$\frac{x^2}{x^2} + \frac{y^2}{x^2} = \frac{r^2}{x^2}$$

$$1 + \left(\frac{y}{x}\right)^2 = \left(\frac{r}{x}\right)^2$$

$$1 + (\tan \theta)^2 = (\sec \theta)^2$$

or $$\tan^2 \theta + 1 = \sec^2 \theta.$$

On the other hand, dividing through by y^2 leads to

$$1 + \cot^2 \theta = \csc^2 \theta.$$

These three identities are called the **Pythagorean identities** since the original equation that led to them, $x^2 + y^2 = r^2$, comes from the Pythagorean theorem.

Pythagorean Identities

$$\sin^2 \theta + \cos^2 \theta = 1 \qquad \tan^2 \theta + 1 = \sec^2 \theta$$
$$1 + \cot^2 \theta = \csc^2 \theta$$

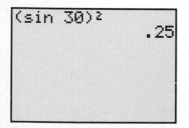

This screen supports the statement found in the Note.

NOTE Although we usually write $\sin^2 \theta$, for example, it should be entered as $(\sin \theta)^2$ in your calculator. To test yourself, verify that $\sin^2 30° = .25$.

As before, we have given only one form of each identity. However, algebraic transformations can be made to get equivalent identities. For example, by subtracting $\sin^2 \theta$ from both sides of $\sin^2 \theta + \cos^2 \theta = 1$ we get the equivalent identity

$$\cos^2 \theta = 1 - \sin^2 \theta.$$

You should be able to transform these identities quickly, and also recognize their equivalent forms.

THE QUOTIENT IDENTITIES   Recall that $\sin \theta = y/r$ and $\cos \theta = x/r$. Consider the quotient of $\sin \theta$ and $\cos \theta$, where $\cos \theta \neq 0$.

$$\frac{\sin \theta}{\cos \theta} = \frac{y/r}{x/r} = \frac{y}{r} \div \frac{x}{r} = \frac{y}{r} \cdot \frac{r}{x} = \frac{y}{x} = \tan \theta$$

Similarly, it can be shown that $(\cos \theta)/(\sin \theta) = \cot \theta$, for $\sin \theta \neq 0$. Thus we have two more identities, called the **quotient identities.**

Quotient Identities

$$\frac{\sin \theta}{\cos \theta} = \tan \theta \qquad \frac{\cos \theta}{\sin \theta} = \cot \theta$$

◀ **EXAMPLE 5**
Finding other function values given one value and the quadrant

Find $\sin \alpha$ and $\tan \alpha$ if $\cos \alpha = -\sqrt{3}/4$ and α is in quadrant II.
Start with $\sin^2 \alpha + \cos^2 \alpha = 1$, and replace $\cos \alpha$ with $-\sqrt{3}/4$.

$$\sin^2 \alpha + \left(-\frac{\sqrt{3}}{4}\right)^2 = 1 \qquad \text{Replace } \cos \alpha \text{ with } -\frac{\sqrt{3}}{4}.$$

$$\sin^2 \alpha + \frac{3}{16} = 1$$

$$\sin^2 \alpha = \frac{13}{16} \qquad \text{Subtract } \frac{3}{16}.$$

$$\sin \alpha = \pm \frac{\sqrt{13}}{4} \qquad \text{Take square roots.}$$

Since α is in quadrant II, $\sin \alpha > 0$, and

$$\sin \alpha = \frac{\sqrt{13}}{4}.$$

To find $\tan \alpha$, use the quotient identity $\tan \alpha = \dfrac{\sin \alpha}{\cos \alpha}$.

$$\tan \alpha = \frac{\sin \alpha}{\cos \alpha} = \frac{\dfrac{\sqrt{13}}{4}}{\dfrac{-\sqrt{3}}{4}} = \frac{\sqrt{13}}{4} \cdot \frac{4}{-\sqrt{3}} = \frac{\sqrt{13}}{-\sqrt{3}}$$

Rationalize the denominator as follows.

$$\frac{\sqrt{13}}{-\sqrt{3}} = \frac{\sqrt{13}}{-\sqrt{3}} \cdot \frac{\sqrt{3}}{\sqrt{3}}$$

$$= \frac{\sqrt{39}}{-3}$$

$$= -\frac{\sqrt{39}}{3}$$

Therefore, $\tan \alpha = -\sqrt{39}/3$. ▶

CAUTION One of the most common errors in problems like those in Examples 4 and 5 involves an incorrect sign choice when square roots are taken. Notice that in Example 5, we chose the positive square root for $\sin \alpha$, since α was in quadrant II, and the sine function is positive there. If the problem had specified that α was in quadrant III, then we would have had to choose the negative square root.

EXAMPLE 6
Finding other function values, given one function value and the quadrant of θ

Find $\sin \theta$ and $\cos \theta$, if $\tan \theta = 4/3$ and θ is in quadrant III.

Since θ is in quadrant III, $\sin \theta$ and $\cos \theta$ will both be negative. It is tempting to say that since $\tan \theta = (\sin \theta)/(\cos \theta)$ and $\tan \theta = 4/3$, then $\sin \theta = -4$ and $\cos \theta = -3$. This is *incorrect,* however, since both $\sin \theta$ and $\cos \theta$ must be in the interval $[-1, 1]$.

Use the Pythagorean identity $\tan^2 \theta + 1 = \sec^2 \theta$ to find $\sec \theta$, and then the reciprocal identity $\cos \theta = 1/\sec \theta$.

$$\tan^2 \theta + 1 = \sec^2 \theta$$

$$\left(\frac{4}{3}\right)^2 + 1 = \sec^2 \theta \qquad \tan \theta = \tfrac{4}{3}$$

$$\frac{16}{9} + 1 = \sec^2 \theta$$

$$\frac{25}{9} = \sec^2 \theta$$

$$-\frac{5}{3} = \sec \theta \qquad \text{Choose the negative square root since } \theta \text{ is in quadrant III.}$$

$$-\frac{3}{5} = \cos \theta \qquad \text{Secant and cosine are reciprocals.}$$

Since $\sin^2 \theta = 1 - \cos^2 \theta$,

$$\sin^2 \theta = 1 - \left(-\frac{3}{5}\right)^2 \qquad \cos \theta = -\tfrac{3}{5}$$

$$= 1 - \frac{9}{25}$$

$$= \frac{16}{25}$$

$$\sin \theta = -\frac{4}{5}. \qquad \text{Choose the negative square root.}$$

Therefore, we have $\sin \theta = -4/5$ and $\cos \theta = -3/5$. ▶

NOTE   Example 6 can also be worked by drawing θ in standard position in quadrant III, finding r to be 5, and then using the definitions of $\sin \theta$ and $\cos \theta$ in terms of x, y, and r.

1.5 Exercises ▼▼▼▼▼▼▼▼▼▼▼▼▼▼▼▼▼▼▼▼▼▼▼▼▼▼▼▼▼▼▼▼

1. What positive number a is its own reciprocal? Find a value of θ for which $\sin \theta = \csc \theta = a$.
2. What negative number a is its own reciprocal? Find a value of θ for which $\cos \theta = \sec \theta = a$.

Use the appropriate reciprocal identity to find each function value. Rationalize denominators when applicable. In Exercises 11 and 12, use a calculator. See Example 1.

3. $\sin \theta$, if $\csc \theta = 3$
4. $\cos \alpha$, if $\sec \alpha = -2.5$
5. $\cot \beta$, if $\tan \beta = -1/5$
6. $\sin \alpha$, if $\csc \alpha = \sqrt{15}$
7. $\csc \alpha$, if $\sin \alpha = \sqrt{2}/4$
8. $\sec \beta$, if $\cos \beta = -1/\sqrt{7}$
9. $\tan \theta$, if $\cot \theta = -\sqrt{5}/3$
10. $\cot \theta$, if $\tan \theta = \sqrt{11}/5$
11. $\sin \theta$, if $\csc \theta = 1.42716321$
12. $\cos \alpha$, if $\sec \alpha = 9.80425133$

13. Can a given angle γ satisfy both $\sin \gamma > 0$ and $\csc \gamma < 0$? Explain.
14. Suppose that the following item appears on a trigonometry test:

$$\text{Find } \sec \theta, \text{ given that } \cos \theta = 3/2.$$

What is wrong with this test item?

15. One form of a particular reciprocal identity is

$$\tan \theta = \frac{1}{\cot \theta}.$$

Give two other equivalent forms of this identity.

16. What is wrong with the following statement? $\tan 90° = \dfrac{1}{\cot 90°}$

Find the tangent of each angle. See Example 1.

17. $\cot \gamma = 2$
18. $\cot \phi = -3$
19. $\cot \omega = \sqrt{3}/3$
20. $\cot \theta = \sqrt{6}/12$
21. $\cot \alpha = -.01$
22. $\cot \beta = .4$

Find a value of the variable.

23. $\tan(3B - 4°) = \dfrac{1}{\cot(5B - 8°)}$
24. $\cos(6A + 5°) = \dfrac{1}{\sec(4A + 15°)}$
25. $\sec(2\alpha + 6°) \cos(5\alpha + 3°) = 1$
26. $\sin(4\theta + 2°) \csc(3\theta + 5°) = 1$
27. $\dfrac{1}{\tan(2k + 1°)} = \cot(4k - 3°)$
28. $\dfrac{1}{\sin(3\theta - 1°)} = \csc(2\theta + 3°)$

Identify the quadrant or quadrants for the angle satisfying the given conditions. See Example 2.

29. $\sin \alpha > 0$, $\cos \alpha < 0$

30. $\cos \beta > 0$, $\tan \beta > 0$

31. $\tan \gamma > 0$, $\cot \gamma > 0$

32. $\sin \beta < 0$, $\cos \beta > 0$

33. $\tan \omega < 0$, $\cot \omega < 0$

34. $\csc \theta < 0$, $\cos \theta < 0$

35. $\cos \beta < 0$

36. $\tan \theta > 0$

Give the signs of the six trigonometric functions for each angle.

37. $74°$ **38.** $129°$ **39.** $183°$ **40.** $298°$ **41.** $302°$

42. $406°$ **43.** $412°$ **44.** $-82°$ **45.** $-14°$ **46.** $-121°$

In Exercises 47–50, without using a calculator, decide which is greater.

47. $\sin 30°$ or $\tan 30°$ **48.** $\sin 20°$ or $\sin 21°$ **49.** $\sin 33°$ or $\sec 33°$ **50.** $\cos 5°$ or $\cos^2 5°$

Decide whether each statement is possible *or* impossible. *See Example 3.*

51. $\sin \theta = 2$ **52.** $\cos \alpha = -1.001$ **53.** $\tan \beta = .92$ **54.** $\cot \omega = -12.1$

55. $\csc \alpha = 1/2$ **56.** $\sec \alpha = 1$ **57.** $\tan \theta = 1$ **58.** $\sin \beta + 1 = .6$

59. $\sec \omega + 1 = 1.3$ **60.** $\csc \theta - 1 = -.2$

61. $\sin \alpha = 1/2$ and $\csc \alpha = 2$ **62.** $\tan \beta = 2$ and $\cot \beta = -2$

Use identities to find the indicated function value. Use a calculator in Exercises 71–74. See Examples 4–6.

63. $\tan \alpha$, if $\sec \alpha = 3$, with α in quadrant IV

64. $\cos \theta$, if $\sin \theta = 2/3$, with θ in quadrant II

65. $\sin \alpha$, if $\cos \alpha = -1/4$, with α in quadrant II

66. $\csc \beta$, if $\cot \beta = -1/2$, with β in quadrant IV

67. $\tan \theta$, if $\cos \theta = 1/3$, with θ in quadrant IV

68. $\sec \theta$, if $\tan \theta = \sqrt{7}/3$, with θ in quadrant III

69. $\cos \beta$, if $\csc \beta = -4$, with β in quadrant III

70. $\sin \theta$, if $\sec \theta = 2$, with θ in quadrant IV

71. $\sin \beta$, if $\cot \beta = 2.40129813$, with β in quadrant I

72. $\cot \alpha$, if $\csc \alpha = -3.5891420$, with α in quadrant III

73. $\csc \alpha$, if $\tan \alpha = .98244655$, with α in quadrant III

74. $\tan \beta$, if $\sin \beta = .49268329$, with β in quadrant II

In Exercises 75 and 76, the given graphing calculator screen is obtained for a particular stored value of X. *What will the screen display for the value of the expression in the final line of the display?*

75.

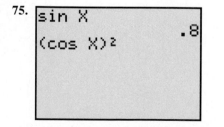

76.

```
tan X
                2
(1/cos X)²
```

77. Does there exist an angle θ with $\cos \theta = -.6$ and $\sin \theta = .8$?

78. Does there exist an angle θ with $\cos \theta = 2/3$ and $\sin \theta = 3/4$?

*Find all the trigonometric function values for each angle. Use a calculator in Exercises 87
and 88. See Examples 4–6.*

79. $\tan \alpha = -15/8$, with α in quadrant II

80. $\cos \alpha = -3/5$, with α in quadrant III

81. $\cot \gamma = 3/4$, with γ in quadrant III

82. $\sin \beta = 7/25$, with β in quadrant II

83. $\tan \beta = \sqrt{3}$, with β in quadrant III

84. $\csc \theta = 2$, with θ in quadrant II

85. $\sin \beta = \sqrt{5}/7$, with $\tan \beta > 0$

86. $\cot \alpha = \sqrt{3}/8$, with $\sin \alpha > 0$

87. $\cot \theta = -1.49586$, with θ in quadrant IV

88. $\sin \alpha = .164215$, with α in quadrant II

89. Derive the identity $1 + \cot^2 \theta = \csc^2 \theta$ by dividing $x^2 + y^2 = r^2$ by y^2.

90. Using a method similar to the one given in this section showing that $(\sin \theta)/(\cos \theta) = \tan \theta$, show that

$$\frac{\cos \theta}{\sin \theta} = \cot \theta.$$

91. True or false: For all angles θ, $\sin \theta + \cos \theta = 1$. If false, give an example showing why it is false.

92. True or false: Since $\cot \theta = \dfrac{\cos \theta}{\sin \theta}$, if $\cot \theta = \dfrac{1}{2}$ with θ in quadrant I, then $\cos \theta = 1$ and $\sin \theta = 2$. If false, explain why.

93. The straight line in the figure here determines both the angle α and the angle β with the positive x-axis. Explain why $\tan \alpha = \tan \beta$.

Chapter 1 Summary

SECTION	KEY IDEAS
1.1 Basic Concepts	
	Pythagorean Theorem If the two shorter sides (the legs) of a right triangle have lengths a and b, respectively, and if the length of the hypotenuse (the longest side, opposite the 90° angle) is c, then $$a^2 + b^2 = c^2.$$

SECTION	KEY IDEAS

1.2 Angles

Types of Angles

Name	Angle Measure	Example
Acute angle	Between 0° and 90°	60° 82°
Right angle	Exactly 90°	90°
Obtuse angle	Between 90° and 180°	97° 138°
Straight angle	Exactly 180°	180°

1.3 Angle Relationships and Similar Triangles

Vertical angles have equal measures.

The sum of the measures of the angles of any triangle is 180°.

When a transversal intersects parallel lines, the following angles formed have equal measure: alternate interior, alternate exterior, and corresponding. Interior angles on the same side of the transversal are supplements.

Similar triangles have corresponding angles with the same measures, and corresponding sides proportional.

1.4 Definitions of the Trigonometric Functions

Definitions of the Trigonometric Functions

Let (x, y) be a point other than the origin on the terminal side of an angle θ in standard position. Let $r = \sqrt{x^2 + y^2}$, the distance from the origin to (x, y). Then

$$\sin \theta = \frac{y}{r} \qquad\qquad \csc \theta = \frac{r}{y} \quad (y \neq 0)$$

$$\cos \theta = \frac{x}{r} \qquad\qquad \sec \theta = \frac{r}{x} \quad (x \neq 0)$$

$$\tan \theta = \frac{y}{x} \quad (x \neq 0) \qquad \cot \theta = \frac{x}{y} \quad (y \neq 0).$$

SECTION	KEY IDEAS

Trigonometric Function Values for Quadrantal Angles

θ	$0°$	$90°$	$180°$	$270°$	$360°$
$\sin \theta$	0	1	0	-1	0
$\cos \theta$	1	0	-1	0	1
$\tan \theta$	0	undefined	0	undefined	0
$\cot \theta$	undefined	0	undefined	0	undefined
$\sec \theta$	1	undefined	-1	undefined	1
$\csc \theta$	undefined	1	undefined	-1	undefined

1.5 Using the Definitions of the Trigonometric Functions

Reciprocal Identities

$$\sin \theta = \frac{1}{\csc \theta} \qquad \csc \theta = \frac{1}{\sin \theta}$$

$$\cos \theta = \frac{1}{\sec \theta} \qquad \sec \theta = \frac{1}{\cos \theta}$$

$$\tan \theta = \frac{1}{\cot \theta} \qquad \cot \theta = \frac{1}{\tan \theta}$$

Pythagorean Identities

$$\sin^2 \theta + \cos^2 \theta = 1$$
$$\tan^2 \theta + 1 = \sec^2 \theta$$
$$1 + \cot^2 \theta = \csc^2 \theta$$

Quotient Identities

$$\frac{\sin \theta}{\cos \theta} = \tan \theta \qquad \frac{\cos \theta}{\sin \theta} = \cot \theta$$

Signs of Trigonometric Functions

y

$x < 0$
$y > 0$
$r > 0$

II
Sine and cosecant
positive

$x > 0$
$y > 0$
$r > 0$

I
All functions
positive

$x < 0$
$y < 0$
$r > 0$

III
Tangent and cotangent
positive

$x > 0$
$y < 0$
$r > 0$

IV
Cosine and secant
positive

x

0

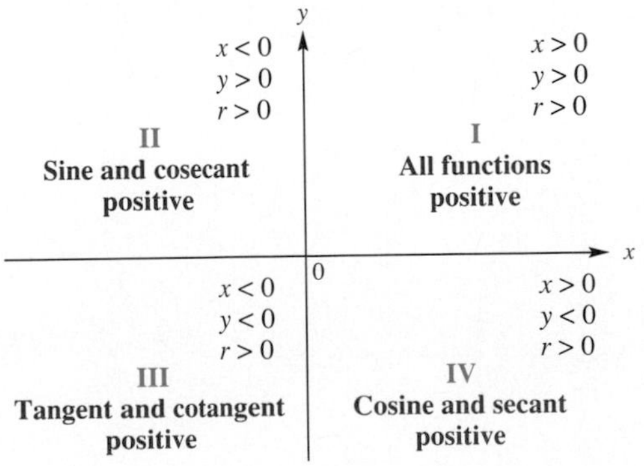

Chapter 1   Review Exercises ▼▼▼▼▼▼▼▼▼▼▼▼▼▼▼▼▼▼▼▼▼▼▼▼▼▼▼▼▼

1. Suppose the point (a, b) lies in the first quadrant. Describe how you would move from the point (a, b) to the point $(-a, -b)$.

2. Find the midpoint of the line segment connecting the two points (a, b) and $(0, 0)$.

Write each of the following sets using interval notation.

3. $\{x \mid x \leq -4\}$

4.

Find the distance between each of the following pairs of points.

5. $(4, -2)$ and $(1, -6)$

6. $(-6, 3)$ and $(-2, -5)$

7. Use the distance formula to determine whether the points $(-2, -2)$, $(8, 4)$, and $(2, 14)$ are the vertices of a right triangle.   *Plotted 3 points, find dist. all 3.   yes on MV.   Pythaum*

8. State in your own words the vertical line test for the graph of a function.

Let $f(x) = -x^2 + 3x + 2$. Find each of the following.

9. $f(0)$

10. $f(-2)$

11. $f(x + 1)$

12. The accompanying table was generated by a graphing calculator for a function $Y_1 = f(x)$. Use the table to answer the following questions.
   (a) What is $f(1)$?
   (b) If $f(x) = -14$, what is the value of x?
   (c) At what point does the graph of $y = f(x)$ cross the y-axis?
   (d) At what point does the graph of $y = f(x)$ cross the x-axis?

X	Y₁	
0	6	
1	6	
2	4	
3	0	
4	-6	
5	-14	
6	-24	

X=0

Find the domain and range. Identify any functions.

13. $y = 9x + 2$

14. $4x - 7y = 1$

15. $y = |x|$

16. $y = \sqrt{x}$

17. $x + 1 = y^2$

18. $x = \sqrt{y + 3}$

19.

20.

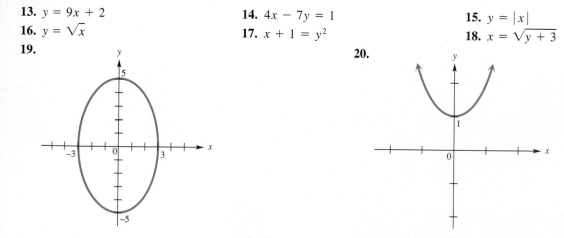

Find the angle of smallest possible positive measure coterminal with each angle.

21. $-51°$

22. $-174°$

23. $792°$

24. Let *n* represent any integer, and write an expression for all angles coterminal with an angle of 270°.

Work each problem.

25. A pulley is rotating 320 times per minute. Through how many degrees does a point on the edge of the pulley move in 2/3 second?

26. The propeller of a speedboat rotates 650 times per minute. Through how many degrees will a point on the edge of the propeller rotate in 2.4 seconds?

Convert decimal degrees to degrees, minutes, seconds, and convert degrees, minutes, seconds to decimal degrees. Round to the nearest second or the nearest thousandth of a degree, as appropriate. Use a calculator as necessary.

27. 47° 25′ 11″

28. 119° 8′ 3″

29. 74.2983°

30. −61.5034°

31. 183.0972°

32. 275.1005°

Find the measure of each marked angle.

33.

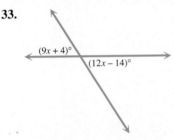

(9x + 4)°
(12x − 14)°

34.

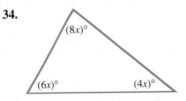

(8x)°
(6x)°
(4x)°

Find all unknown angle measures in each pair of similar triangles.

35.

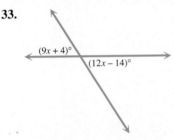

Z
32° T
41°
X   Y V   U

36.

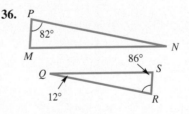

P
82°
M   N
Q   86° S
12°   R

Find the unknown side lengths in each pair of similar triangles.

37.

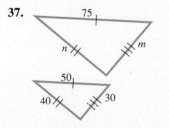

75
n   m
50
40   30

38.

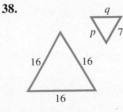

q
p 7
16   16
16

Find the unknown measurement. There are two similar triangles in each figure.

39.

40.

41. Complete the following statement: If two triangles are similar, their corresponding sides are _____ and their corresponding angles are _____ .

42. If a tree 20 feet tall casts a shadow of 8 feet, how long would the shadow of a 30-foot tree be at the same time?

Find the trigonometric function values of each angle. If a value is undefined, say so.

43.

44.

45. 180°

46. 360°

Find the values of all the trigonometric functions for an angle in standard position having the following point on its terminal side.

47. $(-8, 15)$ **48.** $(3, -4)$ **49.** $(1, -5)$

50. $(9, -2)$ **51.** $(6\sqrt{3}, -6)$ **52.** $(-2\sqrt{2}, 2\sqrt{2})$

53. Find the values of all the trigonometric functions for an angle in standard position having its terminal side defined by the equation $5x - 3y = 0$, $x \geq 0$.

54. If the terminal side of a quadrantal angle lies along the y-axis, which of its trigonometric functions are undefined?

In Exercises 55–57, consider an angle θ in standard position whose terminal side has the equation $y = -5x$, with $x \leq 0$.

55. Sketch θ and use an arrow to show the rotation if $0° \leq \theta < 360°$.

56. Find the exact values of $\sin \theta$ and $\cos \theta$.

57. Refer to the equation above, and give the exact value of $\tan \theta$.

Evaluate each expression.

58. $4 \sec 180° - 2 \sin^2 270°$

59. $-\cot^2 90° + 4 \sin 270° - 3 \tan 180°$

60. Explain why, for any value of θ, the point in the plane with coordinates $(\cos \theta, \sin \theta)$ lies on the unit circle.

61. The accompanying graphing calculator screen was obtained for a particular stored value of X. What will the screen display for the value of the expression in the final line of the display?

```
cos X
                        .1
(tan X)²
```

Decide whether each statement is possible *or* impossible.

62. $\sin \theta = 3/4$ and $\csc \theta = 4/3$

63. $\sec \theta = -2/3$

64. $\tan \theta = 1.4$

65. $\cos \theta = .25$ and $\sec \theta = -4$

Find all the trigonometric function values for each angle. Rationalize denominators when applicable.

66. $\sin \theta = \sqrt{3}/5$ and $\cos \theta < 0$

67. $\cos \gamma = -5/8$, with γ in quadrant III

68. $\tan \alpha = 2$, with α in quadrant III

69. $\sec \beta = -\sqrt{5}$, with β in quadrant II

70. $\sin \theta = -2/5$, with θ in quadrant III

71. $\sec \alpha = 5/4$, with α in quadrant IV

72. If, for some particular angle θ, $\sin \theta < 0$ and $\cos \theta > 0$, in what quadrant must θ lie? What is the sign of $\tan \theta$?

73. Explain how you would find the cotangent of an angle θ whose tangent is 1.6778490 using a calculator. Then find $\cot \theta$.

74.  At present, the north star Polaris is located very near the celestial north pole. However, because Earth is inclined 23.5°, the moon's gravitational pull on Earth is uneven. As a result, Earth slowly precesses (moves in) like a spinning top and the direction of the celestial north pole traces out a circular path once every 26,000 years. See the figure. For example, in approximately A.D.14,000 the star Vega will be located at the celestial north pole—and not the star Polaris. As viewed from the center C of this circular path, calculate the angle in seconds that the celestial north pole moves each year. (*Source:* Zeilik, M., S. Gregory, and E. Smith, *Introductory Astronomy and Astrophysics,* Saunders College Publishers, 1992.)

75. The depths of unknown craters on the moon can be approximated by comparing the lengths of their shadows to the shadows of nearby craters with known depths. The crater Aristillus is 11,000 feet deep and its shadow was measured as 1.5 mm on a photograph. Its companion crater, Autolycus, had a shadow of 1.3 mm on the same photograph. Use similar triangles to determine the depth of the crater Autolycus. (*Source:* Webb, T., *Celestial Objects for Common Telescopes,* Dover Publications, Inc., 1962.)

76. The lunar mountain peak Huygens has a height of 21,000 feet. The shadow of Huygens on a photograph was 2.8 mm while the nearby mountain Bradley had a shadow of 1.8 mm on the same photograph. Calculate the height of Bradley. (*Source:* Webb, T., *Celestial Objects for Common Telescopes,* Dover Publications, Inc., 1962.)

Acute Angles and Right Triangles

Highway transportation is critical to the economy of the United States. In 1970 there were 1150 billion vehicle miles traveled and by the year 2000 this will have increased to approximately 2500 billion miles.* The average vehicle in the United States carries 1.13 people. As a result, traffic delays and congestion will continue to increase. Designing highways for safety and efficiency has become a critical issue that saves both lives and time. Many problems like the following can be solved through the use of mathematics and trigonometry.

When an automobile travels around a curve, objects like trees, buildings, and fences situated on the inside of the curve can obstruct a driver's vision. It is expensive to clear land and move buildings that are located on the inside of the curve—and wasteful if it is not necessary. However, it is essential to maintain visibility on curves for a safe stopping distance. If the highway curve shown in the figure has a radius of 600 feet with a speed limit of 55 miles per hour, how far to the inside of the curve should the highway department clear the land?

2.1 Trigonometric Functions of Acute Angles

2.2 Trigonometric Functions of Non-Acute Angles

2.3 Finding Trigonometric Function Values Using a Calculator

2.4 Solving Right Triangles

2.5 Further Applications of Right Triangles

---

* Mannering, F., and W. Kilareski, *Principles of Highway Engineering and Traffic Control,* John Wiley & Sons, Inc., 1990.

The solution of this problem, given in Section 2.4, requires the use of trigonometry. Without trigonometry, our present freeway system would not have been possible. In this chapter we will learn how trigonometry is used to design curves, compute grade resistance, and calculate safe stopping distances.

So far, the definitions of the trigonometric functions have been used only for quadrantal angles such as 0°, 90°, 180°, or 270°. This chapter extends our study to include finding the values of trigonometric functions of other angles. The chapter ends with some applications of trigonometry.

2.1 Trigonometric Functions of Acute Angles ▼▼▼

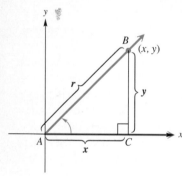

FIGURE 1

Figure 1 shows an acute angle A in standard position. The definitions of the trigonometric function values of angle A require x, y, and r. As drawn in Figure 1, x and y are the lengths of the two legs of the right triangle ABC, and r is the length of the hypotenuse.

The side of length y is called the **side opposite** angle A, and the side of length x is called the **side adjacent** to angle A. The lengths of these sides can be used to replace x and y in the definition of the trigonometric functions, with r replaced with the length of the hypotenuse, to get the following right triangle-based definitions.

Right Triangle-Based Definitions of Trigonometric Functions

For any acute angle A in standard position,

$$\sin A = \frac{y}{r} = \frac{\text{side opposite}}{\text{hypotenuse}} \qquad \csc A = \frac{r}{y} = \frac{\text{hypotenuse}}{\text{side opposite}}$$

$$\cos A = \frac{x}{r} = \frac{\text{side adjacent}}{\text{hypotenuse}} \qquad \sec A = \frac{r}{x} = \frac{\text{hypotenuse}}{\text{side adjacent}}$$

$$\tan A = \frac{y}{x} = \frac{\text{side opposite}}{\text{side adjacent}} \qquad \cot A = \frac{x}{y} = \frac{\text{side adjacent}}{\text{side opposite}}.$$

◀EXAMPLE 1
Finding trigonometric function values of an acute angle in a right triangle

Find the values of the trigonometric functions for angles A and B in the right triangle in Figure 2.

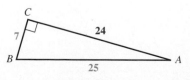

FIGURE 2

The length of the side opposite angle A is 7. The length of the side adjacent to angle A is 24, and the length of the hypotenuse is 25. Using the relationships given above,

$$\sin A = \frac{\text{side opposite}}{\text{hypotenuse}} = \frac{7}{25} \qquad \csc A = \frac{\text{hypotenuse}}{\text{side opposite}} = \frac{25}{7}$$

$$\cos A = \frac{\text{side adjacent}}{\text{hypotenuse}} = \frac{\mathbf{24}}{25} \qquad \sec A = \frac{\text{hypotenuse}}{\text{side adjacent}} = \frac{25}{\mathbf{24}}$$

$$\tan A = \frac{\text{side opposite}}{\text{side adjacent}} = \frac{7}{\mathbf{24}} \qquad \cot A = \frac{\text{side adjacent}}{\text{side opposite}} = \frac{\mathbf{24}}{7}.$$

The length of the side opposite angle B is 24, while the length of the side adjacent to B is 7, making

$$\sin B = \frac{24}{25} \qquad \tan B = \frac{24}{7} \qquad \sec B = \frac{25}{7}$$

$$\cos B = \frac{7}{25} \qquad \cot B = \frac{7}{24} \qquad \csc B = \frac{25}{24}. \quad \blacktriangleright$$

In Example 1, you may have noticed that $\sin A = \cos B$, $\cos A = \sin B$, and so on. Such relationships are always true for the two acute angles of a right triangle. Figure 3 shows a right triangle with acute angles A and B and a right angle at C. (Whenever we use A, B, and C to name the angles in a right triangle, C will be the right angle.) The length of the side opposite angle A is a, and the length of the side opposite angle B is b. The length of the hypotenuse is c.

By the definitions given above, $\sin A = a/c$. Since $\cos B$ is also equal to a/c,

$$\sin A = \frac{a}{c} = \cos B.$$

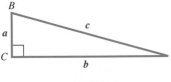

FIGURE 3

In a similar manner,

$$\tan A = \frac{a}{b} = \cot B \qquad \sec A = \frac{c}{b} = \csc B.$$

The sum of the three angles in any triangle is 180°. Since angle C equals 90°, angles A and B must have a sum of $180° - 90° = 90°$. As mentioned in Chapter 1, angles with a sum of 90° are called complementary angles. Since angles A and B are complementary and $\sin A = \cos B$, the functions sine and cosine are called **cofunctions.** Also, tangent and cotangent are cofunctions, as are secant and cosecant. And since the angles A and B are complementary, $A + B = 90°$, or

$$B = 90° - A,$$

giving

$$\sin A = \cos B = \cos(90° - A).$$

Similar results, called the **cofunction identities,** are true for the other trigonometric functions.

Cofunction Identities

For any acute angle A,

$$\sin A = \cos(90° - A) \qquad \csc A = \sec(90° - A)$$
$$\cos A = \sin(90° - A) \qquad \sec A = \csc(90° - A)$$
$$\tan A = \cot(90° - A) \qquad \cot A = \tan(90° - A).$$

(These identities will be extended to *any* angle A, and not just acute angles, in Chapter 5.) It would be wise to memorize all the identities presented in this book.

◀**EXAMPLE 2**
Writing functions in terms of cofunctions

Write each of the following in terms of cofunctions.

(a) $\cos 52°$
Since $\cos A = \sin(90° - A)$,
$$\cos 52° = \sin(90° - 52°) = \sin 38°.$$

(b) $\tan 71° = \cot 19°$

(c) $\sec 24° = \csc 66°$ ▶

◀**EXAMPLE 3**
Solving equations by using the cofunction identities

Find a value of θ satisfying each of the following. Assume that all angles involved are acute angles.

(a) $\cos(\theta + 4°) = \sin(3\theta + 2°)$
Since sine and cosine are cofunctions, this equation is true if the sum of the angles is 90°, or
$$(\theta + 4°) + (3\theta + 2°) = 90°$$
$$4\theta + 6° = 90°$$
$$4\theta = 84°$$
$$\theta = 21°.$$

(b) $\tan(2\theta - 18°) = \cot(\theta + 18°)$
$$(2\theta - 18°) + (\theta + 18°) = 90°$$
$$3\theta = 90°$$
$$\theta = 30°$$ ▶

Figure 4 shows three right triangles. From left to right, the length of each hypotenuse is the same, but angle A increases in measure. As angle A increases in measure from $0°$ to $90°$, the length of the side opposite angle A also increases.

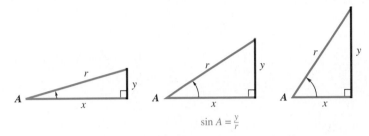

$$\sin A = \frac{y}{r}$$

As A increases, y increases. Since r is fixed, $\sin A$ increases.

FIGURE 4

Since

$$\sin A = \frac{\text{side opposite}}{\text{hypotenuse}},$$

as angle A increases, the numerator of this fraction also increases, while the denominator is fixed. This means that $\sin A$ *increases* as A increases from $0°$ to $90°$.

As angle A increases from $0°$ to $90°$, the length of the side adjacent to A decreases. Since r is fixed, the ratio x/r will decrease. This ratio gives $\cos A$, showing that the values of cosine *decrease* as the angle measure changes from $0°$ to $90°$. Finally, increasing A from $0°$ to $90°$ causes y to increase and x to decrease, making the values of $y/x = \tan A$ increase.

A similar discussion shows that as A increases from $0°$ to $90°$, the values of $\sec A$ increase, while the values of $\cot A$ and $\csc A$ decrease.

◀ **EXAMPLE 4**
Comparing function values of acute angles

Tell whether each of the following is *true* or *false*.

(a) $\sin 21° > \sin 18°$

In the interval from $0°$ to $90°$, as the angle increases, so does the sine of the angle, which makes $\sin 21° > \sin 18°$ a true statement.

(b) $\cos 49° \le \cos 56°$

As the angle increases, the cosine of the angle decreases. The given statement $\cos 49° \le \cos 56°$ is false. ▶

Certain special angles, such as $30°$, $45°$, and $60°$, occur so often in trigonometry and in more advanced mathematics that they deserve special study. We can find the exact trigonometric function values of these angles by using properties of geometry and the Pythagorean theorem.

TRIGONOMETRIC VALUES OF SPECIAL ANGLES To find the trigonometric function values for 30° and 60°, we start with an equilateral triangle, a triangle with all sides of equal length. Each angle of such a triangle has a measure of 60°. While the results we will obtain are independent of the length, for convenience, we choose the length of each side to be 2 units. See Figure 5(a).

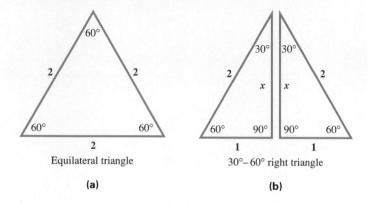

(a) Equilateral triangle

(b) 30°–60° right triangle

FIGURE 5

Bisecting one angle of this equilateral triangle leads to two right triangles, each of which has angles of 30°, 60°, and 90°, as shown in Figure 5(b). Since the hypotenuse of one of these right triangles has a length of 2, the shortest side will have a length of 1. (Why?) If x represents the length of the medium side, then, by the Pythagorean theorem,

$$2^2 = 1^2 + x^2$$
$$4 = 1 + x^2$$
$$3 = x^2$$
$$\sqrt{3} = x.$$

Figure 6 summarizes our results, showing a 30°–60° right triangle.

As shown in the figure, the side opposite the 30° angle has length 1; that is, for the 30° angle,

$$\text{hypotenuse} = 2, \qquad \text{side opposite} = 1, \qquad \text{side adjacent} = \sqrt{3}.$$

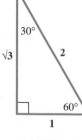

FIGURE 6

Using the definitions of the trigonometric functions,

$$\sin 30° = \frac{\text{side opposite}}{\text{hypotenuse}} = \frac{1}{2} \qquad\qquad \csc 30° = \frac{2}{1} = 2$$

$$\cos 30° = \frac{\text{side adjacent}}{\text{hypotenuse}} = \frac{\sqrt{3}}{2} \qquad\qquad \sec 30° = \frac{2}{\sqrt{3}} = \frac{2\sqrt{3}}{3}$$

$$\tan 30° = \frac{\text{side opposite}}{\text{side adjacent}} = \frac{1}{\sqrt{3}} = \frac{\sqrt{3}}{3} \qquad\qquad \cot 30° = \frac{\sqrt{3}}{1} = \sqrt{3}.$$

The denominator was rationalized for tan 30° and sec 30°.

In a similar manner,

$$\sin 60° = \frac{\sqrt{3}}{2} \qquad \tan 60° = \sqrt{3} \qquad \sec 60° = 2$$

$$\cos 60° = \frac{1}{2} \qquad \cot 60° = \frac{\sqrt{3}}{3} \qquad \csc 60° = \frac{2\sqrt{3}}{3}.$$

The values of the trigonometric functions for 45° can be found by starting with a 45°–45° right triangle, as shown in Figure 7. This triangle is isosceles, and, for convenience, we choose the lengths of the equal sides to be 1 unit. (As before, the results are independent of the length of the equal sides of the right triangle.) Since the shorter sides each have length 1, if r represents the length of the hypotenuse, then

$$1^2 + 1^2 = r^2$$
$$2 = r^2$$
$$\sqrt{2} = r.$$

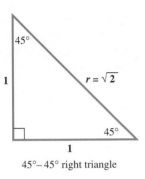

45°–45° right triangle

FIGURE 7

Using the measures indicated on the 45°–45° right triangle in Figure 7, we find

$$\sin 45° = \frac{1}{\sqrt{2}} = \frac{\sqrt{2}}{2} \qquad \tan 45° = \frac{1}{1} = 1 \qquad \sec 45° = \frac{\sqrt{2}}{1} = \sqrt{2}$$

$$\cos 45° = \frac{1}{\sqrt{2}} = \frac{\sqrt{2}}{2} \qquad \cot 45° = \frac{1}{1} = 1 \qquad \csc 45° = \frac{\sqrt{2}}{1} = \sqrt{2}.$$

The importance of these exact trigonometric function values of 30°, 60°, and 45° angles cannot be overemphasized. It is essential to memorize them. They are summarized in the chart that follows.

Function Values of Special Angles

θ	$\sin \theta$	$\cos \theta$	$\tan \theta$	$\cot \theta$	$\sec \theta$	$\csc \theta$
30°	$\dfrac{1}{2}$	$\dfrac{\sqrt{3}}{2}$	$\dfrac{\sqrt{3}}{3}$	$\sqrt{3}$	$\dfrac{2\sqrt{3}}{3}$	2
45°	$\dfrac{\sqrt{2}}{2}$	$\dfrac{\sqrt{2}}{2}$	1	1	$\sqrt{2}$	$\sqrt{2}$
60°	$\dfrac{\sqrt{3}}{2}$	$\dfrac{1}{2}$	$\sqrt{3}$	$\dfrac{\sqrt{3}}{3}$	2	$\dfrac{2\sqrt{3}}{3}$

NOTE You should be able to reproduce the 30°–60° and 45°–45° right triangles on this chart quickly. The latter is not difficult to do if you learn the values of sin 30°, sin 45°, and sin 60°. Then complete the rest of the chart using the reciprocal identities, the cofunction identities, and the quotient identities.

CONNECTIONS A convenient way to quickly produce a chart of the trigonometric values for the special angles is to produce the chart shown below. Write the angles in the first column. In the second column each numerator is a radical with the numbers 0, 1, 2, 3, and 4, in order, placed under it. Each denominator is 2. In the third column each numerator is a radical with the numbers 4, 3, 2, 1, and 0, in order, placed under it. Each denominator is 2. Simplifying these fractions gives the values shown in the chart above for sin θ and cos θ. *Note that this works only for the degree measures shown below and cannot be extended to other values of θ.* The other trigonometric function values are easily found from these basic ones.

θ	$\sin \theta$	$\cos \theta$
0°	$\dfrac{\sqrt{0}}{2}$	$\dfrac{\sqrt{4}}{2}$
30°	$\dfrac{\sqrt{1}}{2}$	$\dfrac{\sqrt{3}}{2}$
45°	$\dfrac{\sqrt{2}}{2}$	$\dfrac{\sqrt{2}}{2}$
60°	$\dfrac{\sqrt{3}}{2}$	$\dfrac{\sqrt{1}}{2}$
90°	$\dfrac{\sqrt{4}}{2}$	$\dfrac{\sqrt{0}}{2}$

FOR DISCUSSION OR WRITING
Verify that the simplified forms of the fractions in the table agree with the values shown earlier.

In Exercises 57 and 58, we generalize the relationships among the sides of a 30°–60° right triangle and a 45°–45° right triangle.

Since a calculator finds trigonometric function values at the touch of a key, you may wonder why we spend so much time in finding values for special angles. We do this because a calculator gives only *approximate* values in most cases, while we often need *exact* values. For example, tan 30° can be found on a scientific calculator by first setting the machine in the *degree mode,* then entering 30 and pressing the tan key to get

$$\tan 30° \approx .57735027.$$

(The symbol ≈ means "is approximately equal to.") Earlier, however, we found the exact value:

$$\tan 30° = \frac{\sqrt{3}}{3}.$$

2.1 Exercises ▼▼▼▼▼▼▼▼▼▼▼▼▼▼▼▼▼▼▼▼▼▼▼▼▼▼▼▼▼▼▼▼

In each exercise, find the values of the six trigonometric functions for angle A. Leave answers as fractions. See Example 1.

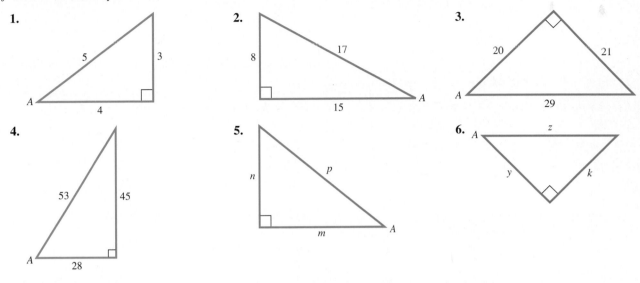

1.   5   3   4 A

2.   8   17   15 A

3.   20   21 A 29

4.   53   45 A 28

5. n p m A

6. A z y k

Suppose ABC is a right triangle with sides of lengths a, b, and c with right angle at C (see Figure 3). Find the unknown side length using the Pythagorean theorem, and then find the values of the six trigonometric functions for angle B. Rationalize denominators when applicable.

7. $a = 5, b = 12$ **8.** $a = 3, b = 5$ **9.** $a = 6, c = 7$ **10.** $b = 7, c = 12$

11. Write a summary of the relationships between cofunctions of complementary angles.

Write each of the following in terms of the cofunction. Assume that all angles in which an unknown appears are acute angles. See Example 2.

12. $\tan 50°$ **13.** $\cot 73°$ **14.** $\csc 47°$ **15.** $\sec 39°$

16. $\cos(\alpha + 20°)$ **17.** $\cot(\beta - 10°)$ **18.** $\tan 25.4°$ **19.** $\sin 38.7°$

20. With a calculator, evaluate $\sin(90° - A)$ and $\cos A$ for various values of A. (Include values greater than 90° and less than zero.) What do you find?

Find a solution for each equation. Assume that all angles in which an unknown appears are acute angles. See Example 3.

21. $\tan \alpha = \cot(\alpha + 10°)$ **22.** $\cos \theta = \sin 2\theta$

23. $\sin(2\gamma + 10°) = \cos(3\gamma - 20°)$ **24.** $\sec(\beta + 10°) = \csc(2\beta + 20°)$

25. $\tan(3B + 4°) = \cot(5B - 10°)$ **26.** $\cot(5\theta + 2°) = \tan(2\theta + 4°)$

Tell whether the statement is true *or* false. *See Example 4.*

27. $\tan 28° \leq \tan 40°$ **28.** $\sin 50° > \sin 40°$ **29.** $\sin 46° < \cos 46°$
(*Hint:* $\cos 46° = \sin 44°$)

30. $\cos 28° < \sin 28°$ **31.** $\tan 41° < \cot 41°$ **32.** $\cot 30° < \tan 40°$

33. Recall that a graphing calculator will return a 1 for a true statement and a 0 for a false statement. What possible values (between 0° and 90°) for A will produce the accompanying graphing calculator screen?

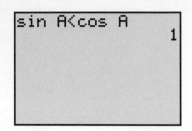

34. Find the angle θ in the interval $[0°, 90°)$ for which $\sin \theta = \cos \theta$.

Refer to the discussion in this section to give the exact *trigonometric function value. Do not use a calculator.*

35. $\tan 30°$ **36.** $\cot 30°$ **37.** $\sin 30°$ **38.** $\cos 30°$

39. $\csc 45°$ **40.** $\sec 45°$ **41.** $\cos 45°$ **42.** $\sin 45°$

43. $\sin 60°$ **44.** $\cos 60°$ **45.** $\tan 60°$ **46.** $\cot 60°$

47. Refer to Table 1. What trigonometric functions are Y_1 and Y_2?

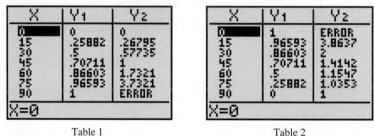

Table 1   Table 2

48. Refer to Table 2. What trigonometric functions are Y_1 and Y_2?

49. What value of A between 0° and 90° will produce the output for the accompanying graphing calculator screen?

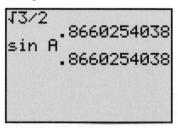

50. A student was asked to give the exact value of $\sin 45°$. Using a calculator, he gave the answer .7071067812. The teacher did not give him credit. What was the teacher's reason for this?

51. With a graphing calculator, find the coordinates of the point of intersection of $y = x$ and $y = \sqrt{1 - x^2}$. These coordinates are the cosine and sine of what angle between 0° and 90°?

52. Find the equation of the line passing through the origin and making a 60° angle with the x-axis.

53. Find the equation of the line passing through the origin and making a 30° angle with the x-axis.

54. What angle does the line $y = \dfrac{\sqrt{3}}{3}x$ make with the positive x-axis?

55. What angle does the line $y = \sqrt{3}x$ make with the positive x-axis?

56. Which pair of trigonometric functions are both reciprocals and cofunctions?

57. Construct an equilateral triangle with each side having length $2k$.
   (a) What is the measure of each angle?
   (b) Label one angle A. Drop a perpendicular from A to the side opposite A. Two 30° angles are formed at A, and two right triangles are formed. What is the length of each side opposite each 30° angle?
   (c) What is the length of the perpendicular constructed in part (b)?
   (d) From the results of parts (a)–(c), complete the following statement: In a 30°–60° right triangle, the hypotenuse is always _____ times as long as the shorter leg, and the longer leg has a length that is _____ times as long as that of the shorter leg. Also, the shorter leg is opposite the _____ angle, and the longer leg is opposite the _____ angle.

58. Construct a square with each side of length k.
   (a) Draw a diagonal of the square. What is the measure of each angle formed by a side of the square and this diagonal?
   (b) What is the length of the diagonal?
   (c) From the results of parts (a) and (b), complete the following statement: In a 45°–45° right triangle, the hypotenuse has a length that is _____ times as long as either leg.

Use the results of Exercises 57 and 58 to find the exact value of each labeled part in each figure.

59.

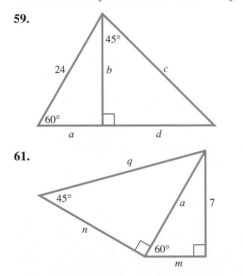

60.

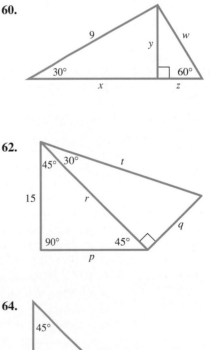

61.

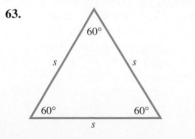

62.

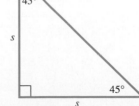

Find a formula for the area of each figure in terms of s.

63.

64.

65. Suppose you know the length of one side and one acute angle of a right triangle. Can you determine the measures of all the sides and angles of the triangle?

66. Refer to the table in the Connections box in this section. Explain why this pattern cannot possibly continue past 90°. (*Hint:* What is the maximum value of the sine ratio?)

67. Construct a table similar to the table in the Connections box in this section for values of the cosine of those angles.

68. If aerodynamic resistance is ignored, the braking distance D (in feet) for an automobile to change its velocity from V_1 to V_2 (feet per second) can be calculated using the equation

$$D = \frac{1.05(V_1{}^2 - V_2{}^2)}{64.4(K_1 + K_2 + \sin \theta)}.$$

K_1 is a constant determined by the efficiency of the brakes and tires, K_2 is a constant determined by the rolling resistance of the automobile, and θ is the grade of the highway. (*Source:* Mannering, F., and W. Kilareski, *Principles of Highway Engineering and Traffic Control*, John Wiley & Sons, Inc., 1990.)

(a) Compute the number of feet required to slow a car from 55 to 30 miles per hour while traveling uphill with a grade of $\theta = 3.5°$. Let $K_1 = .4$ and $K_2 = .02$. (*Hint:* Change miles per hour to feet per second.)

(b) Repeat part (a) with $\theta = -2°$.

(c) How is braking distance affected by the grade θ? Does this agree with your driving experience?

69. Refer to Exercise 68. An automobile is traveling at 90 miles per hour on a highway with a downhill grade of $\theta = -3.5°$. The driver sees a stalled truck in the road 200 feet away and immediately applies the brakes. Assuming that a collision cannot be avoided, how fast (in miles per hour) is the car traveling when it hits the truck? (Use the same values for K_1 and K_2 as in Exercise 68.)

2.2   Trigonometric Functions of Non-Acute Angles ▼▼▼

Now that we have found trigonometric function values for acute angles, we can use those results to find trigonometric function values for other types of angles. Associated with every nonquadrantal angle in standard position is a positive acute angle called its reference angle. A **reference angle** for an angle θ, written θ', is the positive acute angle made by the terminal side of angle θ and the x-axis. Figure 8 shows several angles θ (each less than one complete counterclockwise revolution) in quadrants II, III, and IV, respectively, with the reference angle θ' also shown. In quadrant I, θ and θ' are the same. If an angle θ is negative or has measure greater than 360°, its reference angle is found by first finding its coterminal angle that is between 0° and 360°, and then using the diagrams in Figure 8.

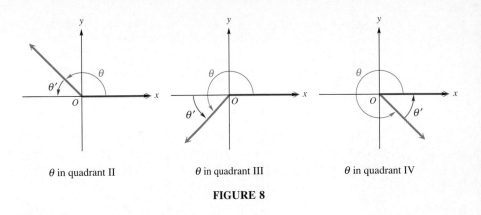

θ in quadrant II   θ in quadrant III   θ in quadrant IV

FIGURE 8

CAUTION   A very common error is to find the reference angle by using the terminal side of θ and the y-axis. *The reference angle is always found with reference to the x-axis.*

◀EXAMPLE 1
Finding reference angles

Find the reference angles for the following angles.

(a) 218°

As shown in Figure 9, the positive acute angle made by the terminal side of this angle and the x-axis is $218° - 180° = 38°$. For $\theta = 218°$, the reference angle $\theta' = 38°$.

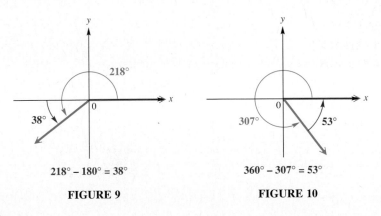

$218° - 180° = 38°$ $360° - 307° = 53°$

FIGURE 9   **FIGURE 10**

(b) 1387°

First find a coterminal angle between 0° and 360°. Divide 1387° by 360° to get a quotient of about 3.9. Begin by subtracting 360° three times (because of the 3 in 3.9):

$$1387° - 3 \cdot 360° = 307°.$$

The reference angle for 307° (and thus for 1387°) is $360° - 307° = 53°$. See Figure 10. ▶

The preceding example suggests the following table for finding the reference angle θ' for any angle θ between $0°$ and $360°$.

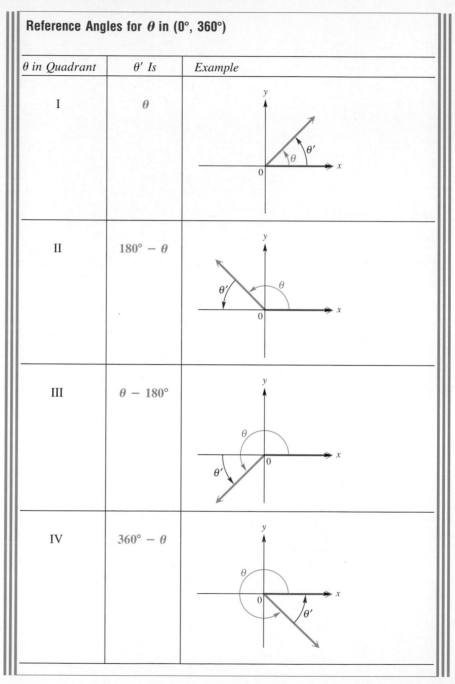

Reference Angles for θ in ($0°$, $360°$)		
θ in Quadrant	θ' Is	Example
I	θ	
II	$180° - \theta$	
III	$\theta - 180°$	
IV	$360° - \theta$	

We can now find exact trigonometric function values of angles with reference angles of 30°, 60°, or 45°. In Example 2 we show how to use these function values to find the trigonometric function values for 210°.

◀**EXAMPLE 2**
Finding trigonometric function values of 210°

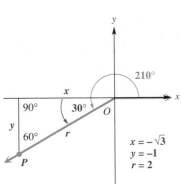

FIGURE 11

Find the values of the trigonometric functions for 210°.

Even though a 210° angle is not an angle of a right triangle, the ideas mentioned earlier can still be used to find the trigonometric function values for this angle. To do so, we draw an angle of 210° in standard position, as shown in Figure 11. We choose point P on the terminal side of the angle so that the distance from the origin O to P is 2. By the results from 30°–60° right triangles, the coordinates of point P become $(-\sqrt{3}, -1)$, with $x = -\sqrt{3}$, $y = -1$, and $r = 2$. Then, by the definitions of the trigonometric functions,

$$\sin 210° = -\frac{1}{2} \qquad \tan 210° = \frac{\sqrt{3}}{3} \qquad \sec 210° = -\frac{2\sqrt{3}}{3}$$

$$\cos 210° = -\frac{\sqrt{3}}{2} \qquad \cot 210° = \sqrt{3} \qquad \csc 210° = -2. \ ◗$$

Notice in Example 2 that the trigonometric function values of 210° correspond in absolute value to those of its reference angle 30°. The signs are different for the sine, cosine, secant, and cosecant functions because 210° is a quadrant III angle. These results suggest a shortcut for finding the trigonometric function values of a non-acute angle, using the reference angle. In Example 2, the reference angle for 210° is 30°, as shown in Figure 11. Simply by using the trigonometric function values of the reference angle, 30°, and choosing the correct signs for a quadrant III angle, we obtain the same results as found in Example 2.

Based on this work, the values of the trigonometric functions for any non-quadrantal angle θ can be determined by finding the function values for an angle between 0° and 90°. To do this, perform the following steps.

Finding Trigonometric Function Values for Any Non-Quadrantal Angle

1. If $\theta > 360°$, or if $\theta < 0°$, find a coterminal angle by adding or subtracting 360° as many times as needed to get an angle greater than 0° but less than 360°.
2. Find the reference angle θ'.
3. Find the necessary values of the trigonometric functions for the reference angle θ'.
4. Determine the correct signs for the values found in Step 3. (Use the table of signs in Section 1.5.) This result gives the value of the trigonometric functions for angle θ.

EXAMPLE 3
Finding trigonometric function values using reference angles

Use reference angles to find the exact value of each of the following.

(a) $\cos(-240°)$

The reference angle is 60°, as shown in Figure 12. Since the cosine is negative in quadrant II,

$$\cos(-240°) = -\cos 60° = -\frac{1}{2}.$$

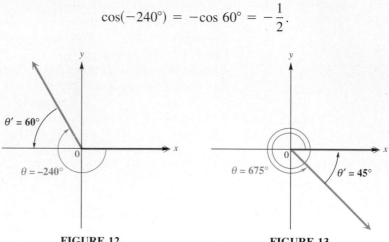

FIGURE 12   FIGURE 13

(b) $\tan 675°$

Begin by subtracting 360° to get a coterminal angle between 0° and 360°.

$$675° - 360° = 315°$$

As shown in Figure 13, the reference angle is $360° - 315° = 45°$. An angle of 315° is in quadrant IV, so the tangent will be negative, and

$$\tan 675° = \tan 315° = -\tan 45° = -1. \ \blacktriangleright$$

The ideas discussed in this section can be reversed to find the measures of certain angles, given a trigonometric function value and an interval in which the angle must lie. We are most often interested in the interval [0°, 360°).

EXAMPLE 4
Finding angle measures given an interval and a function value

Find all values of θ, if θ is in the interval [0°, 360°) and $\cos \theta = -\sqrt{2}/2$.

Since cosine here is negative, θ must lie in either quadrant II or III. Since the absolute value of $\cos \theta$ is $\sqrt{2}/2$, the reference angle θ' must be 45°. The two possible angles θ are sketched in Figure 14. The quadrant II angle θ must equal $180° - 45° = 135°$, and the quadrant III angle θ must equal $180° + 45° = 225°. \ \blacktriangleright$

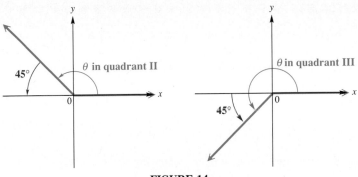

FIGURE 14

Exact values can be used in evaluating expressions, as shown in the next example.

EXAMPLE 5
Evaluating an expression with function values of special angles

Evaluate $\cos 120° + 2 \sin^2 60° - \tan^2 30°$.
   Since $\cos 120° = -1/2$, $\sin 60° = \sqrt{3}/2$, and $\tan 30° = \sqrt{3}/3$,

$$\cos 120° + 2 \sin^2 60° - \tan^2 30° = -\frac{1}{2} + 2\left(\frac{\sqrt{3}}{2}\right)^2 - \left(\frac{\sqrt{3}}{3}\right)^2$$

$$= -\frac{1}{2} + 2\left(\frac{3}{4}\right) - \frac{3}{9}$$

$$= \frac{2}{3}. \quad \blacktriangleright$$

As mentioned earlier, the values of the trigonometric functions of coterminal angles are the same.

EXAMPLE 6
Using coterminal angles to find function values

Evaluate each of the following by first expressing the function in terms of an angle between 0° and 360°.

(a) $\cos 780°$
   Add or subtract 360° as many times as necessary so that the final angle is between 0° and 360°. Subtracting 720°, which is $2 \cdot 360°$, gives

$$\cos 780° = \cos(60° + 2 \cdot 360°)$$

$$= \cos 60°$$

$$= \frac{1}{2}.$$

(b) $\tan(-405°)$
   Add 360° to get $-405° + 360° = -45°$. Following the method of Example 2, we find that $\tan(-45°) = -\tan 45°$, because the reference angle is 45°, and $-45°$ is in quadrant IV where the tangent function is negative. Thus,

$$\tan(-405°) = \tan(-45°) = -\tan 45° = -1. \quad \blacktriangleright$$

2.2 Exercises ▼▼

1. In Example 2(a), why was 2 a good choice for r? Could any other positive number have been used?

2. Explain how the reference angle is used to find values of the trigonometric functions for an angle in quadrant III.

3. Explain why two coterminal angles have the same values for their trigonometric functions.

4. If two angles have the same values for each of the six trigonometric functions, must the angles be coterminal? Explain your reasoning.

Find the reference angle for each of the following. See Example 1.

5. 98°   **6.** 212°   **7.** −135°   **8.** −60°   **9.** 750°   **10.** 480°

Note: *The remaining exercises in this set are* not *to be worked with a calculator.*

Use the methods of this section to find the exact *values of the six trigonometric functions for each of the following angles. Rationalize denominators when applicable. See Examples 2, 3, and 6.*

11. 120°	**12.** 135°	**13.** 150°	**14.** 225°	**15.** 240°
16. 300°	**17.** 315°	**18.** 405°	**19.** 420°	**20.** 480°
21. 495°	**22.** 570°	**23.** 750°	**24.** 1305°	**25.** 1500°
26. 2670°	**27.** −390°	**28.** −510°	**29.** −1020°	**30.** −1290°

Complete the following table with exact trigonometric function values using the methods of this section. See Examples 2 and 3.

θ	$\sin\theta$	$\cos\theta$	$\tan\theta$	$\cot\theta$	$\sec\theta$	$\csc\theta$
31. 30°	$1/2$	$\sqrt{3}/2$	_____	_____	$2\sqrt{3}/3$	2
32. 45°	_____	_____	1	1	_____	_____
33. 60°	_____	$1/2$	$\sqrt{3}$	_____	2	_____
34. 120°	$\sqrt{3}/2$	_____	$-\sqrt{3}$	_____	_____	$2\sqrt{3}/3$
35. 135°	$\sqrt{2}/2$	$-\sqrt{2}/2$	_____	_____	$-\sqrt{2}$	$\sqrt{2}$
36. 150°	_____	$-\sqrt{3}/2$	$-\sqrt{3}/3$	_____	_____	2
37. 210°	$-1/2$	_____	$\sqrt{3}/3$	$\sqrt{3}$	_____	-2
38. 240°	$-\sqrt{3}/2$	$-1/2$	_____	_____	-2	$-2\sqrt{3}/3$

39. Does there exist an angle θ with the function values $\cos\theta = .6$ and $\sin\theta = -.8$?

40. Does there exist an angle θ with the function values $\cos\theta = 2/3$ and $\sin\theta = 3/4$?

Suppose θ is in the interval $(90°, 180°)$. Find the sign of the following.

41. $\sin\dfrac{\theta}{2}$ **42.** $\cos\dfrac{\theta}{2}$ **43.** $\cot(\theta + 180°)$

44. $\sec(\theta + 180°)$ **45.** $\cos(-\theta)$ **46.** $\sin(-\theta)$

47. Explain why $\sin\theta = \sin(\theta + n \cdot 360°)$ for any angle θ and any integer n.

48. Explain why $\cos\theta = \cos(\theta + n \cdot 360°)$ for any angle θ and any integer n.

49. Without using a calculator, determine which of the following numbers is closest to $\cos 115°$: .4, .6, 0, −.4, or −.6.

50. Without using a calculator, determine which of the following numbers is closest to sin 115°: .9, .1, 0, −.9, or −.1.

51. For what angles θ between 0° and 360° does cos θ = −sin θ?

52. For what angles θ between 0° and 360° does cos θ = sin θ?

Tell whether each statement is true *or* false. *If false, tell why. See Example 5.*

53. sin 30° + sin 60° = sin(30° + 60°)

54. sin(30° + 60°) = sin 30° · cos 60° + sin 60° · cos 30°

55. cos 60° = 2 cos² 30° − 1

56. cos 60° = 2 cos 30°

57. sin 120° = sin 150° − sin 30°

58. sin 210° = sin 180° + sin 30°

59. sin 120° = sin 180° · cos 60° − sin 60° · cos 180°

60. cos 300° = cos 240° · cos 60° − sin 240° · sin 60°

61. When a highway goes downhill and then uphill it is said to have a *sag curve*. Sag curves are designed so that at night, headlights shine sufficiently far down the road to allow for a safe stopping distance. See the figure. The minimum length L of a sag curve is determined by the height h of the car's headlights above the pavement, the downhill grade $\theta_1 < 0°$, the uphill grade $\theta_2 > 0°$, and the safe stopping distance S for a given speed limit. In addition, L is dependent on the vertical alignment of the headlights. Headlights are usually pointed upward at a slight angle α above the horizontal of the car. Using these quantities, L can then be computed using the formula $L = \dfrac{(\theta_2 - \theta_1)S^2}{200(h + S \tan \alpha)}$ where $S < L$.

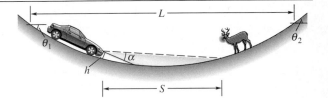

(*Source:* Mannering, F., and W. Kilareski, *Principles of Highway Engineering and Traffic Control,* John Wiley & Sons, Inc., 1990.)

(a) Compute L for a 55 mile per hour speed limit where h = 1.9 ft, α = .9°, θ_1 = −3°, θ_2 = 4°, and S = 336 feet.

(b) Repeat part (a) with α = 1.5°.

(c) How does the alignment of the headlights affect the value of L?

2.3   Finding Trigonometric Function Values Using a Calculator ▼▼▼

With the technological advances of this era in mind, the examples and exercises in this text assume that all students have access to scientific calculators. However, since calculators differ among makes and models, students should always consult their owner's manuals for specific information if questions arise concerning their use.

CAUTION   Thus far in this book, we have studied only one type of measure for angles—degree measure; another type of measure, radians, will be studied in Chapter 3. When evaluating trigonometric functions of angles given in degrees, it is a common error to use the incorrect mode; remember that the calculator must be set in the *degree mode*. One way to avoid this problem is to get in the habit of always starting work by finding sin 90°. If the displayed answer is 1, the calculator is set for degree measure; otherwise it is not.

Almost all calculator values of trigonometric functions are approximations.

EXAMPLE 1
Finding function values with a calculator

Use a calculator to approximate the values of the following trigonometric functions.

(a) sin 49° 12′

Convert 49° 12′ to decimal degrees, as explained in Chapter 1.

$$49° \ 12′ = 49 \ \frac{12°}{60} = 49.2°$$

To eight decimal places,

$$\sin 49° \ 12′ = \sin 49.2° \approx .75699506.$$

(b) sec 97.977°

Calculators do not have secant keys. However,

$$\sec \theta = \frac{1}{\cos \theta}$$

for all angles θ where $\cos \theta \neq 0$. So find sec 97.977° by first finding cos 97.977° and then taking the reciprocal to get

$$\sec 97.977° \approx -7.205879213.$$

(c) cot 51.4283°

Use the identity $\cot \theta = 1/\tan \theta$.

$$\cot 51.4283° \approx .79748114$$

(d) $\sin(-246°) \approx .91354546$

(e) sin 130° 48′

130° 48′ is equal to 130.8°.

$$\sin 130° \ 48′ = \sin 130.8° \approx .75699506 \quad \blacktriangleright$$

Notice that the values found in parts (a) and (e) of Example 1 are the same. The reason for this is that 49° 12′ is the reference angle for 130° 48′ and the sine function is positive for a quadrant II angle.

FINDING ANGLES So far in this section we have used a calculator to find trigonometric function values of angles. This process can be reversed. For now we restrict our attention to angles in the interval [0°, 90°]. The measure of an angle can be found from one of its trigonometric function values as shown in the next example.

EXAMPLE 2
Finding angle measures with a calculator

Use a calculator to find a value of θ in the interval [0°, 90°] satisfying each of the following. Leave answers in decimal degrees.

(a) sin θ = .81815000

We find θ using a key labeled $\boxed{\text{arc}}$ or $\boxed{\text{INV}}$ together with the $\boxed{\text{sin}}$ key. Some calculators may require a key labeled $\boxed{\text{sin}^{-1}}$ instead. Check your owner's manual to see how your calculator handles this. Again, make sure the calculator is set for degree measure. You should get

$$\theta \approx 54.900028°.$$

(b) sec $\theta = 1.0545829$

Use the identity $\cos \theta = 1/\sec \theta$. Enter 1.0545829 and find the reciprocal. This gives $\cos \theta \approx .9482421913$. Now find θ as shown in part (a). The result is

$$\theta \approx 18.514704°. \quad \blacktriangleright$$

CAUTION Compare Examples 1(b) and 2(b). Note that the reciprocal is used *before* the inverse cosine key when finding the angle, but *after* the cosine key when finding the trigonometric function.

 **EXAMPLE 3**
Finding grade resistance

When an automobile travels uphill or downhill on a highway, it experiences a force due to gravity. This force F in lb is called **grade resistance** and is computed using the equation $F = W \sin \theta$ where θ is the grade and W is the weight of the automobile.* If the automobile is moving uphill $\theta > 0°$; if downhill $\theta < 0°$. See Figure 15.

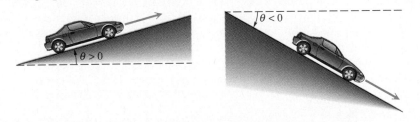

FIGURE 15

(a) Calculate F to two significant digits for a 2500-lb car traveling an uphill grade with $\theta = 2.5°$.

$$F = W \sin \theta = 2500 \sin 2.5° \approx 110 \text{ lb}$$

(b) Calculate F to two significant digits for a 5000-lb truck traveling a downhill grade with $\theta = -6.1°$.

$$F = W \sin \theta = 5000 \sin(-6.1°) \approx -530 \text{ lb}$$

F is negative because the truck is moving downhill.

(c) Calculate F for $\theta = 0°$ and $\theta = 90°$. Do these answers agree with your intuition?

$$F = W \sin \theta = W \sin 0° = W(0) = 0 \text{ lb}$$
$$F = W \sin \theta = W \sin 90° = W(1) = W \text{ lb}$$

This agrees with intuition because if $\theta = 0°$ then there is level ground and gravity does not cause the vehicle to roll. If $\theta = 90°$ the road would be vertical and the full weight of the vehicle would be pulled downward by gravity, so $F = W$. $\quad \blacktriangleright$

---

* Mannering, F., and W. Kilareski, *Principles of Highway Engineering and Traffic Control*, John Wiley & Sons, Inc., 1990.

2.3 Exercises ▼▼▼▼▼▼▼▼▼▼▼▼▼▼▼▼▼▼▼▼▼▼▼▼▼▼▼▼▼▼▼▼▼▼▼▼

Use a calculator to find a decimal approximation for each value. Give as many digits as your calculator displays. In Exercises 17–26, simplify the expression before using the calculator. See Example 1.

1. tan 29° 30′

2. sin 38° 42′

3. cot 41° 24′

4. cos 27° 10′

5. sec 13° 15′

6. csc 44° 30′

7. sin 39° 40′

8. tan 17° 12′

9. csc 145° 45′

10. cot 183° 48′

11. cos 421° 30′

12. sec 312° 12′

13. tan(−80° 6′)

14. sin(−317° 36′)

15. cot(−512° 20′)

16. cos(−15′)

17. $\dfrac{1}{\sec 14.8°}$

18. $\dfrac{1}{\csc 514° 24'}$

19. $\dfrac{1}{\cot 23.4°}$

20. $\dfrac{\sin 33°}{\cos 33°}$

21. $\dfrac{\cos 77°}{\sin 77°}$

22. cos(90° − 3.69°)

23. cot(90° − 4.72°)

24. sin(90° − 17° 12′)

25. sec² 47.8° − 1

26. sin² 17.7° + cos² 17.7°

27. A student, wishing to use a calculator to verify the value of sin 30°, enters the information correctly but gets a display of −.98803162. He knows that the display should be .5, and he also knows that his calculator is in good working order. What do you think is the problem?

28. A certain make of calculator does not allow the input of angles outside of a particular interval when finding trigonometric function values. For example, trying to find cos 2000° using the methods of this section would give an error message, despite the fact that cos 2000° can be evaluated. Explain how you would find cos 2000° using this calculator.

Find a value of θ in [0°, 90°) that satisfies the statement. Leave your answer in decimal degrees. See Example 2.

29. sin θ = .84802194

30. tan θ = 1.4739716

31. sec θ = 1.1606249

32. cot θ = 1.2575516

33. sin θ = .72144101

34. sec θ = 2.7496222

35. tan θ = 6.4358841

36. sin θ = .27843196

37. What value of A between 0° and 90° will produce the output in the accompanying graphing calculator screen?

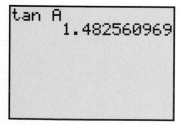

38. What value of A will produce the output in the accompanying graphing calculator screen?

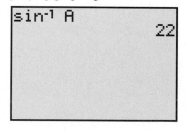

Use a calculator to find each of the following. (As shown later in Chapter 5, all these answers should be integers.)

39. sin 35° cos 55° + cos 35° sin 55°

40. cos 100° cos 80° − sin 100° sin 80°

41. cos 75° 29′ cos 14° 31′ − sin 75° 29′ sin 14° 31′

42. sin 28° 14′ cos 61° 46′ + cos 28° 14′ sin 61° 46′

When a light ray travels from one medium, such as air, to another medium, such as water or glass, the speed of the light changes, and the direction in which the ray is traveling changes. (This is why a fish under water is in a different position than it appears to be.) These changes are given by Snell's law

$$\frac{c_1}{c_2} = \frac{\sin \theta_1}{\sin \theta_2},$$

where c_1 is the speed of light in the first medium, c_2 is the speed of light in the second medium, and θ_1 and θ_2 are the angles shown in the figure. In the following exercises, assume that $c_1 = 3 \times 10^8$ m per sec. Find the speed of light in the second medium.

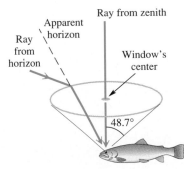

43. $\theta_1 = 46°$, $\theta_2 = 31°$

44. $\theta_1 = 39°$, $\theta_2 = 28°$

Find θ_2 for the following values of θ_1 and c_2. Round to the nearest degree.

45. $\theta_1 = 40°$, $c_2 = 1.5 \times 10^8$ m per sec

46. $\theta_1 = 62°$, $c_2 = 2.6 \times 10^8$ m per sec

The figure here shows a fish's view of the world above the surface of the water. (From "The Amateur Scientist" by Jearl Walker in Scientific American, March 1984. Copyright © 1984 by Scientific American, Inc. Reprinted by permission of Scientific American, Inc. All rights reserved.) Suppose that a light ray comes from the horizon, enters the water, and strikes the fish's eye.

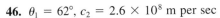

47. Let us assume that this ray gives a value of 90° for angle θ_1 in the formula for Snell's law. (In a practical situation this angle would probably be a little less than 90°.) The speed of light in water is about 2.254×10^8 m per sec. Find angle θ_2.

(Your result should have been about 48.7°. This means that a fish sees the world above the water as a cone, making an angle of 48.7° with the vertical.)

48. Suppose an object is located at a true angle of 29.6° above the horizon. Find the apparent angle above the horizon to a fish.

Use a calculator to decide whether the following statements are true or false. It may be that a true statement will lead to results that differ in the last decimal place, due to rounding error.

49. $\cos 40° = 2 \cos 20°$

50. $\sin 10° + \sin 10° = \sin 20°$

51. $\cos 70° = 2 \cos^2 35° - 1$

52. $\sin 50° = 2 \sin 25° \cdot \cos 25°$

53. $2 \cos 38° 22' = \cos 76° 44'$

54. $\cos 40° = 1 - 2 \sin^2 80°$

55. $\frac{1}{2} \sin 40° = \sin \frac{1}{2}(40°)$

56. $\sin 39° 48' + \cos 39° 48' = 1$

See Example 3 to work Exercises 57–64.

57. What is the grade resistance of a 2400-pound car traveling on a $-2.4°$ downhill grade?

58. What is the grade resistance of a 2100-pound car traveling on a $1.8°$ uphill grade?

59. A 3000-pound car traveling uphill has a grade resistance of 150 pounds. What is the angle of the grade?

60. A car traveling on a $1.5°$ uphill grade has a grade resistance of 120 pounds. What is the weight of the car?

61. A car traveling on a $-3°$ downhill grade has a grade resistance of -145 pounds. What is the weight of the car?

62. A 2600-pound car traveling downhill has a grade resistance of 130 pounds. What is the angle of the grade?

63. Which has the greater grade resistance: a 2200-pound car on a $2°$ uphill grade or a 2000-pound car on a $2.2°$ uphill grade?

64. Complete the table for the values of $\sin \theta$, $\tan \theta$, and $\dfrac{\pi\theta}{180}$ to four decimal places.

θ	$\sin \theta$	$\tan \theta$	$\frac{\pi\theta}{180}$
0°			
.5°			
1°			
1.5°			
2°			
2.5°			
3°			
3.5°			
4°			

(a) How do $\sin \theta$, $\tan \theta$, and $\dfrac{\pi\theta}{180}$ compare for small grades of θ?

(b) Highway grades are usually small. Give two approximations to the grade resistance $F = W \sin \theta$ that do not use the sine function.

(c) A stretch of highway has a 4-ft vertical rise for every 100 ft of horizontal run. Use an approximation from part (a) to estimate the grade resistance for a 2000-lb car on this stretch of highway.

(d) A stretch of highway has a $3.75°$ grade. Without calculating a trigonometric function, estimate the grade resistance for an 1800-lb car on this stretch of highway.

65. When highway curves are designed, the outside of the curve is often slightly elevated or inclined above the inside of the curve. See the figure. This inclination is called **superelevation.** For safety reasons it is important that both the curve's radius and superelevation are correct for a given speed limit. If an automobile is traveling at velocity V (in feet per second), the safe radius R for a curve with superelevation α can be calculated using the formula $R = \dfrac{V^2}{g(f + \tan \alpha)}$ where f and g are constants. (*Source:* Mannering, F., and W. Kilareski, *Principles of Highway Engineering and Traffic Control*, John Wiley & Sons, Inc., 1990.)

(a) A roadway is being designed for automobiles traveling at 45 miles per hour. If $\alpha = 3°$, $g = 32.2$, and $f = .14$, calculate R.

(b) What should the radius of the curve be if the speed limit is increased to 70 miles per hour?

(c) How would increasing the angle α affect the results? Verify your answer by repeating parts (a) and (b) with $\alpha = 4°$.

66. Refer to Exercise 65. A highway curve has a radius of $R = 1150$ ft and a superelevation of $\alpha = 2.1°$. What should the speed limit (in miles per hour) be for this curve?

2.4 Solving Right Triangles ▼▼▼

Many applications require finding a measurement that cannot be measured directly: for example, the height of a tree or a flagpole, or the angle formed between the horizontal and the line of sight to the top of a building. These measurements are often found by solving right triangles.

To **solve a triangle** means to find the measures of all the angles and sides of the triangle. This section and the next discuss methods of solving right triangles. (Methods for solving other triangles are presented in Chapter 7.)

Before we solve triangles, a short discussion concerning accuracy and significant digits is appropriate. Suppose we glance quickly at a room and guess that it is 15 feet by 18 feet. To calculate the length of a diagonal of the room, the Pythagorean theorem can be used.

$$d^2 = 15^2 + 18^2$$
$$d^2 = 549$$
$$d = \sqrt{549}$$

On a calculator, $\sqrt{549} \approx 23.430749$.

Should this answer be given as the length of the diagonal of the room? Of course not. The number 23.430749 contains 6 decimal places, while the original data of 15 feet and 18 feet are only accurate to the nearest foot. Since the results of a problem can be no more accurate than the least accurate number in any calculation, we really should say that the diagonal of the 15- by 18-foot room is 23 feet.

If a wall is measured to the nearest foot and found to be 18 feet long, actually this means that the wall has a length between 17.5 feet and 18.5 feet. If the wall is measured more accurately and found to be 18.3 feet long, then its length is really between 18.25 feet and 18.35 feet. A measurement of 18.00 feet would indicate that the length of the wall is between 17.995 feet and 18.005 feet. The measurement 18 feet is said to have two **significant digits** of accuracy; 18.0 has three significant digits, and 18.00 has four.

What about the measurement 900 meters? We cannot tell whether this represents a measurement to the nearest meter, ten meters, or hundred meters. To avoid this problem the number can be written in scientific notation as 9.00×10^2 to the nearest meter, 9.0×10^2 to the nearest ten meters, or 9×10^2 to the nearest hundred meters. These three cases have 3, 2, and 1 significant digits, respectively.

A significant digit is a digit obtained by actual measurement. A number that represents the result of counting, or a number that results from theoretical work and is not the result of a measurement, is an **exact number.** There are 50 states in the United States, so 50 is an exact number. The number of states is not 49 3/4 or 50 1/4; nor is the number 50 used here to represent "some number between 45 and 55." In the formula for the perimeter of a rectangle, $P = 2L + 2W$, the 2's are obtained from the definition of perimeter, and are exact numbers.

Most values of trigonometric functions are approximations, and virtually all measurements are approximations. To perform calculations on such approximate numbers, follow the rules given below.

Calculation with Significant Digits

For *adding* and *subtracting,* round the answer so that the last digit you keep is in the right-most column in which all the numbers have significant digits.

For *multiplying* or *dividing,* round the answer to the least number of significant digits found in any of the given numbers.

For *powers* and *roots,* round the answer so that it has the same number of significant digits as the number whose power or root you are finding.

When solving triangles, use the following table for deciding on significant digits in angle measure.

Significant Digits for Angles

Number of Significant Digits	Angle Measure to Nearest:
2	Degree
3	Ten minutes, or nearest tenth of a degree
4	Minute, or nearest hundredth of a degree
5	Tenth of a minute, or nearest thousandth of a degree

For example, an angle measuring 52° 30′ has three significant digits (assuming that 30′ is measured to the nearest ten minutes).

In using trigonometry to solve triangles, a labeled sketch is an important aid. It is conventional to use a to represent the length of the side opposite angle A, b for the length of the side opposite angle B, and so on. As mentioned earlier, in a right triangle the letter c is reserved for the hypotenuse. Figure 16 shows the labeling of a typical right triangle.

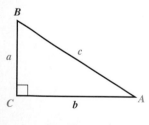

FIGURE 16

EXAMPLE 1
Solving a right triangle given an angle and a side

Solve right triangle ABC, with $A = 34° 30′$ and $c = 12.7$ in. See Figure 17.

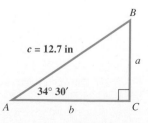

$c = 12.7$ in

34° 30′

FIGURE 17

To solve the triangle, find the measures of the remaining sides and angles. The value of a can be found with a trigonometric function involving the known values of angle A and side c. Since the sine of angle A is given by the quotient of the side opposite A and the hypotenuse, use $\sin A$.

$$\sin A = \frac{a}{c}$$

Substituting known values gives

$$\sin 34° \, 30' = \frac{a}{12.7},$$

or, upon multiplying both sides by 12.7,

$$a = 12.7 \sin 34° \, 30'$$
$$a = 12.7(.56640624) \qquad \text{Use a calculator.}$$
$$a = 7.19 \text{ in.}$$

The value of b could be found with the Pythagorean theorem. It is better, however, to use the information given in the problem rather than a result just calculated. If a mistake were to be made in finding a, then b also would be incorrect. Also, rounding more than once may cause the result to be less accurate. Using $\cos A$ gives

$$\cos A = \frac{\text{side adjacent}}{\text{hypotenuse}} = \frac{b}{c}$$
$$\cos 34° \, 30' = \frac{b}{12.7}$$
$$b = 12.7 \cos 34° \, 30'$$
$$b = 10.5 \text{ in.}$$

Once b has been found, the Pythagorean theorem could be used as a check. All that remains to solve triangle ABC is to find the measure of angle B. Since $A + B = 90°$ and $A = 34° \, 30'$,

$$A + B = 90°$$
$$B = 90° - A$$
$$B = 89° \, 60' - 34° \, 30'$$
$$B = 55° \, 30'. \quad \blacktriangleright$$

NOTE   In Example 1 we could have started by finding the measure of angle B and then used the trigonometric function values of B to find the unknown sides. The process of solving a right triangle (like many problems in mathematics) can usually be done in several ways, each resulting in the correct answer. However, in order to retain as much accuracy as can be expected, always use given information as much as possible, and avoid rounding off in intermediate steps.

◖EXAMPLE 2
Solving a right triangle given two sides

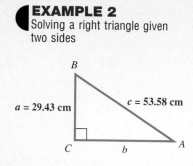

FIGURE 18

Solve right triangle ABC if $a = 29.43$ cm and $c = 53.58$ cm.

Draw a sketch showing the given information, as in Figure 18. One way to begin is to find angle A by using the sine.

$$\sin A = \frac{\text{side opposite}}{\text{hypotenuse}}$$

$$\sin A = \frac{29.43}{53.58}$$

Using $\boxed{\text{INV}}\ \boxed{\text{sin}}$ or $\boxed{\text{sin}^{-1}}$ on a calculator, we find that $A = 33.32°$. The measure of B is $90° - 33.32° = 56.68°$.

We now find b from the Pythagorean theorem, $a^2 + b^2 = c^2$, or $b^2 = c^2 - a^2$. Since $c = 53.58$ and $a = 29.43$,

$$b^2 = 53.58^2 - 29.43^2$$

giving

$$b = 44.77 \text{ cm.} \quad ◗$$

■ PROBLEM SOLVING The process of solving right triangles is easily adapted to solving applied problems. A crucial step in such applications involves sketching the triangle and labeling the given parts correctly. Then we can use the methods described in the earlier examples to find the unknown value or values. ■

Many applications of right triangles involve the angle of elevation or the angle of depression. The **angle of elevation** from point X to point Y (above X) is the acute angle formed by ray XY and a horizontal ray with endpoint at X. The angle of elevation is always measured from the horizontal. See Figure 19. The **angle of depression** from point X to point Y (below X) is the acute angle formed by ray XY and a horizontal ray with endpoint X. Again, see Figure 19.

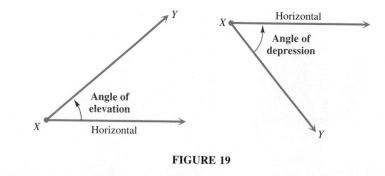

FIGURE 19

CAUTION Errors are often made in interpreting the angle of depression. Remember that both the angle of elevation *and* the angle of depression are measured between the line of sight and the horizontal.

EXAMPLE 3
Finding a length when the angle of elevation is known

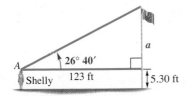

FIGURE 20

Shelly McCarthy knows that when she stands 123 feet from the base of a flagpole, the angle of elevation to the top is 26° 40'. If her eyes are 5.30 feet above the ground, find the height of the flagpole.

The length of the side adjacent to Shelly is known and the length of the side opposite her is to be found. See Figure 20. The ratio that involves these two values is the tangent.

$$\tan A = \frac{\text{side opposite}}{\text{side adjacent}}$$

$$\tan 26° \, 40' = \frac{a}{123}$$

$$a = 123 \tan 26° \, 40'$$

$$a = 61.8 \text{ feet}$$

Since Shelly's eyes are 5.30 feet above the ground, the height of the flagpole is

$$61.8 + 5.30 = 67.1 \text{ feet.} \quad \blacktriangleright$$

EXAMPLE 4
Finding the angle of elevation when lengths are known

The length of the shadow of a building 34.09 meters tall is 37.62 meters. Find the angle of elevation of the sun.

As shown in Figure 21, the angle of elevation of the sun is angle B. Since the side opposite B and the side adjacent to B are known, use the tangent ratio to find B.

$$\tan B = \frac{34.09}{37.62}$$

$$B = 42.18°$$

The angle of elevation of the sun is 42.18°. $\quad \blacktriangleright$

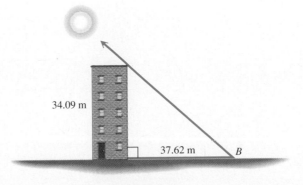

FIGURE 21

2.4 Exercises ▼▼

Refer to the discussion of accuracy and significant digits in this section to work the first ten exercises.

1. What is the difference between a measurement of 23.0 feet and a measurement of 23.00 feet?

2. What number indicates a measurement between 25.95 and 26.05 pounds?

Fill in the blank in Exercises 3 and 4.

3. If *h* is the actual height of a building and the height is measured as 58.6 ft, then $|h - 58.6| \leq$ _____ .

4. If *w* is the actual weight of a car and the weight is measured as 15.00×10^2 pounds, then $|w - 1500| \leq$ _____ .

5. When Mt. Everest was first surveyed, the surveyors obtained a height of 29,000 ft to the nearest foot. State the range represented by this number. (The surveyors thought that no one would believe a measurement of 29,000 ft, so they reported it as 29,002.)

6. New Orleans Saints kicker Tom Dempsey holds the National Football League record for the longest field goal. On November 8, 1970, he kicked a 63-yd field goal against the Detroit Lions to win the game. What range does the number 63 represent here?

7. According to the *Guinness Book of World Records,* the widest long-span bridge is the 1650-ft-long Sydney Harbour Bridge in Australia, which is 160 ft wide. State the ranges represented by these two numbers if they represent accuracy to the nearest foot.

8. At Denny's, a chain of restaurants, the Low-Cal Special is said to have "approximately 472 calories."

What is the range of calories represented by this number? By claiming "approximately 472 calories," they are probably claiming more accuracy than is possible. In your opinion, what might be a better claim?

Find the error in each statement.

9. I have 2 bushel baskets, each containing 65 apples. I know that $2 \times 65 = 130$, but 2 has only one significant figure, so I must write the answer as 1×10^2, or 100. I therefore have 100 apples.

10. The formula for the circumference of a circle is $C = 2\pi r$. My circle has a radius of 54.98 cm, and my calculator has a $\boxed{\pi}$ key, giving fifteen digits of accuracy. Pressing the right buttons gives 345.44953. Because 2 has only one significant digit, however, the answer must be given as 3×10^2, or 300 cm. (What is the correct answer?)

In the remaining exercises in this set, use a calculator as necessary.

Solve each right triangle. See Examples 1 and 2.

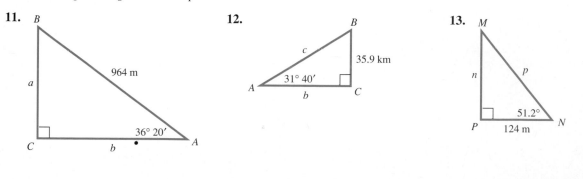

11.   **12.**   **13.**

14.

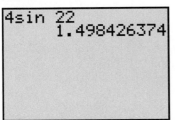

15.

16.

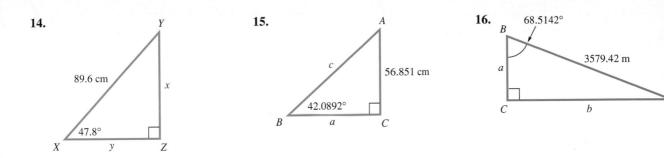

In Exercises 17 and 18, assume the calculator is in degree mode.

17. Make up a right triangle problem whose solution is obtained from the accompanying graphing calculator screen.

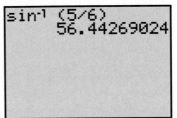

18. Make up a right triangle problem whose solution is obtained from the accompanying graphing calculator screen.

19. Can a right triangle be solved if we are given the measures of its two acute angles and no side lengths? Explain.

20. If we are given an acute angle and a side in a right triangle, what unknown part of the triangle requires the least work to find?

21. Explain why you can always solve a right triangle if you know the measures of one side and one acute angle.

22. Explain why you can always solve a right triangle if you know the lengths of two sides.

Solve each right triangle. In each case, C = 90°. If the angle information is given in degrees and minutes, give the answers in the same way. If given in decimal degrees, do likewise in your answers. When two sides are given, give answers in degrees and minutes.

23. $A = 28.00°$, $c = 17.4$ ft

24. $B = 46.00°$, $c = 29.7$ m

25. $B = 73.00°$, $b = 128$ in

26. $A = 61° 00'$, $b = 39.2$ cm

27. $a = 76.4$ yd, $b = 39.3$ yd

28. $a = 958$ m, $b = 489$ m

29. $a = 18.9$ cm, $c = 46.3$ cm

30. $b = 219$ m, $c = 647$ m

31. $A = 53° 24'$, $c = 387.1$ ft

32. $A = 13° 47'$, $c = 1285$ m

33. $B = 39° 9'$, $c = .6231$ m

34. $B = 82° 51'$, $c = 4.825$ cm

35. When is an angle of elevation equal to 90°?

36. Can an angle of elevation be more than 90°?

37. Use the ideas found in Section 1.3 involving a transversal intersecting parallel lines to explain why the angle of depression *DAB* has the same measure as the angle of elevation *ABC* in the accompanying figure.

38. Why is angle *CAB not* an angle of depression in the accompanying figure?

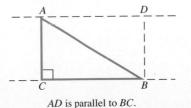

AD is parallel to *BC*.

Solve each problem. See Examples 1–3.

39. A 13.5-m fire-truck ladder is leaning against a wall. Find the distance the ladder goes up the wall if it makes an angle of 43° 50′ with the ground.

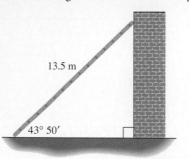

13.5 m

43° 50′

40. A guy wire 77.4 m long is attached to the top of an antenna mast that is 71.3 m high. Find the angle that the wire makes with the ground.

41. Find the length of a guy wire that makes an angle of 45° 30′ with the ground if the wire is attached to the top of a tower 63.0 m high.

42. To measure the height of a flagpole, Donna Garbarino finds that the angle of elevation from a point 24.73 ft from the base to the top is 38° 12′. Find the height of the flagpole.

43. To find the distance *RS* across a lake, a surveyor lays off *RT* = 53.1 m, with angle *T* = 32° 10′, and angle *S* = 57° 50′. Find length *RS*.

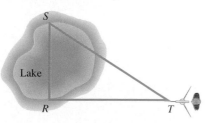

S

Lake

R   T

44. The length of the base of an isosceles triangle is 42.36 in. Each base angle is 38.12°. Find the length of each of the two equal sides of the triangle. (*Hint:* Divide the triangle into two right triangles.)

45. Find the altitude of an isosceles triangle having a base of 184.2 cm if the angle opposite the base is 68° 44′.

Work each problem involving an angle of elevation or depression. See Examples 3 and 4.

46. Suppose the angle of elevation of the sun is 23.4°. Find the length of the shadow cast by Cindy Newman, who is 5.75 ft tall.

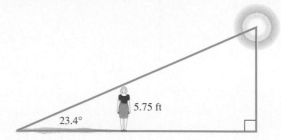

23.4°

5.75 ft

47. The shadow of a vertical tower is 40.6 m long when the angle of elevation of the sun is 34.6°. Find the height of the tower.

48. Find the angle of elevation of the sun if a 48.6-ft flagpole casts a shadow 63.1 ft long.

49. The angle of depression from the top of a building to a point on the ground is 32° 30′. How far is the point on the ground from the top of the building if the building is 252 m high?

50. An airplane is flying 10,500 feet above the level ground. The angle of depression from the plane to the base of a tree is 13° 50′. How far horizontally must the plane fly to be directly over the tree?

10,500 ft

51. The angle of elevation from the top of a small building to the top of a nearby taller building is 46° 40′, while the angle of depression to the bottom is 14° 10′. If the smaller building is 28.0 m high, find the height of the taller building.

52. A video camera is to be mounted on a bank wall so as to have a good view of the head teller (see the figure). Find the angle of depression that the lens should make with the horizontal.

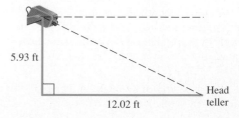

5.93 ft

12.02 ft

Head teller

53. A company safety committee has recommended that a floodlight be mounted in a parking lot so as to illuminate the employee exit. See the figure. Find the angle of depression of the light.

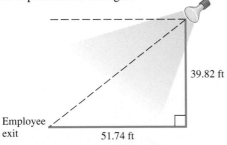

54. To determine the diameter of the sun, an astronomer might sight with a transit (a device used by surveyors for measuring angles) first to one edge of the sun and then to the other, finding that the included angle equals 1° 4′. Assuming that the distance from the Earth to the sun is 92,919,800 mi, calculate the diameter of the sun. See the figure.

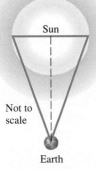

55. The figure shows a magnified view of the threads of a bolt. Find x if d is 2.894 mm.

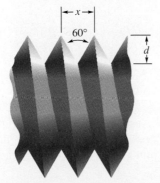

56. A degree may seem like a very small unit, but an error of one degree in measuring an angle may be very significant. For example, suppose a laser beam directed toward the visible center of the moon misses its assigned target by 30 sec. How far is it (in mi) from its assigned target? Take the distance from the surface of the Earth to that of the moon to be 234,000 mi. (From *A Sourcebook of Applications of School Mathematics* by Donald Bushaw et al. Copyright © 1980 by The Mathematical Association of America. Reprinted by permission. The material was prepared with the support of National Science Foundation Grant No. SED72-01123 A05. However, any opinions, findings, conclusions, or recommendations expressed herein are those of the authors and do not necessarily reflect the views of NSF.)

57. The highest mountain peak in the world is Mt. Everest located in the Himalayas. The height of this enormous mountain was determined in 1856 by surveyors using trigonometry long before it was first climbed in 1953. This difficult measurement had to be done from a great distance. At an altitude of 14,545 feet on a different mountain, the straight line distance to the peak of Mt. Everest is 27.0134 miles and its angle of elevation is $\theta = 5.82°$. See the figure. (*Source:* Dunham, W., *The Mathematical Universe,* John Wiley & Sons, Inc., 1994.)

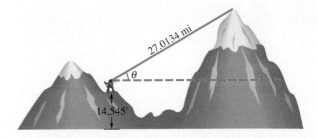

(a) Approximate the height of Mt. Everest.
(b) In the actual measurement, Mt. Everest was over 100 miles away and the curvature of the Earth had to be taken into account. Would the curvature of the Earth make the peak appear taller or shorter than it actually is?

58. 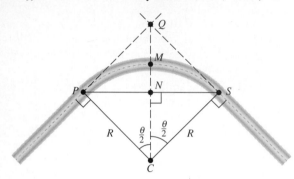 A basic highway curve connecting two straight sections of road is often circular. See the accompanying figure. The points P and S mark the beginning and end of the curve. Let Q be the point of intersection where the two straight sections of highway leading into the curve would meet if extended. The radius of the curve is R and the central angle θ denotes how many degrees the curve turns. (*Source:* Mannering, F., and W. Kilareski, *Principles of Highway Engineering and Traffic Control,* John Wiley & Sons, Inc., 1990.)

These obstructions prevent the driver from seeing sufficiently far down the highway to ensure a safe stopping distance. See the accompanying figure. The *minimum* distance d that should be cleared on the inside of the highway is given by the equation $d = R\left(1 - \cos\dfrac{\beta}{2}\right)$. (*Source:* Mannering, F., and W. Kilareski, *Principles of Highway Engineering and Traffic Control,* John Wiley & Sons, Inc., 1990.)

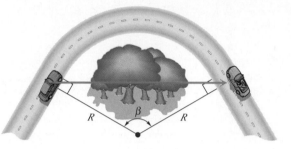

(a) If $R = 965$ ft and $\theta = 37°$, find the distance d between P and Q.

(b) Find an expression in terms of R and θ for the distance between points M and N.

59. 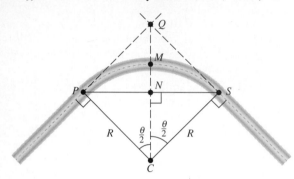 Refer to Exercise 58. When an automobile travels along a circular curve, objects like trees and buildings situated on the inside of the curve can obstruct a driver's vision.

(a) It can be shown that if β is measured in degrees then $\beta \approx \dfrac{57.3S}{R}$ where S is the safe stopping distance for the given speed limit. Compute d for a 55 mile per hour speed limit if $S = 336$ feet and $R = 600$ feet.

(b) Compute d for a 65 mile per hour speed limit if $S = 485$ feet.

(c) How does the speed limit affect the amount of land that should be cleared on the inside of the curve?

2.5 Further Applications of Right Triangles ▼▼▼

Other applications of right triangles involve **bearing,** an important idea in navigation. There are two common ways to express bearing. *When a single angle is given, such as 164°, it is understood that the bearing is measured in a clockwise direction from due north.* Several sample bearings using this first type of system are shown in Figure 22.

NOTE In the following examples and exercises, the problems all result in right triangles, so the methods of the previous section apply. Chapter 7 will include problems involving bearing that result in triangles that are *not* right triangles and require other methods to solve.

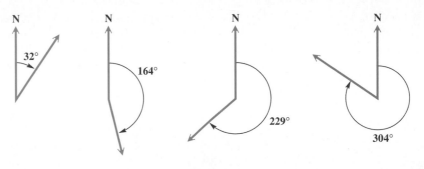

FIGURE 22

EXAMPLE 1
Solving a problem involving bearing (first type)

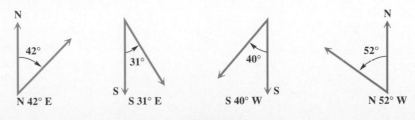

FIGURE 23

Radar stations A and B are on an east-west line, 3.7 kilometers apart. Station A detects a plane at C, on a bearing of 61°. Station B simultaneously detects the same plane, on a bearing of 331°. Find the distance from A to C.

Draw a sketch showing the given information, as in Figure 23. Since a line drawn due north is perpendicular to an east-west line, right angles are formed at A and B, so that angles CAB and CBA can be found. Angle C is a right angle because angles CAB and CBA are complementary. (If C were not a right angle, the methods of Chapter 7 would be needed.) Find distance b by using the cosine function.

$$\cos 29° = \frac{b}{3.7}$$
$$3.7 \cos 29° = b$$
$$b = 3.2 \text{ kilometers}$$
Use a calculator and round to the nearest tenth. ▶

■ PROBLEM SOLVING   It would be foolish to attempt to solve the problem in Example 1 without drawing a sketch. The importance of a correctly labeled sketch in applications such as this cannot be overemphasized, as some of the necessary information is often not given in the problem, and can only be determined from the sketch.   ■

The second common system for expressing bearing starts with a north-south line and uses an acute angle to show the direction, either east or west, from this line. Figure 24 shows several sample bearings using this system. Either N or S always comes first, followed by an acute angle, and then E or W.

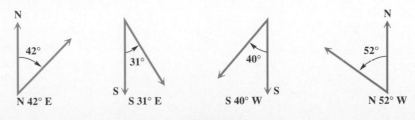

N 42° E   S 31° E   S 40° W   N 52° W

FIGURE 24

EXAMPLE 2
Solving a problem involving bearing (second type)

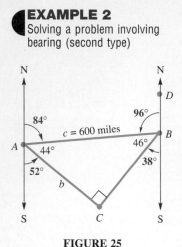

FIGURE 25

The bearing from *A* to *C* is S 52° E. The bearing from *A* to *B* is N 84° E. The bearing from *B* to *C* is S 38° W. A plane flying at 250 miles per hour takes 2.4 hours to go from *A* to *B*. Find the distance from *A* to *C*.

Make a sketch of the situation. First draw the two bearings from point *A*. Choose a point *B* on the bearing N 84° E from *A* and draw the bearing to *C*. Point *C* will be located where the bearing lines from *A* and *B* intersect as shown in Figure 25.

Since the bearing from *A* to *B* is N 84° E, angle *ABD* is 180° − 84° = 96°. Thus, angle *ABC* is 46°. Also, angle *BAC* is 180° − (84° + 52°) = 44°. Angle *C* is 180° − (44° + 46°) = 90°. From the statement of the problem, a plane flying at 250 miles per hour takes 2.4 hours to go from *A* to *B*. The distance from *A* to *B* is the product of rate and time, or

$$c = \text{rate} \times \text{time} = 250(2.4) = 600 \text{ miles.}$$

To find *b*, the distance from *A* to *C*, use the sine. (The cosine could also have been used.)

$$\sin 46° = \frac{b}{c}$$

$$\sin 46° = \frac{b}{600}$$

$$600 \sin 46° = b$$

$$b = 430 \text{ miles} \quad \blacktriangleright$$

The next example uses the idea of the angle of elevation first discussed in the previous section.

EXAMPLE 3
Solving a problem involving angle of elevation

Francisco needs to know the height of a tree. From a given point on the ground he finds that the angle of elevation to the top of the tree is 36.7°. He then moves back 50 feet. From the second point, the angle of elevation to the top of the tree is 22.2°. See Figure 26. Find the height of the tree.

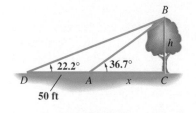

FIGURE 26

The figure shows two unknowns: x, the distance from the center of the trunk of the tree to the point where the first observation was made, and h, the height of the tree. Since nothing is given about the length of the hypotenuse of either triangle ABC or triangle BCD, use a ratio that does not involve the hypotenuse — the tangent.

In triangle ABC, $\quad \tan 36.7° = \dfrac{h}{x} \quad$ or $\quad h = x \tan 36.7°$.

In triangle BCD, $\quad \tan 22.2° = \dfrac{h}{50 + x} \quad$ or $\quad h = (50 + x) \tan 22.2°$.

Since each of these expressions equals h, these expressions must be equal. Thus,

$$x \tan 36.7° = (50 + x) \tan 22.2°.$$

Now use algebra to solve for x.

$$x \tan 36.7° = 50 \tan 22.2° + x \tan 22.2° \qquad \text{Distributive property}$$

$$x \tan 36.7° - x \tan 22.2° = 50 \tan 22.2° \qquad \text{Get } x \text{ terms on one side.}$$

$$x(\tan 36.7° - \tan 22.2°) = 50 \tan 22.2° \qquad \text{Factor out } x \text{ on the left.}$$

$$x = \frac{50 \tan 22.2°}{\tan 36.7° - \tan 22.2°} \qquad \text{Divide by the coefficient of } x.$$

We saw above that $h = x \tan 36.7°$. Substituting for x,

$$h = \left(\frac{50 \tan 22.2°}{\tan 36.7° - \tan 22.2°}\right)(\tan 36.7°).$$

From a calculator,

$$\tan 36.7° = .74537703$$
$$\tan 22.2° = .40809244$$

so

$$\tan 36.7° - \tan 22.2° = .74537703 - .40809244 = .33728459$$

and

$$h = \left(\frac{50(.40809244)}{.33728459}\right)(.74537703) = 45 \text{(rounded)}.$$

The height of the tree is approximately 45 feet. ▶

NOTE In practice we usually do not write down the intermediate calculator approximation steps. However, we have done this in Example 3 so that the reader may follow the steps more easily.

CONNECTIONS An alternative approach to solving the problem in Example 3 uses the intersection of graphs capability of the graphing calculator. This approach is based on a similar solution proposed by a student, John Cree, as explained in a letter to the editor in the January 1995 issue of *Mathematics Teacher* from Cree's teacher, Robert Ruzich.*

In the figure we have superimposed Figure 26 on a coordinate axis with the origin at D. Since the tangent of the angle between the x-axis and the graph of a line with equation $y = mx + b$ is the slope of the line, m, the segment BD lies on the graph of $Y_1 = (\tan 22.2°)x$. (Here, the value of b is 0.) By similar reasoning, we can find an equation of the line containing segment AB. Then we can find the point of intersection of the two graphs. The y-coordinate of this point gives the length of segment BC, which is the height of the tree.

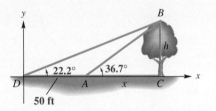

FOR DISCUSSION OR WRITING
1. Find the equation of the line containing segment AB, following the reasoning we used to get Y_1.
2. Use the intersection of graphs method to find the coordinates of the point of intersection. Use a window of [0, 200] by [0, 100]. Does the y-value agree with the solution in Example 3?
3. Explain why the tangent of angle D in the figure gives the slope of the line containing segment BD.

EXAMPLE 4
Using trigonometry to measure a distance

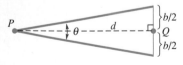

FIGURE 27

A method that surveyors use to determine a small distance d between two points P and Q is called the **subtense bar method.** The subtense bar with length b is centered at Q and situated perpendicular to the line of sight between P and Q. See Figure 27. The angle θ is measured and then the distance d can be determined.

(a) Find d when $\theta = 1°\ 23'\ 12''$ and $b = 2$ meters.

From Figure 27 we see that

$$\cot \frac{\theta}{2} = \frac{d}{b/2}$$

$$d = \frac{b}{2} \cot \frac{\theta}{2}.$$

---

*Reprinted by permission of *Mathematics Teacher,* January 1995.

To evaluate $\theta/2$, we change θ to decimal degrees: $1° \, 23' \, 12'' = 1.386667°$, so

$$d = \frac{2}{2} \cot \frac{1.386667°}{2} \approx 82.6341 \text{ m.}$$

(b) The angle θ usually cannot be measured more accurately than to the nearest $1''$. How much change would there be in the value of d if θ were measured $1''$ larger?
Use $\theta = 1° \, 23' \, 13'' \approx 1.386944°$.

$$d = \frac{2}{2} \cot \frac{1.386944°}{2} \approx 82.6176 \text{ m.}$$

The difference is $82.6341 - 82.6176 \approx .017$ m. ▶   •

2.5 Exercises ▼▼▼▼▼▼▼▼▼▼▼▼▼▼▼▼▼▼▼▼▼▼▼▼▼▼▼▼▼▼▼▼

An observer for a radar station is located at the origin of a coordinate system. For each of the points in Exercises 1–4, find the bearing of an airplane located at that point. Express the bearing using both methods.

1. $(-4, 0)$ **2.** $(-3, -3)$ **3.** $(-5, 5)$ **4.** $(0, -2)$

5. The ray $y = x, x \geq 0$ contains the origin and all points in the coordinate system whose bearing from the origin is $45°$. Determine the equation of a ray consisting of the origin and all points whose bearing from the origin is $240°$.

6. Repeat Exercise 5 for a bearing of $150°$.

Work each problem. In these exercises, assume the course of a plane or ship is on the indicated bearing. See Examples 1 and 2.

7. A plane flies 1.3 hr at 110 mph on a bearing of $40°$. It then turns and flies 1.5 hr at the same speed on a bearing of $130°$. How far is the plane from its starting point?

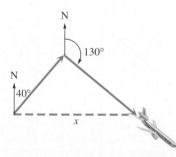

8. A ship travels 50 km on a bearing of $27°$, and then travels on a bearing of $117°$ for 140 km. Find the distance traveled from the starting point to the ending point.

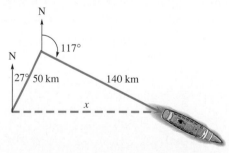

9. Two ships leave a port at the same time. The first ship sails on a bearing of $40°$ at 18 knots (nautical miles per hour) and the second at a bearing of $130°$ at 26 knots. How far apart are they after 1.5 hours?

10. Two lighthouses are located on a north-south line. From lighthouse A the bearing of a ship 3742 m away is $129° \, 43'$. From lighthouse B the bearing of the ship is $39° \, 43'$. Find the distance between the lighthouses.

11. A ship leaves its home port and sails on a bearing of N 28° 10′ E. Another ship leaves the same port at the same time and sails on a bearing of S 61° 50′ E. If the first ship sails at 24.0 mph and the second sails at 28.0 mph, find the distance between the two ships after 4 hr.

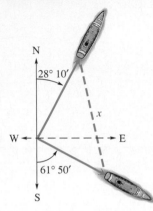

12. Radio direction finders are set up at points A and B, which are 2.50 mi apart on an east-west line. From A

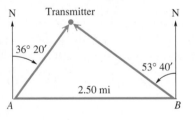

it is found that the bearing of the signal from a radio transmitter is N 36° 20′ E, while from B the bearing of the same signal is N 53° 40′ W. Find the distance of the transmitter from B.

13. The bearing from Winston-Salem, North Carolina, to Danville, Virginia, is N 42° E. The bearing from Danville to Goldsboro, North Carolina, is S 48° E. A car driven by Mark Ferrari, traveling at 60 mph, takes 1 hr to go from Winston-Salem to Danville and 1.8 hr to go from Danville to Goldsboro. Find the distance from Winston-Salem to Goldsboro.

14. The bearing from Atlanta to Macon is S 27° E, and the bearing from Macon to Augusta is N 63° E. An automobile traveling at 60 mph needs 1 1/4 hr to go from Atlanta to Macon and 1 3/4 hr to go from Macon to Augusta. Find the distance from Atlanta to Augusta.

15. Solve the equation $ax = b + cx$ for x in terms of a, b, and c. (*Note:* This is in essence the calculation carried out in Example 3.)

16. Explain why the line $y = (\tan \theta)(x - a)$ passes through the point $(a, 0)$ and makes an angle of $\theta°$ with the x-axis.

17. Find the equation of the line passing through the point $(25, 0)$ that makes an angle of 35° with the x-axis.

18. Find the equation of the line passing through the point $(5, 0)$ that makes an angle of 15° with the x-axis.

In Exercises 19–24 use the method of Example 3 and then check your answer with the graphing method described in the Connections discussion. Drawing a sketch for these problems where one is not given may be helpful.

19. Find h as indicated in the figure.

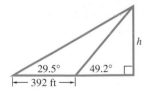

20. Find h as indicated in the figure.

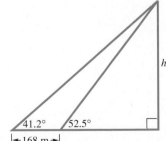

21. The angle of elevation from a point on the ground to the top of a pyramid is 35° 30′. The angle of elevation from a point 135 ft farther back to the top of the pyramid is 21° 10′. Find the height of the pyramid.

22. Debbie Maybury, a whale researcher standing at the top of a tower, is watching a whale approach the tower directly. When she first begins watching the whale, the angle of depression of the whale is 15° 50′. Just as the whale turns away from the tower, the angle of depression is 35° 40′. If the height of the tower is 68.7 m, find the distance traveled by the whale as it approaches the tower.

23. A scanner antenna is on top of the center of a house. The angle of elevation from a point 28.0 m from the center of the house to the top of the antenna is 27° 10′, and the angle of elevation to the bottom of the antenna is 18° 10′. Find the height of the antenna.

24. The angle of elevation from Lone Pine to the top of Mt. Whitney is 10° 50′. Van Dong Le, traveling 7.00 km from Lone Pine along a straight, level road toward Mt. Whitney, finds the angle of elevation to be 22° 40′. Find the height of the top of Mt. Whitney above the level of the road.

25. Refer to Example 4. A variation of the subtense bar method that surveyors use to determine larger distances d between two points P and Q is shown in the figure. In this case the subtense bar with length b is placed between the points P and Q so that the bar is centered on and perpendicular to the line of sight connecting P and Q. The angles α and β are measured from points P and Q, respectively. (*Source:* Mueller, I. and K. Ramsayer, *Introduction to Surveying,* Frederick Ungar Publishing Co., New York, 1979.)

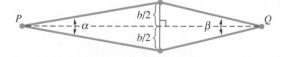

(a) Find a formula for d involving α, β, and b.
(b) Use your formula to determine d if $\alpha = 37'\ 48''$, $\beta = 42'\ 3''$, and $b = 2$ meters.

Solve the following exercises using the techniques of Section 2.4.

26. Find the minimum height h above the surface of the Earth so that a pilot at point A in the figure can see an object on the horizon at C, 125 mi away. Assume that the radius of the Earth is 4.00×10^3 mi.

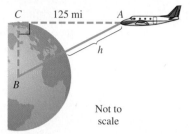

Not to scale

27. In one area, the lowest angle of elevation of the sun in winter is 23° 20′. Find the minimum distance x that a plant needing full sun can be placed from a fence 4.65 ft high. See the figure.

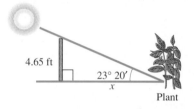

Plant

28. A tunnel is to be dug from A to B (see the figure). Both A and B are visible from C. If AC is 1.4923 mi and BC is 1.0837 mi, and if C is 90°, find the measures of angles A and B.

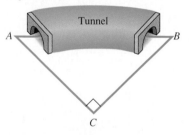

29. A piece of land has the shape shown in the figure. Find x.

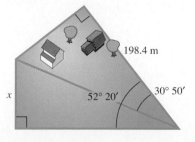

Chapter 2   Summary ▼▼▼▼▼▼▼▼▼▼▼▼▼▼▼▼▼▼▼▼▼▼▼▼▼▼▼▼▼▼▼▼▼▼▼▼

SECTION	KEY IDEAS

2.1 Trigonometric Functions of Acute Angles

Right Triangle-Based Definitions of the Trigonometric Functions

For any acute angle A in standard position,

$$\sin A = \frac{y}{r} = \frac{\text{side opposite}}{\text{hypotenuse}} \qquad \csc A = \frac{r}{y} = \frac{\text{hypotenuse}}{\text{side opposite}}$$

$$\cos A = \frac{x}{r} = \frac{\text{side adjacent}}{\text{hypotenuse}} \qquad \sec A = \frac{r}{x} = \frac{\text{hypotenuse}}{\text{side adjacent}}$$

$$\tan A = \frac{y}{x} = \frac{\text{side opposite}}{\text{side adjacent}} \qquad \cot A = \frac{x}{y} = \frac{\text{side adjacent}}{\text{side opposite}}.$$

Cofunction Identities

For any acute angle A,

$$\sin A = \cos(90° - A) \qquad \cot A = \tan(90° - A)$$
$$\cos A = \sin(90° - A) \qquad \csc A = \sec(90° - A)$$
$$\tan A = \cot(90° - A) \qquad \sec A = \csc(90° - A).$$

Function Values of Special Angles

θ	$\sin \theta$	$\cos \theta$	$\tan \theta$	$\cot \theta$	$\sec \theta$	$\csc \theta$
30°	$\frac{1}{2}$	$\frac{\sqrt{3}}{2}$	$\frac{\sqrt{3}}{3}$	$\sqrt{3}$	$\frac{2\sqrt{3}}{3}$	2
45°	$\frac{\sqrt{2}}{2}$	$\frac{\sqrt{2}}{2}$	1	1	$\sqrt{2}$	$\sqrt{2}$
60°	$\frac{\sqrt{3}}{2}$	$\frac{1}{2}$	$\sqrt{3}$	$\frac{\sqrt{3}}{3}$	2	$\frac{2\sqrt{3}}{3}$

2.2 Trigonometric Functions of Non-Acute Angles

Reference Angles for θ in (0°, 360°)

θ in Quadrant	θ' Is
I	θ
II	$180° - \theta$
III	$\theta - 180°$
IV	$360° - \theta$

Finding Trigonometric Function Values for Any Angle

1. Add or subtract 360° as many times as needed to get an angle of at least 0° but less than 360°.
2. Find the reference angle θ'.
3. Find the trigonometric function value for θ'.
4. Find the correct sign.

SECTION	KEY IDEAS
2.5 Further Applications of Right Triangles	

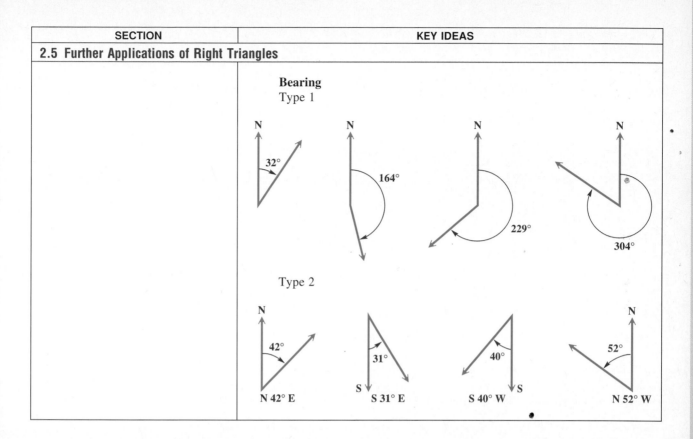

Bearing

Type 1

N

32°

N

164°

N

229°

N

304°

Type 2

N

42°

N 42° E

S

31°

S 31° E

S

40°

S 40° W

N

52°

N 52° W

Chapter 2   Review Exercises ▼▼▼▼▼▼▼▼▼▼▼▼▼▼▼▼▼▼▼▼▼▼▼▼▼▼

Find the values of the trigonometric functions for each angle A.

1.

11

60

A

61

2.

40

58

42

A

Solve each equation. Assume that all angles are acute angles.

3. $\sin 4\beta = \cos 5\beta$

4. $\sec(2\gamma + 10°) = \csc(4\gamma + 20°)$

5. $\tan(5x + 11°) = \cot(6x + 2°)$

6. $\cos\left(\dfrac{3\theta}{5} + 11°\right) = \sin\left(\dfrac{7\theta}{10} + 40°\right)$

Tell whether each statement is true *or* false.

7. $\sin 46° < \sin 58°$

8. $\cos 47° < \cos 58°$

9. $\sec 48° \geq \cos 42°$

10. $\sin 22° \geq \csc 68°$

11. Explain why, in the figure, the cosine of angle A is equal to the sine of angle B.

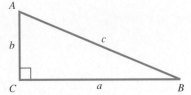

Find the values of the six trigonometric functions for each angle. Give exact values. Do not use a calculator. Rationalize denominators when applicable.

12. $120°$

13. $225°$

14. $300°$

15. $750°$

16. $-225°$

17. $-390°$

Find all values of θ, if θ is in the interval $[0°, 360°)$ and θ has the given function value.

18. $\sin \theta = -\dfrac{1}{2}$

19. $\cos \theta = -\dfrac{1}{2}$

20. $\cot \theta = -1$

21. $\sec \theta = -\dfrac{2\sqrt{3}}{3}$

Evaluate each expression. Give exact values.

22. $\cos 60° + 2 \sin^2 30°$

23. $\tan^2 120° - 2 \cot 240°$

24. $\cot^2 300° + \cos^2 120° - 3 \sin^2 240°$

25. $\sec^2 300° - 2 \cos^2 150° + \tan 45°$

26. If A, B, and C are the three angles of a triangle, then

$$\tan A + \tan B + \tan C = \tan A \tan B \tan C \text{ (where } A, B, C \neq 90°)$$

and

$$\sin^2 A + \sin^2 B + \sin^2 C = 2(1 + \cos A \cos B \cos C).$$

Verify these equations for a $30°–30°–120°$ triangle.

Use a calculator to find the value.

27. $\sin 72° \, 30'$

28. $\sec 222° \, 30'$

29. $\cot 305.6°$

30. $\csc 78° \, 21'$

31. $\sec 58.9041°$

32. $\tan 11.7689°$

33. $\sin 89.0043°$

34. $\cot 1.49783°$

35. A test item read, "Find the exact value of $\sin 60°$." A student gave the answer .8660254038, obtained with his powerful new calculator, yet did not receive credit. Was the teacher justified in not accepting his answer? Explain.

36. Which one of the following cannot be *exactly* determined using the methods of this chapter?
    (a) $\cos 135°$ **(b)** $\cot(-45°)$ **(c)** $\sin 300°$ **(d)** $\tan 140°$

Use a calculator to find the value of θ, where θ is in the interval [0°, 90°). Give the answers in decimal degrees.

37. sin θ = .82584121

38. cot θ = 1.1249386

39. cos θ = .97540415

40. sec θ = 1.2637891

41. tan θ = 1.9633124

42. csc θ = 9.5670466

Find two angles in the interval [0°, 360°) that satisfy each of the following. Leave your answer in decimal degrees.

43. sin θ = .73254290

44. tan θ = 1.3865342

Tell whether each statement is true *or* false. *If false, tell why. Use a calculator for Exercises 45 and 48.*

45. sin 50° + sin 40° = sin 90°

46. cos 210° = cos 180° · cos 30° − sin 180° · sin 30°

47. sin 240° = 2 sin 120° · cos 120°

48. sin 42° + sin 42° = sin 84°

49. A student wants to use a calculator to find the value of cot 25°. However, instead of calculating $\dfrac{1}{\tan 25}$, he calculates $\tan^{-1} 25$. Will this produce the correct answer?

For each angle θ, use a calculator to find cos θ *and* sin θ. *Use your results to decide what quadrant the angle lies in.*

50. θ = 2976°

51. θ = 1997°

52. θ = 4000°

Solve each of the following right triangles. In Exercise 54, give angles to the nearest minute. In Exercises 55 and 56, label the triangle as shown in Figure 16 in Section 2.4.

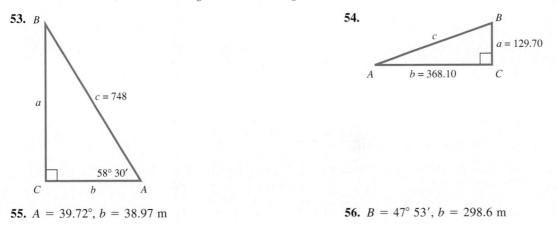

53. triangle with B at top, right angle at C, c = 748, 58° 30′ at A, sides a and b

54. triangle with B at top right, right angle at C, c, a = 129.70, b = 368.10, A at left

55. A = 39.72°, b = 38.97 m

56. B = 47° 53′, b = 298.6 m

Solve each problem.

57. The angle of elevation from a point 93.2 ft from the base of a tower to the top of the tower is 38° 20′. Find the height of the tower.

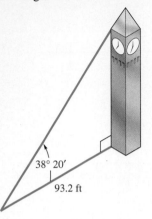

38° 20′

93.2 ft

58. The angle of depression of a television tower to a point on the ground 36.0 m from the bottom of the tower is 29.5°. Find the height of the tower.

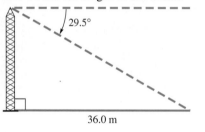

29.5°

36.0 m

59. A rectangle has adjacent sides measuring 10.93 cm and 15.24 cm. The angle between the diagonal and the longer side is 35.65°. Find the length of the diagonal.

60. An isosceles triangle has a base of length 49.28 m. The angle opposite the base is 58.746°. Find the length of each of the two equal sides.

61. The bearing of *B* from *C* is 254°. The bearing of *A* from *C* is 344°. The bearing of *A* from *B* is 32°. The distance from *A* to *C* is 780 m. Find the distance from *A* to *B*.

62. The bearing from point *A* to point *B* is S 55° E, and from point *B* to point *C* is N 35° E. If a ship sails from *A* to *B*, a distance of 80 km, and then from *B* to *C*, a distance of 74 km, how far is it from *A* to *C*?

63. Two cars leave an intersection at the same time. One heads due south at 55 mph. The other travels due west. After two hours, the bearing of the car headed west from the car headed south is 324°. How far apart are they at that time?

64. Find a formula for *h* in terms of *k*, *A*, and *B*. Assume *A* < *B*.

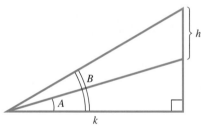

h

B

A

k

65. Make up a right triangle problem whose solution is 3 tan 25°.

66. Make up a right triangle problem whose solution is sin⁻¹(3/4).

In Exercises 67–69, find a line segment in the accompanying figure whose length is equal to the function value. (Hint: Use right triangles.)

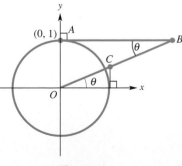

67. cot θ   **68.** sec θ   **69.** csc θ

70. Artificial satellites that orbit Earth often use VHF signals to communicate with the ground. VHF signals travel in straight lines. The height h of the satellite above Earth and the time T that the satellite can communicate with a fixed location on the ground are related by the equation

$$h = R\left(\frac{1}{\cos(180T/P)} - 1\right)$$

where $R = 3955$ miles is the radius of Earth and P is the period for the satellite to orbit Earth. (*Source:* Schlosser, W., T. Schmidt-Kaler, and E. Milone, *Challenges of Astronomy,* Springer-Verlag, 1991.)

(a) Find h when $T = 25$ min and $P = 140$ min. (Evaluate the cosine function in degree mode.)

(b) What is the value of h if T is increased to 30?

71. The first fundamental problem of surveying is to determine the coordinates of a point Q given the coordinates of a point P, the distance between P and Q, and the bearing θ from P to Q. See the figure. (*Source:* Mueller, I., and K. Ramsayer, *Introduction to Surveying,* Frederick Ungar Publishing Co., New York, 1979.)

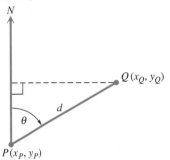

(a) Find a formula for the coordinates (x_Q, y_Q) of the point Q given θ, the coordinates (x_P, y_P) of P, and the distance d between P and Q.

(b) Use your formula to determine (x_Q, y_Q) if $(x_P, y_P) = (123.62, \ 337.95)$, $\theta = 17° \ 19' \ 22''$, and $d = 193.86$ ft.

3

Radian Measure and the Circular Functions

Finding sources of energy has been an important concern since the beginning of civilization. During the past 100 years, people have relied on fossil fuels for a large portion of their energy requirements. Fossils fuels are finite and limited. The heavy use of fossil fuels has caused irreversible damage to our environment and may be accelerating a greenhouse effect. Nuclear energy as an alternative has a potential for providing almost unlimited amounts of energy. Unfortunately, it creates health risks and dangerous nuclear wastes. Currently there is no completely safe disposal method for nuclear wastes. As a result, no new nuclear power plants have been ordered in the United States since 1978.

Over the past twenty-five years the production of solar energy has evolved from a mere kilowatt of electricity to hundreds of megawatts. Solar energy has many advantages over traditional energy sources in that it does not pollute and has the potential of being an unlimited, cheap source of energy. Its use and production is not limited to a small number of countries but is readily available throughout the United States and the world. The North American Southwest has some of the brightest sunlight in the world with a potential to provide up to 2500 kilowatt-hours per square meter.

In the design of solar power plants, engineers need to position solar panels perpendicular to the sun's rays so that maximum energy can be collected. Understanding the movement and position of the sun at any time and

Source: Winter, C., R. Sizmann, and Vant-Hunt (Editors), *Solar Power Plants,* Springer-Verlag, 1991.

date are fundamental concepts for solar energy collection. How high in the sky will the sun be in Sacramento or in New Orleans on February 29, 2000, at 3 P.M.? How many hours of daylight will there be in Hartford, Connecticut, on August 12, 2001? Answers to questions like these are necessary for generating electricity from sunlight. (See Section 3.3 for the answers.) Because Earth moves in a nearly circular orbit around the sun, the circular functions will be essential for answering these and many other questions.

In most work involving applications of trigonometry, angles are measured in degrees.* In more advanced work in mathematics, the use of *radian measure* of angles is preferred. Radian measure also allows us to treat our familiar trigonometric functions as functions with domains of *real numbers,* rather than angles. These ideas are introduced in this chapter.

3.1 Radian Measure ▼▼▼

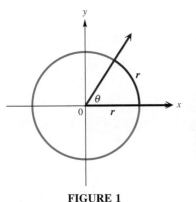

FIGURE 1

Figure 1 shows an angle θ in standard position along with a circle of radius r. The vertex of θ is at the center of the circle. Angle θ intercepts an arc on the circle equal in length to the radius of the circle. Because of this, angle θ is said to have a measure of one radian.

> **Radian**
>
> An angle that has its vertex at the center of a circle and that intercepts an arc on the circle equal in length to the radius of the circle has a measure of **one radian.**

It follows that an angle of measure 2 radians intercepts an arc equal in length to twice the radius of the circle, an angle of measure 1/2 radian intercepts an arc equal in length to half the radius of the circle, and so on.

The circumference of a circle, the distance around the circle, is given by $C = 2\pi r$, where r is the radius of the circle. The formula $C = 2\pi r$ shows that the radius can be laid off 2π times around a circle. Therefore, an angle of 360°, which corresponds to a complete circle, intercepts an arc equal in length to 2π

---

* This method of angle measure dates back to the Babylonians, who were the first to subdivide the circumference of a circle into 360 parts. There are various theories as to why the number 360 was chosen. One is that it is approximately the number of days in a year, and it has many divisors which makes it convenient to work with. Another involves a roundabout theory dealing with the length of a Babylonian mile.

times the radius of the circle. Because of this, an angle of 360° has a measure of 2π radians:

$$360° = 2\pi \text{ radians.}$$

An angle of 180° is half the size of an angle of 360°, so an angle of 180° has half the radian measure of an angle of 360°.

$$180° = \frac{1}{2}(2\pi) \text{ radians}$$

Degree/Radian Relationship

$$180° = \pi \text{ radians}$$

We can use this simple relationship to convert from degree measure to radian measure or from radians to degrees. It helps to remember that radians are associated with π and degrees with 180°. We convert by using ratios as shown in the next example.

EXAMPLE 1
Converting degrees to radians

Convert each degree measure to radians.

(a) 45°

Let r represent the radian measure. Use the proportion

$$\frac{r}{\pi} = \frac{45}{180}.$$

Solve for r and reduce the fraction.

$$r = \frac{45\pi}{180} = \frac{\pi}{4}$$

(b) 249.8°

$$\frac{r}{\pi} = \frac{249.8}{180}$$

$$r = \frac{249.8\pi}{180} \approx 4.360 \text{ radians}$$

to the nearest thousandth. Note that in the second step of each example the radian measure r equals the degree measure multiplied by $\pi/180°$. ▸

EXAMPLE 2
Converting radians to degrees

Convert each of the following radian measures to degrees.

(a) $\dfrac{9\pi}{4}$

Let d represent the degree measure.

$$\frac{\frac{9\pi}{4}}{\pi} = \frac{d}{180°}$$

$$\frac{9\pi}{4} \cdot \frac{1}{\pi} = \frac{d}{180°}$$

$$\frac{9}{4} = \frac{d}{180°}$$

$$d = \frac{9}{4} \cdot 180° = 405°$$

As the last step indicates, since π radians $= 180°$, when converting from radian measure given as a multiple of π, simply replace π with $180°$ and simplify.

(b) 4.25 (Give the answer in decimal degrees.)

$$\frac{4.25}{\pi} = \frac{d}{180°}$$

$$d = \frac{(4.25)180°}{\pi} \approx 243.5°$$

In the last step we used the π key on a calculator to complete the computation.   ▶

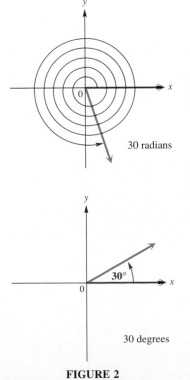

30 radians

30 degrees

FIGURE 2

Converting Between Degrees and Radians

Use the proportion

$$\frac{\text{radian measure}}{\pi} = \frac{\text{degree measure}}{180°}$$

or the following shortcuts:

1. If a radian measure involves a multiple of π, replace π with $180°$ and simplify to convert to degrees.
2. Multiply a degree measure by $\pi/180°$ to convert to radians.

Some calculators have a key that automatically converts between decimal degrees and radians. Check your owner's manual to see if your calculator has this feature. If not, you may wish to write a program to accomplish this.

There is a general agreement among mathematicians that if no unit of measure is specified for an angle, radian measure is understood.

CAUTION Figure 2 shows angles measuring 30 radians and 30°. These angle measures are not at all close, so be careful not to confuse them.

Trigonometric function values for angles measured in radians can be found by first converting the radian measure to degrees. (You should try to skip this intermediate step as soon as possible, and find the function values directly from the radian measure.)*

EXAMPLE 3
Finding a function value of an angle in radian measure

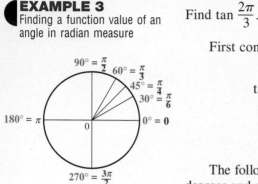

FIGURE 3

Find $\tan \dfrac{2\pi}{3}$.

First convert $\dfrac{2\pi}{3}$ radians to degrees.

$$\tan \frac{2\pi}{3} = \tan\left(\frac{2}{3} \cdot 180°\right) \qquad \text{Substitute } 180° \text{ for } \pi.$$

$$= \tan 120°$$

$$= -\sqrt{3} \qquad \text{Use the methods of Chapter 2.} \quad \blacktriangleright$$

The following table and Figure 3 give some equivalent angles measured in degrees and radians. It will be useful to memorize these equivalent values. Keep in mind that $180° = \pi$ radians. Then it will be easy to reproduce the rest of the table.

Equivalent Angle Measures in Degrees and Radians

Degrees	Radians		Degrees	Radians	
	Exact	Approximate		Exact	Approximate
0°	0	0	90°	$\dfrac{\pi}{2}$	1.57
30°	$\dfrac{\pi}{6}$	.52	180°	π	3.14
45°	$\dfrac{\pi}{4}$	.79	270°	$\dfrac{3\pi}{2}$	4.71
60°	$\dfrac{\pi}{3}$	1.05	360°	2π	6.28

EXAMPLE 4
Finding a function value of an angle in radian measure

Find $\sin \dfrac{3\pi}{2}$.

From the table, $\dfrac{3\pi}{2}$ radians $= 270°$, so

$$\sin \frac{3\pi}{2} = \sin 270° = -1. \quad \blacktriangleright$$

*A table giving the values of the trigonometric functions for common radian and degree measures is given inside the front cover of this book.

3.1 Exercises ▼▼▼▼▼▼▼▼▼▼▼▼▼▼▼▼▼▼▼▼▼▼▼▼▼▼▼▼▼▼▼▼▼▼▼▼

In Exercises 1–4, each angle θ is a whole number when measured in radians. Give the radian measure of the angle.

1.

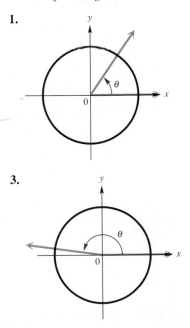

2.

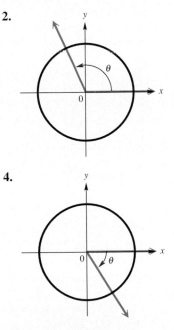

3.

4.

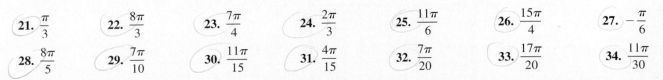

Convert each degree measure to radians. Leave answers as multiples of π. See Example 1.

5. 60°	**6.** 30°	**7.** 90°	**8.** 120°	**9.** 150°
10. 270°	**11.** 300°	**12.** 315°	**13.** 450°	**14.** 480°

15. In your own words, explain how to convert degree measure to radian measure.

16. In your own words, explain how to convert radian measure to degree measure.

17. In your own words, explain the meaning of radian measure.

18. Explain the difference between degree measure and radian measure.

19. Show that you can convert from radian measure to degree measure by multiplying by $180°/\pi$.

20. Explain why an angle of radian measure t in standard position intercepts an arc of length t on the circle of radius 1.

Convert each radian measure to degrees. See Example 2.

21. $\dfrac{\pi}{3}$	**22.** $\dfrac{8\pi}{3}$	**23.** $\dfrac{7\pi}{4}$	**24.** $\dfrac{2\pi}{3}$	**25.** $\dfrac{11\pi}{6}$	**26.** $\dfrac{15\pi}{4}$	**27.** $-\dfrac{\pi}{6}$
28. $\dfrac{8\pi}{5}$	**29.** $\dfrac{7\pi}{10}$	**30.** $\dfrac{11\pi}{15}$	**31.** $\dfrac{4\pi}{15}$	**32.** $\dfrac{7\pi}{20}$	**33.** $\dfrac{17\pi}{20}$	**34.** $\dfrac{11\pi}{30}$

Convert each degree measure to radians. See Example 1.

35. 39°   **36.** 74°   **37.** 42° 30′   **38.** 53° 40′

39. 139° 10′   **40.** 174° 50′   **41.** 64.29°   **42.** 85.04°

43. 56° 25′   **44.** 122° 37′   **45.** 47.6925°   **46.** 23.0143°

Convert each radian measure to degrees. Write answers to the nearest minute. See Example 2.

47. 2   **48.** 5   **49.** 1.74   **50.** 3.06

51. .3417   **52.** 9.84763   **53.** 5.01095   **54.** −3.47189

55. The value of sin 30 is not 1/2. Explain why.

56. Explain in your own words what is meant by an angle of one radian.

Find the exact value of each of the following without using a calculator. See Examples 3 and 4.

57. $\sin \dfrac{\pi}{3}$ **58.** $\cos \dfrac{\pi}{6}$ **59.** $\tan \dfrac{\pi}{4}$ **60.** $\cot \dfrac{\pi}{3}$ **61.** $\sec \dfrac{\pi}{6}$ **62.** $\csc \dfrac{\pi}{4}$

63. $\sin \dfrac{\pi}{2}$ **64.** $\csc \dfrac{\pi}{2}$ **65.** $\tan \dfrac{2\pi}{3}$ **66.** $\cot \dfrac{2\pi}{3}$ **67.** $\sin \dfrac{5\pi}{6}$ **68.** $\tan \dfrac{5\pi}{6}$

69. $\cos 3\pi$ **70.** $\sec \pi$ **71.** $\sin \dfrac{4\pi}{3}$ **72.** $\cot\left(-\dfrac{2\pi}{3}\right)$ **73.** $\sin\left(-\dfrac{7\pi}{6}\right)$ **74.** $\cos\left(-\dfrac{\pi}{6}\right)$

75. The figure shows the same angles measured in both degrees and radians. Complete the missing measures.

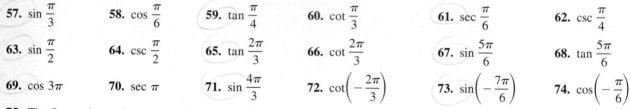
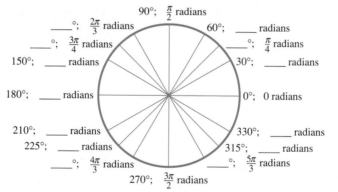

76. Find the measure (in both degrees and radians) of the angle θ formed in the accompanying screen by the line passing through the origin and the positive part of the x-axis. Use the displayed values of x and y at the bottom of the screen.

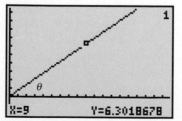

77. In the accompanying screen, was the calculator in degree or radian mode?

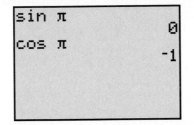

78. A circular pulley is rotating about its center. Through how many radians would it turn in (**a**) 8 rotations, and (**b**) 30 rotations?

79. Through how many radians will the hour hand on a clock rotate in (**a**) 24 hours, and (**b**) 4 hours?

80. A space vehicle is orbiting the Earth in a circular orbit. What radian measure corresponds to (**a**) 2.5 orbits, and (**b**) 4/3 of an orbit?

81. The **solar constant** S is the amount of energy per unit area that reaches Earth's atmosphere from the sun. It is equal to 1367 watts per square meter but varies slightly throughout the seasons. This fluctuation ΔS in S can be calculated using the formula $\Delta S = .034S \sin\left[\dfrac{2\pi(82.5 - N)}{365.25}\right]$. In this formula, N is the day number covering a four-year period where $N = 1$ corresponds to January 1 of a leap year and $N = 1461$ corresponds to December 31 of the fourth year. (*Source:* Winter, C., R. Sizmann, and Vant-Hunt (Editors), *Solar Power Plants,* Springer-Verlag, 1991.)
(**a**) Calculate ΔS for $N = 80$ which is the spring equinox in the first year.
(**b**) Calculate ΔS for $N = 1268$ which is the summer solstice in the fourth year.
(**c**) What is the maximum value of ΔS?
(**d**) Find a value for N where ΔS is equal to zero.

3.2 Applications of Radian Measure ▼▼▼

Radian measure is used to simplify certain formulas. Two of these formulas are discussed in this section. Both would be more complicated if expressed in degrees.

ARC LENGTH OF A CIRCLE The first of these formulas is used to find the length of an arc of a circle. The formula comes from the fact (proven in plane geometry) that the length of an arc is proportional to the measure of its central angle.

In Figure 4, angle QOP has a measure of 1 radian and intercepts an arc of length r on the circle. Angle ROT has a measure of θ radians and intercepts an arc of length s on the circle. Since the lengths of the arcs are proportional to the measure of their central angles,

$$\frac{s}{r} = \frac{\theta}{1}.$$

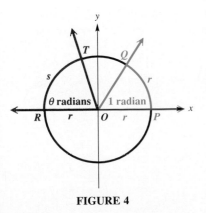

FIGURE 4

Multiplying both sides by r gives the following result.

Length of Arc

The length s of the arc intercepted on a circle of radius r by a central angle of measure θ radians is given by the product of the radius and the radian measure of the angle, or

$$s = r\theta, \qquad \theta \text{ in radians.}$$

This formula is a good example of the usefulness of radian measure. To see why, write the equivalent formula for an angle measured in degrees.

CAUTION When applying the formula $s = r\theta$, the value of θ *must be* expressed in *radians*.

◖**EXAMPLE 1**
Finding arc length using
$s = r\theta$

A circle has a radius of 18.2 centimeters. Find the length of the arc intercepted by a central angle having each of the following measures.

(a) $\dfrac{3\pi}{8}$ radians

Here $r = 18.2$ cm and $\theta = \dfrac{3\pi}{8}$. Since $s = r\theta$,

$$s = 18.2\left(\frac{3\pi}{8}\right) \text{ centimeters}$$

$$s = \frac{54.6\pi}{8} \text{ centimeters} \qquad \text{The exact answer}$$

or $\qquad s \approx 21.4$ centimeters. $\qquad$ Calculator approximation

(b) $144°$

The formula $s = r\theta$ requires that θ be measured in radians. First, convert θ to radians by the methods of the previous section. Using the shortcut, we multiply $144°$ by $\pi/180°$.

$$144° = 144°\left(\frac{\pi}{180°}\right) \text{ radians} \qquad \text{Change from degrees to radians.}$$

$$144° = \frac{4\pi}{5} \text{ radians}$$

Now $\qquad s = 18.2\left(\dfrac{4\pi}{5}\right)$ centimeters $\qquad$ Use $s = r\theta$.

$$s = \frac{72.8\pi}{5} \text{ centimeters,}$$

or $\qquad s \approx 45.7$ centimeters. ◗

EXAMPLE 2
Finding the distance between two cities using latitudes

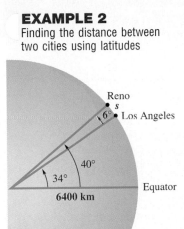

FIGURE 5

Reno, Nevada, is approximately due north of Los Angeles. The latitude of Reno is 40° N, while that of Los Angeles is 34° N. (The N in 34° N means *north* of the equator.) If the radius of Earth is 6400 kilometers, find the north-south distance between the two cities.

Latitude gives the measure of a central angle with vertex at Earth's center whose initial side goes through the equator and whose terminal side goes through the given location. As shown in Figure 5, the central angle for Reno and Los Angeles is 6°. The distance between the two cities can thus be found by the formula $s = r\theta$, after 6° is first converted to radians.

$$6° = 6°\left(\frac{\pi}{180°}\right) = \frac{\pi}{30} \text{ radians}$$

The distance between the two cities is

$$s = r\theta$$

$$s = 6400\left(\frac{\pi}{30}\right) \text{ kilometers} \qquad r = 6400, s = \frac{\pi}{30}$$

$$\approx 670 \text{ kilometers.} \quad \blacktriangleright$$

EXAMPLE 3
Finding a length using $s = r\theta$

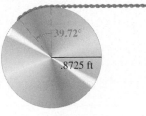
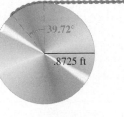

FIGURE 6

A rope is being wound around a drum with radius .8725 feet. (See Figure 6.) How much rope will be wound around the drum if the drum is rotated through an angle of 39.72°?

The length of rope wound around the drum is just the arc length for a circle of radius .8725 feet and a central angle of 39.72°. Use the formula $s = r\theta$, with the angle converted to radian measure.

$$s = r\theta$$

$$s = (.8725)\left[(39.72°)\left(\frac{\pi}{180°}\right)\right] \qquad \text{Convert to radians.}$$

$$\approx .6049$$

The length of the rope wound around the drum is approximately .6049 foot. $\blacktriangleright$

EXAMPLE 4
Finding an angle measure using $s = r\theta$

FIGURE 7

Two gears are adjusted so that the smaller gear drives the larger one as shown in Figure 7. If the smaller gear rotates through 225°, through how many degrees will the larger gear rotate?

First find the radian measure of the angle, which will give the arc length on the smaller gear that determines the motion of the larger gear. Since 225° = $5\pi/4$ radians, for the smaller gear,

$$s = r\theta = 2.5\left(\frac{5\pi}{4}\right) \approx 9.8 \text{ centimeters.}$$

An arc length of 9.8 centimeters on the larger gear corresponds to an angle measure θ, in radians, of

$$s = r\theta$$
$$9.8 = 4.8\theta$$
$$2.0 \approx \theta.$$

Changing back to degrees shows that the larger gear rotates through

$$2.0\left(\frac{180°}{\pi}\right) \approx 110°,$$

to two significant figures. ▶

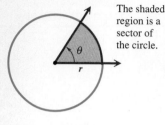

The shaded region is a sector of the circle.

FIGURE 8

SECTOR OF A CIRCLE   The other useful formula given in this section is used to find the area of a "piece of pie," or sector. A **sector of a circle** is the portion of the interior of a circle intercepted by a central angle. See Figure 8.

To find the area of a sector, assume that the radius of the circle is r. A complete circle can be thought of as an angle with a measure of 2π radians. If a central angle for the sector has measure θ radians, then the sector makes up a fraction $\theta/(2\pi)$ of a complete circle. The area of a complete circle is $A = \pi r^2$. Therefore, the area of the sector is given by the product of the fraction $\theta/(2\pi)$ and the total area, πr^2, or

$$\text{area of sector} = \frac{\theta}{2\pi}(\pi r^2) = \frac{1}{2}r^2\theta, \qquad \theta \text{ in radians.}$$

This discussion is summarized as follows.

Area of a Sector

The area of a sector of a circle of radius r and central angle θ is given by

$$A = \frac{1}{2}r^2\theta, \quad \theta \text{ in radians.}$$

CAUTION   As in the formula for arc length, the measure of θ must be in radians when using this formula for the area of a sector.

▶**EXAMPLE 5**
Finding area using
$A = (1/2)r^2\theta$

FIGURE 9

Figure 9 shows a field in the shape of a sector of a circle. The central angle is 15° and the radius of the circle is 321 meters. Find the area of the field.

First, convert 15° to radians.

$$15° = 15°\left(\frac{\pi}{180°}\right) = \frac{\pi}{12} \text{ radians}$$

Now use the formula for the area of a sector.

$$A = \frac{1}{2}r^2\theta$$

$$= \frac{1}{2}(321)^2\left(\frac{\pi}{12}\right)$$

$$\approx 13,500 \text{ m}^2. \quad ▶$$

3.2 Exercises ▼▼▼▼▼▼▼▼▼▼▼▼▼▼▼▼▼▼▼▼▼▼▼▼▼▼▼▼▼▼▼▼▼▼

In Exercises 1–2, find the exact length of the arc intercepted by the given central angle.

1.

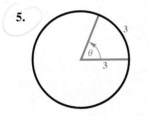

2.

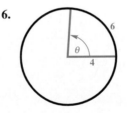

In Exercises 3–4, find the radius of the circle. The measure of a central angle (in radians) and length of its intercepted arc are given.

3.

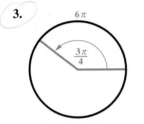

4.

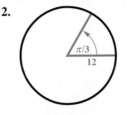

In Exercises 5–6, find the measure of the central angle (in radians).

5.

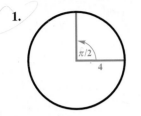

6.

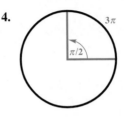

Unless otherwise directed, give calculator approximations in your answers in this exercise set.

Find the length of the arc intercepted by a central angle θ in a circle of radius r. See Example 1.

7. $r = 12.3$ cm, $\theta = \dfrac{2\pi}{3}$ radians

8. $r = .892$ cm, $\theta = \dfrac{11\pi}{10}$ radians

9. $r = 253$ m, $\theta = \dfrac{2\pi}{5}$ radians

10. $r = 120$ mm, $\theta = \dfrac{\pi}{9}$ radians

11. $r = 4.82$ m, $\theta = 60°$

12. $r = 71.9$ cm, $\theta = 135°$

13. Find the measure (in radians) of a central angle that intercepts an arc of length 5 inches in a circle of radius 2 inches.

14. Find the radius of a circle in which a central angle of 2 radians intercepts an arc of length 3 feet.

15. Find the radius of a circle in which a central angle of $\pi/5$ radians intercepts an arc of length 4 inches.

16. Find the measure (in radians) of a central angle that intercepts an arc of length 30 cm in a circle of radius 5 cm.

17. If the radius of a circle is doubled, how is the length of the arc intercepted by a fixed central angle changed?

18. In your own words, explain what is meant by a *sector* of a circle.

Find the distance in kilometers between each pair of cities, assuming they lie on the same north-south line. See Example 2.

19. Panama City, Panama, 9° N, and Pittsburgh, Pennsylvania, 40° N

20. Farmersville, California, 36° N, and Penticton, British Columbia, 49° N

21. New York City, New York, 41° N, and Lima, Peru, 12° S

22. Halifax, Nova Scotia, 45° N, and Buenos Aires, Argentina, 34° S

23. Madison, South Dakota, and Dallas, Texas, are 1200 km apart and lie on the same north-south line. The latitude of Dallas is 33° N. What is the latitude of Madison?

24. Charleston, South Carolina, and Toronto, Canada, are 1100 km apart and lie on the same north-south line. The latitude of Charleston is 33° N. What is the latitude of Toronto?

Work each applied problem. See Examples 3 and 4.

25. (a) How many inches will the weight in the figure rise if the pulley is rotated through an angle of 71°50′?
(b) Through what angle, to the nearest minute, must the pulley be rotated to raise the weight 6 in?

26. Find the radius of the pulley in the figure if a rotation of 51.6° raises the weight 11.4 cm.

27. The rotation of the smaller wheel in the figure causes the larger wheel to rotate. Through how many degrees will the larger wheel rotate if the smaller one rotates through 60.0°?

28. Find the radius of the larger wheel in the figure if the smaller wheel rotates 80.0° when the larger wheel rotates 50.0°.

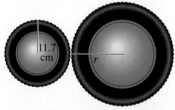

29. The figure shows the chain drive of a bicycle. How far will the bicycle move if the pedals are rotated through 180°? Assume the radius of the bicycle wheel is 13.6 in.

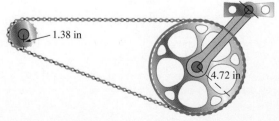

30. The speedometer of a small pickup truck is designed to be accurate with tires of radius 14 in.

(a) Find the number of rotations of a tire in 1 hr if the truck is driven at 55 mph.

(b) Suppose that oversize tires of radius 16 in are placed on the truck. If the truck is now driven for 1 hr with the speedometer reading 55 mph, how far has the truck gone? If the speed limit is 55 mph, does the driver deserve a speeding ticket?

*If a central angle is very small, there is little difference in length between an arc and the inscribed chord. See the figure. Approximate each of the following lengths by finding the necessary arc length. (*Note: *When a central angle intercepts* an arc, the arc is said to subtend *the angle.*)*

Arc length ≈ length of inscribed chord

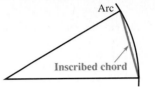

31. A railroad track in the desert is 3.5 km away. A train on the track subtends (horizontally) an angle of 3°20′. Find the length of the train.

32. An oil tanker 2.3 km at sea subtends a 1°30′ angle horizontally. Find the length of the ship.

33. The full moon subtends an angle of 1/2°. The moon is 240,000 mi away. Find the diameter of the moon.

34. The mast of Brent Simon's boat is 32 ft high. If it subtends an angle of 2°10′, how far away is it?

In Exercises 35–36, find the area of the sector.

35.

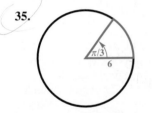

36.

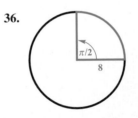

In Exercises 37–38, find the measure (in radians) of the central angle. The number inside the sector is the area.

37.

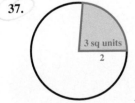

38.

Find the area of a sector of a circle having radius r and central angle θ. See Example 5.

39. $r = 29.2$ m, $\theta = \dfrac{5\pi}{6}$ radians

40. $r = 59.8$ km, $\theta = \dfrac{2\pi}{3}$ radians

41. $r = 52$ cm, $\theta = \dfrac{3\pi}{10}$ radians

42. $r = 25$ mm, $\theta = \dfrac{\pi}{15}$ radians

43. $r = 12.7$ cm, $\theta = 81°$

44. $r = 18.3$ m, $\theta = 125°$

45. Find the measure (in radians) of a central angle of a sector of area 16 square inches in a circle of radius 3.0 inches.

46. Find the radius of a circle in which a central angle of $\pi/4$ radians determines a sector of area 36 square ft.

47. Find the radius of a circle in which a central angle of $\pi/6$ radians determines a sector of area 64 square meters.

48. Find the measure (in radians) of a central angle of a sector of area 25 square inches in a circle of radius 10 inches.

49. Consider the area-of-a-sector formula $A = (1/2)r^2\theta$. What well-known formula corresponds to the special case $\theta = 2\pi$?

50. If the radius of a circle is doubled and the central angle of a sector is unchanged, how is the area of the sector changed?

---

51. The sector in the accompanying graphing calculator screen is bounded above by the line $y = (\sqrt{3}/3)x$, below by the x-axis, and on the right by the circle $x^2 + y^2 = 4$. What is the area of the sector?

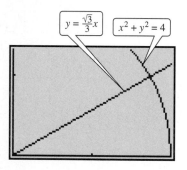

$y = \frac{\sqrt{3}}{3}x$ $x^2 + y^2 = 4$

52. The figure shows Medicine Wheel, a Native American structure in northern Wyoming. This circular structure is perhaps 200 years old. There are 32 spokes in the wheel, all equally spaced.

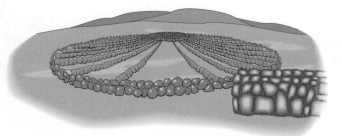

(a) Find the measure of each central angle in degrees and in radians.

(b) If the radius of the wheel is 76 ft, find the circumference.

(c) Find the length of each arc intercepted by consecutive pairs of spokes.

(d) Find the area of each sector formed by consecutive spokes.

53. The unusual corral in the figure is separated into 26 areas, many of which approximate sectors of a circle. Assume that the corral has a diameter of 50 m.

(a) Find the central angle for each region, assuming that the 26 regions are all equal sectors, with the fences meeting at the center.

(b) What is the area of each sector?

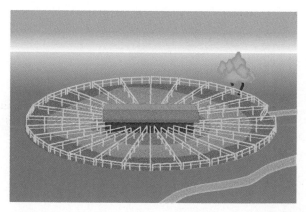

54. Eratosthenes (*ca.* 230 B.C.) made a famous measurement of the Earth. He observed at Syene (the modern Aswan) at noon and at the summer solstice that a vertical stick had no shadow, while at Alexandria (on the same meridian as Syene) the sun's rays were inclined 1/50 of a complete circle to the vertical. See the figure. He then calculated the circumference of the Earth from the known distance of 5000 stades between Alexandria and Syene. Obtain Eratosthenes' result of 250,000 stades for the circumference of the Earth. There is reason to suppose that a stade is about equal to 516.7 ft. Assuming this, use Eratosthenes' result to calculate the polar diameter of the Earth in miles. (The actual polar diameter of the Earth, to the nearest mile, is 7900 mi.) (*Source:* Eves, Howard, *A Survey of Geometry,* Vol. 1. Reprinted by permission of the author.)

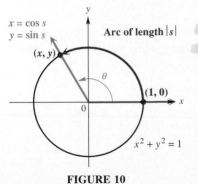

Multiply the area of the base times the height to find the volume of each solid.

55.

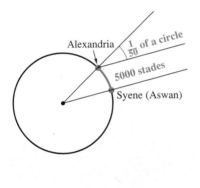

56.

Outside radius is r_1, inside radius is r_2

57. A 300-megawatt solar-power plant requires approximately 950,000 square meters of land area in order to collect the required amount of energy from sunlight.
(a) If this land area is circular, what is its radius?
(b) If this land area is a 35° sector of a circle, what is its radius?

3.3 Circular Functions of Real Numbers ▼▼▼

So far we have defined the six trigonometric functions for *angles.* The angles can be measured either in degrees or in radians. While the domain of the trigonometric functions is a set of angles, the range is a set of real numbers. In advanced work, such as calculus, it is necessary to modify the trigonometric functions so that the domain contains not angles, but real numbers. To do this we use the relationship between an angle θ and an arc of length s on a circle.

Look at Figure 10. In the figure, starting at the point (1, 0), we lay off an arc of length $|s|$ along the circle. We go counterclockwise if s is positive and clockwise if s is negative. The endpoint of the arc is the point (x, y). The circle in Figure 10 is a **unit circle,** a circle with center at the origin and a radius of one unit (hence the name *unit circle*). Recall from algebra that the equation of this circle is

$$x^2 + y^2 = 1.$$

FIGURE 10

$x = \cos s$
$y = \sin s$
(x, y)
Arc of length $|s|$
θ
$(1, 0)$
$x^2 + y^2 = 1$

We saw earlier that the radian measure of θ is related to the arc length s. In fact, for θ measured in radians, we know that $s = r\theta$. Here, $r = 1$, so s, which is measured in linear units such as inches or centimeters, is numerically equal to θ, measured in radians. Thus, the trigonometric functions of angle θ in radians found by choosing a point (x, y) on the unit circle can be rewritten as functions of the arc length s, a real number. To distinguish these from the trigonometric functions of angles, they are called **circular functions.**

Circular Functions

$$\sin s = y \qquad \tan s = \frac{y}{x}, x \neq 0 \qquad \sec s = \frac{1}{x}, x \neq 0$$

$$\cos s = x \qquad \cot s = \frac{x}{y}, y \neq 0 \qquad \csc s = \frac{1}{y}, y \neq 0$$

NOTE Since $\sin s = y$ and $\cos s = x$, we can replace x and y in the equation $x^2 + y^2 = 1$ and obtain the familiar Pythagorean identity

$$\cos^2 s + \sin^2 s = 1.$$

Since the ordered pair (x, y) represents a point on the unit circle,

$$-1 \leq x \leq 1 \qquad \text{and} \qquad -1 \leq y \leq 1,$$

making $\qquad -1 \leq \cos s \leq 1 \qquad$ and $\qquad -1 \leq \sin s \leq 1.$

For any value of s, both $\sin s$ and $\cos s$ exist, so the domain of these functions is the set of all real numbers. For $\tan s$, defined as y/x, $x \neq 0$. The only way x can equal 0 is when the arc length s is $\pi/2$, $-\pi/2$, $3\pi/2$, $-3\pi/2$, and so on. To avoid a zero denominator, the domain of tangent must be restricted to those values of s satisfying

$$s \neq \frac{\pi}{2} + n\pi, \quad n \text{ any integer.}$$

The definition of secant also has x in the denominator, making the domain of secant the same as the domain of tangent. Both cotangent and cosecant are defined with a denominator of y. To guarantee that $y \neq 0$, the domain of these functions must be the set of all values of s satisfying

$$s \neq 0 + n\pi, \quad n \text{ any integer.}$$

The domains of the circular functions are summarized in the following box. Compare these with the domains of the trigonometric functions.

Domains of the Circular Functions

The domains of the circular functions are as follows. Assume that n is any integer and s is a real number.

Sine and Cosine Functions: $(-\infty, \infty)$

Tangent and Secant Functions: $\left\{s \mid s \neq \dfrac{\pi}{2} + n\pi\right\}$

Cotangent and Cosecant Functions: $\{s \mid s \neq n\pi\}$

As mentioned at the beginning of this section, in Figure 10, s is the radian measure of angle θ and the radius r equals 1. Using the definition of the trigonometric functions,

$$\sin \theta = \frac{y}{r} = \frac{y}{1} = y = \sin s$$

$$\cos \theta = \frac{x}{r} = \frac{x}{1} = x = \cos s.$$

Similar results hold for the other four functions.

As shown above, the trigonometric functions and the circular functions lead to the same function values. Because of this, a value such as $\sin \pi/2$ can be found without worrying about whether $\pi/2$ is a real number or the radian measure of an angle. In either case, $\sin \pi/2 = 1$. All the formulas developed in this book are valid for either angles or real numbers. For example, $\sin \theta = 1/\csc \theta$ is equally valid for θ as the measure of an angle in degrees or radians or for θ as a real number.

We can use the ideas of the preceding discussion to find exact circular function values of numbers expressed as rational multiples of π. Example 1 illustrates this.

EXAMPLE 1
Finding an exact function value
of a rational multiple of π

Find the exact circular function value for each of the following.

(a) $\cos \dfrac{2\pi}{3}$

As mentioned earlier, we can consider $2\pi/3$ as the radian measure of an angle. Since

$$\frac{2\pi}{3} = \frac{2}{3} \cdot 180° = 120°,$$

$$\cos \frac{2\pi}{3} = \cos 120° = -\frac{1}{2}.$$

(b) $\tan\left(-\dfrac{\pi}{4}\right)$

$$\tan\left(-\frac{\pi}{4}\right) = \tan(-45°) = -1$$

(c) $\csc \dfrac{10\pi}{3}$

Because $10\pi/3$ is not between 0 and 2π, we must first find a number between 0 and 2π with which it is coterminal. To do this subtract 2π as many times as needed. Here, 2π must be subtracted only once.

$$\frac{10\pi}{3} - 2\pi = \frac{10\pi}{3} - \frac{6\pi}{3} = \frac{4\pi}{3}$$

$$\csc \frac{10\pi}{3} = \csc \frac{4\pi}{3} = \csc 240° = -\frac{2\sqrt{3}}{3} \quad \blacktriangleright$$

NOTE   The values found in Example 1 can also be determined without converting to degrees by using reference angles measured in radians.

In order to use a calculator to find an approximation of a circular function of a real number, we must first set the calculator to *radian* mode. The next example shows how to find such approximations.

**EXAMPLE 2**
Finding approximate circular function values

Use a calculator to find an approximation for each of the following circular function values.

(a) $\cos .5149$

$$\cos .5149 \approx .87034197$$

(b) $\cot 1.3209$
Because calculators do not have keys for cotangent, secant, and cosecant, to find these values we must use the appropriate reciprocal function. To find $\cot 1.3209$, we first find $\tan 1.3209$ and then find the reciprocal.

$$\cot 1.3209 = \frac{1}{\tan 1.3209} \approx .25523149$$

(c) $\sec(-2.9234)$

$$\sec(-2.9234) = \frac{1}{\cos(-2.9234)} \approx -1.0242855 \quad \blacktriangleright$$

CAUTION   One of the most common errors in trigonometry involves using calculators in degree mode when radian mode should be used. Remember that if you are finding a circular function value of a real number, the calculator *must* be in radian mode.

EXAMPLE 3
Finding a number with a given circular function value

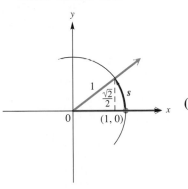

FIGURE 11

(a) Find the value of s in the interval $[0, \pi/2]$ that has $\cos s = .96854556$.

The value of s can be found with a calculator set for radian mode. Recall from Section 2.3 how we found an angle measure given a trigonometric function value of the angle. The same procedure is repeated here with the calculator set to radian mode to find that

$$\cos .25147856 = .96854556,$$

and $0 < .25147856 < \pi/2$, so $s = .25147856$.

(b) Find the exact value of s in the interval $[0, \pi/2]$ for which $\sin s = \sqrt{2}/2$.

Sketch a triangle in quadrant I and use the definition of $\sin s$ to label the sides as shown in Figure 11. To relate it to the definition of the trigonometric function $\sin \theta$, multiply the lengths of each side by 2. We recognize this as a right triangle with the two acute angles of 45°. To find s, convert 45° to radians, to get $s = \pi/4$. ◗

The next example answers the first question posed in the application presented at the beginning of the chapter.

EXAMPLE 4
Finding the angle of elevation of the sun

Knowing the position of the sun in the sky is essential for solar-power plants. Solar panels need to be positioned perpendicular to the sun's rays for maximum efficiency. The angle of elevation θ of the sun in the sky at any latitude L can be calculated using the formula

$$\sin \theta = \cos D \cos L \cos \omega + \sin D \sin L$$

where $\theta = 0$ corresponds to sunrise and $\theta = \pi/2$ occurs if the sun is directly overhead. ω is the number of radians that the Earth has rotated through since noon when $\omega = 0$. D is the declination of the sun which varies because the Earth is tilted on its axis. (*Source:* Winter, C., R. Sizmann, and Vant-Hunt (Editors), *Solar Power Plants,* Springer-Verlag, 1991.)

Sacramento, California, has a latitude of $L = 38.5°$ or .6720 radians. Find the angle of elevation θ of the sun at 3 P.M. on February 29, 2000, where at that time, $D \approx -.1425$ and $\omega \approx .7854$.

Use the formula for $\sin \theta$.

$$\sin \theta = \cos D \cos L \cos \omega + \sin D \sin L$$
$$= \cos(-.1425) \cos(.6720) \cos(.7854) + \sin(-.1425) \sin(.6720)$$
$$\approx .4593$$

Thus, $\theta \approx .4773$ radians or 27.3°. ◗

▨ At the beginning of this section we introduced the concept of circular functions. We showed how the unit circle $x^2 + y^2 = 1$ can be used to find the cosine and sine of a real number s. To illustrate this concept, we can use a graphing calculator to graph the circle. We must first solve the equation for y, getting the two functions

$$Y_1 = \sqrt{1 - x^2} \quad \text{and} \quad Y_2 = -\sqrt{1 - x^2}.$$

With your calculator in function mode, graph these two equations in the same window to get the graph of the circle. To get an undistorted figure, a *square window* must be used. (See your instruction manual for details.)

The TRACE function of the calculator allows us to find coordinates of points on the graph. Experiment with this feature, and notice that the x and y coordinates are displayed below the graph. The x-coordinate represents the cosine of the length of the arc from the point $(1, 0)$ to the point indicated by the cursor. For example, one such point is

$$x = .9 \qquad y = .43588989.$$

This point is shown in the figure.

While the calculator does not give the arc length, it can be found by setting the calculator in radian mode and finding either $\cos^{-1} .9$ or $\sin^{-1} .43588989$. By doing this, we find the arc length is approximately .4510268.

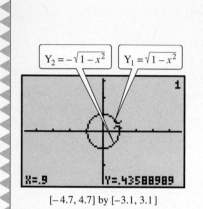

$[-4.7, 4.7]$ by $[-3.1, 3.1]$

In Figure 12 you will find a graph of the unit circle $x^2 + y^2 = 1$ with a great deal of important information. Degree and radian measures are given for the first counterclockwise revolution, and the coordinates of the points on the circle are also given. Remember that the first coordinate is the cosine of the angle (or number) and the second coordinate is the sine. This figure should help you in the rest of your study of trigonometry.

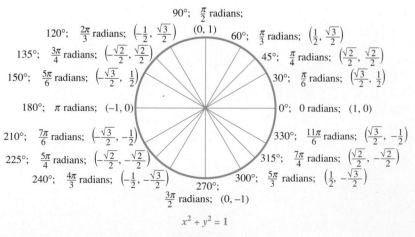

FIGURE 12

3.3 Exercises ▼▼▼▼▼▼▼▼▼▼▼▼▼▼▼▼▼▼▼▼▼▼▼▼▼▼▼▼▼▼▼▼▼▼

1. So far, we can find exact values of trigonometric functions for quadrantal angles and for angles having $\pi/3$, $\pi/4$, or $\pi/6$ as reference angles. If we consider s to be a radian-measured angle, $0 \le s \le 2\pi$, how many such exact values can be determined?

2. In your own words, explain how the cosine function associates a value with the real number s.

Find the exact circular function value for each of the following. See Example 1. Evaluate

3-17

3. $\sin \dfrac{7\pi}{6}$

4. $\cos \dfrac{5\pi}{3}$

5. $\tan \dfrac{3\pi}{4}$

6. $\sin \dfrac{5\pi}{3}$

7. $\cos \dfrac{7\pi}{6}$

8. $\tan \dfrac{4\pi}{3}$

9. $\sec \dfrac{2\pi}{3}$

10. $\csc \dfrac{11\pi}{6}$

11. $\cot \dfrac{5\pi}{6}$

12. $\cos\left(-\dfrac{4\pi}{3}\right)$

13. $\sin\left(-\dfrac{5\pi}{6}\right)$

14. $\tan \dfrac{17\pi}{3}$

15. $\sec \dfrac{23\pi}{6}$

16. $\csc \dfrac{13\pi}{3}$

17. $\cos \dfrac{13\pi}{4}$

18. Suppose a student attempts to work Exercises 3–17 above using a calculator. Can the student expect to get the correct results, according to the directions given for those exercises? Explain.

Use a calculator to find an approximation for each circular function value. Be sure that your calculator is set in radian mode. See Example 2.

19. $\sin .8203$

20. $\cot .6632$

21. $\cos .6429$

22. $\tan .9047$

23. $\sin 1.5097$

24. $\cot .0465$

25. $\csc 1.3875$

26. $\tan 1.3032$

27. $\sin 7.5835$

28. $\tan 6.4752$

29. $\cot 7.4526$

30. $\cos 6.6701$

31. $\tan 4.0230$

32. $\cot 3.8426$

33. $\cos 4.2528$

34. $\sin 3.4645$

35. $\sin(-2.2864)$

36. $\cot(-2.4871)$

37. $\cos(-3.0602)$

38. $\tan(-1.7861)$

Find the value of s in the interval $[0, \pi/2]$ that makes each statement true. See Example 3(a).

39. $\tan s = .21264138$ Arc Arc Tan

40. $\cos s = .78269876$

41. $\sin s = .99184065$

42. $\cot s = .29949853$

43. $\cot s = .62084613$

44. $\tan s = 2.6058440$

45. $\cos s = .57834328$

46. $\sin s = .98771924$

47. $\cot s = .09637041$

48. $\csc s = 1.0219553$

49. $\tan s = 1.6213129$

50. $\cos s = .92728460$

51. What (exact) value of S between 0 and $\pi/2$ produces the output for the accompanying graphing calculator screen?

```
sin S
         .8660254038
√3/2
         .8660254038
```

In Exercises 52–57, find the exact value of s in the given interval that has the given circular function value. Do not use a calculator. See Example 3(b).

52. $\left[\dfrac{\pi}{2}, \pi\right]$; $\sin s = \dfrac{1}{2}$

53. $\left[\dfrac{\pi}{2}, \pi\right]$; $\cos s = -\dfrac{1}{2}$

54. $\left[\pi, \dfrac{3\pi}{2}\right]$; $\tan s = \sqrt{3}$

55. $\left[\pi, \dfrac{3\pi}{2}\right]$; $\sin s = -\dfrac{1}{2}$

56. $\left[\dfrac{3\pi}{2}, 2\pi\right]$; $\tan s = -1$

57. $\left[\dfrac{3\pi}{2}, 2\pi\right]$; $\cos s = \dfrac{\sqrt{3}}{2}$

The graphing calculator screen shows a point on the unit circle. What is the length of the shortest arc of the circle from (1, 0) to the point?

58.

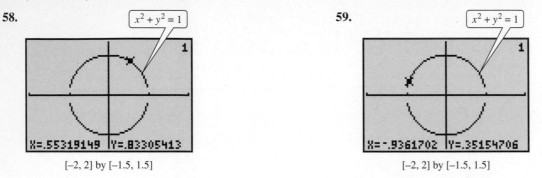

$x^2 + y^2 = 1$

X=.55319149 Y=.83305413

[−2, 2] by [−1.5, 1.5]

59.

$x^2 + y^2 = 1$

X=−.9361702 Y=.35154706

[−2, 2] by [−1.5, 1.5]

Suppose an arc of length s lies on the unit circle $x^2 + y^2 = 1$, starting at the point (1, 0) and terminating at the point (x, y). (See Figure 10.) Use a calculator to find the approximate coordinates for (x, y). (Hint: $x = \cos s$ and $y = \sin s$.)

60. $s = 2.5$ **61.** $s = 3.4$ **62.** $s = -7.4$ **63.** $s = -3.9$

For each value of s, use a calculator to find sin s and cos s and then use the results to decide in which quadrant an angle of s radians lies.

64. $s = 51$ **65.** $s = 49$ **66.** $s = 65$ **67.** $s = 79$

---

68. The values of the circular functions repeat every 2π. For this reason, circular functions are used to describe things that repeat periodically. For example, the maximum afternoon temperature in a given city might be approximated by

$$t = 60 - 30 \cos \frac{x\pi}{6},$$

where t represents the maximum afternoon temperature in month x, with $x = 0$ representing January, $x = 1$ representing February, and so on. Find the maximum afternoon temperature for each of the following months.
(a) January   (b) April
(c) May   (d) June
(e) August   (f) October

69. The temperature in Fairbanks is approximated by

$$T(x) = 37 \sin\left[\frac{2\pi}{365}(x - 101)\right] + 25,$$

where $T(x)$ is the temperature in degrees Fahrenheit on day x, with $x = 1$ corresponding to January 1 and $x = 365$ corresponding to December 31. Use a calculator to estimate the temperature on the following days. (*Source:* Lando, Barbara and Clifton Lando, "Is the Graph of Temperature Variation a Sine Curve?" *The Mathematics Teacher,* 70 (September 1977): 534–37.)
(a) March 1 (day 60)   (b) April 1 (day 91)
(c) Day 150   (d) June 15
(e) September 1   (f) October 31

Solve the equation for $0 \le x \le 2\pi$. (Hint: Use the unit circle definition of cosine and sine discussed in this section.)

70. $\sin x = \sin(x + 2)$ **71.** $\cos x = \cos(x + 1)$

72. Refer to Example 4.
(a) Repeat the example for New Orleans, which has a latitude of $L = 30°$.
(b) Compare your answers. Do they agree with your intuition?

73. The ability to calculate the number of daylight hours H at any location is important in estimating the potential solar energy production. H can be calculated using the formula

$$\cos(.1309H) = -\tan D \tan L$$

where D and L are defined in Example 4. Use this trigonometric equation to calculate the shortest and longest days in Minneapolis, Minnesota, if its latitude is $L = 44.88°$, the shortest day occurs when $D = -23.44°$, and the longest day occurs when $D = 23.44°$. (*Source:* Winter, C., R. Sizmann, and Vant-Hunt (Editors), *Solar Power Plants,* Springer-Verlag, 1991.)

74. Refer to Exercise 73. Calculate the number of daylight hours at Hartford, Connecticut, on August 12, 2001. (*Hint:* $D \approx .26724$ and the latitude is $L = 41.93°$. Remember to convert degrees to radians.)

3.4 Linear and Angular Velocity ▼▼▼

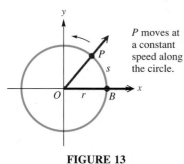

P moves at a constant speed along the circle.

FIGURE 13

In many situations it is necessary to know at what speed a point on a circular disk is moving or how fast the central angle of such a disk is changing. Some examples occur with machinery involving gears or pulleys or the speed of a car around a curved portion of highway. Suppose that point P moves at a constant speed along a circle of radius r and center O. See Figure 13. The measure of how fast the position of P is changing is called **linear velocity.** If v represents linear velocity, then

$$\text{velocity} = \frac{\text{distance}}{\text{time}}$$

$$v = \frac{s}{t},$$

where s is the length of the arc traced by point P at time t. (This formula is just a restatement of the familiar result $d = rt$ with s as distance, v as the rate, and t as time.)

Look at Figure 13 again. As point P moves along the circle, ray OP rotates around the origin. Since the ray OP is the terminal side of angle POB, the measure of the angle changes as P moves along the circle. The measure of how fast angle POB is changing is called **angular velocity.** Angular velocity, written ω (the greek letter *omega*), can be given as

$$\omega = \frac{\theta}{t}, \quad \theta \text{ in radians,}$$

where θ is the measure of angle POB at time t. As with the earlier formulas in this chapter, θ must be measured in radians, with ω expressed as radians per unit of time. Angular velocity is used in physics and engineering, among other applications.

In Section 3.2 the length s of the arc intercepted on a circle of radius r by a central angle of measure θ radians was found to be $s = r\theta$. Using this formula, the formula for linear velocity, $v = s/t$, becomes

$$v = \frac{r\theta}{t}$$

$$= r \cdot \frac{\theta}{t}$$

or $\qquad\qquad\qquad v = r\omega. \qquad \omega = \frac{\theta}{t}$

This last formula relates linear and angular velocity.

A radian is a "pure number," with no units associated with it. This is why the product of the length r, measured in units such as centimeters, and ω, measured in units such as radians per second, is velocity, v, measured in units such as centimeters per second.

All the formulas given in this section are summarized below.

Angular and Linear Velocity

Angular Velocity	Linear Velocity
$\omega = \dfrac{\theta}{t}$	$v = \dfrac{s}{t}$
(ω in radians per unit time, θ in radians)	$v = \dfrac{r\theta}{t}$
	$v = r\omega$

◖EXAMPLE 1
Using the linear and angular velocity formulas

Suppose that point P is on a circle with a radius of 10 centimeters, and ray OP is rotating with angular velocity of $\pi/18$ radians per second.

(a) Find the angle generated by P in 6 seconds.

The velocity of ray OP is $\omega = \pi/18$ radians per second. Since $\omega = \theta/t$, then in 6 seconds

$$\frac{\pi}{18} = \frac{\theta}{6},$$

or $\theta = 6(\pi/18) = \pi/3$ radians.

(b) Find the distance traveled by P along the circle in 6 seconds.

In 6 seconds P generates an angle of $\pi/3$ radians. Since $s = r\theta$,

$$s = 10\left(\frac{\pi}{3}\right) = \frac{10\pi}{3} \text{ centimeters.}$$

(c) Find the linear velocity of P.

Since $v = s/t$, in 6 seconds

$$v = \frac{\dfrac{10\pi}{3}}{6} = \frac{5\pi}{9} \text{ centimeters per second.} \quad \blacktriangleright$$

■ **PROBLEM SOLVING** In practical applications, angular velocity is often given as revolutions per unit of time, which must be converted to radians per unit of time before using the formulas given in this section. ■

EXAMPLE 2
Using the linear and angular velocity formulas

A belt runs a pulley of radius 6 centimeters at 80 revolutions per minute.

(a) Find the angular velocity of the pulley in radians per second.

In one minute, the pulley makes 80 revolutions. Each revolution is 2π radians, for a total of

$$80(2\pi) = 160\pi \text{ radians per minute.}$$

Since there are 60 seconds in a minute, ω, the angular velocity in radians per second, is found by dividing 160π by 60.

$$\omega = \frac{160\pi}{60} = \frac{8\pi}{3} \text{ radians per second}$$

(b) Find the linear velocity of the belt in centimeters per second.

The linear velocity of the belt will be the same as that of a point on the circumference of the pulley. Thus,

$$v = r\omega$$
$$v = 6\left(\frac{8\pi}{3}\right)$$
$$v = 16\pi \text{ centimeters per second}$$
$$v \approx 50.3 \text{ centimeters per second.} \quad \blacktriangleright$$

EXAMPLE 3
Finding the linear velocity and distance traveled by a satellite

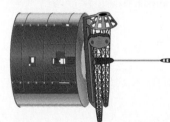

A satellite traveling in a circular orbit 1600 kilometers above the surface of the Earth takes two hours to make an orbit. Assume that the radius of the Earth is 6400 kilometers.

(a) Find the linear velocity of the satellite.

The distance of the satellite from the center of the Earth is

$$r = 1600 + 6400 = 8000 \text{ kilometers.}$$

For one orbit $\theta = 2\pi$, and

$$s = r\theta = 8000(2\pi) \text{ kilometers.}$$

Since it takes 2 hours to complete an orbit, the linear velocity is

$$v = \frac{s}{t} = \frac{8000(2\pi)}{2}$$
$$= 8000\pi$$
$$\approx 25,000 \text{ kilometers per hour.}$$

(b) Find the distance traveled in 4.5 hours.

$$s = vt = (8000\pi)(4.5)$$
$$= 36,000\pi$$
$$\approx 110,000 \text{ kilometers.} \quad \blacktriangleright$$

3.4 Exercises ▼▼▼▼▼▼▼▼▼▼▼▼▼▼▼▼▼▼▼▼▼▼▼▼▼▼▼▼▼▼▼▼▼▼▼▼

1. If a point moves around the circumference of the unit circle at an angular velocity of 1 radian per second, how long will it take for the point to move around the entire circle?

2. If a point moves around the circumference of the unit circle at the speed of one unit per second, how long will it take for the point to move around the entire circle?

3. What is the difference between linear velocity and angular velocity?

4. Explain why linear velocity is affected by the radius of the circle, whereas angular velocity is not.

Use the formula $\omega = \theta/t$ to find the value of the missing variable.

5. $\omega = 2\pi/3$ radians per sec, $t = 3$ sec

6. $\omega = \pi/4$ radians per min, $t = 5$ min

7. $\theta = 3\pi/4$ radians, $t = 8$ sec

8. $\theta = 2\pi/5$ radians, $t = 10$ sec

9. $\theta = 2\pi/9$ radians, $\omega = 5\pi/27$ radians per min

10. $\theta = 3\pi/8$ radians, $\omega = \pi/24$ radians per min

11. $\theta = 3.871142$ radians, $t = 21.4693$ sec

12. $\omega = .90674$ radians per min, $t = 11.876$ min

Use the formula $v = r\omega$ to find the value of the missing variable.

13. $r = 12$ m, $\omega = 2\pi/3$ radians per sec

14. $r = 8$ cm, $\omega = 9\pi/5$ radians per sec

15. $v = 9$ m per sec, $r = 5$ m

16. $v = 18$ ft per sec, $r = 3$ ft

17. $v = 107.692$ m per sec, $r = 58.7413$ m

18. $r = 24.93215$ cm, $\omega = .372914$ radians per sec

The formula $\omega = \theta/t$ can be rewritten as $\theta = \omega t$. Using ωt for θ changes $s = r\theta$ to $s = r\omega t$. Use the formula $s = r\omega t$ to find the value of the missing variables.

19. $r = 6$ cm, $\omega = \pi/3$ radians per sec, $t = 9$ sec

20. $r = 9$ yd, $\omega = 2\pi/5$ radians per sec, $t = 12$ sec

21. $s = 6\pi$ cm, $r = 2$ cm, $\omega = \pi/4$ radians per sec

22. $s = 12\pi/5$ m, $r = 3/2$ m, $\omega = 2\pi/5$ radians per sec

23. $s = 3\pi/4$ km, $r = 2$ km, $t = 4$ sec

24. $s = 8\pi/9$ m, $r = 4/3$ m, $t = 12$ sec

25. Explain the similarities between the familiar $d = rt$ formula and the formula $s = vt$.

26. Suppose you must convert k radians per second to degrees per minute. Explain how you would do this.

Find ω for each of the following.

27. The hour hand of a clock

28. The minute hand of a clock

29. The second hand of a clock

30. A line from the center to the edge of a phonograph record revolving 33 1/3 times per minute

Find v for each of the following.

31. The tip of the minute hand of a clock, if the hand is 7 cm long

32. The tip of the second hand of a clock, if the hand is 28 mm long

33. A point on the edge of a flywheel of radius 2 m, rotating 42 times per min

34. A point on the tread of a tire of radius 18 cm, rotating 35 times per min

35. The tip of an airplane propeller 3 m long, rotating 500 times per min (*Hint: r* = 1.5 m)

36. A point on the edge of a gyroscope of radius 83 cm, rotating 680 times per minute

---

Solve the following problems, which review the ideas of this section. See Examples 1–3.

37. Two pulleys of diameters 4 m and 2 m, respectively, are connected by a belt. The larger pulley rotates 80 times per min. Find the speed of the belt in meters per second and the angular velocity of the smaller pulley.

38. The Earth travels about the sun in an orbit that is almost circular. Assume that the orbit is a circle, with a radius of 93,000,000 mi. (See the figure.)
   (a) Assume that a year is 365 days, and find θ, the angle formed by the Earth's movement in one day.
   (b) Give the angular velocity in radians per hour.
   (c) Find the linear velocity of the Earth in miles per hour.

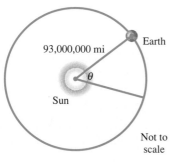

39. The pulley shown has a radius of 12.96 cm. Suppose it takes 18 sec for 56 cm of belt to go around the pulley. Find the angular velocity of the pulley in radians per sec.

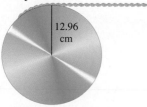

40. The two pulleys in the figure have radii of 15 cm and 8 cm, respectively. The larger pulley rotates 25 times in 36 sec. Find the angular velocity of each pulley in radians per sec.

41. A gear is driven by a chain that travels 1.46 m per sec. Find the radius of the gear if it makes 46 revolutions per min.

42. A thread is being pulled off a spool at the rate of 59.4 cm per sec. Find the radius of the spool if it makes 152 revolutions per min.

43. The Earth revolves on its axis once every 24 hr. Assuming that the Earth's radius is 6400 km, find the following.
   (a) Angular velocity of the Earth in radians per day and radians per hr
   (b) Linear velocity at the North Pole or South Pole
   (c) Linear velocity at Quito, Ecuador, a city on the equator
   (d) Linear velocity at Salem, Oregon (halfway from the equator to the North Pole)

44. A railroad track is laid along the arc of a circle of radius 1800 ft. The circular part of the track subtends a central angle of 40°. How long (in seconds) will it take a point on the front of a train traveling 30 mph to go around this portion of the track?

45. A 90-horsepower outboard motor at full throttle will rotate its propeller at 5000 revolutions per minute. Find the angular velocity of the propeller in radians per second.

Chapter 3   Summary ▼▼▼▼▼▼▼▼▼▼▼▼▼▼▼▼▼▼▼▼▼▼▼▼▼▼▼▼▼▼▼▼

SECTION	KEY IDEAS
3.1 Radian Measure	
	Radian An angle that has its vertex at the center of a circle and that intercepts an arc on the circle equal in length to the radius of the circle has a measure of **one radian.** **Degree/Radian Relationship** $$180° = \pi \text{ radians}$$ **Converting Between Degrees and Radians** Use the proportion $$\frac{\text{radian measure}}{\pi} = \frac{\text{degree measure}}{180°}$$ or the following shortcuts: **1.** If a radian measure involves a multiple of π, replace π with $180°$ and simplify to convert to degrees. **2.** Multiply a degree measure by $\pi/180°$ to convert to radians.
3.2 Applications of Radian Measure	
	Length of Arc The length s of the arc intercepted on a circle of radius r by a central angle of measure θ radians is given by the product of the radius and the radian measure of the angle, or $$s = r\theta, \quad \theta \text{ in radians.}$$ **Area of Sector** The area of a sector of a circle of radius r and central angle θ is given by $$A = \frac{1}{2}r^2\theta, \quad \theta \text{ in radians.}$$
3.3 Circular Functions of Real Numbers	
	Circular Functions Start at the point $(1, 0)$ on the unit circle $x^2 + y^2 = 1$ and lay off an arc of length $|s|$ along the circle, going counterclockwise if s is positive, and clockwise if s is negative. Let the endpoint of the arc be at the point (x, y). The six circular functions of s are defined as follows. (Assume that no denominators are zero.) $$\sin s = y \qquad \tan s = \frac{y}{x} \qquad \sec s = \frac{1}{x}$$ $$\cos s = x \qquad \cot s = \frac{x}{y} \qquad \csc s = \frac{1}{y}$$

SECTION	KEY IDEAS
3.3 Circular Functions of Real Numbers	
	Domains of the Circular Functions The domains of the circular functions are as follows. Assume that n is any integer. **Sine and Cosine Functions** $$(-\infty, \infty)$$ **Tangent and Secant Functions** $$\left\{ s \mid s \neq \frac{\pi}{2} + n\pi \right\}$$ **Cotangent and Cosecant Functions** $$\{ s \mid s \neq n\pi \}$$
3.4 Linear and Angular Velocity	
	Angular and Linear Velocity

	Angular Velocity	*Linear Velocity*
	$\omega = \dfrac{\theta}{t}$	$v = \dfrac{s}{t}$
	(ω in radians per unit time, θ in radians)	$v = \dfrac{r\theta}{t}$
		$v = r\omega$

Chapter 3   Review Exercises ▼▼▼▼▼▼▼▼▼▼▼▼▼▼▼▼▼▼▼▼▼▼▼▼▼▼▼▼

1. Which is bigger—an angle of $1°$ or an angle of 1 radian? Discuss and justify your observations.

2. Find three angles coterminal to an angle of 1 radian.

3. Find three angles coterminal to an angle of 2 radians.

4. Give an expression that generates all angles coterminal with an angle of $\pi/6$ radians. Let n represent any integer.

Convert each of the following degree measures to radians. Leave answers as multiples of π.

5. $45°$ 6. $120°$ 7. $80°$ 8. $175°$

9. $330°$ 10. $800°$ 11. $1020°$ 12. $2000°$

Convert each of the following radian measures to degrees.

13. $\dfrac{5\pi}{4}$ 14. $\dfrac{9\pi}{10}$ 15. $\dfrac{8\pi}{3}$ 16. $-\dfrac{6\pi}{5}$ 17. $-\dfrac{11\pi}{18}$ 18. $\dfrac{21\pi}{5}$ 19. $\dfrac{14\pi}{15}$ 20. $\dfrac{33\pi}{5}$

Evaluate in degrees

Convert each radian measure to degrees in order to find the exact function value. Do not use a calculator.

21. $\tan \dfrac{\pi}{3}$

22. $\cos \dfrac{2\pi}{3}$

23. $\sin\left(-\dfrac{5\pi}{6}\right)$

24. $\cot \dfrac{11\pi}{6}$

25. $\tan\left(-\dfrac{7\pi}{3}\right)$

26. $\sec \dfrac{\pi}{3}$

27. $\csc\left(-\dfrac{11\pi}{6}\right)$

28. $\cot\left(-\dfrac{17\pi}{3}\right)$

Radius · Radian

$S = R\theta$

Solve the problem. Use a calculator as necessary.

29. The radius of a circle is 15.2 cm. Find the length of an arc of the circle intercepted by a central angle of $3\pi/4$ radians.

30. Find the length of an arc intercepted by a central angle of .769 radians on a circle with a radius of 11.4 cm.

31. A circle has a radius of 8.973 cm. Find the length of an arc on this circle intercepted by a central angle of 49.06°.

32. A central angle of $7\pi/4$ radians forms a sector of a circle. Find the area of the sector if the radius of the circle is 28.69 in.

33. Find the area of a sector of a circle having a central angle of 21° 40′ in a circle of radius 38.0 m.

34. A tree 2000 yd away subtends an angle of 1° 10′. Find the height of the tree to two significant digits.

Assume that the radius of Earth is 6400 km in the next two exercises.

35. Find the distance in kilometers between cities on a north-south line that are on latitudes 28° N and 12° S, respectively.

36. Two cities on the equator have longitudes of 72° E and 35° W, respectively. Find the distance between the cities.

In Exercises 37 and 38, find the measure of the central angle θ (in radians) and the area of the sector.

37.

38.

39. The hour hand of a wall clock measures six inches from its tip to the center of the clock.
  (a) Through what angle (in radians) does the hour hand pass between 1 o'clock and 3 o'clock?
  (b) What distance does the tip of the hour hand travel during the time period from 1 o'clock to 3 o'clock?

40. Describe what would happen to the central angle for a given arc length of a circle if the circle's radius were doubled. (Assume everything else is unchanged.)

Use a calculator to find an approximation for the circular function values. Be sure your calculator is set in radian mode.

41. $\sin 1.0472$

42. $\tan 1.2275$

43. $\cos(-.2443)$

44. $\cot 3.0543$

45. $\tan 7.3159$

46. $\sin 4.8386$

47. $\sec .4864$

48. $\csc(-.8385)$

Find the value of s in the interval $[0, \pi/2]$ *that makes each of the following true.*

49. cos *s* = .92500448

50. tan *s* = 4.0112357

51. sin *s* = .49244294

52. csc *s* = 1.2361343

53. cot *s* = .50221761

54. sec *s* = 4.5600039

Find the exact value of s in the given interval that has the given circular function value. Do not use a calculator.

55. $\left[0, \dfrac{\pi}{2}\right]$; cos $s = \dfrac{\sqrt{2}}{2}$

56. $\left[\dfrac{\pi}{2}, \pi\right]$; tan $s = -\sqrt{3}$

57. $\left[\pi, \dfrac{3\pi}{2}\right]$; sec $s = -\dfrac{2\sqrt{3}}{3}$

58. $\left[\dfrac{3\pi}{2}, 2\pi\right]$; sin $s = -\dfrac{1}{2}$

59. Without using a calculator, determine which of the following numbers is closest to cos 2: .4, .6, 0, −.4, or −.6.

60. Without using a calculator, determine which of the following numbers is closest to sin 2: .9, .1, 0, −.9, or −.1.

61. Find the measure (in both degrees and radians) of the angle *θ* formed in the accompanying screen by the line passing through the origin and the positive part of the *x*-axis. Use the displayed values of *x* and *y* at the bottom of the screen.

62. The graphing calculator screen shows a point on the unit circle. What is the length of the shortest arc of the circle from (1, 0) to the point?

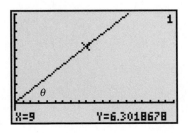

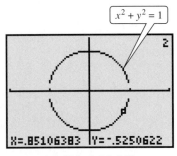

[−2, 2] by [−1.5, 1.5]

Solve each problem.

63. Find *t* if $\theta = 5\pi/12$ radians and $\omega = 8\pi/9$ radians per sec.

64. Find *θ* if *t* = 12 sec and $\omega = 9$ radians per sec.

65. Find *ω* if *t* = 8 sec and $\theta = 2\pi/5$ radians.

66. Find *ω* if $s = 12\pi/25$ ft, $r = 3/5$ ft, and *t* = 15 sec.

67. Find *s* if *r* = 11.46 cm, $\omega = 4.283$ radians per sec, and *t* = 5.813 sec.

68. Find the linear velocity of a point on the edge of a flywheel of radius 7 m if the flywheel is rotating 90 times per sec.

69. The shortest path for the sun's rays through the Earth's atmosphere occurs when the sun is directly overhead. Disregarding the curvature of the Earth, as the sun moves lower on the horizon, the distance that sunlight passes through the atmosphere increases by a factor of $\csc \theta$ where θ is the angle of elevation of the sun. This increased distance reduces both the intensity of the sun and the amount of ultraviolet light that reaches the Earth's surface. See the accompanying figure. (*Source:* Winter, C., R. Sizmann, and Vant-Hunt (Editors), *Solar Power Plants,* Springer-Verlag, 1991.)

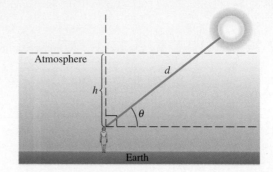

(a) Verify that $d = h \csc \theta$.

(b) Determine θ when $d = 2h$.

(c) The atmosphere filters out the ultraviolet light that causes skin to burn. Compare the difference between sunbathing when $\theta = \pi/2$ and $\pi/3$. Which measure gives less ultraviolet light?

4

Graphs of the Circular Functions

Many cities throughout the world experience seasons. Although each city has its own unique weather patterns, there are also similarities between cities as to when seasons occur and how corresponding temperatures vary. Temperature changes are a primary cause of seasons. The table lists the average monthly temperatures in °F in Vancouver, Canada.

Month	Temperature
Jan	36
Feb	39
Mar	43
Apr	48
May	55
June	59
July	64
Aug	63
Sept	57
Oct	50
Nov	43
Dec	39

Source: Miller, A., and J. Thompson, *Elements of Meteorology,* Charles E. Merrill Publishing Company, Columbus, Ohio, 1975.

Do these temperatures have a pattern? Can we use graphs of the circular functions to model average monthly temperatures in Vancouver? Can temperatures for other cities like Phoenix, Arizona, and Buenos Aires, Argentina, be modeled using a common mathematical technique? The answers to these questions will be found in the examples and exercises of Section 4.2.

The average temperatures in Vancouver are coldest in January and warmest in July. These temperatures cycle yearly and may change only slightly over many years. Seasonal temperature changes occur periodically because Earth's axis is tilted and its orbit around the sun is nearly circular. When a phenomenon such as temperature results from circular periodic motion, the circular functions are often used to mathematically model the data. The graphs of these functions are essential in describing things like world temperatures and seasonal carbon dioxide levels. Their graphs will provide us with both a picture and a better understanding of periodic phenomena.

4.1 Graphs of the Sine and Cosine Functions ▼▼▼

Many things in daily life repeat with a predictable pattern: in warm areas electricity use goes up in the summer and down in the winter, the price of fresh fruit goes down in the summer and up in the winter, and attendance at amusement parks increases in the summer and declines in autumn. Because the sine and cosine functions repeat their values over and over in a regular pattern, they are examples of *periodic functions*. Figure 1 shows a sinusoid (sine graph) that represents a normal heartbeat.

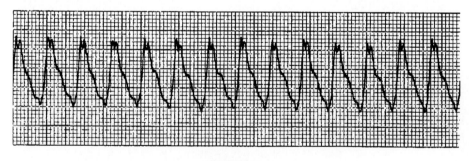

FIGURE 1

Periodic Function

A **periodic function** is a function f such that

$$f(x) = f(x + p),$$

for every real number x in the domain of f and for some positive real number p. The smallest possible positive value of p is the **period** of the function.

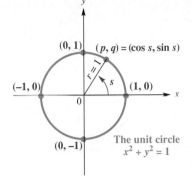

FIGURE 2

The circumference of the unit circle is 2π, so the smallest value of p for which the sine and cosine functions repeat is 2π. Therefore, the sine and cosine functions are periodic functions with period 2π.

In Section 3.3 we saw that if an arc of length s is traced along the unit circle $x^2 + y^2 = 1$, starting at the point $(1, 0)$, the terminal point of the arc has coordinates $(\cos s, \sin s)$. Let us examine the function $y = \sin s$. We saw that this function has domain $(-\infty, \infty)$ and range $[-1, 1]$. Since the sine function has period 2π, one period of $y = \sin s$ can be graphed using values of s from 0 to 2π. To graph this period, look at Figure 2, which shows a unit circle with a point (p, q) marked on it. Based on the definitions for circular functions given in Section 3.3, for any angle s, in radians, $p = \cos s$ and $q = \sin s$. As s increases from 0 to $\pi/2$ (or 90°), q (or $\sin s$) increases from 0 to 1, while p (or $\cos s$) decreases from 1 to 0.

As s increases from $\pi/2$ to π (or 180°), q decreases from 1 to 0, while p decreases from 0 to -1. Similar results can be found for the other quadrants, as shown in the following table.

As s Increases from	sin s	cos s
0 to $\pi/2$	increases from 0 to 1	decreases from 1 to 0
$\pi/2$ to π	decreases from 1 to 0	decreases from 0 to -1
π to $3\pi/2$	decreases from 0 to -1	increases from -1 to 0
$3\pi/2$ to 2π	increases from -1 to 0	increases from 0 to 1

Any letter could be used instead of s for the arc length. In the rest of this chapter we will use circular functions of x (rather than s), so that we are graphing on the familiar xy-coordinate system.

Selecting key values of x and finding the corresponding values of $\sin x$ give the following results. (Decimals are rounded to the nearest tenth.)

x	0	$\pi/4$	$\pi/2$	$3\pi/4$	π	$5\pi/4$	$3\pi/2$	$7\pi/4$	2π
$\sin x$	0	.7	1	.7	0	$-.7$	-1	$-.7$	0

Plotting the points from the table of values and connecting them with a smooth curve gives the solid portion of the graph in Figure 3. Since $y = \sin x$

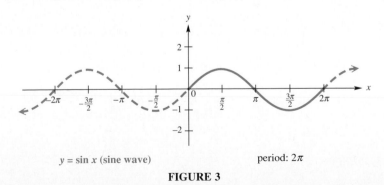

$y = \sin x$ (sine wave)   period: 2π

FIGURE 3

is periodic and has $(-\infty, \infty)$ as its domain, the graph continues in both directions indefinitely, as indicated by the arrows. This graph is sometimes called a **sine wave** or **sinusoid.** You should learn the shape of this graph and be able to sketch it quickly. The key points of the graph are $(0, 0)$, $(\pi/2, 1)$, $(\pi, 0)$, $(3\pi/2, -1)$, and $(2\pi, 0)$. By plotting these five points and connecting them with the characteristic sine wave, you can quickly sketch the graph.

The same scales are used on both the x and y axes of Figure 3 so as not to distort the graph. Since the period of $y = \sin x$ is 2π, it is convenient to use subdivisions of 2π on the x-axis. The more familiar x-values, 1, 2, 3, 4, and so on, are still present, but are usually not shown to avoid cluttering the graph. These values are shown in Figure 4.

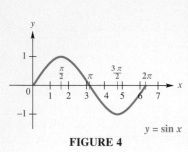

FIGURE 4

Sine graphs occur in many different practical applications. For one application, look back at Figure 2 and assume that the line from the origin to the point (p, q) is part of the pedal of a bicycle wheel, with a foot placed at (p, q). As mentioned earlier, q is equal to $\sin x$, showing that the height of the pedal from the horizontal axis in Figure 2 is given by $\sin x$. By choosing various angles for the pedal and calculating q for each angle, the height of the pedal leads to the sine curve shown in Figure 5. Two sample points are shown in Figure 5.

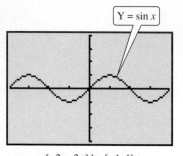

$[-2\pi, 2\pi]$ by $[-4, 4]$

$\text{Xscl} = \dfrac{\pi}{2}$ $\text{Yscl} = 1$

The graph of $Y = \sin x$ is shown in the standard *trig window*, having the dimensions and scales indicated.

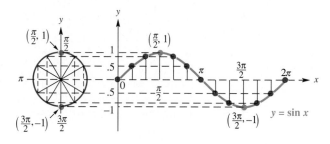

FIGURE 5

The graph of $y = \cos x$ can be found in much the same way as the graph of $y = \sin x$ was found. The domain of cosine is $(-\infty, \infty)$, and the range of $y = \cos x$ is $[-1, 1]$. A table of values is shown below for $y = \cos x$.

x	0	$\pi/4$	$\pi/2$	$3\pi/4$	π	$5\pi/4$	$3\pi/2$	$7\pi/4$	2π
$\cos x$	1	.7	0	$-.7$	-1	$-.7$	0	.7	1

Here the key points are $(0, 1)$, $(\pi/2, 0)$, $(\pi, -1)$, $(3\pi/2, 0)$, and $(2\pi, 1)$.

The graph of $y = \cos x$, in Figure 6, has the same shape as the graph of $y = \sin x$. In fact, it is the graph of the sine function, shifted $\pi/2$ units to the left.

The examples in the rest of this section show graphs that are "stretched" either vertically, horizontally, or both when compared with the graphs of $y = \sin x$ or $y = \cos x$.

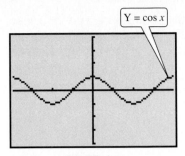

Trig Window
This is a calculator-generated graph of the cosine function.

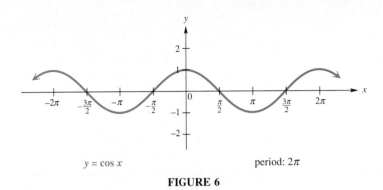

$y = \cos x$ period: 2π

FIGURE 6

◖EXAMPLE 1
Graphing $y = a \sin x$

Graph $y = 2 \sin x$.

For a given value of x, the value of y is twice as large as it would be for $y = \sin x$, as shown in the table of values. The only change in the graph is the range, which becomes $[-2, 2]$. See Figure 7, which also shows a graph of $y = \sin x$ for comparison.

x	0	$\pi/2$	π	$3\pi/2$	2π
$\sin x$	0	1	0	-1	0
$2 \sin x$	0	2	0	-2	0

◗

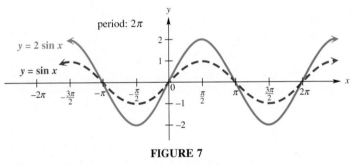

FIGURE 7

NOTE When graphing periodic functions, it is customary to first graph over one period. Then more of the graph can be sketched using the fact that the graph repeats the cycle over and over.

Generalizing from Example 1 gives the following.

Amplitude of Sine and Cosine

The graph of $y = a \sin x$ or $y = a \cos x$, with $a \neq 0$, will have the same shape as the graph of $y = \sin x$ or $y = \cos x$, respectively, except with range $[-|a|, |a|]$. The number $|a|$ is called the **amplitude.** (The amplitude of a periodic function can be interpreted as half the difference between its maximum and minimum values.)

No matter what the value of the amplitude, the periods of $y = a \sin x$ and $y = a \cos x$ are still 2π. Now suppose $y = \sin 2x$. We can complete a table of values for the interval $[0, 2\pi]$.

x	0	$\pi/4$	$\pi/2$	$3\pi/4$	π	$5\pi/4$	$3\pi/2$	$7\pi/4$	2π
$\sin 2x$	0	1	0	-1	0	1	0	-1	0

The period here is π, which equals $2\pi/2$. What about $y = \sin 4x$? Look at the table below.

x	0	$\pi/8$	$\pi/4$	$3\pi/8$	$\pi/2$	$5\pi/8$	$3\pi/4$	$7\pi/8$	π
$\sin 4x$	0	1	0	-1	0	1	0	-1	0

These values suggest that a complete cycle is achieved in $\pi/2$ units, which is reasonable since

$$\sin\left(4 \cdot \frac{\pi}{2}\right) = \sin 2\pi = 0.$$

Throughout this chapter we assume $b > 0$. If a function has $b < 0$, then the identities of the next chapter can be used to change the function to one in which $b > 0$. In general, the graph of a function of the form $y = \sin bx$ or $y = \cos bx$, for $b > 0$, will have a period different from 2π when $b \neq 1$. To see why this is so, remember that the values of $\sin bx$ or $\cos bx$ will take on all possible values as bx ranges from 0 to 2π. Therefore, to see what the period of either of these will be, we must solve the compound inequality

$$0 \leq bx \leq 2\pi.$$

Dividing by the positive number b gives

$$0 \leq x \leq \frac{2\pi}{b}.$$

Therefore, the period is $2\pi/b$. By dividing the interval $[0, 2\pi/b]$ into four equal parts, we obtain the values for which $\sin bx$ or $\cos bx$ is $-1, 0,$ or 1. These will give minimum points, x-intercepts, and maximum points on the graph. Once these points are determined, the graph can be completed by joining the points with a smooth sinusoidal curve.

NOTE To divide an interval into four equal parts, we find the three middle values as follows:

1. Find the midpoint of the interval.
2. Find the two midpoints of the intervals found in Step 1.

EXAMPLE 2
Graphing $y = \sin bx$

Graph $y = \sin 2x$.

For this function, $b = 2$, so the period is $2\pi/2 = \pi$. Therefore, the graph will complete one period over the interval $[0, \pi]$.

The endpoints are 0 and π, and the three middle points are

$$\frac{1}{4}(0 + \pi), \frac{1}{2}(0 + \pi), \text{and} \frac{3}{4}(0 + \pi),$$

which give the following x-values.

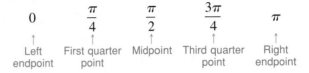

0	$\dfrac{\pi}{4}$	$\dfrac{\pi}{2}$	$\dfrac{3\pi}{4}$	π
↑	↑	↑	↑	↑
Left endpoint	First quarter point	Midpoint	Third quarter point	Right endpoint

We now plot the points from the table of values given earlier, and join them with a smooth sinusoidal curve. More of the graph can be sketched by repeating this cycle over and over, as shown in Figure 8. Notice that the amplitude is not changed. The graph of $y = \sin x$ is included for comparison. ▶

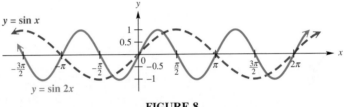

FIGURE 8

Generalizing from Example 2 leads to the following result.

> **Period of Sine and Cosine Functions**
>
> For $b > 0$, the graph of $y = \sin bx$ will look like that of $y = \sin x$, but with a period of $2\pi/b$. Also, the graph of $y = \cos bx$ will look like that of $y = \cos x$, but with a period of $2\pi/b$.

EXAMPLE 3
Graphing $y = \cos bx$

Graph $y = \cos(2/3)x$.

For this function the period is

$$\frac{2\pi}{2/3} = 3\pi.$$

Divide the interval $[0, 3\pi]$ into four equal parts to get the x-values that will yield minimum points, maximum points, and x-intercepts. These x-values are as follows.

$$0 \qquad \frac{3\pi}{4} \qquad \frac{3\pi}{2} \qquad \frac{9\pi}{4} \qquad 3\pi$$

These values are used to get a table of key points for one period.

x	0	$3\pi/4$	$3\pi/2$	$9\pi/4$	3π
$(2/3)x$	0	$\pi/2$	π	$3\pi/2$	2π
$\cos(2/3)x$	1	0	-1	0	1

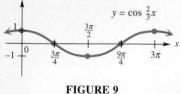

FIGURE 9

The amplitude is 1, because the maximum value is 1, the minimum value is -1, and half of $1 - (-1) = (1/2)(2) = 1$.

Now plot these points and join them with a smooth curve. The graph is shown in Figure 9. ▶

NOTE Look at the middle row of the table in Example 3. The method of dividing the interval $[0, 2\pi/b]$ into four equal parts will always give the values 0, $\pi/2$, π, $3\pi/2$, and 2π for this row, resulting in values of -1, 0, or 1 for the circular function. These lead to key points on the graph, which can then be easily sketched.

The steps used to graph $y = a \sin bx$ or $y = a \cos bx$, where $b > 0$, are given below.

Graphing the Sine and Cosine Functions

To graph $y = a \sin bx$ or $y = a \cos bx$, with $b > 0$:

1. Find the period, $2\pi/b$. Start at 0 on the x-axis and lay off a distance of $2\pi/b$.
2. Divide the interval into four equal parts.
3. Evaluate the function for each of the five x-values resulting from Step 2. The points will be maximum points, minimum points, and x-intercepts.
4. Plot the points found in Step 3, and join them with a sinusoidal curve with amptitude $|a|$.
5. Draw additional cycles of the graph, to the right and to the left, as needed.

The functions in Examples 4 and 5 have both amplitude and period affected by constants.

EXAMPLE 4
Graphing $y = a \sin bx$

Graph $y = -2 \sin 3x$.

Step 1   For this function, $b = 3$, so the period is $2\pi/3$. We will first graph the function over the interval $[0, 2\pi/3]$.

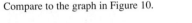

FIGURE 10

$\left[0, \frac{2\pi}{3}\right]$ by $[-2.5, 2.5]$

$\text{Xscl} = \frac{\pi}{6}$ $\text{Yscl} = 1$

Compare to the graph in Figure 10.

Step 2 Dividing the interval $[0, 2\pi/3]$ into four equal parts gives the *x*-values 0, $\pi/6$, $\pi/3$, $\pi/2$, and $2\pi/3$.

Step 3 Make a table of points determined by the *x*-values resulting from Step 2.

x	0	$\pi/6$	$\pi/3$	$\pi/2$	$2\pi/3$
$3x$	0	$\pi/2$	π	$3\pi/2$	2π
$\sin 3x$	0	1	0	-1	0
$-2 \sin 3x$	0	-2	0	2	0

Step 4 Plot the points $(0, 0)$, $(\pi/6, -2)$, $(\pi/3, 0)$, $(\pi/2, 2)$, and $(2\pi/3, 0)$, and join them with a sinusoidal curve with amplitude 2. See Figure 10.

Step 5 If necessary, the graph in Figure 10 can be extended by repeating the cycle over and over.

Notice the effect of the negative value of *a*. When *a* is negative, the graph of $y = a \sin bx$ will be the reflection about the *x*-axis of the graph of $y = |a| \sin bx$. The amplitude here is $|-2| = 2$. ▶

EXAMPLE 5
Graphing $y = a \cos bx$

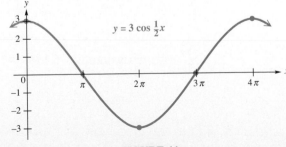

$[0, 4\pi]$ by $[-3.5, 3.5]$
$\text{Xscl} = \pi$ $\text{Yscl} = 1$

Compare to the graph in Figure 11.

Graph $y = 3 \cos \frac{1}{2}x$.

The period is $2\pi/(1/2) = 4\pi$. The key points have *x*-values of

$$0, \qquad \frac{1}{4}(4\pi) = \pi, \qquad \frac{1}{2}(4\pi) = 2\pi,$$

$$\frac{3}{4}(4\pi) = 3\pi, \qquad \text{and} \qquad 4\pi.$$

Evaluating the function for these *x*-values gives the following points.

$$(0, 3) \qquad (\pi, 0) \qquad (2\pi, -3) \qquad (3\pi, 0) \qquad (4\pi, 3)$$

Figure 11 shows these points joined with a smooth curve. The amplitude is 3. ▶

FIGURE 11

The graphing calculator can provide an enlightening look at how the constants a and b affect the graph of a function of the form $y = a \sin bx$ or $y = a \cos bx$. Be sure the calculator is set for radians. Some graphing calculators have a built-in "Trig" window with domain $[-2\pi, 2\pi]$, range $[-4, 4]$, Xscl $= \pi/4$, and Yscl $= 1$. Of course, other settings may be preferable in some cases. By graphing pairs of functions such as $y = \sin x$ and $y = 2 \sin x$, or $y = \sin x$ and $y = -2 \sin x$, we can observe the effect of the coefficient a. See the figures below.

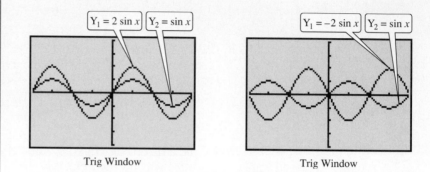

Trig Window   Trig Window

To determine the effect of the positive constant b on the graph of $y = \cos bx$ or $y = \sin bx$, keep the graph of $y = \sin x$ as Y_1 and enter $Y_2 = \sin 2x$, then replace Y_2 with $Y_3 = -\sin 2x$. Notice how the graphs have the same basic shape, but the periods of Y_2 and Y_3 are half that of Y_1. See the figures below.

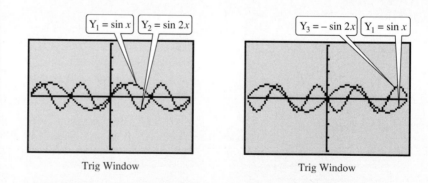

Trig Window   Trig Window

We can also use the graph of $y = \sin x$, for example, to find ordered pairs that satisfy the function, by using the TRACE feature. One such pair is

$$X = 1.1243595, \quad Y = .90199124.$$

This means that $\sin 1.1243595 \approx .90199124$.

EXAMPLE 6
Interpreting a sine
function model

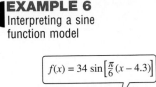

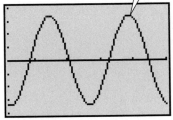

[1, 25] by [−40, 40]
Xscl = 5   Yscl = 10

FIGURE 12

The average temperature (in °F) at Mould Bay, Canada, can be approximated by the circular function

$$f(x) = 34 \sin\left[\frac{\pi}{6}(x - 4.3)\right],$$

where x is the month and $x = 1$ corresponds to January.

(a) Graph f over the interval $1 \le x \le 25$. Determine the amplitude and period of the graph.

The graph is shown in Figure 12. Its amplitude is 34 and the period is

$$\frac{2\pi}{\frac{\pi}{6}} = 12.$$

The function f has a period of 12 months or 1 year which agrees with the changing of the seasons.

(b) What is the average temperature during the month of May?

May is the 5th month, so the average temperature during the month of May is

$$f(5) = 34 \sin\left[\frac{\pi}{6}(5 - 4.3)\right] \approx 12°F.$$

(c) What would be an approximation for the average *yearly* temperature in Mould Bay?

From the graph it appears that the average yearly temperature is about 0°F, since the graph is centered vertically about the line $y = 0$. ▶

4.1 Exercises ▼▼▼▼▼▼▼▼▼▼▼▼▼▼▼▼▼▼▼▼▼▼▼▼▼▼▼▼▼▼▼▼▼▼▼

1. Describe the graph of $y = \sin x$ as if you were explaining it to a friend on the phone.

2. Describe the graph of $y = \cos x$ as if you were explaining it to a friend on the phone.

3. Explain why a graphing calculator will return a 1 for the expression shown in the accompanying screen, no matter what value is stored for X.

4. Give two values in the interval $[0, 2\pi)$ that can be stored in X for the accompanying calculator screen to be obtained. Support your result with your own calculator.

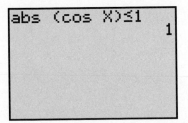

Simulate the accompanying screens on your graphing calculator and then work Exercises 5–8. Place your calculator in parametric mode (see your calculator manual) and set the window shown at the left below. (The values for Tmax *and* Xmax *were entered as* 2π *and the value of* Tstep *was entered as* $\pi/40$.) *Set the functions as shown in the second screen and then graph the functions. The screen will appear as shown at the right below.*

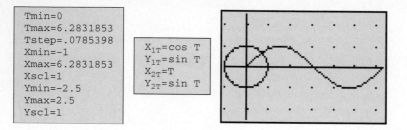

```
Tmin=0
Tmax=6.2831853
Tstep=.0785398
Xmin=-1
Xmax=6.2831853
Xscl=1
Ymin=-2.5
Ymax=2.5
Yscl=1
```

```
X₁ₜ=cos T
Y₁ₜ=sin T
X₂ₜ=T
Y₂ₜ=sin T
```

5. Invoke the TRACE feature and then press the right-arrow key five times. The value of T will be $\pi/8$, since $5 \cdot \pi/40$ is $\pi/8$. What are $\cos \pi/8$ and $\sin \pi/8$?

6. Press the up-arrow key once to move the point on the circle to a point on the graph of $y = \sin x$. How are the two points related?

7. What point on the unit circle corresponds to the highest point of the graph of $y = \sin x$?

8. In the second screen above, change $Y_{2T} = \sin T$ to $Y_{2T} = \cos T$. Graph the functions to obtain the circle and the cosine function. Invoke the TRACE feature, press the right-arrow key a few times, and then press the up-arrow key once to move the point on the circle to a point on the graph of $y = \cos x$. How are the two points related?

Graph the defined function over the interval $[-2\pi, 2\pi]$. *Give the amplitude. See Example 1.*

9. $y = 2 \cos x$ 10. $y = 3 \sin x$ 11. $y = \dfrac{2}{3} \sin x$ 12. $y = \dfrac{3}{4} \cos x$

13. $y = -\cos x$ 14. $y = -\sin x$ 15. $y = -2 \sin x$ 16. $y = -3 \cos x$

Graph each defined function over a two-period interval. Give the period and the amplitude. See Examples 2–5.

17. $y = \sin \dfrac{1}{2}x$ 18. $y = \sin \dfrac{2}{3}x$ 19. $y = \cos \dfrac{1}{3}x$ 20. $y = \cos \dfrac{3}{4}x$

21. $y = \sin 3x$ 22. $y = \cos 2x$ 23. $y = 2 \sin \dfrac{1}{4}x$ 24. $y = 3 \sin 2x$

25. $y = -2 \cos 3x$ 26. $y = -5 \cos 2x$ 27. $y = \cos \pi x$ 28. $y = -\sin \pi x$

Match each defined function with its graph in Exercises 29–36.

29. $y = \sin x$

30. $y = \cos x$

31. $y = -\sin x$

32. $y = -\cos x$

33. $y = \sin 2x$

34. $y = \cos 2x$

35. $y = 2 \sin x$

36. $y = 2 \cos x$

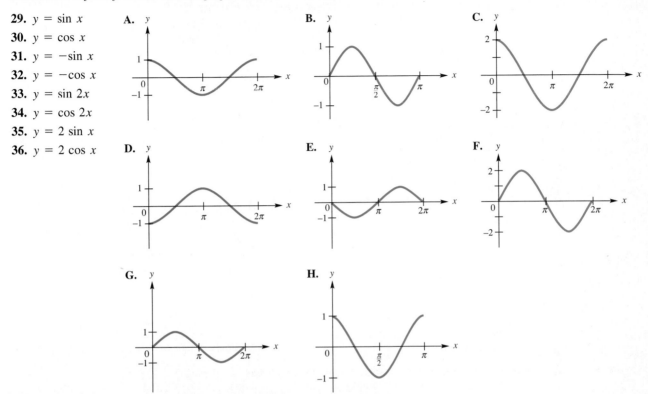

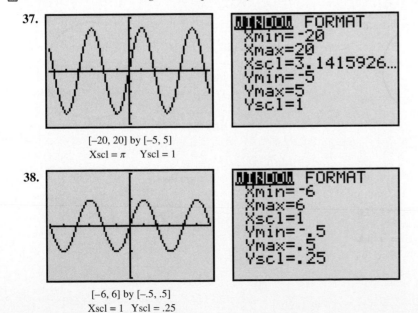

In Exercises 37 and 38, give the equation of a sine function having the given graph.

37.

[−20, 20] by [−5, 5]
Xscl = π Yscl = 1

38.

[−6, 6] by [−.5, .5]
Xscl = 1   Yscl = .25

Exercises 39 and 40 refer to the table. The numbers in the X column are 0, $\pi/6$, $\pi/3$, $\pi/2$, . . . (that is, 0°, 30°, 60°, 90°, . . .). The functions determined by Y_1 and Y_2 have the form $y = a \sin bx$ or $y = a \cos bx$.

X	Y₁	Y₂
0	3	0
.5236	1.5	2
1.0472	-1.5	2
1.5708	-3	0
2.0944	-1.5	-2
2.618	1.5	0
3.1416	3	2

X=0

39. Determine a possible expression for the function determined by Y_1.

40. Determine a possible expression for the function determined by Y_2.

41. Compare the graphs of $y = \sin 2x$ and $y = 2 \sin x$ over the interval $[0, 2\pi]$. Can we say that, in general, $\sin bx = b \sin x$? Explain.

42. Compare the graphs of $y = \cos 3x$ and $y = 3 \cos x$ over the interval $[0, 2\pi]$. Can we say that, in general, $\cos bx = b \cos x$? Explain.

43. Refer to the graph of $y = \sin x$ in Figure 4. The graph completes one cycle between $x = 0$ and $x = 2\pi$. Consider the statement, "The function $y = \sin(bx)$ completes b cycles between 0 and 2π." Use your graphing calculator to confirm the statement for some positive integer values of b, such as 3, 4, and 5. Interpret and confirm the statement for $b = 1/2$ and $b = 3/2$.

44. The graph shown gives the variation in blood pressure for a typical person. Systolic and diastolic pressures are the upper and lower limits of the periodic changes in pressure that produce the pulse. The length of time between peaks is called the period of the pulse.

 (a) Find the amplitude of the graph.

 (b) Find the pulse rate (the number of pulse beats in one minute) for this person.

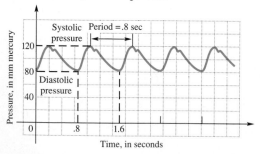

45. Scientists believe that the average annual temperature in a given location is periodic. The overall temperature at a given place during a given season fluctuates as time goes on, from colder to warmer, and back to colder. The graph shows an idealized description of the tem-

perature for the last few thousand years of a location at the same latitude as Anchorage.

 (a) Find the highest and lowest temperatures recorded.

 (b) Use these two numbers to find the amplitude. (*Hint:* An alternative definition of the amplitude is half the difference between the y-values of the highest and lowest points on the graph.)

 (c) Find the period of the graph.

 (d) What is the trend of the temperature now?

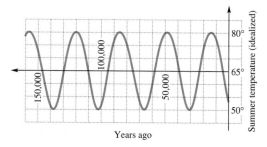

46. Many of the activities of living organisms are periodic. For example, the graph below shows the time that a certain nocturnal animal begins its evening activity.

 (a) Find the amplitude of this graph.

 (b) Find the period.

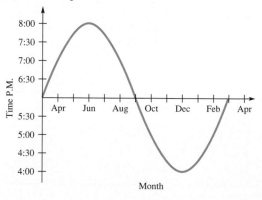

47. The figure shows schematic diagrams of a rhythmically moving arm. The upper arm *RO* rotates back and forth about the point *R*; the position of the arm is measured by the angle *y* between the actual position and the downward vertical position. (*Source:* De Sapio, Rodolfo, *Calculus for the Life Sciences.* Copyright © 1978 by W. H. Freeman and Company. Reprinted by permission.)

(a) Find an equation of the form $y = a \sin kt$ for the graph shown.

(b) How long does it take for a complete movement of the arm?

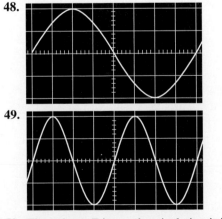

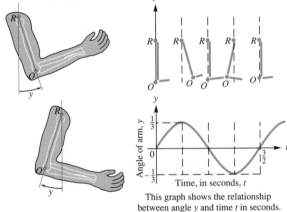

This graph shows the relationship between angle *y* and time *t* in seconds.

Pure sounds produce single sine waves on an oscilloscope. Find the amplitude and period of each sine wave in the following photographs. On the vertical scale, each square represents .5, and on the horizontal scale each square represents 30° or π/6.

48.

49.

50. The voltage *E* in an electrical circuit is given by

$$E = 5 \cos 120\pi t,$$

where *t* is time measured in seconds.

(a) Find the amplitude and the period.

(b) How many cycles are completed in one second? (The number of cycles (periods) completed in one second is the **frequency** of the function.)

(c) Find *E* when $t = 0, .03, .06, .09, .12$.

(d) Graph *E* for $0 \le t \le 1/30$.

51. For another electrical circuit, the voltage *E* is given by

$$E = 3.8 \cos 40\pi t,$$

where *t* is time measured in seconds.

(a) Find the amplitude and the period.

(b) Find the frequency. See Exercise 50(b).

(c) Find *E* when $t = .02, .04, .08, .12, .14$.

(d) Graph one period of *E*.

52. At Mauna Loa, Hawaii, atmospheric carbon dioxide levels in parts per million (ppm) have been measured regularly since 1958. The function defined by

$$L(x) = .022x^2 + .55x + 316 + 3.5 \sin(2\pi x)$$

can be used to model these levels, where *x* is in years and $x = 0$ corresponds to 1960. (*Source:* Nilsson, A., *Greenhouse Earth,* John Wiley & Sons, New York, 1992.)

(a) Graph *L* for $15 \le x \le 35$. (*Hint:* Use $325 \le y \le 365$.)

(b) When do the seasonal maximum and minimum carbon dioxide levels occur?

(c) *L* is the sum of a quadratic function and a sine function. What is the significance of each of these functions? Discuss what physical phenomena may be responsible for each function.

53. Refer to the previous exercise. The carbon dioxide content in the atmosphere at Barrow, Alaska, in parts per million (ppm) can be modeled using the function defined by

$$C(x) = .04x^2 + .6x + 330 + 7.5 \sin(2\pi x),$$

where $x = 0$ corresponds to 1970. (*Source:* Zeilik, M., S. Gregory, and E. Smith, *Introductory Astronomy and Astrophysics,* Saunders College Publishing, 1992.)

(a) Graph *C* for $5 \le x \le 25$. (*Hint:* Use $320 \le y \le 380$.)

(b) Discuss possible reasons why the amplitude of the oscillations in the graph of *C* are larger than the amplitude of the oscillations in the graph of *L* in Exercise 52, which models Hawaii.

(c) Define a new *C* function that is valid if *x* represents the actual year where $1970 \le x \le 1995$.

4.2 Translating Graphs of the Sine and Cosine Functions ▼▼▼

In Section 4.1 we studied the basic graphs of $y = \sin x$ and $y = \cos x$, and observed that the constants a and b affect the graphs of

$$y = a \sin bx \qquad \text{and} \qquad y = a \cos bx$$

by changing the amplitude and period, respectively. In this section we will see how the constants c and d affect the graphs of

$$y = c + a \sin b(x - d) \qquad \text{and} \qquad y = c + a \cos b(x - d).$$

HORIZONTAL TRANSLATIONS In general, the graph of the function $y = f(x - d)$ is translated *horizontally* when compared to the graph of $y = f(x)$. The translation is d units to the right if $d > 0$ and $|d|$ units to the left if $d < 0$. See Figure 13. With circular functions, a horizontal translation is called a **phase shift**. In the function $y = f(x - d)$, the expression $x - d$ is called the **argument**.

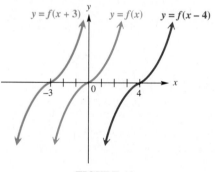

FIGURE 13

In the first example, we show two methods that can be used to graph a circular function involving a phase shift.

◀**EXAMPLE 1**
Graphing $y = \sin(x - d)$

Graph $y = \sin\left(x - \dfrac{\pi}{3}\right)$.

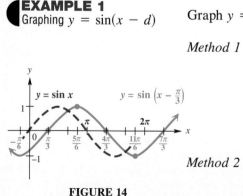

FIGURE 14

Method 1 The argument $x - \pi/3$ indicates that the graph will be translated $\pi/3$ units to the *right* (the phase shift) as compared to the graph of $y = \sin x$. Notice that in Figure 14 the graph of $y = \sin x$ is shown as a dashed curve, and the graph of $y = \sin(x - \pi/3)$ is shown as a solid curve. Therefore, to graph a function using this method, first graph the basic circular function, and then graph the desired function by using the appropriate translation.

Method 2 For the argument $x - \pi/3$ to result in all possible values throughout one period, it must take on all values between 0 and 2π, inclusive. Therefore, to find an interval of one period, we solve the compound inequality

$$0 \le x - \frac{\pi}{3} \le 2\pi.$$

Add $\pi/3$ to each expression to find the interval

$$\frac{\pi}{3} \le x \le \frac{7\pi}{3} \qquad \text{or} \qquad \left[\frac{\pi}{3}, \frac{7\pi}{3}\right].$$

As first shown in Section 4.1, divide this interval into four equal parts, getting the following values.

$$\frac{\pi}{3} \qquad \frac{5\pi}{6} \qquad \frac{4\pi}{3} \qquad \frac{11\pi}{6} \qquad \frac{7\pi}{3}$$

Make a table of points using the x-values above.

x	$\pi/3$	$5\pi/6$	$4\pi/3$	$11\pi/6$	$7\pi/3$
$x - \pi/3$	0	$\pi/2$	π	$3\pi/2$	2π
$\sin(x - \pi/3)$	0	1	0	-1	0

Join these points to get the graph shown in Figure 14. The period is 2π and the amplitude is 1. ▶

EXAMPLE 2
Graphing $y = a \cos(x - d)$

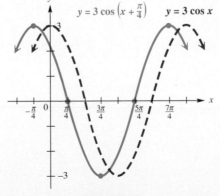

$\left[-\frac{\pi}{4}, \frac{7\pi}{4}\right]$ by $[-3.5, 3.5]$

$Xscl = \frac{\pi}{4}$ $Yscl = 1$

The graph of Y_2 is obtained by shifting the graph of Y_1 $\frac{\pi}{4}$ units to the left.

Graph $y = 3 \cos\left(x + \frac{\pi}{4}\right)$.

Start by writing $3 \cos(x + \pi/4)$ in the form $a \cos(x - d)$.

$$3 \cos\left(x + \frac{\pi}{4}\right) = 3 \cos\left[x - \left(-\frac{\pi}{4}\right)\right]$$

This result shows that $d = -\pi/4$. Since $-\pi/4$ is negative, the phase shift is $|-\pi/4| = \pi/4$ to the left. The period is 2π and the amplitude is 3. The graph is the same as that of $y = 3 \cos x$, except that it is shifted $\pi/4$ units to the left. See Figure 15.

FIGURE 15

Alternatively, the graph can be sketched by first solving the inequality

$$0 \le x + \frac{\pi}{4} \le 2\pi.$$

Adding $-\pi/4$ to each expression gives

$$-\frac{\pi}{4} \le x \le \frac{7\pi}{4},$$

an interval over which one period of the function can be graphed. Dividing this interval into four equal parts gives x-values of $-\pi/4$, $\pi/4$, $3\pi/4$, $5\pi/4$, and $7\pi/4$. A table of points for these x-values once again leads to maximum points, minimum points, and x-intercepts.

x	$-\pi/4$	$\pi/4$	$3\pi/4$	$5\pi/4$	$7\pi/4$
$x + \pi/4$	0	$\pi/2$	π	$3\pi/2$	2π
$3 \cos(x + \pi/4)$	3	0	-3	0	3

This alternative method produces the same graph as the one shown in Figure 15. ▶

The next example shows a function of the form $y = a \cos b(x - d)$. Such functions have both a phase shift (if $d \ne 0$) and a period different from 2π (if $|b| \ne 1$).

EXAMPLE 3
Graphing $y = a \cos b(x - d)$

Graph $y = -2 \cos(3x + \pi)$.

First write the expression in the form $a \cos b(x - d)$ by factoring 3 out of the argument as follows.

$$y = -2 \cos(3x + \pi) = -2 \cos 3\left(x + \frac{\pi}{3}\right)$$

Then $a = -2$, $b = 3$, and $d = -\pi/3$. The amplitude is $|-2| = 2$, and the period is $2\pi/3$ (since the value of b is 3). The phase shift is $|-\pi/3| = \pi/3$ to the left as compared to the graph of $y = -2 \cos 3x$.

The function can be sketched over one period by solving the compound inequality

$$0 \le 3\left(x + \frac{\pi}{3}\right) \le 2\pi$$

to get the interval $[-\pi/3, \pi/3]$. Divide this interval into four equal parts to get the following points.

$$\left(-\frac{\pi}{3}, -2\right) \qquad \left(-\frac{\pi}{6}, 0\right) \qquad (0, 2) \qquad \left(\frac{\pi}{6}, 0\right) \qquad \left(\frac{\pi}{3}, -2\right)$$

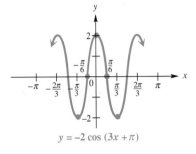

The points found in Example 3 can be obtained using the TABLE feature of a graphing calculator.

Plot these points and then join them with a smooth curve. By graphing an additional half period to the left and to the right, we obtain the sketch shown in Figure 16. ◗

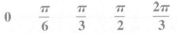

FIGURE 16

VERTICAL TRANSLATIONS   The graph of a function of the form $y = c + f(x)$ is shifted *vertically* as compared with the graph of $y = f(x)$. See Figure 17. The function $y = c + f(x)$ is called a **vertical translation** of $y = f(x)$. The next example illustrates a vertical translation of a circular function.

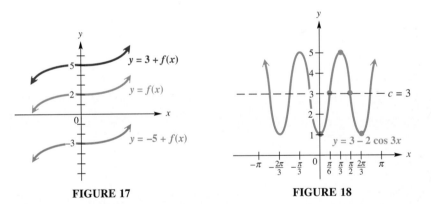

FIGURE 17   **FIGURE 18**

◖**EXAMPLE 4**

Graph $y = 3 - 2 \cos 3x$.

The values of y will be 3 greater than the corresponding values of y in $y = -2 \cos 3x$. This means that the graph of $y = 3 - 2 \cos 3x$ is the same as the graph of $y = -2 \cos 3x$, except with a vertical translation of 3 units upward. Since the period of $y = -2 \cos 3x$ is $2\pi/3$, the key points have the following x-values.

$$0 \quad \frac{\pi}{6} \quad \frac{\pi}{3} \quad \frac{\pi}{2} \quad \frac{2\pi}{3}$$

The key points are shown on the graph in Figure 18, along with more of the graph, which can be sketched by keeping in mind the fact that it is a periodic function. ◗

▧ The graphing calculator also helps to reinforce the concepts of horizontal and vertical translations as seen in Figures 14 and 18 earlier. Again, let Y_1 remain as sin x, and enter Y_2 as sin$(x − \pi/3)$, taking care to insert parentheses as necessary. The graph of Y_2 should be the same as the graph of Y_1 shifted $\pi/3$ units to the right. See the figure on the left.

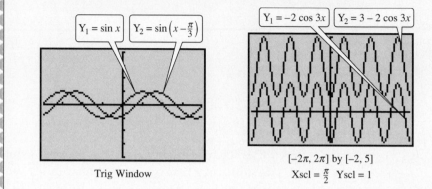

$[-2\pi, 2\pi]$ by $[-2, 5]$
Xscl $= \frac{\pi}{2}$ Yscl $= 1$

Trig Window

To see the effect of a vertical translation, enter $Y_1 = −2 \cos 3x$ and $Y_2 = 3 − 2 \cos 3x$. The constant $c = 3$ in Y_2 will have the effect of shifting the graph of Y_1 three units upward. Adjust the range so that the minimum and maximum values of y are $−2$ and 5, respectively. Now graph these two functions and observe the results, shown in the figure on the right.

In the next example we graph a function that involves all the types of stretching, compressing, and shifting studied in the previous section and this one.

EXAMPLE 5
Graphing $y = c +$
$a \sin b(x − d)$

Graph $y = −1 + 2 \sin 4\left(x + \dfrac{\pi}{4}\right)$.

Here, the amplitude is 2, the period is $2\pi/4 = \pi/2$, the graph is translated down 1 unit and $\pi/4$ units to the left as compared to the graph of $y = \sin x$. Since the graph is translated $\pi/4$ units to the left, start at the x-value $0 − \pi/4 = −\pi/4$. The first period will end at $−\pi/4 + \pi/2 = \pi/4$. The maximum y-values will be $2 − 1 = 1$ and the minimum y-values will be $−2 − 1 = −3$. Sketch the graph using the typical sine curve. See Figure 19.

Alternatively, start by finding an interval over which the graph will complete one cycle. To do this, use the argument $4(x + \pi/4)$ in a compound in-

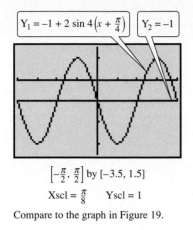

$\left[-\frac{\pi}{2}, \frac{\pi}{2}\right]$ by $[-3.5, 1.5]$

Xscl = $\frac{\pi}{8}$ Yscl = 1

Compare to the graph in Figure 19.

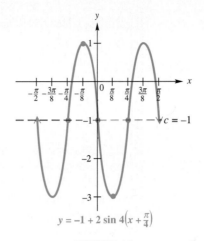

$$y = -1 + 2 \sin 4\left(x + \frac{\pi}{4}\right)$$

FIGURE 19

equality, with 0 as one endpoint and 2π as the other. Then solve the inequality for x.

$$0 \le 4\left(x + \frac{\pi}{4}\right) \le 2\pi$$

$$0 \le x + \frac{\pi}{4} < \frac{\pi}{2} \qquad \text{Divide by 4.}$$

$$-\frac{\pi}{4} \le x \le \frac{\pi}{4} \qquad \text{Subtract } \frac{\pi}{4}.$$

Divide the interval $[-\pi/4, \pi/4]$ into four equal parts to find the key points on the graph, as shown in the following table.

x	$-\pi/4$	$-\pi/8$	0	$\pi/8$	$\pi/4$
$x + \pi/4$	0	$\pi/8$	$\pi/4$	$3\pi/8$	$\pi/2$
$4(x + \pi/4)$	0	$\pi/2$	π	$3\pi/2$	2π
$-1 + 2 \sin 4(x + \pi/4)$	-1	1	-1	-3	-1

Join the key points with a smooth curve as shown in Figure 19. This function has period $\pi/2$ and amplitude $[1 - (-3)]/2 = 2$. The phase shift is $\pi/4$ units to the left, and there is a vertical translation of 1 unit down. ▶

Graphing General Sine and Cosine Functions

To graph the general function $y = c + a \sin b(x - d)$ or $y = c + a \cos b(x - d)$, where $b > 0$, follow these steps.

1. Find an interval whose length is one period $(2\pi/b)$ by solving the compound inequality

$$0 \le b(x - d) \le 2\pi.$$

2. Divide the interval into four equal parts.
3. Evaluate the function for each of the five x-values resulting from Step 2. The points will be maximum points, minimum points, and points that intersect the line $y = c$ ("middle" points of the wave).
4. Plot the points found in Step 3, and join them with a sinusoidal curve.
5. Draw the graph over additional periods, to the right and to the left, as needed.

The amplitude of the function is $|a|$. The vertical translation is c units up if $c > 0$, $|c|$ units down if $c < 0$. The horizontal translation (phase shift) is d units to the right if $d > 0$, and $|d|$ units to the left if $d < 0$.

CONNECTIONS You have probably noticed that the graphs of sin x and cos x are the same shape and each is a horizontal translation (or phase shift) of the other. Because of this we can rewrite any cosine function as a sine function or any sine function as a cosine function. The table of values below was produced by a graphing calculator. It shows that the corresponding y-values for $Y_1 = 2 \sin 2x$ and $Y_2 = 2 \cos 2(x - \pi/4)$ are the same in the intervals shown.

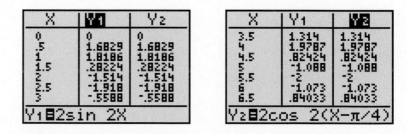

FOR DISCUSSION OR WRITING
1. Graph these two functions and compare their graphs. What do you expect to find?
2. Find a sine function that is equivalent to $y = .5 \cos(x - \pi)$.

EXAMPLE 6
Modeling temperature with a
sine function

The maximum average monthly temperature in New Orleans is 82°F and the minimum is 54°F. The table shows the average monthly temperatures.

Month	Temperature
Jan	54
Feb	55
Mar	61
Apr	69
May	73
June	79
July	82
Aug	81
Sept	77
Oct	71
Nov	59
Dec	55

(a) Using only the maximum and minimum temperatures, determine a function of the form $f(x) = a \sin b(x - d) + c$, where a, b, c, and d are constants, that models the average monthly temperature in New Orleans. Let x represent the month, with January corresponding to $x = 1$.

We can use the maximum and minimum average monthly temperatures to find the amplitude a.

$$a = \frac{82 - 54}{2} = 14$$

The average of the maximum and minimum temperatures is a good choice for c. The average is

$$\frac{82° + 54°}{2} = 68°F.$$

Since the coldest month is January, when $x = 1$, and the hottest month is July, when $x = 7$, we should choose d to be about 4. The table shows that temperatures are actually a little warmer after July than before, so we try $d = 4.2$. Since temperatures repeat every 12 months, b is $2\pi/12 = \pi/6$. Thus,

$$f(x) = a \sin b(x - d) + c = 14 \sin\left[\frac{\pi}{6}(x - 4.2)\right] + 68.$$

(b) On the same coordinate axes, graph f for a two-year period together with the actual data values found in the table.

We show a graphing calculator graph of f together with the data points in Figure 20. The function models the data quite accurately. ▶

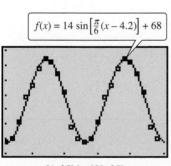

$f(x) = 14 \sin\left[\frac{\pi}{6}(x - 4.2)\right] + 68$

[1, 25] by [50, 85]
Xscl = 5   Yscl = 5

FIGURE 20

4.2 Exercises ▼▼▼▼▼▼▼▼▼▼▼▼▼▼▼▼▼▼▼▼▼▼▼▼▼▼▼▼▼▼▼▼▼▼▼▼▼▼

1. What is the minimum value of $y = 2 + \cos 4x$?
2. What is the maximum value of $y = 5 - 4 \sin 3x$?
3. What positive phase shift converts the graph of $y = \sin x$ to the graph of $y = \cos x$?
4. What negative phase shift converts the graph of $y = \sin x$ to the graph of $y = \cos x$?

Match each function with its graph in Exercises 5–12.

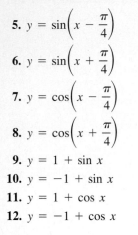

5. $y = \sin\left(x - \dfrac{\pi}{4}\right)$

6. $y = \sin\left(x + \dfrac{\pi}{4}\right)$

7. $y = \cos\left(x - \dfrac{\pi}{4}\right)$

8. $y = \cos\left(x + \dfrac{\pi}{4}\right)$

9. $y = 1 + \sin x$

10. $y = -1 + \sin x$

11. $y = 1 + \cos x$

12. $y = -1 + \cos x$

A.

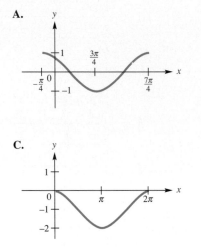

B.

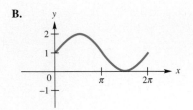

C.

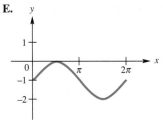

D.

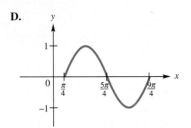

E.

F.

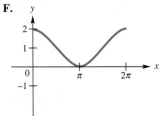

G.

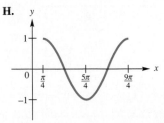

H.

In Exercises 13 and 14, fill in the blanks.

13. The function $y = -4 + \sin 4(x + \pi/2)$ has amplitude _____, period _____, phase shift _____ units to the _____, and has vertical translation _____ units _____.

14. If the graph of $y = \cos x$ is shifted $\pi/2$ units horizontally to the _____, it will coincide with the graph of $y = \sin x$.

Find the amplitude, the period, any vertical translation, and any phase shift of each graph. See Examples 1–5.

15. $y = 2 \sin(x - \pi)$

16. $y = \dfrac{2}{3} \sin\left(x + \dfrac{\pi}{2}\right)$

17. $y = 4 \cos\left(\dfrac{x}{2} + \dfrac{\pi}{2}\right)$

18. $y = -\cos \dfrac{2}{3}\left(x - \dfrac{\pi}{3}\right)$

19. $y = 3 \cos 2\left(x - \dfrac{\pi}{4}\right)$

20. $y = \dfrac{1}{2} \sin\left(\dfrac{x}{2} + \pi\right)$

21. $y = 2 - \sin\left(3x - \dfrac{\pi}{5}\right)$

22. $y = -1 + \dfrac{1}{2} \cos(2x - 3\pi)$

Graph the defined function over a two-period interval. See Examples 1 and 2.

23. $y = \cos\left(x - \dfrac{\pi}{2}\right)$

24. $y = \sin\left(x - \dfrac{\pi}{4}\right)$

25. $y = \sin\left(x + \dfrac{\pi}{4}\right)$

26. $y = \cos\left(x - \dfrac{\pi}{3}\right)$

27. $y = 2 \cos\left(x - \dfrac{\pi}{3}\right)$

28. $y = 3 \sin\left(x - \dfrac{3\pi}{2}\right)$

Graph the defined function over a one-period interval. See Example 3.

29. $y = \dfrac{3}{2} \sin 2\left(x + \dfrac{\pi}{4}\right)$

30. $y = -\dfrac{1}{2} \cos 4\left(x + \dfrac{\pi}{2}\right)$

31. $y = -4 \sin(2x - \pi)$

32. $y = 3 \cos(4x + \pi)$

33. $y = \dfrac{1}{2} \cos\left(\dfrac{1}{2}x - \dfrac{\pi}{4}\right)$

34. $y = -\dfrac{1}{4} \sin\left(\dfrac{3}{4}x + \dfrac{\pi}{8}\right)$

Graph the defined function over a two-period interval. See Example 4.

35. $y = -3 + 2 \sin x$

36. $y = 2 - 3 \cos x$

37. $y = 1 - \dfrac{2}{3} \sin \dfrac{3}{4}x$

38. $y = -1 - 2 \cos 5x$

39. $y = 1 - 2 \cos \dfrac{1}{2}x$

40. $y = -3 + 3 \sin \dfrac{1}{2}x$

41. $y = -2 + \dfrac{1}{2} \sin 3x$

42. $y = 1 + \dfrac{2}{3} \cos \dfrac{1}{2}x$

Graph the defined function over a one-period interval. See Example 5.

43. $y = -3 + 2 \sin\left(x + \dfrac{\pi}{2}\right)$

44. $y = 4 - 3 \cos(x - \pi)$

45. $y = \dfrac{1}{2} + \sin 2\left(x + \dfrac{\pi}{4}\right)$

46. $y = -\dfrac{5}{2} + \cos 3\left(x - \dfrac{\pi}{6}\right)$

47. Consider the function defined by $y = -4 - 3 \sin 2(x - \pi/6)$. Without actually graphing the function, write an explanation of how the constants -4, -3, 2, and $\pi/6$ affect the graph, using the graph of $y = \sin x$ as a basis for comparison.

48. The graph of $y = .5 \sin x + .866 \cos x$ is the same as the graph of a function of the form $y = a \sin(x + \alpha)$. Graph the function with a graphing calculator and then estimate the values of a and α.

For each graph in Exercises 49 and 50, give the equation of a sine function having that graph.

49. (*Note:* Xscl is $\pi/4$.)

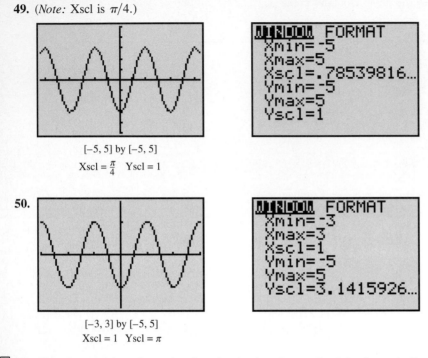

[−5, 5] by [−5, 5]

Xscl = $\frac{\pi}{4}$ Yscl = 1

```
WINDOW FORMAT
 Xmin=-5
 Xmax=5
 Xscl=.78539816…
 Ymin=-5
 Ymax=5
 Yscl=1
```

50.

[−3, 3] by [−5, 5]

Xscl = 1   Yscl = π

```
WINDOW FORMAT
 Xmin=-3
 Xmax=3
 Xscl=1
 Ymin=-5
 Ymax=5
 Yscl=3.1415926…
```

51. Give the equation of a cosine function having the graph of the figure in Exercise 49.

52. Give the equation of a cosine function having the graph of the figure in Exercise 50.

Exercises 53 and 54 refer to the table below. The numbers in the X column are $\pi/2$, $3\pi/4$, π, $\cdots$ (that is, 90°, 135°, 180°, . . .). The functions Y_1 and Y_2 have the form $y = c + a \sin b(x - d)$.

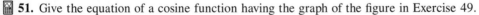

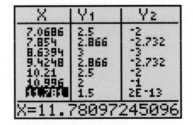

X	Y₁	Y₂
1.5708	2	-1
2.3562	1.5	1E-13
3.1416	1.134	.73205
3.927	1	1
4.7124	1.134	.73205
5.4978	1.5	0
6.2832	2	-1

X=6.28318530718

X	Y₁	Y₂
7.0686	2.5	-2
7.854	2.866	-2.732
8.6394	3	-3
9.4248	2.866	-2.732
10.21	2.5	-2
10.996	2	-1
11.781	1.5	2E-13

X=11.78097245096

53. Determine a possible expression for the function Y_1.

54. Determine a possible expression for the function Y_2.

55. The average temperature (in °F) in Austin, Texas, can be modeled using the trigonometric function

$$f(x) = 17.5 \sin\left[\frac{\pi}{6}(x - 4)\right] + 67.5$$

where x is the month and $x = 1$ corresponds to January. (*Source:* Miller, A., and J. Thompson, *Elements of Meteorology,* Charles E. Merrill Publishing Company, Columbus, Ohio, 1975.)

(a) Graph f over the interval $1 \le x \le 25$. Determine the amplitude, period, phase shift, and vertical translation of f.

(b) What is the average monthly temperature for the month of December?

(c) Determine the maximum and minimum average monthly temperatures and the months when they occur.

(d) What would be an approximation for the average *yearly* temperature in Austin? How is this related to the vertical translation of the sine function in the formula of f?

56. The average monthly temperature (in °F) in Vancouver, Canada, is shown in the table. (*Source:* Miller, A., and J. Thompson, *Elements of Meteorology,* Charles E. Merrill Publishing Company, Columbus, Ohio, 1975.)

Month	Temperature
Jan	36
Feb	39
Mar	43
Apr	48
May	55
June	59
July	64
Aug	63
Sept	57
Oct	50
Nov	43
Dec	39

(a) Plot the average monthly temperature over a two-year period by letting $x = 1$ correspond to the month of January during the first year. Do the data seem to indicate a translated sine graph?

(b) The highest average monthly temperature is 64°F in July and the lowest average monthly temperature is 36°F in January. Their average is 50°F. Graph the data together with the line $y = 50$. What does this line represent with regard to temperature in Vancouver?

(c) Approximate the amplitude, period, and phase shift of the translated sine wave indicated by the data.

(d) Determine a function of the form $f(x) = a \sin b(x - d) + c$, where a, b, c, and d are constants, that models the data.

(e) Graph f together with the data on the same coordinate axes. Comment on how well f models the given data.

57. The average monthly temperature (in °F) in Phoenix, Arizona, is shown in the table. (*Source:* Miller, A., and J. Thompson, *Elements of Meteorology,* Charles E. Merrill Publishing Company, Columbus, Ohio, 1975.)

Month	Temperature
Jan	51
Feb	55
Mar	63
Apr	67
May	77
June	86
July	90
Aug	90
Sept	84
Oct	71
Nov	59
Dec	52

(a) Predict the average yearly temperature and compare it to the actual value of 70°F.

(b) Plot the average monthly temperature over a two-year period by letting $x = 1$ correspond to January of the first year.

(c) Determine a function of the form $f(x) = a \cos b(x - d) + c$, where a, b, c, and d are constants, that models the data.

(d) Graph f together with the data on the same coordinate axes.

4.3 Graphs of the Other Circular Functions ▼▼▼

In this section we discuss the graphs of the four remaining circular functions: cosecant, secant, tangent, and cotangent.

GRAPHS OF COSECANT AND SECANT Since cosecant values are reciprocals of the corresponding sine values, the period of the function $y = \csc x$ is 2π, the same as for $y = \sin x$. The following table shows several values for $y = \sin x$ and the corresponding values of $y = \csc x$.

x	0	$\pi/4$	$\pi/2$	$3\pi/4$	π	$5\pi/4$	$3\pi/2$	2π
$\sin x$	0	$\sqrt{2}/2$	1	$\sqrt{2}/2$	0	$-\sqrt{2}/2$	-1	0
$\csc x$	undefined	$\sqrt{2}$	1	$\sqrt{2}$	undefined	$-\sqrt{2}$	-1	undefined

When $\sin x = 1$, the value of $\csc x$ is also 1, and when $0 < \sin x < 1$, then $\csc x > 1$. Also, if $-1 < \sin x < 0$, then $\csc x < -1$. As x approaches 0, $\sin x$ approaches 0, and $|\csc x|$ gets larger and larger. The graph of $\csc x$ approaches the vertical line $x = 0$ but never touches it. The line $x = 0$ is called a **vertical asymptote.** In fact, the lines $x = n\pi$, where n is any integer, are all vertical asymptotes. Using this information and plotting a few points shows that the graph takes the shape of the solid curve shown in Figure 21. To show how the two graphs are related, the graph of $y = \sin x$ is also shown, as a dashed curve. The domain of the function $y = \csc x$ is $\{x \mid x \neq n\pi,$ where n is any integer$\}$, and the range is $(-\infty, -1] \cup [1, \infty)$.

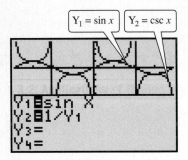

To graph $y = \csc x$, use the fact that $\csc x = \frac{1}{\sin x}$. The graphs of both $Y_1 = \sin x$ and $Y_2 = \csc x$ are shown. The calculator is in split-screen and connected modes.

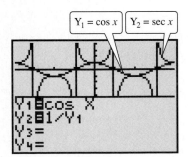

To graph $Y_2 = \sec x$, use the fact that $\sec x = \frac{1}{\cos x}$.

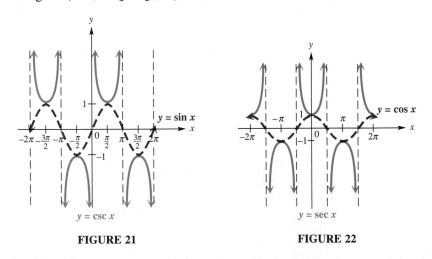

FIGURE 21   FIGURE 22

The graph of $y = \sec x$, shown in Figure 22, is related to the cosine graph in the same way that the graph of $y = \csc x$ is related to the sine graph, because $\sec x = 1/\cos x$. The domain of the function $y = \sec x$ is $\{x \mid x \neq \pi/2 + n\pi,$ where n is any integer$\}$, and the range is $(-\infty, -1] \cup [1, \infty)$.

In order to graph functions based on the cosecant and secant, see the summary that follows.

Graphing the Cosecant and Secant Functions

To graph $y = a \csc bx$ or $y = a \sec bx$, with $b > 0$, follow these steps.

1. Graph the corresponding reciprocal function as a guide, using a dashed curve. That is,

To Graph	Use as a Guide
$y = a \csc bx$	$y = a \sin bx$
$y = a \sec bx$	$y = a \cos bx.$

2. Sketch the vertical asymptotes. They will have equations of the form $x = k$, where k is an x-intercept of the graph of the guide function.
3. Sketch the graph of the desired function by drawing the typical U-shaped branches between the adjacent asymptotes. The branches will be above the graph of the guide function when the guide function values are positive, and below the graph of the guide function when the guide function values are negative. The graph will resemble the graphs in Figures 21 and 22.

Like the sine and cosine functions, the secant and cosecant function graphs may be translated vertically and horizontally. The period of both functions is 2π.

◀EXAMPLE 1
Graphing $y = a \sec bx$

Graph $y = 2 \sec \dfrac{1}{2} x$.

Use the guidelines above.

Step 1 This function involves the secant, so the corresponding reciprocal function will involve the cosine. The function we will graph as a guide is

$$y = 2 \cos \frac{1}{2}x.$$

Using the guidelines of Section 4.1, we find that one period of the graph lies along the interval that satisfies the inequality

$$0 \le \frac{1}{2}x \le 2\pi,$$

or $[0, 4\pi]$. Dividing this interval into four equal parts gives the following key points.

$$(0, 2) \qquad (\pi, 0) \qquad (2\pi, -2) \qquad (3\pi, 0) \qquad (4\pi, 2)$$

These are joined with a smooth curve; it is dashed to indicate that this graph is only a guide. An additional period is graphed as seen in Figure 23(a).

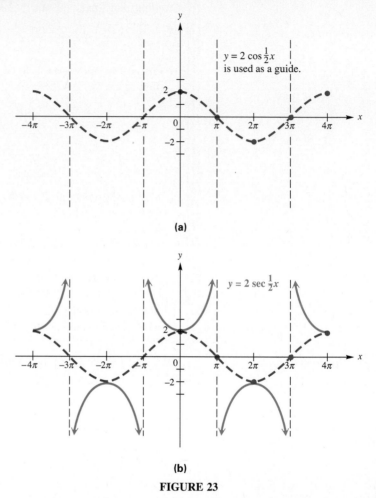

(a)

(b)

FIGURE 23

Step 2 Sketch the vertical asymptotes. These occur at *x*-values for which the guide function equals 0. A few of them have these equations:

$$x = -3\pi \qquad x = -\pi \qquad x = \pi \qquad x = 3\pi.$$

See Figure 23(a).

Step 3 Sketch the graph of $y = 2 \sec(1/2)x$ by drawing in the typical U-shaped branches, approaching the asymptotes. See Figure 23(b). ▶

EXAMPLE 2
Graphing $y = a \csc(x - d)$

Graph $y = \dfrac{3}{2} \csc\left(x - \dfrac{\pi}{2}\right)$.

This function can be graphed using the method of Example 1, by first graphing the corresponding reciprocal function $y = (3/2) \sin(x - \pi/2)$. We can alternatively analyze the function as follows. Compared with the graph of $y = \csc x$, this graph has a phase shift of $\pi/2$ units to the right. Thus, the asymptotes are the lines $x = \pi/2, 3\pi/2$, and so on. Also, there are no values of y between $-3/2$ and $3/2$. As shown in Figure 24, this is related to the increased amplitude of $y = (3/2) \sin x$ compared with $y = \sin x$. (Amplitude does not apply to the secant or cosecant functions; it enters only indirectly from the corresponding cosine or sine graphs.) This means that the graph goes through the points $(\pi, 3/2), (2\pi, -3/2)$, and so on. Two periods are shown in Figure 24. (The graph of the "guide" function, $y = (3/2) \sin(x - \pi/2)$, is shown in red.) ▶

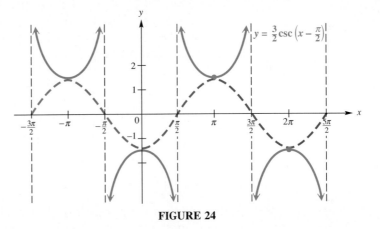

FIGURE 24

To graph the functions discussed in this section with a graphing calculator, we need to use the reciprocal functions of $\sin x$, $\cos x$, and $\tan x$ to get $\csc x$, $\sec x$, and $\cot x$, respectively. Because the graphs of all circular functions except $\sin x$ and $\cos x$ have asymptotes, you may prefer to use dot mode for their graphs. The figures below show the graph of $y = (3/2) \csc(x - \pi/2)$. As shown in the figures, the connected mode will draw vertical lines appearing between the portions of the graph, while the dot mode does not. Compare these graphs with Figure 24.

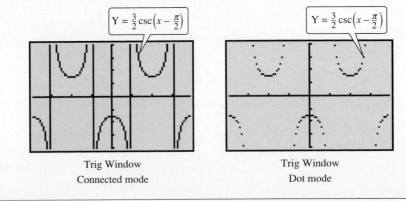

Trig Window
Connected mode

Trig Window
Dot mode

GRAPHS OF TANGENT AND COTANGENT We now study the graphs of the two remaining circular functions, $y = \tan x$ and $y = \cot x$.

Since the values of $y = \tan x$ are positive in quadrants I and III, and negative in quadrants II and IV,

$$\tan(x + \pi) = \tan x,$$

so the period of $y = \tan x$ is π. Thus, the tangent function need be investigated only within an interval of π units. A convenient interval for this purpose is $(-\pi/2, \pi/2)$ because, although the endpoints $-\pi/2$ and $\pi/2$ are not in the domain of $y = \tan x$ (why?), $\tan x$ exists for all other values in the interval. In the interval $(0, \pi/2)$, $\tan x$ is positive. As x goes from 0 to $\pi/2$, a calculator shows that $\tan x$ gets larger and larger without bound. As x goes from $-\pi/2$ to 0, the values of $\tan x$ approach 0 through negative values. These results are summarized in the following table.

As x Increases from	tan x
0 to $\pi/2$	Increases from 0, without bound
$-\pi/2$ to 0	Increases to 0

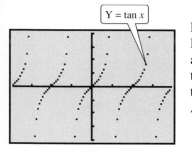

Y = tan x

Trig Window
Dot mode

This is a calculator-generated graph of
$Y = \tan x$.

Based on these results, the graph of $y = \tan x$ will approach the vertical line $x = \pi/2$ but never touch it, so the line $x = \pi/2$ is a vertical asymptote. The lines $x = \pi/2 + n\pi$, where n is any integer, are all vertical asymptotes. These asymptotes are indicated with light dashed lines on the graph in Figure 25. In the interval $(-\pi/2, 0)$, which corresponds to quadrant IV on the unit circle, $\tan x$ is negative, and as x goes from 0 to $-\pi/2$, $\tan x$ gets smaller and smaller. A table of values for $\tan x$, where $-\pi/2 < x < \pi/2$, follows.

x	$-\pi/3$	$-\pi/4$	$-\pi/6$	0	$\pi/6$	$\pi/4$	$\pi/3$
$\tan x$	-1.7	-1	$-.6$	0	$.6$	1	1.7

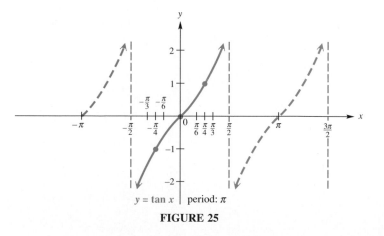

$y = \tan x$ period: π

FIGURE 25

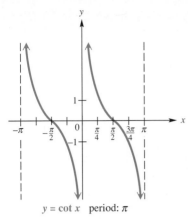

$y = \cot x$ period: π

FIGURE 26

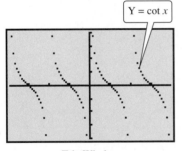

Trig Window
Dot mode

To graph Y = cot x, use the identity
$\cot x = \dfrac{1}{\tan x}$ or the identity $\cot x = \dfrac{\cos x}{\sin x}$.

Plotting the points from the table and letting the graph approach the asymptotes at $x = \pi/2$ and $x = -\pi/2$ gives the portion of the graph shown with a solid curve in Figure 25. More of the graph can be sketched by repeating the same curve, also as shown in the figure. This graph, like the graphs for the sine and cosine functions, should be learned well enough so that a quick sketch can easily be made. Convenient key points are $(-\pi/4, -1)$, $(0, 0)$, and $(\pi/4, 1)$. These points are shown in Figure 25. The lines $x = \pi/2$ and $x = -\pi/2$ are vertical asymptotes. (The idea of *amplitude,* discussed earlier, applies only to the sine and cosine functions. However, here it means that each y-value of $y = a \tan x$ is a times the corresponding y-value of $y = \tan x$.) The domain of the tangent function is $\{x \,|\, x \neq \pi/2 + n\pi$, where n is any integer$\}$. The range is $(-\infty, \infty)$.

The definition $\cot x = 1/(\tan x)$ can be used to find the graph of $y = \cot x$. The period of the cotangent, like that of the tangent, is π. The domain of $y = \cot x$ excludes $0 + n\pi$, where n is any integer, since $1/\tan x$ is undefined for these values of x. Thus, the vertical lines $x = n\pi$ are asymptotes. Values of x that lead to asymptotes for $\tan x$ will make $\cot x = 0$, so $\cot(-\pi/2) = 0$, $\cot \pi/2 = 0$, $\cot 3\pi/2 = 0$, and so on. The values of $\tan x$ increase as x goes from $-\pi/2$ to $\pi/2$, so the values of $\cot x$ will *decrease* as x goes from $-\pi/2$ to $\pi/2$. A table of values for $\cot x$, where $0 < x < \pi$, is shown below.

x	$\pi/6$	$\pi/4$	$\pi/3$	$\pi/2$	$2\pi/3$	$3\pi/4$	$5\pi/6$
$\cot x$	1.7	1	.6	0	$-.6$	-1	-1.7

Plotting these points and using the information discussed above gives the graph of $y = \cot x$ shown in Figure 26. (The graph shows two periods.) The domain of the cotangent function is $\{x \,|\, x \neq n\pi$, where n is any integer$\}$. The range is $(-\infty, \infty)$.

Graphing the Tangent and Cotangent Functions

To graph $y = a \tan bx$ or $y = a \cot bx$, with $b > 0$:

1. The period is π/b. To locate two adjacent vertical asymptotes, solve the following equations for x:

$$\text{For } y = a \tan bx: \quad bx = -\frac{\pi}{2} \quad \text{and} \quad bx = \frac{\pi}{2}$$

$$\text{For } y = a \cot bx: \quad bx = 0 \quad \text{and} \quad bx = \pi.$$

2. Sketch the two vertical asymptotes found in Step 1.
3. Divide the interval formed by the vertical asymptotes into four equal parts.
4. Evaluate the function for the first-quarter point, midpoint, and third-quarter point, using the x-values found in Step 3.
5. Join the points with a smooth curve, approaching the vertical asymptotes. Draw additional asymptotes and periods of the graph as necessary.

Like the other circular functions, the graphs of the tangent and cotangent functions may be shifted horizontally as well as vertically.

EXAMPLE 3
Graphing $y = \tan bx$

Graph $y = \tan 2x$.

Step 1 The period of this function is $\pi/2$. To locate two adjacent vertical asymptotes, solve $2x = -\pi/2$ and $2x = \pi/2$ (since this is a tangent function). We find that the two asymptotes have equations

$$x = -\frac{\pi}{4} \quad \text{and} \quad x = \frac{\pi}{4}.$$

Step 2 Sketch the two vertical asymptotes $x = \pm\pi/4$, as shown in Figure 27.

Step 3 Divide the interval $(-\pi/4, \pi/4)$ into four equal parts. This gives the following key x-values:

first-quarter value: $-\dfrac{\pi}{8}$

middle value: **0**

third-quarter value: $\dfrac{\pi}{8}$.

Step 4 Evaluate the function for the x-values found in Step 3, as shown in the following table.

x	$-\pi/8$	0	$\pi/8$
$2x$	$-\pi/4$	0	$\pi/4$
$\tan 2x$	-1	**0**	1

Step 5 Join these points with a smooth curve, approaching the vertical asymptotes. See Figure 27. Another period has been graphed as well, one-half period to the left and one-half period to the right. ▶

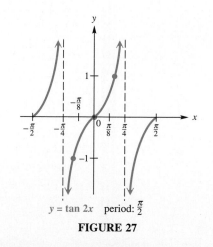

$y = \tan 2x$ period: $\dfrac{\pi}{2}$

FIGURE 27

EXAMPLE 4
Graphing $y = a \tan bx$

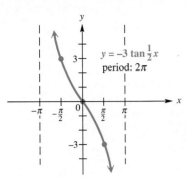

$y = -3 \tan \frac{1}{2}x$
period: 2π

FIGURE 28

Graph $y = -3 \tan \frac{1}{2} x$.

The period is $\pi/(1/2) = 2\pi$. Adjacent asymptotes are at $x = -\pi$ and $x = \pi$. Dividing the interval $-\pi < x < \pi$ into four equal parts gives key x-values of $-\pi/2$, 0, and $\pi/2$. Evaluating the function at these x values gives these key points.

$$\left(-\frac{\pi}{2}, 3\right) \qquad (0, 0) \qquad \left(\frac{\pi}{2}, -3\right)$$

Plotting these points and joining them with a smooth curve gives the graph shown in Figure 28. Notice that, because the coefficient -3 is negative, the graph is reflected about the x-axis compared to the graph of $y = 3 \tan(1/2)x$. ▶

NOTE The function $y = -3 \tan (1/2)x$ of Example 4, graphed in Figure 28, has a graph that compares to the graph of $y = \tan x$ as follows:

1. The period is larger, because $b = 1/2$, and $1/2 < 1$.
2. The graph is "stretched," because $a = -3$, and $|-3| > 1$.
3. Each branch of the graph goes down from left to right (that is, the function decreases) between each pair of adjacent asymptotes, because $a = -3 < 0$. When $a < 0$, the graph is reflected about the x-axis.

Graphing a cotangent function is done in a manner similar to graphing a tangent function, as shown in the next example.

EXAMPLE 5
Graphing $y = a \cot bx$

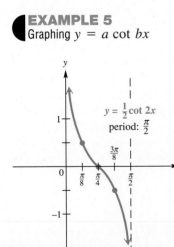

$y = \frac{1}{2}\cot 2x$
period: $\frac{\pi}{2}$

FIGURE 29

Graph $y = \frac{1}{2} \cot 2x$.

Because this function involves the cotangent, we can locate two adjacent asymptotes by solving the equations $2x = 0$ and $2x = \pi$. We find that the lines $x = 0$ (the y-axis) and $x = \pi/2$ are two such asymptotes. Divide the interval $0 < x < \pi/2$ into four equal parts, getting key x-values of $\pi/8$, $\pi/4$, and $3\pi/8$. Evaluating the function at these x-values gives the following key points.

$$\left(\frac{\pi}{8}, \frac{1}{2}\right) \qquad \left(\frac{\pi}{4}, 0\right) \qquad \left(\frac{3\pi}{8}, -\frac{1}{2}\right)$$

Joining these points with a smooth curve approaching the asymptotes gives the graph shown in Figure 29. ▶

As stated earlier, tangent and cotangent function graphs may be translated vertically, horizontally, or both, as shown in the next two examples.

◗EXAMPLE 6
Graphing a tangent function with a vertical translation

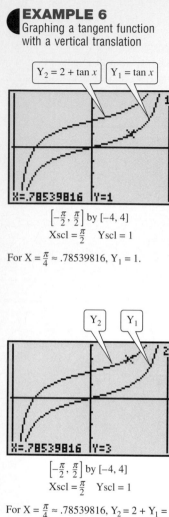

$\left[-\frac{\pi}{2}, \frac{\pi}{2}\right]$ by $[-4, 4]$
$\mathrm{Xscl} = \frac{\pi}{2}$ $\mathrm{Yscl} = 1$

For $\mathrm{X} = \frac{\pi}{4} \approx .78539816$, $\mathrm{Y}_1 = 1$.

$\left[-\frac{\pi}{2}, \frac{\pi}{2}\right]$ by $[-4, 4]$
$\mathrm{Xscl} = \frac{\pi}{2}$ $\mathrm{Yscl} = 1$

For $\mathrm{X} = \frac{\pi}{4} \approx .78539816$, $\mathrm{Y}_2 = 2 + \mathrm{Y}_1 = 2 + 1 = 3$. Notice from the two screens above, $\mathrm{Y}_2 = 2 + \mathrm{Y}_1$. The graph of Y_2 is obtained by translating the graph of Y_1 2 units upward.

Graph $y = 2 + \tan x$.

Every value of y for this function will be 2 units more than the corresponding value of y in $y = \tan x$, causing the graph of $y = 2 + \tan x$ to be translated 2 units upward as compared with the graph of $y = \tan x$. See Figure 30.   ◗

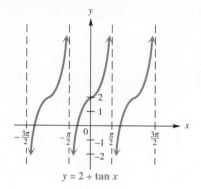

$y = 2 + \tan x$

FIGURE 30

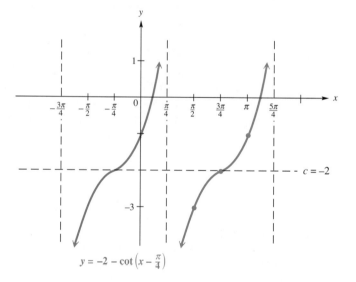

$y = -2 - \cot\left(x - \frac{\pi}{4}\right)$

FIGURE 31

◗EXAMPLE 7
Graphing a cotangent function with vertical and horizontal translations

Graph $y = -2 - \cot\left(x - \dfrac{\pi}{4}\right)$.

Here $b = 1$, so the period is π. The graph will be translated down 2 units (because $c = -2$), reflected about the x-axis (because of the negative sign in front of the cotangent) and will have a phase shift (horizontal translation) of $\pi/4$ units to the right (because of the argument $(x - \pi/4)$). To locate adjacent asymptotes, since this function involves the cotangent, we solve the following equations:

$$x - \frac{\pi}{4} = 0 \qquad x - \frac{\pi}{4} = \pi$$

$$x = \frac{\pi}{4} \qquad x = \frac{5\pi}{4}.$$

Dividing the interval $\pi/4 < x < 5\pi/4$ into four equal parts and evaluating the function at the three key x-values within the interval gives these key points.

$$\left(\frac{\pi}{2}, -3\right) \qquad \left(\frac{3\pi}{4}, -2\right) \qquad (\pi, -1)$$

These points are joined by a smooth curve. This period of the graph, along with one in the interval $-3\pi/4 < x < \pi/4$, is shown in Figure 31. ▶

New functions are often formed by adding or subtracting other functions. A function formed by combining two other functions, such as

$$y = \cos x + \sin x,$$

has historically been graphed using a method known as *addition of ordinates*. (The ordinate of a point is its y-coordinate.) To apply this method to this function, we would graph the functions $y = \cos x$ and $y = \sin x$. Then, for selected values of x, we would add $\cos x$ and $\sin x$, and plot the points $(x, \cos x + \sin x)$. Connecting the selected points with a typical circular function-type curve would give the graph of the desired function. While this method illustrates some valuable concepts involving the arithmetic of functions, it is very time-consuming.

With the technology of the graphing calculator, such an exercise can easily be accomplished. Let $Y_1 = \cos x$, $Y_2 = \sin x$, and $Y_3 = Y_1 + Y_2$. Then graph these three functions in a "Trig" window and observe carefully as the calculator plots them in order. See the figure.

Use TRACE to locate points on each graph with the same x value. Add the resulting coordinates Y_1 and Y_2. The sum should agree with Y_3. (There may be a slight discrepancy in the last decimal place.) Do you see how the term "addition of ordinates" originated? If your calculator has a TABLE feature, you can use it to get corresponding values of Y_1, Y_2, and Y_3.

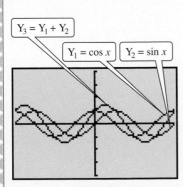

$Y_3 = Y_1 + Y_2$

$Y_1 = \cos x$ $Y_2 = \sin x$

Trig Window

FOR DISCUSSION OR WRITING

Use a graphing calculator to graph the following functions directly and then by graphing each term and the sum in one window.

1. $y = \sin x + \sin 2x$

2. $y = \cos x + \sec x$

4.3 Exercises ▼▼▼

1. Describe the graph of $y = \sec x$ as if you were explaining it to a friend on the phone.

2. Describe the graph of $y = \tan x$ as if you were explaining it to a friend on the phone.

3. True or false: The graph of $y = 3 \csc x$ is the same as the graph of $y = \dfrac{1}{3 \sin x}$. If false, explain why.

4. If c is any number, how many solutions does the equation $c = \tan x$ have in the interval $(-2\pi, 2\pi]$?

5. True or false: The graph of $y = \tan x$ in Figure 25 suggests that $\tan(-x) = -\tan x$ for all x in the domain of $\tan x$.

6. True or false: The graph of $y = \sec x$ in Figure 22 suggests that $\sec(-x) = \sec x$ for all x in the domain of $\sec x$.

Match each equation with its graph in Exercises 7–12.

7. $y = -\csc x$

8. $y = -\sec x$

9. $y = -\tan x$

10. $y = -\cot x$

11. $y = \tan\left(x - \dfrac{\pi}{4}\right)$

12. $y = \cot\left(x - \dfrac{\pi}{4}\right)$

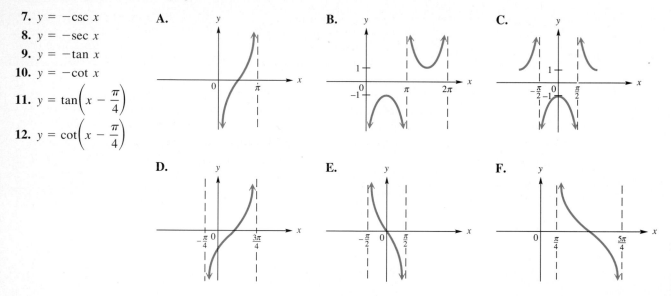

Graph each defined function over a one-period interval. See Examples 1 and 2.

13. $y = \csc\left(x - \dfrac{\pi}{4}\right)$

14. $y = \sec\left(x + \dfrac{3\pi}{4}\right)$

15. $y = \sec\left(x + \dfrac{\pi}{4}\right)$

16. $y = \csc\left(x + \dfrac{\pi}{3}\right)$

17. $y = \sec\left(\dfrac{1}{2}x + \dfrac{\pi}{3}\right)$

18. $y = \csc\left(\dfrac{1}{2}x - \dfrac{\pi}{4}\right)$

19. $y = 2 + 3\sec(2x - \pi)$

20. $y = 1 - 2\csc\left(x + \dfrac{\pi}{2}\right)$

21. $y = 1 - \dfrac{1}{2}\csc\left(x - \dfrac{3\pi}{4}\right)$

22. $y = 2 + \dfrac{1}{4}\sec\left(\dfrac{1}{2}x - \pi\right)$

Graph each defined function over a one-period interval. See Examples 3–5.

23. $y = 2 \tan x$

24. $y = 2 \cot x$

25. $y = \frac{1}{2} \cot x$

26. $y = 2 \tan \frac{1}{4}x$

27. $y = \cot 3x$

28. $y = -\cot \frac{1}{2}x$

Graph each defined function over a two-period interval. See Examples 6 and 7.

29. $y = \tan(2x - \pi)$

30. $y = \tan\left(\frac{x}{2} + \pi\right)$

31. $y = \cot\left(3x + \frac{\pi}{4}\right)$

32. $y = \cot\left(2x - \frac{3\pi}{2}\right)$

33. $y = 1 + \tan x$

34. $y = -2 + \tan x$

35. $y = 1 - \cot x$

36. $y = -2 - \cot x$

37. $y = -1 + 2 \tan x$

38. $y = 3 + \frac{1}{2} \tan x$

39. $y = -1 + \frac{1}{2} \cot(2x - 3\pi)$

40. $y = -2 + 3 \tan(4x + \pi)$

41. $y = \frac{2}{3} \tan\left(\frac{3}{4}x - \pi\right) - 2$

42. $y = 1 - 2 \cot 2\left(x + \frac{\pi}{2}\right)$

43. Consider the function $f(x) = -4 \tan(2x + \pi)$. What is the domain of f? What is its range?

44. Consider the function $g(x) = -2 \csc(4x + \pi)$. What is the domain of g? What is its range?

---

45. A rotating beacon is located at point A next to a long wall. (See the figure.) The beacon is 4 m from the wall. The distance d is given by

$$d = 4 \tan 2\pi t,$$

where t is time measured in seconds since the beacon started rotating. (When $t = 0$, the beacon is aimed at point R. When the beacon is aimed to the right of R, the value of d is positive; d is negative if the beacon is aimed to the left of R.) Find d for the following times.

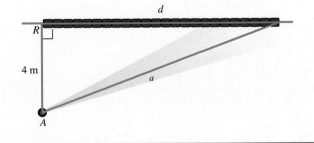

(a) $t = 0$ **(b)** $t = .4$ **(c)** $t = .8$ **(d)** $t = 1.2$
(e) Why is .25 a meaningless value for t?

46. In the figure for Exercise 45, the distance a is given by

$$a = 4|\sec 2\pi t|.$$

Find a for the following times.
(a) $t = 0$ **(b)** $t = .86$ **(c)** $t = 1.24$

47. Simultaneously sketch the graphs of $y = \tan x$ and $y = x$ for $-1 \leq x \leq 1$ and $-1 \leq y \leq 1$ with a graphing calculator. Write a sentence or two describing the relationship of $\tan x$ and x for small x-values.

48. Between each pair of successive asymptotes, a portion of the graph of $y = \sec x$ or $y = \csc x$ resembles a parabola. Can each of these portions actually be a parabola? Explain.

Chapter 4 Summary ▼▼▼▼▼▼▼▼▼▼▼▼▼▼▼▼▼▼▼▼▼▼▼▼▼▼▼▼▼▼▼▼▼

SECTION	KEY IDEAS

4.1 Graphs of the Sine and Cosine Functions   4.2 Translating Graphs of the Sine and Cosine Functions

Cosine and Sine Functions

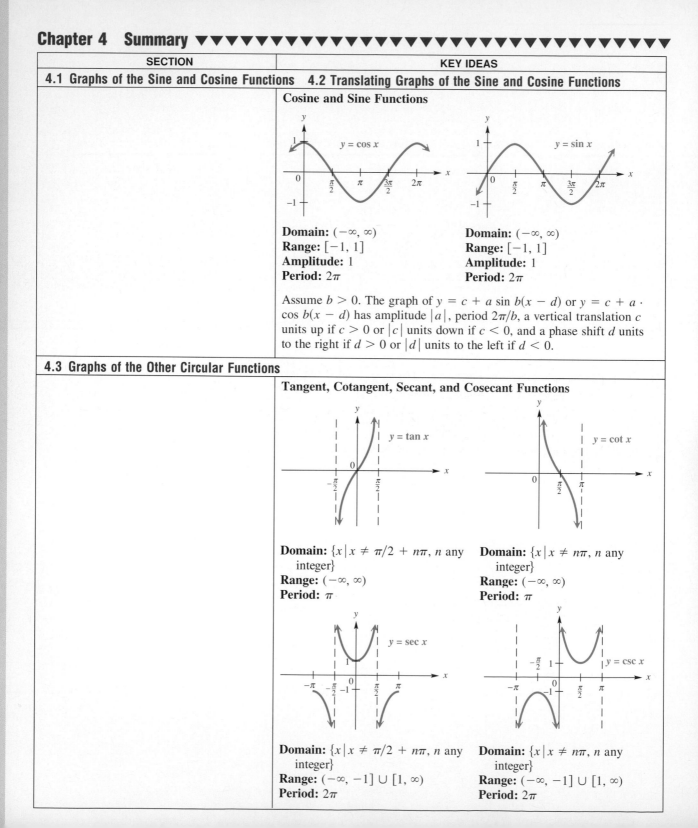

Domain: $(-\infty, \infty)$
Range: $[-1, 1]$
Amplitude: 1
Period: 2π

Domain: $(-\infty, \infty)$
Range: $[-1, 1]$
Amplitude: 1
Period: 2π

Assume $b > 0$. The graph of $y = c + a \sin b(x - d)$ or $y = c + a \cdot \cos b(x - d)$ has amplitude $|a|$, period $2\pi/b$, a vertical translation c units up if $c > 0$ or $|c|$ units down if $c < 0$, and a phase shift d units to the right if $d > 0$ or $|d|$ units to the left if $d < 0$.

4.3 Graphs of the Other Circular Functions

Tangent, Cotangent, Secant, and Cosecant Functions

Domain: $\{x \mid x \neq \pi/2 + n\pi, n \text{ any integer}\}$
Range: $(-\infty, \infty)$
Period: π

Domain: $\{x \mid x \neq n\pi, n \text{ any integer}\}$
Range: $(-\infty, \infty)$
Period: π

Domain: $\{x \mid x \neq \pi/2 + n\pi, n \text{ any integer}\}$
Range: $(-\infty, -1] \cup [1, \infty)$
Period: 2π

Domain: $\{x \mid x \neq n\pi, n \text{ any integer}\}$
Range: $(-\infty, -1] \cup [1, \infty)$
Period: 2π

Chapter 4   Review Exercises ▼▼▼▼▼▼▼▼▼▼▼▼▼▼▼▼▼▼▼▼▼▼▼▼▼▼▼▼

1. Give all basic trigonometric functions that satisfy the condition $f(x + \pi) = -f(x)$.

2. Give all basic trigonometric functions that satisfy the condition $f(x - \pi) = f(x)$.

3. Which of the basic trigonometric functions can attain the value $1/2$?

4. Which of the basic trigonometric functions can attain the value 2?

For each defined function, give the amplitude, period, vertical translation, and phase shift, as applicable.

5. $y = 2 \sin x$

6. $y = \tan 3x$

7. $y = -\dfrac{1}{2} \cos 3x$

8. $y = 2 \sin 5x$

9. $y = 1 + 2 \sin \dfrac{1}{4}x$

10. $y = 3 - \dfrac{1}{4} \cos \dfrac{2}{3}x$

11. $y = 3 \cos\left(x + \dfrac{\pi}{2}\right)$

12. $y = -\sin\left(x - \dfrac{3\pi}{4}\right)$

13. $y = \dfrac{1}{2} \csc\left(2x - \dfrac{\pi}{4}\right)$

14. $y = 2 \sec(\pi x - 2\pi)$

15. $y = \dfrac{1}{3} \tan\left(3x - \dfrac{\pi}{3}\right)$

16. $y = \cot\left(\dfrac{x}{2} + \dfrac{3\pi}{4}\right)$

Of the six basic circular functions, identify the one that satisfies the description.

17. Period is π, x intercepts are of the form $n\pi$, where n is an integer

18. Period is 2π, passes through the origin

19. Period is 2π, passes through the point $(\pi/2, 0)$

20. Period is 2π, domain is $\{x \mid x \neq n\pi$, where n is an integer$\}$

21. Period is π, function is decreasing on the interval $0 < x < \pi$

22. Period is 2π, has vertical asymptotes of the form $x = \pi/2 + n\pi$, where n is an integer

23. Suppose that f is a sine function with period 10 and $f(5) = 2$. Explain why $f(25) = 2$.

24. Suppose that f is a sine function with period π and $f(6\pi/5) = 1$. Explain why $f(-4\pi/5) = 1$.

Graph each defined function over a one-period interval.

25. $y = 3 \sin x$

26. $y = \dfrac{1}{2} \sec x$

27. $y = -\tan x$

28. $y = -2 \cos x$

29. $y = 2 + \cot x$

30. $y = -1 + \csc x$

31. $y = \sin 2x$

32. $y = \tan 3x$

33. $y = 3 \cos 2x$

34. $y = \dfrac{1}{2} \cot 3x$

35. $y = \cos\left(x - \dfrac{\pi}{4}\right)$

36. $y = \tan\left(x - \dfrac{\pi}{2}\right)$

37. $y = \sec\left(2x + \dfrac{\pi}{3}\right)$

38. $y = \sin\left(3x + \dfrac{\pi}{2}\right)$

39. $y = 1 + 2 \cos 3x$

40. $y = -1 - 3 \sin 2x$

41. $y = 2 \sin \pi x$

42. $y = -\dfrac{1}{2} \cos(\pi x - \pi)$

43. $y = 1 - 2 \sec\left(x - \dfrac{\pi}{4}\right)$

44. $y = -\csc(2x - \pi) + 1$

45. Let a person h_1 ft tall stand d ft from an object h_2 ft tall, where $h_2 > h_1$. Let θ be the angle of elevation to the top of the object. (See the figure.)
(a) Show that $d = (h_2 - h_1) \cot \theta$.
(b) Let $h_2 = 55$ and $h_1 = 5$. Graph d for the interval $0 < \theta \le \pi/2$.

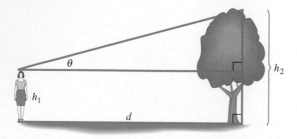

46. The amount of pollution in the air fluctuates with the seasons. It is lower after heavy spring rains and higher after periods of little rain. In addition to this seasonal fluctuation, the long-term trend is upward. An idealized graph of this situation is shown in the figure. Circular functions can be used to describe the fluctuating part of the pollution levels. Powers of the number e (e is the base of natural logarithms; to six decimal places, $e = 2.718282$) can be used to show the long-term growth. In fact, the pollution level in a certain area might be given by

$$P(t) = 7(1 - \cos 2\pi t)(t + 10) + 100e^{.2t},$$

where t is time in years, with $t = 0$ representing January 1 of the base year. Thus, July 1 of the same year would be represented by $t = .5$, and October 1 of the following year would be represented by $t = 1.75$. Find the pollution levels on the following dates.
(a) January 1, base year
(b) July 1, base year
(c) January 1, following year
(d) July 1, following year

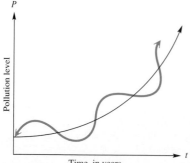

47. The figure shows the population of lynx and hares in Canada for the years 1847–1903. The hares are food for the lynx. An increase in hare population causes an increase in lynx population some time later. The increasing lynx population then causes a decline in hare population.
(a) Estimate the length of one period.
(b) Estimate maximum and minimum hare populations.

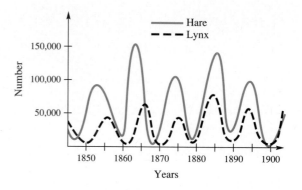

48. Explain how the graph of $y = 2 \cos(3x + 1)$ differs from the graph of $y = 2 \cos 3x + 1$.

49. The average monthly temperature (in °F) in Chicago, Illinois, is shown in the table. (*Source:* Miller, A., and J. Thompson, *Elements of Meteorology,* Charles E. Merrill Publishing Company, Columbus, Ohio, 1975.)

Month	Temperature
Jan	25
Feb	28
Mar	36
Apr	48
May	61
June	72
July	74
Aug	75
Sept	66
Oct	55
Nov	39
Dec	28

(a) Plot the average monthly temperature over a two-year period by letting $x = 1$ correspond to January of the first year.

(b) Determine a model function of the form $f(x) = a \sin b(x - d) + c$, where a, b, c, and d are constants.

(c) Explain the significance of each constant.

(d) Graph f together with the data on the same coordinate axes.

5

Trigonometric Identities

In 1831 Michael Faraday discovered that when a wire is passed by a magnet, a small electric current is produced in that wire. This phenomenon became known as Faraday's Law. Since then people have used this property to generate electric current for homes and businesses by simultaneously rotating thousands of wires near large electromagnets in order to produce massive amounts of electricity. In 1992 U.S. utilities generated 2796 billion kilowatt-hours of electricity. This amount of electricity could power a 100-watt lightbulb for about 3.2 billion years! Electricity has become an important modern convenience in our society.

The electricity supplied to most homes is produced by electric generators that rotate at 60 cycles per second. Because of this rotation, electric current alternates its direction in electrical wires and can be modeled accurately using either the sine or cosine functions. Household current is often rated at 115 volts. If the current is alternating direction in the wires, is the voltage always 115 volts or does it actually vary with time? Electric companies charge customers according to the wattage that an electrical device uses and how long it is turned on. Given both the voltage and current supplied to a lightbulb, how can its wattage be determined? The answers to these questions are found in Sections 5.3 and 5.5.

In order to understand phenomena such as electric current, sound waves, or stress on your back muscles when you bend at the waist, we will need to use not only trigonometric functions but also the many identities that relate the trigonometric functions to each other. The study of electricity, noise control, and biophysics are all fascinating subjects that require knowledge of the trigonometric identities.

Sources: Weidner, R. and R. Sells, *Elementary Classical Physics, Vol. 2,* Allyn and Bacon, 1968. Wright, J. (editor), *The Universal Almanac 1995,* Universal Press Syndicate Company, Kansas City, 1994.

An identity is an equation that is true for *every* value in the domain of its variable. Examples of identities include

$$5(x + 3) = 5x + 15 \quad \text{and} \quad (a + b)^2 = a^2 + 2ab + b^2.$$

This chapter discusses identities involving trigonometric and circular functions. The variables in these functions represent either angles or real numbers. The domain of the variable is assumed to be all values for which a given function is defined.

5.1 Fundamental Identities ▼▼▼

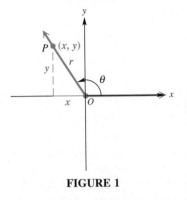

FIGURE 1

In this section we discuss some of the ways we can use the fundamental trigonometric identities first introduced in Chapter 1.* We repeat the basic definitions of the trigonometric functions of an angle θ in standard position. See Figure 1.

Trigonometric Functions

Let (x, y) be a point other than the origin on the terminal side of an angle θ in standard position. Then $r = \sqrt{x^2 + y^2}$ is the distance from the point to the origin. The six trigonometric functions of θ are defined as follows.

$$\sin \theta = \frac{y}{r} \qquad\qquad \csc \theta = \frac{r}{y} \quad (y \neq 0)$$

$$\cos \theta = \frac{x}{r} \qquad\qquad \sec \theta = \frac{r}{x} \quad (x \neq 0)$$

$$\tan \theta = \frac{y}{x} \quad (x \neq 0) \qquad \cot \theta = \frac{x}{y} \quad (y \neq 0)$$

In Chapter 1, these definitions were used to derive the following **reciprocal identities,** which are true for all suitable replacements of the variable.

$$\cot \theta = \frac{1}{\tan \theta} \qquad \csc \theta = \frac{1}{\sin \theta} \qquad \sec \theta = \frac{1}{\cos \theta}$$

Each of these reciprocal identities leads to other forms of the identity. For example, $\csc \theta = 1/\sin \theta$ gives $\sin \theta = 1/\csc \theta$.

From the definitions of the trigonometric functions,

$$\frac{\sin \theta}{\cos \theta} = \frac{y/r}{x/r} = \frac{y}{x} = \tan \theta$$

or

$$\tan \theta = \frac{\sin \theta}{\cos \theta}.$$

---

*All the identities given in this chapter are summarized at the end of the chapter and inside the back cover.

In a similar manner,

$$\cot \theta = \frac{\cos \theta}{\sin \theta}.$$

These last two identities, also derived in Chapter 1, are called the **quotient identities.**

We also saw in Chapter 1 that the definitions of the trigonometric functions were used to derive the identity

$$\sin^2 \theta + \cos^2 \theta = 1.$$

Dividing both sides by $\cos^2 \theta$ then leads to

$$\tan^2 \theta + 1 = \sec^2 \theta,$$

while dividing through by $\sin^2 \theta$ gives

$$1 + \cot^2 \theta = \csc^2 \theta.$$

These last three identities are the **Pythagorean identities.**

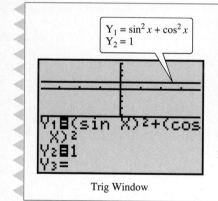

$Y_1 = \sin^2 x + \cos^2 x$
$Y_2 = 1$

Trig Window

A graphing calculator can be used to support an identity—that is, to decide whether two functions are identical. For example, to support the identity $\sin^2 x + \cos^2 x = 1$, let $Y_1 = \sin^2 x + \cos^2 x$ and let $Y_2 = 1$. (Be sure your calculator is set in radian mode.) Now, use the "trig" window and graph the two functions. If the identity is true, you should see no difference in the two graphs. If the equation is not an identity, the graphs of Y_1 and Y_2 will not coincide. As a check, to guard against the possibility that the graphs are different, but one of them is not showing in the window being used, repeat the process, graphing Y_2 first.

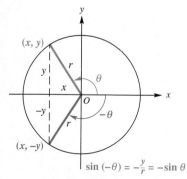

$\sin(-\theta) = -\dfrac{y}{r} = -\sin \theta$

FIGURE 2

As suggested by the circle shown in Figure 2, an angle θ having the point (x, y) on its terminal side has a corresponding angle $-\theta$ with a point $(x, -y)$ on its terminal side. From the definition of sine,

$$\sin(-\theta) = \frac{-y}{r} \qquad \text{and} \qquad \sin \theta = \frac{y}{r},$$

so that $\sin(-\theta)$ and $\sin \theta$ are negatives of each other, or

$$\sin(-\theta) = -\sin \theta.$$

Figure 2 shows an angle θ in quadrant II, but the same result holds for θ in any quadrant. Also, by definition,

$$\cos(-\theta) = \frac{x}{r} \quad \text{and} \quad \cos \theta = \frac{x}{r},$$

so that

$$\cos(-\theta) = \cos\theta.$$

These formulas for $\sin(-\theta)$ and $\cos(-\theta)$ can be used to find $\tan(-\theta)$ in terms of $\tan\theta$:

$$\tan(-\theta) = \frac{\sin(-\theta)}{\cos(-\theta)} = \frac{-\sin\theta}{\cos\theta} = -\frac{\sin\theta}{\cos\theta}$$

or

$$\tan(-\theta) = -\tan\theta.$$

The preceding three identities are **negative-angle identities.**

CONNECTIONS We can derive the negative-angle identities by using the properties of even and odd functions. An **even function** has the property that $f(-x) = f(x)$ for all x. Because of this property, the graph of an even function is symmetric with respect to the y-axis—that is, if folded along the y-axis, the two halves would match. The graph of $y = \cos x$ is symmetric about the y-axis, so cosine is an even function and therefore

$$\cos(-x) = \cos x.$$

A function is an **odd function** if $f(-x) = -f(x)$ for all x. The graph of an odd function is symmetric about the origin, which means that if (x, y) belongs to the function, then $(-x, -y)$ also belongs to the function. The sine graph exhibits this property: for example, $(\pi/2, 1)$ and $(-\pi/2, -1)$ are points on the graph of $y = \sin x$. Thus,

$$\sin(-x) = -\sin x.$$

By recalling (from algebra) the type of symmetry shown by the graph of a trigonometric function, we can remember the three negative-angle identities.

FOR DISCUSSION OR WRITING
Is the tangent function even or odd or neither? Does your answer tell you the correct negative-angle identity for tangent?

The identities given in this section are summarized on the following page. As a group, these are called the **fundamental identities.**

Fundamental Identities

Reciprocal Identities

$$\cot \theta = \frac{1}{\tan \theta} \qquad \sec \theta = \frac{1}{\cos \theta} \qquad \csc \theta = \frac{1}{\sin \theta}$$

Quotient Identities

$$\tan \theta = \frac{\sin \theta}{\cos \theta} \qquad \cot \theta = \frac{\cos \theta}{\sin \theta}$$

Pythagorean Identities

$$\sin^2 \theta + \cos^2 \theta = 1 \qquad \tan^2 \theta + 1 = \sec^2 \theta \qquad 1 + \cot^2 \theta = \csc^2 \theta$$

Negative-Angle Identities

$$\sin(-\theta) = -\sin \theta \qquad \cos(-\theta) = \cos \theta \qquad \tan(-\theta) = -\tan \theta$$

NOTE   The forms of the identities given above are the most commonly recognized forms. Throughout this chapter it will be necessary to recognize alternative forms of these identities as well. For example, two other forms of $\sin^2 \theta + \cos^2 \theta = 1$ are

$$\sin^2 \theta = 1 - \cos^2 \theta$$

and

$$\cos^2 \theta = 1 - \sin^2 \theta.$$

You should be able to transform the basic identities using algebraic transformations.

The fundamental identities are used extensively in trigonometry and in calculus, so it is important to be very familiar with them. One use for these identities is to find the values of other trigonometric functions from the value of a given trigonometric function. Of course, we could find these values by using a right triangle instead, but this is a good way to practice using the fundamental identities. For example, given a value of $\tan \theta$, the value of $\cot \theta$ can be found from the identity $\cot \theta = 1/\tan \theta$. In fact, given any trigonometric function value and the quadrant in which θ lies, the values of all the other trigonometric functions can be found by using identities, as in the following example.

EXAMPLE 1
Finding all trigonometric function values, given one value and the quadrant

If $\tan \theta = -5/3$ and θ is in quadrant II, find the values of the other trigonometric functions using fundamental identities.

The identity $\cot \theta = 1/\tan \theta$ leads to $\cot \theta = -3/5$. Next, find $\sec \theta$ from the identity $\tan^2 \theta + 1 = \sec^2 \theta$.

$$\left(-\frac{5}{3}\right)^2 + 1 = \sec^2 \theta$$

$$\frac{25}{9} + 1 = \sec^2 \theta$$

$$\frac{34}{9} = \sec^2 \theta$$

$$-\sqrt{\frac{34}{9}} = \sec \theta$$

$$-\frac{\sqrt{34}}{3} = \sec \theta$$

We choose the negative square root since $\sec \theta$ is negative in quadrant II. Now find $\cos \theta$:

$$\cos \theta = \frac{1}{\sec \theta} = \frac{-3}{\sqrt{34}} = -\frac{3\sqrt{34}}{34},$$

after rationalizing the denominator. Find $\sin \theta$ by using the identity $\sin^2 \theta + \cos^2 \theta = 1$, with $\cos \theta = -3/\sqrt{34}$.

$$\sin^2 \theta + \left(\frac{-3}{\sqrt{34}}\right)^2 = 1$$

$$\sin^2 \theta = 1 - \frac{9}{34}$$

$$\sin^2 \theta = \frac{25}{34}$$

$$\sin \theta = \frac{5}{\sqrt{34}}$$

$$\sin \theta = \frac{5\sqrt{34}}{34} \qquad \text{Rationalize.}$$

The positive square root is used since $\sin \theta$ is positive in quadrant II. Finally, since $\csc \theta$ is the reciprocal of $\sin \theta$,

$$\csc \theta = \frac{\sqrt{34}}{5}. \qquad \blacktriangleright$$

CAUTION Several comments can be made concerning Example 1.

1. We are given $\tan \theta = -5/3$. Although $\tan \theta = (\sin \theta)/(\cos \theta)$, we should *not* assume that $\sin \theta = -5$ and $\cos \theta = 3$. (Why can these values not possibly be correct?)
2. Problems of this type can usually be worked in more than one way. For example, after finding $\cot \theta = -3/5$, we could have then found $\csc \theta$ using the identity $1 + \cot^2 \theta = \csc^2 \theta$. The remaining function values could then be found as well.
3. The most common error made in problems like this is an incorrect sign choice for the functions. When taking the square root, be sure to choose the sign based on the quadrant of θ and the function being found.

Each of $\tan \theta$, $\cot \theta$, $\sec \theta$, and $\csc \theta$ can easily be expressed in terms of $\sin \theta$ and/or $\cos \theta$. For this reason, we often make such substitutions in an expression so that the expression can be simplified. The next example shows such substitutions.

◖**EXAMPLE 2**
Simplifying an expression by writing in terms of sine and cosine

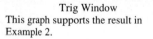

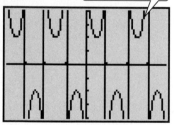

Trig Window
This graph supports the result in Example 2.

Use the fundamental identities to write $\tan \theta + \cot \theta$ in terms of $\sin \theta$ and $\cos \theta$, and then simplify the expression.

From the fundamental identities,

$$\tan \theta + \cot \theta = \frac{\sin \theta}{\cos \theta} + \frac{\cos \theta}{\sin \theta}.$$

Simplify this expression by adding the two fractions on the right side, using the common denominator $\cos \theta \sin \theta$.

$$\tan \theta + \cot \theta = \frac{\sin^2 \theta}{\cos \theta \sin \theta} + \frac{\cos^2 \theta}{\cos \theta \sin \theta}$$

$$= \frac{\sin^2 \theta + \cos^2 \theta}{\cos \theta \sin \theta}$$

Now substitute **1** for $\mathbf{\sin^2 \theta + \cos^2 \theta}$.

$$\tan \theta + \cot \theta = \frac{1}{\cos \theta \sin \theta} \quad ▶$$

Every trigonometric function of an angle θ or a number x can be expressed in terms of every other function. One such case is shown in the next example.

◖**EXAMPLE 3**
Expressing one function in terms of another

Express $\cos x$ in terms of $\tan x$.

Since $\sec x$ is related to both $\cos x$ and $\tan x$ by identities, start with $\tan^2 x + 1 = \sec^2 x$. Then take reciprocals to get

$$\frac{1}{\tan^2 x + 1} = \frac{1}{\sec^2 x}$$

$$\text{or} \qquad \frac{1}{\tan^2 x + 1} = \cos^2 x$$

$$\pm\sqrt{\frac{1}{\tan^2 x + 1}} = \cos x \qquad \text{Take the square root of both sides.}$$

$$\cos x = \frac{\pm 1}{\sqrt{\tan^2 x + 1}}.$$

Rationalize the denominator to get

$$\cos x = \frac{\pm\sqrt{\tan^2 x + 1}}{\tan^2 x + 1}.$$

We choose the + sign or the − sign, depending on the quadrant of x.　▶

CAUTION When working with trigonometric expressions and identities, be sure to write the argument of the function. For example, we would *not* write $\sin^2 + \cos^2 = 1$; an argument such as θ is necessary in this identity.

5.1 Exercises ▼▼▼▼▼▼▼▼▼▼▼▼▼▼▼▼▼▼▼▼▼▼▼▼▼▼▼▼▼▼▼▼▼▼▼▼▼

In Exercises 1–4, the given graphing calculator screen is obtained for a particular stored value of X. What will the screen display for the value of the expression in the final line of the display?

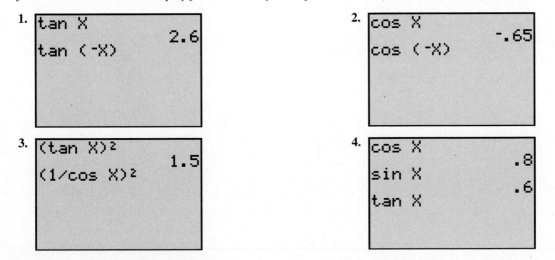

1.
```
tan X
            2.6
tan (-X)
```

2.
```
cos X
           -.65
cos (-X)
```

3.
```
(tan X)²
           1.5
(1/cos X)²
```

4.
```
cos X
            .8
sin X
            .6
tan X
```

In Exercises 5–10, find $\sin s$. *See Example 1.*

5. $\cos s = \dfrac{3}{4}$, s in quadrant I

6. $\cot s = -\dfrac{1}{3}$, s in quadrant IV

7. $\cos s = \dfrac{\sqrt{5}}{5}$, $\tan s < 0$

8. $\tan s = -\dfrac{\sqrt{7}}{2}$, $\sec s > 0$

9. $\sec s = \dfrac{11}{4}$, $\tan s < 0$

10. $\csc s = -\dfrac{8}{5}$

11. Why is it unnecessary to give the quadrant of s in Exercise 10?

12. What is wrong with this problem? "Find sin s if csc $s = -9/5$ and s is in quadrant II."

13. Find tan θ if cos $\theta = -2/5$, and sin $\theta < 0$.

14. Find csc α if tan $\alpha = 6$, and cos $\alpha > 0$.

Use the fundamental identities to find the remaining five trigonometric functions of θ. See Example 1.

15. $\sin \theta = \dfrac{2}{3}$, θ in quadrant II

16. $\cos \theta = \dfrac{1}{5}$, θ in quadrant I

17. $\tan \theta = -\dfrac{1}{4}$, θ in quadrant IV

18. $\csc \theta = -\dfrac{5}{2}$, θ in quadrant III

19. $\cot \theta = \dfrac{4}{3}$, $\sin \theta > 0$

20. $\sin \theta = -\dfrac{4}{5}$, $\cos \theta < 0$

21. $\sec \theta = \dfrac{4}{3}$, $\sin \theta < 0$

22. $\cos \theta = -\dfrac{1}{4}$, $\sin \theta > 0$

For each trigonometric expression in Column I, choose the expression from Column II that completes a fundamental identity.

Column I	Column II
23. $\dfrac{\cos x}{\sin x}$	**(a)** $\sin^2 x + \cos^2 x$
24. $\tan x$	**(b)** $\cot x$
25. $\cos(-x)$	**(c)** $\sec^2 x$
26. $\tan^2 x + 1$	**(d)** $\dfrac{\sin x}{\cos x}$
27. 1	**(e)** $\cos x$

For each expression in Column I, choose the expression from Column II that completes an identity. You will have to rewrite one or both expressions, using a fundamental identity, to recognize the matches.

Column I	Column II
28. $-\tan x \cos x$	**(a)** $\dfrac{\sin^2 x}{\cos^2 x}$
29. $\sec^2 x - 1$	**(b)** $\dfrac{1}{\sec^2 x}$
30. $\dfrac{\sec x}{\csc x}$	**(c)** $\sin(-x)$
31. $1 + \sin^2 x$	**(d)** $\csc^2 x - \cot^2 x + \sin^2 x$
32. $\cos^2 x$	**(e)** $\tan x$

33. A student writes "$1 + \cot^2 = \csc^2$." Comment on this student's work.

34. Another student makes the following claim: "Since $\sin^2 \theta + \cos^2 \theta = 1$, I should be able to also say $\sin \theta + \cos \theta = 1$ if I take the square root of both sides." Comment on this student's statement.

35. Suppose that $\cos \theta = x/(x + 1)$. Find $\sin \theta$.

36. Find $\tan \alpha$ if $\sec \alpha = (p + 4)/p$.

Use the fundamental identities to get an equivalent expression involving only sines and cosines, and then simplify it. See Example 2.

37. $\cot \theta \sin \theta$

38. $\sec \theta \cot \theta \sin \theta$

39. $\cos \theta \csc \theta$

40. $\cot^2 \theta(1 + \tan^2 \theta)$

41. $\sin^2 \theta(\csc^2 \theta - 1)$

42. $(\sec \theta - 1)(\sec \theta + 1)$

43. $(1 - \cos \theta)(1 + \sec \theta)$

44. $\dfrac{\cos \theta + \sin \theta}{\sin \theta}$

45. $\dfrac{\cos^2 \theta - \sin^2 \theta}{\sin \theta \cos \theta}$

46. $\dfrac{1 - \sin^2 \theta}{1 + \cot^2 \theta}$

47. $\tan \theta + \cot \theta$

48. $(\sec \theta + \csc \theta)(\cos \theta - \sin \theta)$

49. $\sin \theta(\csc \theta - \sin \theta)$

50. $\dfrac{1 + \tan^2 \theta}{1 + \cot^2 \theta}$

51. $\sin^2 \theta + \tan^2 \theta + \cos^2 \theta$

52. $\dfrac{\tan(-\theta)}{\sec \theta}$

Complete this chart, so that each trigonometric function in the column at the left is expressed in terms of the functions given across the top. See Example 3.

	$\sin \theta$	$\cos \theta$	$\tan \theta$	$\cot \theta$	$\sec \theta$	$\csc \theta$
53. $\sin \theta$	$\sin \theta$	$\pm\sqrt{1 - \cos^2 \theta}$	$\dfrac{\pm\tan \theta \sqrt{1 + \tan^2 \theta}}{1 + \tan^2 \theta}$			$\dfrac{1}{\csc \theta}$
54. $\cos \theta$		$\cos \theta$	$\dfrac{\pm\sqrt{\tan^2 \theta + 1}}{\tan^2 \theta + 1}$		$\dfrac{1}{\sec \theta}$	
55. $\tan \theta$			$\tan \theta$	$\dfrac{1}{\cot \theta}$		
56. $\cot \theta$			$\dfrac{1}{\tan \theta}$	$\cot \theta$	$\dfrac{\pm\sqrt{\sec^2 \theta - 1}}{\sec^2 \theta - 1}$	
57. $\sec \theta$		$\dfrac{1}{\cos \theta}$			$\sec \theta$	
58. $\csc \theta$	$\dfrac{1}{\sin \theta}$					$\csc \theta$

59. Let $\cos x = \dfrac{1}{5}$. Find all possible values for $\dfrac{\sec x - \tan x}{\sin x}$.

60. Let $\csc x = -3$. Find all possible values for $\dfrac{\sin x + \cos x}{\sec x}$.

61. Look at the graphs of cot x, csc x, and sec x and determine which of them are odd and which are even.

62. Write the functions cot x, csc x, and sec x in terms of the tangent, sine, and cosine functions, respectively, and determine which of them are odd and which are even.

▼▼▼▼▼▼▼▼▼▼▼▼▼ **DISCOVERING CONNECTIONS** (Exercises 63–67) ▼▼▼▼▼▼▼▼▼▼▼▼▼

In Chapter 4 we graphed functions of the form $y = c + a \cdot f[b(x - d)]$ with the assumption that $b > 0$. To see what happens when $b < 0$, work Exercises 63–67 in order.

63. Use a negative-angle identity to write $y = \sin(-2x)$ as a function of $2x$.

64. How does your answer to Exercise 63 relate to $y = \sin(2x)$?

65. Use a negative-angle identity to write $y = \cos(-4x)$ as a function of $4x$.

66. How does your answer to Exercise 65 relate to $y = \cos(4x)$?

67. Use your results from Exercises 63–66 to rewrite the following with a positive value of b.
   (a) $y = \sin(-4x)$ **(b)** $y = \cos(-2x)$ **(c)** $y = -5 \sin(-3x)$

---

68. How does the graph of $y = \sec(-x)$ compare to that of $y = \sec x$?

69. How does the graph of $y = \csc(-x)$ compare to that of $y = \csc x$?

70. How does the graph of $y = \tan(-x)$ compare to that of $y = \tan x$?

71. How does the graph of $y = \cot(-x)$ compare to that of $y = \cot x$?

Show that each of the following is not an identity by replacing the variables with numbers that show the result to be false.

72. $2 \sin s = \sin 2s$ **73.** $\sin x = \sqrt{1 - \cos^2 x}$ **74.** $\sin(x + y) = \sin x + \sin y$

75. Show that $\sin 1° + \sin 2° + \sin 3° + \cdots + \sin 358° + \sin 359° = 0$. (*Hint:* Pair the first term with the last term, the second term with the next to last, and so on. Then use a negative-angle identity.)

76. Refer to Exercise 75. Can you determine the sum $\cos 1° + \cos 2° + \cdots + \cos 358° + \cos 359°$ in a similar manner? Explain why or why not.

5.2 Verifying Trigonometric Identities ▼▼▼

One of the skills required for more advanced work in mathematics (and especially in calculus) is the ability to use the trigonometric identities to write trigonometric expressions in alternate forms. This skill is developed by using the fundamental identities to verify that a trigonometric equation is an identity (for those values of the variable for which it is defined). Here are some hints that may help you get started.

Verifying Identities

1. Learn the fundamental identities given in the last section. Whenever you see either side of a fundamental identity, the other side should come to mind. Also, be aware of equivalent forms of the fundamental identities. For example $\sin^2 \theta = 1 - \cos^2 \theta$ is an alternative form of $\sin^2 \theta + \cos^2 \theta = 1$.

2. Try to rewrite the more complicated side of the equation so that it is identical to the simpler side.

3. It is often helpful to express all trigonometric functions in the equation in terms of sine and cosine and then simplify the result.

4. Usually any factoring or indicated algebraic operations should be performed. For example, the expression $\sin^2 x + 2 \sin x + 1$ can be factored as follows: $(\sin x + 1)^2$. The sum or difference of two trigonometric expressions, such as

$$\frac{1}{\sin \theta} + \frac{1}{\cos \theta},$$

can be added or subtracted in the same way as any other rational expressions:

$$\frac{1}{\sin \theta} + \frac{1}{\cos \theta} = \frac{\cos \theta}{\sin \theta \cos \theta} + \frac{\sin \theta}{\sin \theta \cos \theta}$$
$$= \frac{\cos \theta + \sin \theta}{\sin \theta \cos \theta}.$$

5. As you select substitutions, keep in mind the side you are not changing, because it represents your goal. For example, to verify the identity

$$\tan^2 x + 1 = \frac{1}{\cos^2 x},$$

try to think of an identity that relates $\tan x$ to $\cos x$. Here, since $\sec x = 1/\cos x$ and $\sec^2 x = \tan^2 x + 1$, the secant function is the best link between the two sides.

6. If an expression contains $1 + \sin x$, multiplying both numerator and denominator by $1 - \sin x$ would give $1 - \sin^2 x$, which could be replaced with $\cos^2 x$. Similar results for $1 - \sin x$, $1 + \cos x$, and $1 - \cos x$ may be useful.

These hints are used in the examples of this section.

CAUTION   Verifying identities is not the same as solving equations. Techniques used in solving equations, such as adding the same terms to both sides, or multiplying both sides by the same term, are not valid when working with identities since you are starting with a statement (to be verified) that may not be true.

EXAMPLE 1
Verifying an identity (working with one side)

Verify that

$$\cot s + 1 = \csc s(\cos s + \sin s)$$

is an identity.

Use the fundamental identities to rewrite one side of the equation so that it is identical to the other side. Since the right side is more complicated, it is probably a good idea to work with it. Here we use the method of changing all the trigonometric functions to sine or cosine.

Steps	**Reasons**
$\csc s(\cos s + \sin s) = \dfrac{1}{\sin s}(\cos s + \sin s)$	$\csc s = \frac{1}{\sin s}$
$= \dfrac{\cos s}{\sin s} + \dfrac{\sin s}{\sin s}$	Distributive property
$= \cot s + 1$	$\frac{\cos s}{\sin s} = \cot s; \frac{\sin s}{\sin s} = 1$

The given equation is an identity since the right side equals the left side.   ▶

EXAMPLE 2
Verifying an identity (working with one side)

Verify that

$$\tan^2 \alpha(1 + \cot^2 \alpha) = \frac{1}{1 - \sin^2 \alpha}$$

is an identity.

Working with the left side gives the following.

$\tan^2 \alpha(1 + \cot^2 \alpha) = \tan^2 \alpha + \tan^2 \alpha \cot^2 \alpha$	Distributive property
$= \tan^2 \alpha + \tan^2 \alpha \cdot \dfrac{1}{\tan^2 \alpha}$	$\cot^2 \alpha = \frac{1}{\tan^2 \alpha}$
$= \tan^2 \alpha + 1$	$\tan^2 \alpha \cdot \frac{1}{\tan^2 \alpha} = 1$
$= \sec^2 \alpha$	$\tan^2 \alpha + 1 = \sec^2 \alpha$
$= \dfrac{1}{\cos^2 \alpha}$	$\sec^2 \alpha = \frac{1}{\cos^2 \alpha}$
$= \dfrac{1}{1 - \sin^2 \alpha}$	$\cos^2 \alpha = 1 - \sin^2 \alpha$

Since the left side equals the right side, the given equation is an identity.   ▶

As mentioned earlier, we can use a graphing calculator to support our algebraic verification of an identity. In Example 2, if we let

$$Y_1 = \tan^2 x(1 + \cot^2 x) \qquad \text{and} \qquad Y_2 = \frac{1}{1 - \sin^2 x},$$

and graph these functions in the same window, the graphs coincide, supporting our analytic work in Example 2. See the figure on the left. (Enter $\tan^2 x$ as $(\tan x)^2$.)

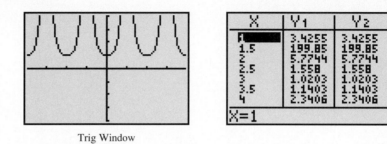

Trig Window

A table of values for Y_1 and Y_2 is shown on the right. The table also supports the identity for selected values in the interval $[1, 4]$.

EXAMPLE 3
Verifying an identity (working with one side)

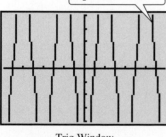

$$Y_1 = \frac{\tan x - \cot x}{\sin x \cos x}$$
$$Y_2 = \sec^2 x - \csc^2 x$$

Trig Window
This graph supports the result in Example 3.

Verify that

$$\frac{\tan t - \cot t}{\sin t \cos t} = \sec^2 t - \csc^2 t$$

is an identity.

Since the left side is the more complicated one, tranform the left side to equal the right side.

$$\frac{\tan t - \cot t}{\sin t \cos t} = \frac{\tan t}{\sin t \cos t} - \frac{\cot t}{\sin t \cos t} \qquad \frac{a-b}{c} = \frac{a}{c} - \frac{b}{c}$$

$$= \tan t \cdot \frac{1}{\sin t \cos t} - \cot t \cdot \frac{1}{\sin t \cos t} \qquad \frac{a}{b} = a \cdot \frac{1}{b}$$

$$= \frac{\sin t}{\cos t} \cdot \frac{1}{\sin t \cos t} - \frac{\cos t}{\sin t} \cdot \frac{1}{\sin t \cos t} \qquad \tan t = \frac{\sin t}{\cos t};$$
$$\cot t = \frac{\cos t}{\sin t}$$

$$= \frac{1}{\cos^2 t} - \frac{1}{\sin^2 t} \qquad \frac{1}{\cos^2 t} = \sec^2 t;$$

$$= \sec^2 t - \csc^2 t \qquad \frac{1}{\sin^2 t} = \csc^2 t$$

Here, writing in terms of sine and cosine only was used in the third line. ▶

EXAMPLE 4
Verifying an identity (working with one side)

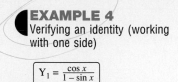

Trig Window
This graph supports the result in Example 4.

Verify that

$$\frac{\cos x}{1 - \sin x} = \frac{1 + \sin x}{\cos x}$$

is an identity.

This time we will work on the right side. Use the suggestion given at the beginning of the section to multiply the numerator and denominator on the right by $1 - \sin x$.

$$\frac{1 + \sin x}{\cos x} = \frac{(1 + \sin x)(1 - \sin x)}{\cos x(1 - \sin x)} \qquad \text{Multiply by 1.}$$

$$= \frac{1 - \sin^2 x}{\cos x(1 - \sin x)}$$

$$= \frac{\cos^2 x}{\cos x(1 - \sin x)} \qquad 1 - \sin^2 x = \cos^2 x$$

$$= \frac{\cos x}{1 - \sin x} \qquad \text{Reduce to lowest terms.} \quad \blacktriangleright$$

If both sides of an identity appear to be equally complex, the identity can be verified by working independently on the left side and on the right side, until each side is changed into some common third result. *Each step, on each side, must be reversible.* With all steps reversible, the procedure is as follows.

left = right
common third
expression

The left side leads to the third expression, which leads back to the right side. This procedure is just a shortcut for the procedure used in the first examples of this section: the left side is changed into the right side, but by going through an intermediate step.

EXAMPLE 5
Verifying an identity (working with both sides)

Verify that

$$\frac{\sec \alpha + \tan \alpha}{\sec \alpha - \tan \alpha} = \frac{1 + 2 \sin \alpha + \sin^2 \alpha}{\cos^2 \alpha}$$

is an identity.

Both sides appear equally complex, so verify the identity by changing each side into a common third expression. Work first on the left, multiplying numerator and denominator by $\cos \alpha$.

$$\frac{\sec \alpha + \tan \alpha}{\sec \alpha - \tan \alpha} = \frac{(\sec \alpha + \tan \alpha) \cos \alpha}{(\sec \alpha - \tan \alpha) \cos \alpha}$$

$\frac{\cos \alpha}{\cos \alpha} = 1;$
multiplicative identity

$$= \frac{\sec \alpha \cos \alpha + \tan \alpha \cos \alpha}{\sec \alpha \cos \alpha - \tan \alpha \cos \alpha}$$

Distributive property

$$= \frac{1 + \tan \alpha \cos \alpha}{1 - \tan \alpha \cos \alpha}$$

$\sec \alpha \cos \alpha = 1$

$$= \frac{1 + \dfrac{\sin \alpha}{\cos \alpha} \cdot \cos \alpha}{1 - \dfrac{\sin \alpha}{\cos \alpha} \cdot \cos \alpha}$$

$\tan \alpha = \frac{\sin \alpha}{\cos \alpha}$

$$= \frac{1 + \sin \alpha}{1 - \sin \alpha}$$

On the right side of the original statement, begin by factoring.

$$\frac{1 + 2 \sin \alpha + \sin^2 \alpha}{\cos^2 \alpha} = \frac{(1 + \sin \alpha)^2}{\cos^2 \alpha}$$

$a^2 + 2ab + b^2 = (a + b)^2$

$$= \frac{(1 + \sin \alpha)^2}{1 - \sin^2 \alpha}$$

$\cos^2 \alpha = 1 - \sin^2 \alpha$

$$= \frac{(1 + \sin \alpha)^2}{(1 + \sin \alpha)(1 - \sin \alpha)}$$

$1 - \sin^2 \alpha =$
$(1 + \sin \alpha)(1 - \sin \alpha)$

$$= \frac{1 + \sin \alpha}{1 - \sin \alpha}$$

Reduce to lowest terms.

We now have shown that

$$\frac{\sec \alpha + \tan \alpha}{\sec \alpha - \tan \alpha} = \frac{1 + \sin \alpha}{1 - \sin \alpha} = \frac{1 + 2 \sin \alpha + \sin^2 \alpha}{\cos^2 \alpha},$$

verifying that the original equation is an identity. ▶

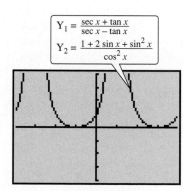

$Y_1 = \frac{\sec x + \tan x}{\sec x - \tan x}$

$Y_2 = \frac{1 + 2 \sin x + \sin^2 x}{\cos^2 x}$

Trig Window
This graph supports the result in Example 5.

CAUTION   This method should be used *only* if the steps are reversible. A good check is to reverse the steps used on the right side to show that you could continue from the last step on the left to get the original statement on the right.

There are usually several ways to verify a given identity. You may wish to go through the examples of this section and verify each using a method different from the one given. For instance, another way to begin verifying the identity in Example 5 is to work on the left as follows.

$$\frac{\sec \alpha + \tan \alpha}{\sec \alpha - \tan \alpha} = \frac{\dfrac{1}{\cos \alpha} + \dfrac{\sin \alpha}{\cos \alpha}}{\dfrac{1}{\cos \alpha} - \dfrac{\sin \alpha}{\cos \alpha}} \qquad \text{Fundamental identities}$$

$$= \frac{\dfrac{1 + \sin \alpha}{\cos \alpha}}{\dfrac{1 - \sin \alpha}{\cos \alpha}} \qquad \text{Add fractions; subtract fractions.}$$

$$= \frac{1 + \sin \alpha}{1 - \sin \alpha} \qquad \text{Divide fractions.}$$

Now we compare this with the result shown in Example 5 for the right side to see that the two sides agree.

CONNECTIONS Much of our work with identities in this chapter is preparation for calculus, where many of the identities we verify here are used. Some calculus problems are simplified by making an appropriate trigonometric substitution. We show one example here where trigonometric substitution and the use of an identity makes it possible to replace an expression such as $\sqrt{9 + x^2}$ with a trigonometric expression without a radical. To do this, we choose $x = 3 \tan \theta$. The reason for this choice will become clear as we continue.

Letting $x = 3 \tan \theta$ gives

$$\sqrt{9 + x^2} = \sqrt{9 + (3 \tan \theta)^2}$$
$$= \sqrt{9 + 9 \tan^2 \theta}$$
$$= \sqrt{9(1 + \tan^2 \theta)}$$
$$= 3\sqrt{1 + \tan^2 \theta}$$
$$= 3\sqrt{\sec^2 \theta}.$$

In the interval $(0, \pi/2)$, the value of $\sec \theta$ is positive, giving

$$\sqrt{9 + x^2} = 3 \sec \theta.$$

FOR DISCUSSION OR WRITING

Substitute $\cos \theta$ for x in $\sqrt{(1 - x^2)^3}$ and simplify. Why is $\cos \theta$ an appropriate choice here?

5.2 Exercises ▼▼▼▼▼▼▼▼▼▼▼▼▼▼▼▼▼▼▼▼▼▼▼▼▼▼▼▼▼▼▼▼▼▼

Perform the indicated operations and simplify the result.

1. $\tan \theta + \dfrac{1}{\tan \theta}$

2. $\dfrac{\cos x}{\sin x} + \dfrac{\sin x}{\cos x}$

3. $\cot s(\tan s + \sin s)$

4. $\sec \beta(\cos \beta + \sin \beta)$

5. $\dfrac{1}{\csc^2 \theta} + \dfrac{1}{\sec^2 \theta}$

6. $\dfrac{1}{\sin \alpha - 1} - \dfrac{1}{\sin \alpha + 1}$

7. $\dfrac{\cos x}{\sec x} + \dfrac{\sin x}{\csc x}$

8. $\dfrac{\cos \gamma}{\sin \gamma} + \dfrac{\sin \gamma}{1 + \cos \gamma}$

9. $(1 + \sin t)^2 + \cos^2 t$

10. $(1 + \tan s)^2 - 2 \tan s$

11. $\dfrac{1}{1 + \cos x} - \dfrac{1}{1 - \cos x}$

12. $(\sin \alpha - \cos \alpha)^2$

Factor each trigonometric expression.

13. $\sin^2 \gamma - 1$

14. $\sec^2 \theta - 1$

15. $(\sin x + 1)^2 - (\sin x - 1)^2$

16. $(\tan x + \cot x)^2 - (\tan x - \cot x)^2$

17. $2 \sin^2 x + 3 \sin x + 1$

18. $4 \tan^2 \beta + \tan \beta - 3$

19. $\cos^4 x + 2 \cos^2 x + 1$

20. $\cot^4 x + 3 \cot^2 x + 2$

21. $\sin^3 x - \cos^3 x$

22. $\sin^3 \alpha + \cos^3 \alpha$

Each expression simplifies to a constant, a single circular function, or a power of a circular function. Use the fundamental identities to simplify each expression.

23. $\tan \theta \cos \theta$

24. $\cot \alpha \sin \alpha$

25. $\sec r \cos r$

26. $\cot t \tan t$

27. $\dfrac{\sin \beta \tan \beta}{\cos \beta}$

28. $\dfrac{\csc \theta \sec \theta}{\cot \theta}$

29. $\sec^2 x - 1$

30. $\csc^2 t - 1$

31. $\dfrac{\sin^2 x}{\cos^2 x} + \sin x \csc x$

32. $\dfrac{1}{\tan^2 \alpha} + \cot \alpha \tan \alpha$

Given a trigonometric identity such as those discussed in this section, it is possible to form another identity by replacing each function with its cofunction. Therefore, for example, since the equation

$$\frac{\cot \theta}{\csc \theta} = \cos \theta,$$

in Exercise 33, represents an identity, the equation

$$\frac{\tan \theta}{\sec \theta} = \sin \theta$$

is also an identity. After verifying the identities in Exercises 33–68, you may wish to construct other identities using this method, in order to provide more practice exercises for yourself.

In Exercises 33–68, verify each trigonometric identity. See Examples 1–5.

33. $\dfrac{\cot \theta}{\csc \theta} = \cos \theta$

34. $\dfrac{\tan \alpha}{\sec \alpha} = \sin \alpha$

35. $\dfrac{1 - \sin^2 \beta}{\cos \beta} = \cos \beta$

Prove

Simplify left side

36. $\dfrac{\tan^2 \gamma + 1}{\sec \gamma} = \sec \gamma$

37. $\cos^2 \theta(\tan^2 \theta + 1) = 1$

38. $\sin^2 \beta(1 + \cot^2 \beta) = 1$

39. $\cot s + \tan s = \sec s \csc s$

40. $\sin^2 \alpha + \tan^2 \alpha + \cos^2 \alpha = \sec^2 \alpha$

41. $\dfrac{\cos \alpha}{\sec \alpha} + \dfrac{\sin \alpha}{\csc \alpha} = \sec^2 \alpha - \tan^2 \alpha$

42. $\dfrac{\sin^2 \gamma}{\cos \gamma} = \sec \gamma - \cos \gamma$

43. $\sin^4 \theta - \cos^4 \theta = 2 \sin^2 \theta - 1$

44. $\dfrac{\cos \theta}{\sin \theta \cot \theta} = 1$

45. $(1 - \cos^2 \alpha)(1 + \cos^2 \alpha) = 2 \sin^2 \alpha - \sin^4 \alpha$

46. $\tan^2 \gamma \sin^2 \gamma = \tan^2 \gamma + \cos^2 \gamma - 1$

47. $\dfrac{\cos \theta + 1}{\tan^2 \theta} = \dfrac{\cos \theta}{\sec \theta - 1}$

48. $\dfrac{(\sec \theta - \tan \theta)^2 + 1}{\sec \theta \csc \theta - \tan \theta \csc \theta} = 2 \tan \theta$

49. $\dfrac{1}{1 - \sin \theta} + \dfrac{1}{1 + \sin \theta} = 2 \sec^2 \theta$

50. $\dfrac{1}{\sec \alpha - \tan \alpha} = \sec \alpha + \tan \alpha$

51. $\dfrac{\tan s}{1 + \cos s} + \dfrac{\sin s}{1 - \cos s} = \cot s + \sec s \csc s$

52. $\dfrac{1 - \cos x}{1 + \cos x} = (\cot x - \csc x)^2$

53. $\dfrac{\cot \alpha + 1}{\cot \alpha - 1} = \dfrac{1 + \tan \alpha}{1 - \tan \alpha}$

54. $\dfrac{1}{\tan \alpha - \sec \alpha} + \dfrac{1}{\tan \alpha + \sec \alpha} = -2 \tan \alpha$

55. $\sin^2 \alpha \sec^2 \alpha + \sin^2 \alpha \csc^2 \alpha = \sec^2 \alpha$

56. $\dfrac{\csc \theta + \cot \theta}{\tan \theta + \sin \theta} = \cot \theta \csc \theta$

57. $\sec^4 x - \sec^2 x = \tan^4 x + \tan^2 x$

58. $\dfrac{1 - \sin \theta}{1 + \sin \theta} = \sec^2 \theta - 2 \sec \theta \tan \theta + \tan^2 \theta$

59. $\sin \theta + \cos \theta = \dfrac{\sin \theta}{1 - \dfrac{\cos \theta}{\sin \theta}} + \dfrac{\cos \theta}{1 - \dfrac{\sin \theta}{\cos \theta}}$

60. $\dfrac{\sin \theta}{1 - \cos \theta} - \dfrac{\sin \theta \cos \theta}{1 + \cos \theta} = \csc \theta(1 + \cos^2 \theta)$

61. $\dfrac{\sec^4 s - \tan^4 s}{\sec^2 s + \tan^2 s} = \sec^2 s - \tan^2 s$

62. $\dfrac{\cot^2 t - 1}{1 + \cot^2 t} = 1 - 2 \sin^2 t$

63. $\dfrac{\tan^2 t - 1}{\sec^2 t} = \dfrac{\tan t - \cot t}{\tan t + \cot t}$

64. $(1 + \sin x + \cos x)^2 = 2(1 + \sin x)(1 + \cos x)$

65. $\dfrac{1 + \cos x}{1 - \cos x} - \dfrac{1 - \cos x}{1 + \cos x} = 4 \cot x \csc x$

66. $(\sec \alpha - \tan \alpha)^2 = \dfrac{1 - \sin \alpha}{1 + \sin \alpha}$

67. $(\sec \alpha + \csc \alpha)(\cos \alpha - \sin \alpha) = \cot \alpha - \tan \alpha$

68. $\dfrac{\sin^4 \alpha - \cos^4 \alpha}{\sin^2 \alpha - \cos^2 \alpha} = 1$

69. A student claims that the equation

$$\cos \theta + \sin \theta = 1$$

is an identity, since by letting $\theta = 90°$ (or $\pi/2$ radians) we get $0 + 1 = 1$, a true statement. Comment on this student's reasoning.

70. Explain why the method described in the text involving working on both sides of an identity to show that each side is equal to the same expression is a valid method of verifying an identity. When using this method, what must be true about each step taken? (*Hint:* See the discussion preceding Example 5.)

In Exercises 71–74, graph each function and conjecture an identity. Then prove your conjecture.

71. $(\sec \theta + \tan \theta)(1 - \sin \theta)$

72. $(\csc \theta + \cot \theta)(\sec \theta - 1)$

73. $\dfrac{\cos \theta + 1}{\sin \theta + \tan \theta}$

74. $\tan \theta \sin \theta + \cos \theta$

📱 *In Exercises 75–84, graph the functions on each side of the equals sign to determine if the equation might be an identity. (Note: Use a domain whose length is at least 2π.) If the equation looks like an identity, prove it algebraically.*

75. $\dfrac{2 + 5 \cos s}{\sin s} = 2 \csc s + 5 \cot s$

76. $1 + \cot^2 s = \dfrac{\sec^2 s}{\sec^2 s - 1}$

77. $\dfrac{\tan s - \cot s}{\tan s + \cot s} = 2 \sin^2 s$

78. $\dfrac{1}{1 + \sin s} + \dfrac{1}{1 - \sin s} = \sec^2 s$

79. $\dfrac{1 - \tan^2 s}{1 + \tan^2 s} = \cos^2 s - \sin s$

80. $\dfrac{\sin^3 s - \cos^3 s}{\sin s - \cos s} = \sin^2 s + 2 \sin s \cos s + \cos^2 s$

81. $\sin^2 s + \cos^2 s = \dfrac{1}{2}(1 - \cos 4s)$

82. $\cos 3s = 3 \cos s + 4 \cos^3 s$

83. $\tan^2 x - \sin^2 x = (\tan x \sin x)^2$

84. $\dfrac{\cot \theta}{\csc \theta + 1} = \sec \theta - \tan \theta$

By substituting a number for s or t, show that the equation is not an identity for all *real numbers s and t.*

85. $\sin(\csc s) = 1$ **86.** $\sqrt{\cos^2 s} = \cos s$ **87.** $\csc t = \sqrt{1 + \cot^2 t}$ **88.** $\cos t = \sqrt{1 - \sin^2 t}$

89. Let $\tan \theta = t$ and show that

$$\sin \theta \cos \theta = \frac{t}{t^2 + 1}.$$

90. When does $\sin x = \sqrt{1 - \cos^2 x}$?

91. An equation that is an identity has an infinite number of solutions. If an equation has an infinite number of solutions, is it necessarily an identity? Discuss.

5.3   Sum and Difference Identities for Cosine ▼▼▼

Several examples presented throughout this book should have convinced you by now that $\cos(A - B)$ does *not* equal $\cos A - \cos B$. For example, if $A = \pi/2$ and $B = 0$,

$$\cos(A - B) = \cos\left(\frac{\pi}{2} - 0\right) = \cos \frac{\pi}{2} = 0,$$

while

$$\cos A - \cos B = \cos \frac{\pi}{2} - \cos 0 = 0 - 1 = -1.$$

The actual formula for $\cos(A - B)$ is derived in this section. Start by locating angles A and B in standard position on a unit circle, with $B < A$. Let S and Q be the points where angles A and B, respectively, intersect the circle.

Locate point R on the unit circle so that angle POR equals the difference $A - B$. See Figure 3.

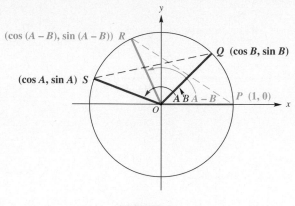

FIGURE 3

Point Q is on the unit circle, so by the work with circular functions in Chapter 3, the x-coordinate of Q is given by the cosine of angle B, while the y-coordinate of Q is given by the sine of angle B:

$$Q \text{ has coordinates } (\cos B, \sin B).$$

In the same way,

$$S \text{ has coordinates } (\cos A, \sin A),$$

and

$$R \text{ has coordinates } (\cos(A - B), \sin(A - B)).$$

Angle SOQ also equals $A - B$. Since the central angles SOQ and POR are equal, chords PR and SQ are equal. By the distance formula, since $PR = SQ$,

$$\sqrt{[\cos(A - B) - 1]^2 + [\sin(A - B) - 0]^2}$$
$$= \sqrt{(\cos A - \cos B)^2 + (\sin A - \sin B)^2}.$$

Squaring both sides and clearing parentheses gives

$$\cos^2(A - B) - 2\cos(A - B) + 1 + \sin^2(A - B)$$
$$= \cos^2 A - 2\cos A \cos B + \cos^2 B + \sin^2 A - 2\sin A \sin B + \sin^2 B.$$

Since $\sin^2 x + \cos^2 x = 1$ for any value of x, rewrite the equation as

$$2 - 2\cos(A - B) = 2 - 2\cos A \cos B - 2\sin A \sin B$$
$$\cos(A - B) = \cos A \cos B + \sin A \sin B.$$

This is the identity for $\cos(A - B)$. Although Figure 3 shows angles A and B in the second and first quadrants, respectively, it can be shown that this result is the same for any values of these angles.

To find a similar expression for $\cos(A + B)$, rewrite $A + B$ as $A - (-B)$ and use the identity for $\cos(A - B)$ found above, along with the fact that $\cos(-B) = \cos B$ and $\sin(-B) = -\sin B$.

$$\cos(A + B) = \cos[A - (-B)]$$
$$= \cos A \cos(-B) + \sin A \sin(-B)$$
$$= \cos A \cos B + \sin A (-\sin B)$$
$$\cos(A + B) = \cos A \cos B - \sin A \sin B$$

The two formulas we have just derived are summarized as follows.

> **Cosine of Sum or Difference**
>
> $$\cos(A - B) = \cos A \cos B + \sin A \sin B$$
> $$\cos(A + B) = \cos A \cos B - \sin A \sin B$$

These identities are important in calculus and other areas of mathematics and useful in certain applications. Although a calculator can be used to find an approximation for $\cos 15°$, for example, the method shown below can be applied to give practice using the sum and difference identities, as well as to get an exact value.

EXAMPLE 1
Using the cosine sum and difference identities to find exact values

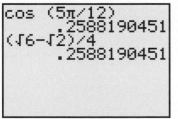

See Example 1(b). The calculator provides support by giving the same approximation for both $\cos \frac{5\pi}{12}$ and $\frac{\sqrt{6} - \sqrt{2}}{4}$.

Find the *exact* value of the following.

(a) $\cos 15°$

To find $\cos 15°$, write $15°$ as the sum or difference of two angles with known function values. Since we know the exact trigonometric function values of both $45°$ and $30°$, write $15°$ as $45° - 30°$. (We could also use $60° - 45°$.) Then use the identity for the cosine of the difference of two angles.

$$\cos 15° = \cos(45° - 30°)$$
$$= \cos 45° \cos 30° + \sin 45° \sin 30° \quad \text{Use the cosine of the difference identity.}$$
$$= \frac{\sqrt{2}}{2} \cdot \frac{\sqrt{3}}{2} + \frac{\sqrt{2}}{2} \cdot \frac{1}{2}$$
$$= \frac{\sqrt{6} + \sqrt{2}}{4}$$

(b) $\cos \frac{5}{12}\pi = \cos\left(\frac{\pi}{6} + \frac{\pi}{4}\right)$

$$= \cos\frac{\pi}{6} \cos\frac{\pi}{4} - \sin\frac{\pi}{6} \sin\frac{\pi}{4} \quad \text{Use the cosine of the sum identity.}$$
$$= \frac{\sqrt{3}}{2} \cdot \frac{\sqrt{2}}{2} - \frac{1}{2} \cdot \frac{\sqrt{2}}{2}$$
$$= \frac{\sqrt{6} - \sqrt{2}}{4}$$

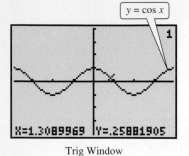

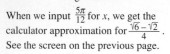

Trig Window

When we input $\frac{5\pi}{12}$ for x, we get the calculator approximation for $\frac{\sqrt{6} - \sqrt{2}}{4}$. See the screen on the previous page.

(c) $\cos 87° \cos 93° - \sin 87° \sin 93°$

From the identity for the cosine of the sum of two angles,

$$\cos 87° \cos 93° - \sin 87° \sin 93° = \cos(87° + 93°)$$
$$= \cos(180°)$$
$$= -1. \quad \blacktriangleright$$

NOTE In Example 1(b) we used the fact that $5\pi/12 = \pi/6 + \pi/4$. At first glance, this sum may not be obvious; but think of the values $\pi/6$ and $\pi/4$ in terms of fractions with denominator 12: $\pi/6 = 2\pi/12$ and $\pi/4 = 3\pi/12$. The list below may help you with problems of this type.

$$\frac{\pi}{3} = \frac{4\pi}{12}$$

$$\frac{\pi}{4} = \frac{3\pi}{12}$$

$$\frac{\pi}{6} = \frac{2\pi}{12}$$

Using this list, for example, we see that $\pi/12 = \pi/3 - \pi/4$ (or $\pi/4 - \pi/6$).

The identities for the cosine of the sum and difference of two angles can be used to derive other identities. Recall the *cofunction identities,* which were presented earlier for values of θ in the interval $[0°, 90°]$.

Cofunction Identities

$$\cos(90° - \theta) = \sin \theta \qquad \cot(90° - \theta) = \tan \theta$$
$$\sin(90° - \theta) = \cos \theta \qquad \sec(90° - \theta) = \csc \theta$$
$$\tan(90° - \theta) = \cot \theta \qquad \csc(90° - \theta) = \sec \theta$$

Similar identities can be obtained for a real number domain by replacing $90°$ by $\pi/2$.

These identities now can be generalized for any angle θ, not just those between $0°$ and $90°$. For example, substituting $90°$ for A and θ for B in the identity given above for $\cos(A - B)$ gives

$$\cos(90° - \theta) = \cos 90° \cos \theta + \sin 90° \sin \theta$$
$$= 0 \cdot \cos \theta + 1 \cdot \sin \theta$$
$$= \sin \theta.$$

This result is true for *any* value of θ since the identity for $\cos(A - B)$ is true for any values of A and B. For the derivations of other cofunction identities, see Exercises 77 and 78.

EXAMPLE 2
Using the cofunction identities

Find an angle θ that satisfies each of the following.

(a) $\cot \theta = \tan 25°$
Since tangent and cotangent are cofunctions,

$$\cot \theta = \tan(90° - \theta).$$

This means that

$$\tan(90° - \theta) = \tan 25°,$$

or
$$90° - \theta = 25°$$
$$\theta = 65°.$$

(b) $\sin \theta = \cos(-30°)$
In the same way,

$$\sin \theta = \cos(90° - \theta) = \cos(-30°),$$

giving
$$90° - \theta = -30°$$
$$\theta = 120°.$$

(c) $\csc \dfrac{3\pi}{4} = \sec \theta$

Cosecant and secant are cofunctions, so

$$\csc \frac{3\pi}{4} = \sec\left(\frac{\pi}{2} - \frac{3\pi}{4}\right) = \sec \theta$$

$$\sec\left(-\frac{\pi}{4}\right) = \sec \theta$$

$$-\frac{\pi}{4} = \theta. \quad \blacktriangleright$$

NOTE   Because trigonometric (and circular) functions are periodic, the solutions in Example 2 are not unique. In each case, we give only one of infinitely many possibilities.

If one of the angles A or B in the identities for $\cos(A + B)$ and $\cos(A - B)$ is a quadrantal angle, then the identity allows us to write the expression in terms of a single function of A or B. The next example illustrates this.

EXAMPLE 3
Reducing $\cos(A - B)$ to a function of a single variable

Write $\cos(180° - \theta)$ as a trigonometric function of θ.
Use the difference identity. Replace A with $180°$ and B with θ.

$$\cos(180° - \theta) = \cos 180° \cos \theta + \sin 180° \sin \theta$$
$$= (-1) \cos \theta + (0) \sin \theta$$
$$= -\cos \theta \quad \blacktriangleright$$

EXAMPLE 4
Finding $\cos(s + t)$ given information about s and t

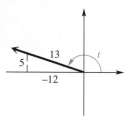

FIGURE 4

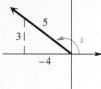

FIGURE 5

Suppose that $\sin s = 3/5$, $\cos t = -12/13$, and both s and t are in quadrant II. Find $\cos(s + t)$.

By the identity above, $\cos(s + t) = \cos s \cos t - \sin s \sin t$. The values of $\sin s$ and $\cos t$ are given, so that $\cos(s + t)$ can be found if $\cos s$ and $\sin t$ are known. We can find $\cos s$ and $\sin t$ by sketching two angles in the second quadrant, one with $\sin s = 3/5$ and the other with $\cos t = -12/13$. See Figures 4 and 5.

In Figure 4, since $\sin s = 3/5 = y/r$, let $y = 3$ and $r = 5$. Substituting gives $\sqrt{x^2 + 3^2} = 5^2$; solve to get $x = -4$. Thus, $\cos s = -4/5$. In Figure 5, $\cos t = -12/13 = x/r$, so we let $x = -12$ and $r = 13$. Then $\sqrt{(-12)^2 + y^2} = 13^2$; solve to get $y = 5$. Thus, $\sin t = 5/13$.

Now find $\cos(s + t)$.

$$\cos(s + t) = \cos s \cos t - \sin s \sin t$$
$$= -\frac{4}{5} \cdot \left(-\frac{12}{13}\right) - \frac{3}{5} \cdot \frac{5}{13}$$
$$= \frac{48}{65} - \frac{15}{65}$$
$$= \frac{33}{65}$$

NOTE In Example 4, the values of $\cos s$ and $\sin t$ could also be found by using the Pythagorean identities. The problem could then be solved using the identity for $\cos(s + t)$ in the same way as shown in the example.

EXAMPLE 5
Applying the cosine difference identity to voltage

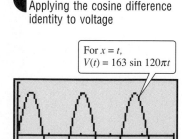

For $x = t$,
$V(t) = 163 \sin 120\pi t$

[0, .05] by [−200, 200]
Xscl = .01   Yscl = 50

FIGURE 6

Common household electrical current is called alternating current because the current alternates direction within the wires. The voltage V in a typical 115-volt outlet can be expressed using the equation $V = 163 \sin \omega t$, where ω is the angular velocity (in radians per second) of the rotating generator at the electrical plant and t is time measured in seconds.*

(a) It is essential for electrical generators to rotate at precisely 60 cycles per second so that household appliances and computers will function properly. Determine ω for these electrical generators. (Alternating current that cycles 60 times per second is often listed as 60 hertz. 1 hertz (Hz) is equal to 1 cycle per second.)

Since each cycle is 2π radians, at 60 cycles per second, $\omega = 60(2\pi) = 120\pi$ radians per second.

(b) Graph V on the interval $0 \leq t \leq .05$.

$V = 163 \sin \omega t = 163 \sin 120\pi t$. Because the amplitude is 163 here, we choose $-200 \leq V \leq 200$ for the range, as shown in Figure 6.

---

*Source: Bell, D., *Fundamentals of Electric Circuits 2e,* Reston Publishing Company, Inc., Reston, Virginia, 1981.

   (c) For what value of ϕ will the graph of $V = 163 \cos(\omega t - \phi)$ be the same as the graph of $V = 163 \sin \omega t$?

      Since $\cos(x - \pi/2) = \sin x$, choose $\phi = \pi/2$. In the exercises, you will be asked to use the cosine of the difference identity to show that $V = 163 \sin \omega t$ and $V = 163 \cos(\omega t - \pi/2)$ are equivalent equations. ▶

5.3 Exercises ▼▼▼▼▼▼▼▼▼▼▼▼▼▼▼▼▼▼▼▼▼▼▼▼▼▼▼▼▼▼▼▼▼▼▼

1. Compare the formulas for $\cos(A - B)$ and $\cos(A + B)$. How do they differ? How are they alike?

2. What does the cofunction identity $\cos(\pi/2 - \theta) = \sin \theta$ imply about the graphs of the cosine and sine functions? [*Hint:* First observe that $\cos(\pi/2 - \theta)$ is the same as $\cos(\theta - \pi/2)$.]

Use the sum and difference identities for cosine to find the exact value. (Do not use a calculator.) See Example 1.

3. $\cos 75°$

4. $\cos(-15°)$

5. $\cos(-75°)$

6. $\cos(105°)$ [*Hint:* $105° = 60° + 45°$]

7. $\cos(-105°)$ [*Hint:* $-105° = -60° + (-45°)$]

8. $\cos\left(\dfrac{7\pi}{12}\right)$

9. $\cos\left(-\dfrac{\pi}{12}\right)$

10. $\cos\left(-\dfrac{5\pi}{12}\right)$

11. $\cos 40° \cos 50° - \sin 40° \sin 50°$

12. $\cos(-10°) \cos 35° + \sin(-10°) \sin 35°$

13. $\cos \dfrac{2\pi}{5} \cos \dfrac{\pi}{10} - \sin \dfrac{2\pi}{5} \sin \dfrac{\pi}{10}$

14. $\cos \dfrac{7\pi}{9} \cos \dfrac{2\pi}{9} - \sin \dfrac{7\pi}{9} \sin \dfrac{2\pi}{9}$

Write each of the following in terms of the cofunction of a complementary angle. See Example 2.

15. $\tan 87°$

16. $\sin 15°$

17. $\cos \dfrac{\pi}{12}$

18. $\sin \dfrac{2\pi}{5}$

19. $\csc(-14° \, 24')$

20. $\sin 142° \, 14'$

21. $\sin \dfrac{5\pi}{8}$

22. $\cot \dfrac{9\pi}{10}$

23. $\sec 146° \, 42'$

24. $\tan 174° \, 3'$

25. $\cot 176.9814°$

26. $\sin 98.0142°$

Use the cofunction identities to fill in the blank with the appropriate trigonometric function name. See Example 2.

27. $\cot \dfrac{\pi}{3} = \underline{\hspace{1.2cm}} \dfrac{\pi}{6}$

28. $\sin \dfrac{2\pi}{3} = \underline{\hspace{1.2cm}} \left(-\dfrac{\pi}{6}\right)$

29. $\underline{\hspace{1.2cm}} 33° = \sin 57°$

30. $\underline{\hspace{1.2cm}} 72° = \cot 18°$

31. $\cos 70° = \dfrac{1}{\underline{\hspace{0.8cm}} 20°}$

32. $\tan 24° = \dfrac{1}{\underline{\hspace{0.8cm}} 66°}$

Use the cofunction identities to find an angle θ that makes each statement true. See Example 2.

33. $\tan \theta = \cot(45° + 2\theta)$

34. $\sin \theta = \cos(2\theta - 10°)$

35. $\sec \theta = \csc\left(\dfrac{\theta}{2} + 20°\right)$

36. $\cos \theta = \sin(3\theta + 10°)$

37. $\sin(3\theta - 15°) = \cos(\theta + 25°)$

38. $\cot(\theta - 10°) = \tan(2\theta + 20°)$

Use the identities for the cosine of a sum or a difference to reduce each expression to a single function of θ. See Example 3.

39. $\cos(0° - \theta)$ **40.** $\cos(90° - \theta)$ **41.** $\cos(180° - \theta)$ **42.** $\cos(270° - \theta)$

43. $\cos(0° + \theta)$ **44.** $\cos(90° + \theta)$ **45.** $\cos(180° + \theta)$ **46.** $\cos(270° + \theta)$

Find $\cos(s + t)$ and $\cos(s - t)$. See Example 4.

47. $\cos s = -1/5$ and $\sin t = 3/5$, s and t in quadrant II

48. $\sin s = 2/3$ and $\sin t = -1/3$, s in quadrant II and t in quadrant IV

49. $\sin s = 3/5$ and $\sin t = -12/13$, s in quadrant I and t in quadrant III

50. $\cos s = -8/17$ and $\cos t = -3/5$, s and t in quadrant III

51. $\sin s = \sqrt{5}/7$ and $\sin t = \sqrt{6}/8$, s and t in quadrant I

52. $\cos s = \sqrt{2}/4$ and $\sin t = -\sqrt{5}/6$, s and t in quadrant IV

Tell whether each statement is true *or* false.

53. $\cos 42° = \cos(30° + 12°)$ **54.** $\cos(-24°) = \cos 16° - \cos 40°$

55. $\cos 74° = \cos 60° \cos 14° + \sin 60° \sin 14°$ **56.** $\cos 140° = \cos 60° \cos 80° - \sin 60° \sin 80°$

57. $\cos \dfrac{\pi}{3} = \cos \dfrac{\pi}{12} \cos \dfrac{\pi}{4} - \sin \dfrac{\pi}{12} \sin \dfrac{\pi}{4}$ **58.** $\cos \dfrac{2\pi}{3} = \cos \dfrac{11\pi}{12} \cos \dfrac{\pi}{4} + \sin \dfrac{11\pi}{12} \sin \dfrac{\pi}{4}$

59. $\cos 70° \cos 20° - \sin 70° \sin 20° = 0$ **60.** $\cos 85° \cos 40° + \sin 85° \sin 40° = \dfrac{\sqrt{2}}{2}$

61. $\tan\left(\theta - \dfrac{\pi}{2}\right) = \cot \theta$ **62.** $\sin\left(\theta - \dfrac{\pi}{2}\right) = \cos \theta$

Verify the identity.

63. $\cos\left(\dfrac{\pi}{2} + x\right) = -\sin x$ **64.** $\sec(\pi - x) = -\sec x$

65. $\cos 2x = \cos^2 x - \sin^2 x$ [*Hint:* $\cos 2x = \cos(x + x)$.]

66. $1 + \cos 2x - \cos^2 x = \cos^2 x$ [*Hint:* Use the result in Exercise 65.]

67. $\cos(\pi + s - t) = -\sin s \sin t - \cos s \cos t$ **68.** $\cos\left(\dfrac{\pi}{2} + s - t\right) = \sin(t - s)$

69. $\cos(\alpha + \beta) \cos(\alpha - \beta) = 1 - \sin^2 \alpha - \sin^2 \beta$ **70.** $\cos 4x \cos 7x - \sin 4x \sin 7x = \cos 11x$

71. Suppose a fellow student tells you that the cosine of the sum of two angles is the sum of their cosines. Write in your own words how you would correct this student's statement.

72. By a cofunction identity, $\cos 20° = \sin 70°$. What are some values other than 70° that make $\cos 20° = \sin \theta$ a true statement?

▼▼▼▼▼▼▼▼▼▼▼▼▼▼ **DISCOVERING CONNECTIONS** (Exercises 73–76) ▼▼▼▼▼▼▼▼▼▼▼▼▼▼

The identities for $\cos(A + B)$ and $\cos(A - B)$ can be used to find exact values of expressions like $\cos 195°$ and $\cos 255°$, where the angle is not in the first quadrant. Work Exercises 73–76 in order to see how this is done.

73. By writing 195° as 180° + 15°, use the identity for $\cos(A + B)$ to express $\cos 195°$ as $-\cos 15°$.

74. Use the identity for $\cos(A - B)$ to find $-\cos 15°$.

75. By the results of Exercises 73 and 74, cos 195° = _____ .

76. Find the exact value of each of the following using the method shown in Exercises 73–75.

   (a) cos 255°   **(b)** cos $\dfrac{11\pi}{12}$

---

77. Use the identity cos(90° − θ) = sin θ, and replace θ with 90° − A, to derive the identity cos A = sin(90° − A).

78. Use the identities in Exercise 77 to derive the identity tan A = cot(90° − A).

79. Let $f(x) = \cos x$. Prove that $\dfrac{f(x+h) - f(x)}{h} = \cos x\left(\dfrac{\cos h - 1}{h}\right) - \sin x\left(\dfrac{\sin h}{h}\right)$.

80. Without using a calculator, show that cos 140° + cos 100° + cos 20° = 0.

   Exercises 81 and 82 refer to Example 5.

81. How many times does the current oscillate in .05 second?

82. What are the maximum and minimum voltages in this outlet? Is the voltage always equal to 115 volts?

83.   Sound is a result of waves applying pressure to a person's eardrum. For a pure sound wave radiating outward in a spherical shape, the trigonometric function $P = \dfrac{a}{r}\cos\left[\dfrac{2\pi r}{\lambda} - ct\right]$ can be used to express the sound pressure at a radius of r feet from the source: t is the time in seconds, λ is the length of the sound wave in feet, c is the speed of sound in feet per second, and a is the maximum sound pressure at the source measured in pounds per square foot. (*Source:* Beranek, L., *Noise and Vibration Control,* Institute of Noise Control Engineering, Washington, D.C., 1988.)

   (a) Let $a = .4$ lb per ft², $\lambda = 4.9$ ft, and $c = 1026$ ft per sec. Graph the sound pressure at a distance of $r = 10$ feet from its source over the interval $0 \le t \le .05$. Describe P at this distance.

   (b) Now let $a = 3$ and $t = 10$. Graph the sound pressure for $0 \le r \le 20$. What happens to the pressure P as the radius r increases?

   (c) Suppose a person stands at a radius r so that $r = n\lambda$ where n is a positive integer. Use the difference identity for cosine to simplify P in this situation.

5.4   Sum and Difference Identities For Sine and Tangent ▼▼▼

Formulas for sin(A + B) and sin(A − B) can be developed from the results in Section 5.3. Start with the cofunction relationship

$$\sin \theta = \cos(90° - \theta).$$

Replace θ with A + B.

$$\sin(A + B) = \cos[90° - (A + B)]$$
$$= \cos[(90° - A) - B]$$

Using the formula for cos(A − B) from the previous section gives

$$\sin(A + B) = \cos(90° - A)\cos B + \sin(90° - A)\sin B$$

or $\sin(A + B) = \sin A \cos B + \cos A \sin B$.

(The cofunction relationships were used in the last step.)

Now write $\sin(A - B)$ as $\sin[A + (-B)]$ and use the identity for $\sin(A + B)$ to get

$$\sin(A - B) = \sin[A + (-B)]$$
$$= \sin A \cos(-B) + \cos A \sin(-B)$$
$$= \sin A \cos B - \cos A \sin B$$

since $\cos(-B) = \cos B$ and $\sin(-B) = -\sin B$. In summary,

$$\sin(A - B) = \sin A \cos B - \cos A \sin B.$$

Using the identities for $\sin(A + B)$, $\cos(A + B)$, $\sin(A - B)$, and $\cos(A - B)$, and the identity $\tan \theta = \sin \theta / \cos \theta$, gives the following identities.

$$\tan(A + B) = \frac{\tan A + \tan B}{1 - \tan A \tan B}$$

$$\tan(A - B) = \frac{\tan A - \tan B}{1 + \tan A \tan B}$$

We show the proof for the first of these two identities. The proof for the other is very similar. Start with

$$\tan(A + B) = \frac{\sin(A + B)}{\cos (A + B)}$$
$$= \frac{\sin A \cos B + \cos A \sin B}{\cos A \cos B - \sin A \sin B}.$$

To express this result in terms of the tangent function, multiply both numerator and denominator by $1/(\cos A \cos B)$.

$$\tan(A + B) = \frac{\dfrac{\sin A \cos B + \cos A \sin B}{1}}{\dfrac{\cos A \cos B - \sin A \sin B}{1}} \cdot \frac{\dfrac{1}{\cos A \cos B}}{\dfrac{1}{\cos A \cos B}}$$

$$= \frac{\dfrac{\sin A \cos B}{\cos A \cos B} + \dfrac{\cos A \sin B}{\cos A \cos B}}{\dfrac{\cos A \cos B}{\cos A \cos B} - \dfrac{\sin A \sin B}{\cos A \cos B}}$$

$$= \frac{\dfrac{\sin A}{\cos A} + \dfrac{\sin B}{\cos B}}{1 - \dfrac{\sin A}{\cos A} \cdot \dfrac{\sin B}{\cos B}}$$

Using the identity $\tan \theta = \sin \theta / \cos \theta$,

$$\tan(A + B) = \frac{\tan A + \tan B}{1 - \tan A \tan B}.$$

The identities given in this section are summarized below.

Sine and Tangent of Sum or Difference

$$\sin(A + B) = \sin A \cos B + \cos A \sin B$$
$$\sin(A - B) = \sin A \cos B - \cos A \sin B$$
$$\tan(A + B) = \frac{\tan A + \tan B}{1 - \tan A \tan B}$$
$$\tan(A - B) = \frac{\tan A - \tan B}{1 + \tan A \tan B}$$

Again, the following examples and the corresponding exercises are given primarily to offer practice in using these new identities.

EXAMPLE 1
Using the sine and tangent sum and difference identities to find exact values

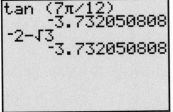

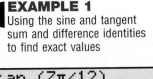

See Example 1(b). The calculator provides support by giving the same approximation for both tan $\frac{7\pi}{12}$ and $-2 - \sqrt{3}$.

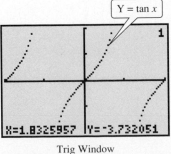

Trig Window
Dot mode

When we input $\frac{7\pi}{12}$ for x, we get the calculator approximation for $-2 - \sqrt{3}$. See the screen above this one.

Find the *exact* value of the following.

(a) $\sin 75° = \sin(45° + 30°)$
$$= \sin 45° \cos 30° + \cos 45° \sin 30°$$
$$= \frac{\sqrt{2}}{2} \cdot \frac{\sqrt{3}}{2} + \frac{\sqrt{2}}{2} \cdot \frac{1}{2}$$
$$= \frac{\sqrt{6}}{4} + \frac{\sqrt{2}}{4} = \frac{\sqrt{6} + \sqrt{2}}{4}$$

(b) $\tan \dfrac{7\pi}{12} = \tan\left(\dfrac{\pi}{3} + \dfrac{\pi}{4}\right)$
$$= \frac{\tan \dfrac{\pi}{3} + \tan \dfrac{\pi}{4}}{1 - \tan \dfrac{\pi}{3} \tan \dfrac{\pi}{4}}$$
$$= \frac{\sqrt{3} + 1}{1 - \sqrt{3} \cdot 1}$$
$$= \frac{\sqrt{3} + 1}{1 - \sqrt{3}} \cdot \frac{1 + \sqrt{3}}{1 + \sqrt{3}}$$
$$= \frac{\sqrt{3} + 3 + 1 + \sqrt{3}}{1 - 3}$$
$$= \frac{4 + 2\sqrt{3}}{-2}$$
$$= -2 - \sqrt{3}$$

(c) $\sin 40° \cos 160° - \cos 40° \sin 160° = \sin(40° - 160°)$
$$= \sin(-120°)$$
$$= -\sin 120°$$
$$= -\frac{\sqrt{3}}{2}$$ ▶

◖EXAMPLE 2
Writing a function as an expression involving functions of θ

Write each of the following as an expression involving functions of θ.

(a) $\sin(30° + \theta)$

Using the identity for $\sin(A + B)$,

$$\sin(30° + \theta) = \sin 30° \cos \theta + \cos 30° \sin \theta$$

$$= \frac{1}{2} \cos \theta + \frac{\sqrt{3}}{2} \sin \theta.$$

(b) $\tan(45° - \theta) = \dfrac{\tan 45° - \tan \theta}{1 + \tan 45° \tan \theta} = \dfrac{1 - \tan \theta}{1 + \tan \theta}$

(c) $\sin(180° + \theta) = \sin 180° \cos \theta + \cos 180° \sin \theta$
$$= 0 \cdot \cos \theta + (-1) \sin \theta$$
$$= -\sin \theta \quad ▶$$

◖EXAMPLE 3
Finding functions and the quadrant of $A + B$ given information about A and B

If $\sin A = 4/5$ and $\cos B = -5/13$, where A is in quadrant II and B is in quadrant III, find each of the following.

(a) $\sin(A + B)$

The identity for $\sin(A + B)$ requires $\sin A$, $\cos A$, $\sin B$, and $\cos B$. Two of these values are given. The two missing values, $\cos A$ and $\sin B$, must be found first. These values can be found with the identity $\sin^2 x + \cos^2 x = 1$. To find $\cos A$, use

$$\sin^2 A + \cos^2 A = 1$$

$$\frac{16}{25} + \cos^2 A = 1$$

$$\cos^2 A = \frac{9}{25}$$

$$\cos A = -\frac{3}{5}. \qquad \text{Since } A \text{ is in quadrant II, } \cos A < 0.$$

In the same way, $\sin B = -12/13$. Now use the formula for $\sin(A + B)$.

$$\sin(A + B) = \frac{4}{5}\left(-\frac{5}{13}\right) + \left(-\frac{3}{5}\right)\left(-\frac{12}{13}\right)$$

$$= -\frac{20}{65} + \frac{36}{65} = \frac{16}{65}$$

(b) $\tan(A + B)$

Use the values of sine and cosine from part (a) to get $\tan A = -4/3$ and $\tan B = 12/5$. Then

$$\tan(A + B) = \frac{-\dfrac{4}{3} + \dfrac{12}{5}}{1 - \left(-\dfrac{4}{3}\right)\left(\dfrac{12}{5}\right)} = \frac{\dfrac{16}{15}}{1 + \dfrac{48}{15}} = \frac{\dfrac{16}{15}}{\dfrac{63}{15}} = \frac{16}{63}.$$

(c) the quadrant of $A + B$

From the results of parts (a) and (b), we find that $\sin(A + B)$ is positive and $\tan(A + B)$ is also positive. Therefore, $A + B$ must be in quadrant I, since it is the only quadrant in which both sine and tangent are positive. ▶

The final example shows how to verify identities using the results of this section and the previous one.

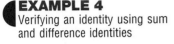

EXAMPLE 4
Verifying an identity using sum and difference identities

Verify the identity

$$\sin\left(\frac{\pi}{6} + s\right) + \cos\left(\frac{\pi}{3} + s\right) = \cos s.$$

Work on the left side, using the identities for $\sin(A + B)$ and $\cos(A + B)$.

$$\sin\left(\frac{\pi}{6} + s\right) + \cos\left(\frac{\pi}{3} + s\right)$$

$$= \left(\sin\frac{\pi}{6} \cos s + \cos\frac{\pi}{6} \sin s\right) \qquad \sin(A + B) = \sin A \cos B + \cos A \sin B$$

$$+ \left(\cos\frac{\pi}{3} \cos s - \sin\frac{\pi}{3} \sin s\right) \qquad \cos(A + B) = \cos A \cos B - \sin A \sin B$$

$$= \left(\frac{1}{2} \cos s + \frac{\sqrt{3}}{2} \sin s\right) \qquad \sin\frac{\pi}{6} = \frac{1}{2}; \quad \cos\frac{\pi}{6} = \frac{\sqrt{3}}{2}$$

$$+ \left(\frac{1}{2} \cos s - \frac{\sqrt{3}}{2} \sin s\right) \qquad \cos\frac{\pi}{3} = \frac{1}{2}; \quad \sin\frac{\pi}{3} = \frac{\sqrt{3}}{2}$$

$$= \frac{1}{2} \cos s + \frac{1}{2} \cos s$$

$$= \cos s \quad ▶$$

5.4 Exercises ▼▼▼▼▼▼▼▼▼▼▼▼▼▼▼▼▼▼▼▼▼▼▼▼▼▼▼▼▼▼▼▼▼▼

1. Compare the formulas for $\sin(A - B)$ and $\sin(A + B)$. How do they differ? How are they alike?

2. Compare the formulas for $\tan(A - B)$ and $\tan(A + B)$. How do they differ? How are they alike?

Use the identities of this section to find the exact value of each of the following. See Example 1.

3. $\sin 15°$ **4.** $\sin 105°$ **5.** $\tan 15°$ **6.** $\tan 105°$ **7.** $\sin(-105°)$ **8.** $\tan(-105°)$

9. $\sin\dfrac{5\pi}{12}$ **10.** $\tan\dfrac{5\pi}{12}$ **11.** $\tan\dfrac{\pi}{12}$ **12.** $\sin\dfrac{\pi}{12}$ **13.** $\sin\left(-\dfrac{7\pi}{12}\right)$ **14.** $\tan\left(-\dfrac{7\pi}{12}\right)$

15. $\sin 76° \cos 31° - \cos 76° \sin 31°$ **16.** $\sin 40° \cos 50° + \cos 40° \sin 50°$

17. $\dfrac{\tan 80° + \tan 55°}{1 - \tan 80° \tan 55°}$ **18.** $\dfrac{\tan 80° - \tan(-55°)}{1 + \tan 80° \tan(-55°)}$

19. $\dfrac{\tan 100° + \tan 80°}{1 - \tan 100° \tan 80°}$

20. $\sin 100° \cos 10° - \cos 100° \sin 10°$

21. $\sin \dfrac{\pi}{5} \cos \dfrac{3\pi}{10} + \cos \dfrac{\pi}{5} \sin \dfrac{3\pi}{10}$

22. $\dfrac{\tan \dfrac{5\pi}{12} + \tan \dfrac{\pi}{4}}{1 - \tan \dfrac{5\pi}{12} \tan \dfrac{\pi}{4}}$

Use the identities of this section and the previous one to express each of the following as an expression involving functions of x or θ. See Example 2.

23. $\cos(30° + \theta)$ **24.** $\cos(45° - \theta)$ **25.** $\cos(60° + \theta)$ **26.** $\cos(\theta - 30°)$ **27.** $\cos\left(\dfrac{3\pi}{4} - x\right)$

28. $\sin(45° + \theta)$ **29.** $\tan(\theta + 30°)$ **30.** $\tan\left(\dfrac{\pi}{4} + x\right)$ **31.** $\sin\left(\dfrac{\pi}{4} + x\right)$ **32.** $\sin(180° - \theta)$

33. $\sin(270° - \theta)$ **34.** $\tan(180° + \theta)$ **35.** $\tan(360° - \theta)$ **36.** $\sin(\pi + \theta)$ **37.** $\tan(\pi - \theta)$

38. Why is it not possible to follow Example 2 and find a formula for $\tan(270° - \theta)$?

39. What happens when you try to evaluate $\dfrac{\tan 65.902° + \tan 24.098°}{1 - \tan 65.902° \tan 24.098°}$?

40. Show that if A, B, and C are the angles of a triangle, then $\sin(A + B + C) = 0$.

For each of the following, find $\sin(s + t)$, $\sin(s - t)$, $\tan(s + t)$, $\tan(s - t)$, the quadrant of $s + t$, and the quadrant of $s - t$. See Example 3.

41. $\cos s = 3/5$ and $\sin t = 5/13$, s and t in quadrant I

42. $\cos s = -1/5$ and $\sin t = 3/5$, s and t in quadrant II

43. $\sin s = 2/3$ and $\sin t = -1/3$, s in quadrant II and t in quadrant IV

44. $\sin s = 3/5$ and $\sin t = -12/13$, s in quadrant I and t in quadrant III

45. $\cos s = -8/17$ and $\cos t = -3/5$, s and t in quadrant III

46. $\cos s = -15/17$ and $\sin t = 4/5$, s in quadrant II and t in quadrant I

47. $\sin s = -4/5$ and $\cos t = 12/13$, s in quadrant III and t in quadrant IV

48. $\sin s = -5/13$ and $\sin t = 3/5$, s in quadrant III and t in quadrant II

49. $\cos s = -\sqrt{7}/4$ and $\sin t = \sqrt{3}/5$, s and t in quadrant II

50. $\cos s = \sqrt{11}/5$ and $\cos t = \sqrt{2}/6$, s and t in quadrant IV

🖩 *In Exercises 51–54, graph each expression and use the graph to conjecture an identity. Then verify your conjecture as in Example 4.*

51. $\sin\left(\dfrac{\pi}{2} + x\right)$ **52.** $\sin\left(\dfrac{3\pi}{2} + x\right)$ **53.** $\tan\left(\dfrac{\pi}{2} + x\right)$ **54.** $\dfrac{1 + \tan x}{1 - \tan x}$

Verify that each statement is an identity. See Example 4.

55. $\sin 2x = 2 \sin x \cos x$ [*Hint:* $\sin 2x = \sin(x + x)$]

56. $\sin(x + y) + \sin(x - y) = 2 \sin x \cos y$

57. $\tan(x - y) - \tan(y - x) = \dfrac{2(\tan x - \tan y)}{1 + \tan x \tan y}$

58. $\sin(210° + x) - \cos(120° + x) = 0$

59. $\dfrac{\cos(\alpha - \beta)}{\cos \alpha \sin \beta} = \tan \alpha + \cot \beta$

60. $\dfrac{\sin(s + t)}{\cos s \cos t} = \tan s + \tan t$

61. $\dfrac{\sin(x - y)}{\sin(x + y)} = \dfrac{\tan x - \tan y}{\tan x + \tan y}$

62. $\dfrac{\sin(x + y)}{\cos(x - y)} = \dfrac{\cot x + \cot y}{1 + \cot x \cot y}$

63. $\dfrac{\sin(s - t)}{\sin t} + \dfrac{\cos(s - t)}{\cos t} = \dfrac{\sin s}{\sin t \cos t}$

64. $\dfrac{\tan(\alpha + \beta) - \tan \beta}{1 + \tan(\alpha + \beta) \tan \beta} = \tan \alpha$

Find the exact value. (See the technique developed in Exercises 73–76 of Section 5.3.)

65. $\sin 165°$ **66.** $\tan 165°$ **67.** $\tan 255°$ **68.** $\sin 255°$ **69.** $\sin 285°$

70. $\tan 285°$ **71.** $\sin \dfrac{11\pi}{12}$ **72.** $\tan \dfrac{11\pi}{12}$ **73.** $\tan\left(-\dfrac{13\pi}{12}\right)$ **74.** $\sin\left(-\dfrac{13\pi}{12}\right)$

75. Derive the identity for $\tan(A - B)$ using the identity for $\tan(A + B)$, and the fact that $A - B = A + (-B)$.

76. Derive the identity for $\tan(A - B)$ using the identities for $\sin(A - B)$ and $\cos(A - B)$, and the fact that $\tan(A - B) = \dfrac{\sin(A - B)}{\cos(A - B)}$.

Derive a formula for each statement.

77. $\sin(A + B + C)$ **78.** $\cos(A + B + C)$

79. Let $f(x) = \sin x$. Show that $\dfrac{f(x + h) - f(x)}{h} = \sin x\left(\dfrac{\cos h - 1}{h}\right) + \cos x\left(\dfrac{\sin h}{h}\right)$.

▼▼▼▼▼▼▼▼▼▼▼▼▼▼ **DISCOVERING CONNECTIONS** (Exercises 80–85) ▼▼▼▼▼▼▼▼▼▼▼▼▼▼

Refer to the sketch at left below. By the definition of $\tan \theta$, $m = \tan \theta$, where m is the slope and θ is the angle of inclination of the line. The following exercises, which depend on properties of triangles, refer to the triangle ABC in the sketch at right below. Assume that all angles are measured in degrees.

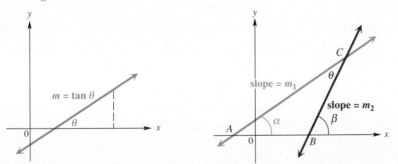

80. In terms of β, what is the measure of angle ABC?

81. Use the fact that the sum of the angles in a triangle is $180°$ to express θ in terms of α and β.

82. Apply the formula for $\tan(A - B)$ to obtain an expression for $\tan \theta$ in terms of $\tan \alpha$ and $\tan \beta$.

83. Replace $\tan \alpha$ by m_1 and $\tan \beta$ by m_2 to obtain $\tan \theta = \dfrac{m_2 - m_1}{1 + m_1 m_2}$.

▦ *In Exercises 84 and 85, use the result from Exercise 83 to find the angle between pairs of lines. Use a calculator and round to the nearest tenth of a degree.*

84. $x + y = 9,\ 2x + y = -1$ **85.** $5x - 2y + 4 = 0,\ 3x + 5y = 6$

86. When the two voltages $V_1 = 30 \sin 120\pi t$ and $V_2 = 40 \cos 120\pi t$ are applied to the same circuit, the resulting voltage V will be equal to their sum. (*Source:* Bell, D., *Fundamentals of Electric Circuits 2e,* Reston Publishing Company Inc., Reston, Virginia, 1981.)

(a) Graph $V = V_1 + V_2$ over the interval $0 \le t \le .05$.

(b) Use the graph and our work in Chapter 4 to estimate values for a and ϕ so that the voltage $V = a \sin(120\pi t + \phi)$.

(c) Use identities to verify that your expression for V is valid.

87. If a person bends at the waist with a straight back making an angle of θ degrees with the horizontal, then the force F exerted on the back muscles can be approximated by the equation $F = \dfrac{.6W \sin(\theta + 90°)}{\sin 12°}$, where W is the weight of the person. (*Source:* Metcalf, H., *Topics in Classical Biophysics,* Prentice-Hall, Inc., Englewood Cliffs, New Jersey, 1980.)

(a) Calculate F when $W = 170$ lb and $\theta = 30°$.

(b) Use an identity to show that F is approximately equal to $2.9W \cos \theta$.

(c) For what value of θ is F maximum?

5.5 Double-Angle Identities ▼▼▼

Some special cases of the identities for the sum of two angles are used often enough to be expressed as separate identities. These are the identities that result from the addition identities when $A = B$, so that $A + B = 2A$. These identities, called the **double-angle identities,** are derived in this section.

In the identity $\cos(A + B) = \cos A \cos B - \sin A \sin B$, let $B = A$ to derive an expression for $\cos 2A$.

$$\cos 2A = \cos(A + A)$$
$$= \cos A \cos A - \sin A \sin A$$
$$\cos 2A = \cos^2 A - \sin^2 A$$

Two other useful forms of this identity can be obtained by substituting either $\cos^2 A = 1 - \sin^2 A$ or $\sin^2 A = 1 - \cos^2 A$. Replace $\cos^2 A$ with $1 - \sin^2 A$ to get

$$\cos 2A = \cos^2 A - \sin^2 A$$
$$= (1 - \sin^2 A) - \sin^2 A$$
$$\cos 2A = 1 - 2 \sin^2 A,$$

and replace $\sin^2 A$ with $1 - \cos^2 A$ to get

$$\cos 2A = \cos^2 A - (1 - \cos^2 A)$$
$$= \cos^2 A - 1 + \cos^2 A$$
$$\cos 2A = 2 \cos^2 A - 1.$$

We find $\sin 2A$ with the identity $\sin(A + B) = \sin A \cos B + \cos A \sin B$, letting $B = A$.

$$\sin 2A = \sin(A + A)$$
$$= \sin A \cos A + \cos A \sin A$$
$$\sin 2A = 2 \sin A \cos A$$

Using the identity for $\tan(A + B)$, we find $\tan 2A$.

$$\tan 2A = \tan(A + A)$$

$$= \frac{\tan A + \tan A}{1 - \tan A \tan A}$$

$$\tan 2A = \frac{2 \tan A}{1 - \tan^2 A}$$

A summary of the double-angle identities follows.

Double-Angle Identities

$$\cos 2A = \cos^2 A - \sin^2 A \qquad \cos 2A = 1 - 2 \sin^2 A$$

$$\cos 2A = 2 \cos^2 A - 1 \qquad \sin 2A = 2 \sin A \cos A$$

$$\tan 2A = \frac{2 \tan A}{1 - \tan^2 A}$$

EXAMPLE 1
Using the double-angle identities

Given $\cos \theta = 3/5$ and $\sin \theta < 0$, use identities to find $\sin 2\theta$, $\cos 2\theta$, and $\tan 2\theta$.

In order to find $\sin 2\theta$, we must first find the value of $\sin \theta$. From the identity $\sin^2 \theta + \cos^2 \theta = 1$, we obtain

$$\sin^2 \theta + \left(\frac{3}{5}\right)^2 = 1$$

$$\sin^2 \theta = \frac{16}{25}$$

$$\sin \theta = -\frac{4}{5}. \qquad \text{Choose the negative square root, since } \sin \theta < 0.$$

Using the double-angle identity for sine, we get

$$\sin 2\theta = 2 \sin \theta \cos \theta = 2\left(-\frac{4}{5}\right)\left(\frac{3}{5}\right) = -\frac{24}{25}.$$

Now find $\cos 2\theta$, using the first form of the identity. (Any form may be used.)

$$\cos 2\theta = \cos^2 \theta - \sin^2 \theta = \frac{9}{25} - \frac{16}{25} = -\frac{7}{25}$$

The value of $\tan 2\theta$ can be found in either of two ways. We can use the double-angle identity, and the fact that $\tan \theta = (\sin \theta)/(\cos \theta) = (-4/5)/(3/5) = -4/3$.

$$\tan 2\theta = \frac{2 \tan \theta}{1 - \tan^2 \theta} = \frac{2\left(-\frac{4}{3}\right)}{1 - \frac{16}{9}} = \frac{-\frac{8}{3}}{-\frac{7}{9}} = \frac{24}{7}$$

As an alternative method, we can find tan 2θ by finding the quotient of sin 2θ and cos 2θ.

$$\tan 2\theta = \frac{\sin 2\theta}{\cos 2\theta} = \frac{-24/25}{-7/25} = \frac{24}{7} \quad \blacktriangleright$$

◖EXAMPLE 2
Finding functions of θ given
information about 2θ

Find the values of the six trigonometric functions of θ if cos $2\theta = 4/5$ and $90° < \theta < 180°$.

Use one of the double-angle identities for cosine to get a trigonometric function of θ.

$$\cos 2\theta = 1 - 2\sin^2\theta$$

$$\frac{4}{5} = 1 - 2\sin^2\theta$$

$$-\frac{1}{5} = -2\sin^2\theta$$

$$\frac{1}{10} = \sin^2\theta$$

$$\sin\theta = \sqrt{\frac{1}{10}} = \frac{\sqrt{10}}{10}$$

We choose the positive square root since θ terminates in quadrant II. Values of cos θ and tan θ can now be found by using the fundamental identities or by sketching and labeling a right triangle in quadrant II. Using a triangle as in Figure 7, we have

$$\cos\theta = \frac{-3}{\sqrt{10}} = -\frac{3\sqrt{10}}{10},$$

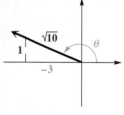

FIGURE 7

and tan $\theta = 1/(-3) = -1/3$. Find the other three functions using reciprocals.

$$\csc\theta = \frac{1}{\sin\theta} = \sqrt{10}, \quad \sec\theta = \frac{1}{\cos\theta} = -\frac{\sqrt{10}}{3}, \quad \cot\theta = \frac{1}{\tan\theta} = -3 \quad \blacktriangleright$$

◖EXAMPLE 3
Simplifying expressions using
double-angle identities

Simplify each of the following using double-angle identities.

(a) $\cos^2 7x - \sin^2 7x$

This expression suggests one of the identities for cos $2A$: cos $2A = \cos^2 A - \sin^2 A$. Substituting $7x$ for A gives

$$\cos^2 7x - \sin^2 7x = \cos 2(7x) = \cos 14x.$$

(b) $\sin 15° \cos 15°$

If this expression were 2 sin 15° cos 15°, then we could apply the identity for sin $2A$ directly, since sin $2A = 2\sin A \cos A$. We can still apply the identity with $A = 15°$ by writing the multiplicative identity element 1 as $(1/2)(2)$:

$$\sin 15° \cos 15° = \left(\frac{1}{2}\right)(2) \sin 15° \cos 15° \qquad \text{Multiply by 1 in the form } \tfrac{1}{2}(2).$$

$$= \frac{1}{2}(2 \sin 15° \cos 15°) \qquad \text{Associative property}$$

$$= \frac{1}{2}(\sin 2 \cdot 15°) \qquad 2 \sin A \cos A = \sin 2A, \text{ with } A = 15°$$

$$= \frac{1}{2} \sin 30°$$

$$= \frac{1}{2} \cdot \frac{1}{2} \qquad \sin 30° = \tfrac{1}{2}$$

$$= \frac{1}{4}. \quad \blacktriangleright$$

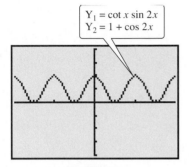

$Y_1 = \cot x \sin 2x$
$Y_2 = 1 + \cos 2x$

Trig Window
This graph supports the result in Example 4.

◖EXAMPLE 4
Verifying an identity

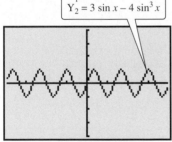

$Y_1 = \sin 3x$
$Y_2 = 3 \sin x - 4 \sin^3 x$

Trig Window
This graph supports the result in Example 5.

◖EXAMPLE 5
Deriving a multiple-angle identity

Double-angle identities can be used to verify certain identities, as shown in the next example.

Verify the identity

$$\cot x \sin 2x = 1 + \cos 2x.$$

Work on the left side.

$$\cot x \sin 2x = \frac{\cos x}{\sin x} \cdot \sin 2x \qquad \cot x = \tfrac{\cos x}{\sin x}$$

$$= \frac{\cos x}{\sin x}(2 \sin x \cos x) \qquad \sin 2x = 2 \sin x \cos x$$

$$= 2 \cos^2 x$$

$$= 1 + \cos 2x \qquad 2 \cos^2 x - 1 = \cos 2x \quad \blacktriangleright$$

The methods used earlier to derive the identities for double angles can also be used to find identities for expressions such as $\sin 3s$.

Write $\sin 3s$ in terms of $\sin s$.

$$\sin 3s = \sin (2s + s)$$

$$= \sin 2s \cos s + \cos 2s \sin s$$

$$= (2 \sin s \cos s) \cos s + (\cos^2 s - \sin^2 s) \sin s$$

$$= 2 \sin s \cos^2 s + \cos^2 s \sin s - \sin^3 s$$

$$= 2 \sin s(1 - \sin^2 s) + (1 - \sin^2 s) \sin s - \sin^3 s$$

$$= 2 \sin s - 2 \sin^3 s + \sin s - \sin^3 s - \sin^3 s$$

$$= 3 \sin s - 4 \sin^3 s \quad \blacktriangleright$$

CONNECTIONS The identities for cos $(A + B)$ and cos $(A - B)$ can be used to derive a group of identities that are useful in calculus because they make it possible to rewrite a product as a sum.

Adding the identities for cos $(A + B)$ and cos $(A - B)$ gives

$$\cos(A + B) = \cos A \cos B - \sin A \sin B$$
$$\cos(A - B) = \cos A \cos B + \sin A \sin B$$
$$\overline{\cos(A + B) + \cos(A - B) = 2 \cos A \cos B}$$

or

$$\cos A \cos B = \frac{1}{2}[\cos(A + B) + \cos(A - B)].$$

Similarly, subtracting cos$(A + B)$ from cos$(A - B)$ gives the following identity.

$$\sin A \sin B = \frac{1}{2}[\cos(A - B) - \cos(A + B)]$$

Using the identities for sin $(A + B)$ and sin $(A - B)$ in the same way, we get two more identities.

$$\sin A \cos B = \frac{1}{2}[\sin(A + B) + \sin(A - B)]$$

$$\cos A \sin B = \frac{1}{2}[\sin(A + B) - \sin(A - B)]$$

These last two identities make it possible to rewrite an expression involving both sine and cosine as an expression with just one of these functions. In solving a conditional equation, this conversion can be very useful.

FOR DISCUSSION OR WRITING
1. Show that the double-angle identity sin $2A = 2 \sin A \cos A$ is a special case of the identity for sin $A \cos B$ given above.
2. Show that the double-angle identity cos $2A = 2 \cos^2 A - 1$ is a special case of the identity for cos $A \cos B$ given above.

The final example in this section answers one of the questions posed in the introduction to this chapter.

EXAMPLE 6
Determining wattage consumption

If a toaster is plugged into a common household outlet, the wattage consumed is not constant. Instead, it varies at a high frequency according to the equation $W = V^2/R$ where V is the voltage and R is a constant that measures the resistance of the toaster in ohms.* Graph the wattage W

*Source: Bell, D., *Fundamentals of Electric Circuits,* Second Edition, Reston Publishing Company, Inc., Reston, Virginia, 1981.

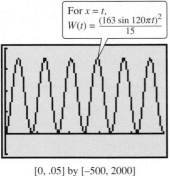

For $x = t$,
$$W(t) = \frac{(163 \sin 120\pi t)^2}{15}$$

[0, .05] by [−500, 2000]
Xscl = .01   Yscl = 500

FIGURE 8

consumed by a typical toaster with $R = 15$ and $V = 163 \sin 120\pi t$ over the interval $0 \le t \le .05$. How many oscillations are there?

By substituting the given values into the wattage equation, we get

$$W = \frac{V^2}{R} = \frac{(163 \sin 120\pi t)^2}{15}.$$

The graph is shown in Figure 8. To determine the range for W, we note that $\sin 120\pi t$ has a maximum value of 1, so the expression for W has a maximum value of $163^2/15 \approx 1771$. The minimum value is 0. The graph shows that there are 6 oscillations. In Exercise 91 you will be asked to show that the equation given above for W is equivalent to the equation $W = a \cos(\omega t) + c$, for specific values of a, c, and ω. ▶

5.5 Exercises ▼▼

In Exercises 1 and 2, the given graphing calculator screen is obtained for a particular stored value of X. *What will the screen display for the value of the expression in the final line of the display?*

1.
```
sin (2X)
                      .4
(sin X)(cos X)
```

2.
```
cos (2X)
                      .3
2(sin X)²
```

Use the identities in this section to find values of the six trigonometric functions for each of the following. See Examples 1 and 2.

3. θ, given $\cos 2\theta = 3/5$ and θ terminates in quadrant I

4. α, given $\cos 2\alpha = 3/4$ and α terminates in quadrant III

5. x, given $\cos 2x = -5/12$ and $\pi/2 < x < \pi$

6. t, given $\cos 2t = 2/3$ and $\pi/2 < t < \pi$

7. 2θ, given $\sin \theta = 2/5$ and $\cos \theta < 0$

8. 2β, given $\cos \beta = -12/13$ and $\sin \beta > 0$

9. $2x$, given $\tan x = 2$ and $\cos x > 0$

10. $2x$, given $\tan x = 5/3$ and $\sin x < 0$

11. 2α, given $\sin \alpha = -\sqrt{5}/7$ and $\cos \alpha > 0$

12. 2α, given $\cos \alpha = \sqrt{3}/5$ and $\sin \alpha > 0$

Use an identity to write each expression as a single trigonometric function or as a single number. See Example 3.

13. $2 \cos^2 15° - 1$

14. $\cos^2 15° - \sin^2 15°$

15. $\dfrac{2 \tan 15°}{1 - \tan^2 15°}$

16. $1 - 2 \sin^2 15°$

17. $2 \sin \dfrac{\pi}{3} \cos \dfrac{\pi}{3}$

18. $\dfrac{2 \tan \dfrac{\pi}{3}}{1 - \tan^2 \dfrac{\pi}{3}}$

19. $1 - 2 \sin^2 22\dfrac{1}{2}°$

20. $2 \sin 22\dfrac{1}{2}° \cos 22\dfrac{1}{2}°$

21. $2 \cos^2 67\frac{1}{2}° - 1$

22. $\dfrac{2 \tan 67\frac{1}{2}°}{1 - \tan^2 67\frac{1}{2}°}$

23. $\sin \dfrac{\pi}{8} \cos \dfrac{\pi}{8}$

24. $\cos^2 \dfrac{\pi}{8} - \dfrac{1}{2}$

25. $\dfrac{\tan 51°}{1 - \tan^2 51°}$

26. $\dfrac{\tan 34°}{2(1 - \tan^2 34°)}$

27. $\dfrac{1}{4} - \dfrac{1}{2} \sin^2 47.1°$

28. $\dfrac{1}{8} \sin 29.5° \cos 29.5°$

29. $\sin^2 \dfrac{2\pi}{5} - \cos^2 \dfrac{2\pi}{5}$

30. $\cos^2 2\alpha - \sin^2 2\alpha$

31. $2 \sin 5x \cos 5x$

32. $2 \cos^2 6\alpha - 1$

33. State in your own words each of the following identities. (*Example:* The sine of twice an angle is two times the sine of the angle times the cosine of the angle.)
(a) cos 2A (first form derived in this section)   (b) cos 2A (second form)
(c) cos 2A (third form)   (d) tan 2A

34. Specific identities for sec 2A, csc 2A, and cot 2A are not usually studied in detail. Why do you think this is so? Give these identities in terms of cos A, sin A, and tan A.

Find the exact value of each of the following in two ways: evaluate the expression directly, and use an appropriate identity. Do not use a calculator.

35. $\sin 2(45°)$

36. $\cos 2(45°)$

37. $\cos 2(60°)$

38. $\tan 2(60°)$

39. $\cos 2\left(\dfrac{5\pi}{3}\right)$

40. $\sin 2\left(\dfrac{\pi}{3}\right)$

41. $\tan 2\left(-\dfrac{\pi}{3}\right)$

42. $\cos 2\left(-\dfrac{9\pi}{4}\right)$

43. $\tan 2\left(-\dfrac{4\pi}{3}\right)$

44. $\tan 2\left(-\dfrac{13\pi}{6}\right)$

45. $\sin 2\left(-\dfrac{11\pi}{2}\right)$

46. $\sin 2\left(-\dfrac{17\pi}{2}\right)$

In Exercises 47–50, graph each expression and use the graph to conjecture an identity. Then verify your conjecture as in Example 4.

47. $\cos^4 x - \sin^4 x$

48. $\dfrac{4 \tan x \cos^2 x - 2 \tan x}{1 - \tan^2 x}$

49. $\dfrac{\cot^2 x - 1}{2 \cot x}$

50. $\dfrac{2 \tan x}{2 - \sec^2 x}$

Verify each identity. See Example 4.

51. $(\sin \gamma + \cos \gamma)^2 = \sin 2\gamma + 1$

52. $\sec 2x = \dfrac{\sec^2 x + \sec^4 x}{2 + \sec^2 x - \sec^4 x}$

53. $\tan 8k - \tan 8k \tan^2 4k = 2 \tan 4k$

54. $\sin 2\gamma = \dfrac{2 \tan \gamma}{1 + \tan^2 \gamma}$

55. $\cos 2y = \dfrac{2 - \sec^2 y}{\sec^2 y}$

56. $-\tan 2\theta = \dfrac{2 \tan \theta}{\sec^2 \theta - 2}$

57. $\sin 4\alpha = 4 \sin \alpha \cos \alpha \cos 2\alpha$

58. $\dfrac{1 + \cos 2x}{\sin 2x} = \cot x$

59. $\tan(\theta - 45°) + \tan(\theta + 45°) = 2 \tan 2\theta$

60. $\cot 4\theta = \dfrac{1 - \tan^2 2\theta}{2 \tan 2\theta}$

61. $\dfrac{2 \cos 2\alpha}{\sin 2\alpha} = \cot \alpha - \tan \alpha$

62. $\sin 4\gamma = 4 \sin \gamma \cos \gamma - 8 \sin^3 \gamma \cos \gamma$

63. $\sin 2\alpha \cos 2\alpha = \sin 2\alpha - 4 \sin^3 \alpha \cos \alpha$

64. $\cos 2x = \dfrac{1 - \tan^2 x}{1 + \tan^2 x}$

65. $\tan s + \cot s = 2 \csc 2s$

66. $\dfrac{\cot \alpha - \tan \alpha}{\cot \alpha + \tan \alpha} = \cos 2\alpha$

67. $1 + \tan x \tan 2x = \sec 2x$

68. $\cot \theta \tan(\theta + \pi) - \sin(\pi - \theta) \cos\left(\dfrac{\pi}{2} - \theta\right) = \cos^2 \theta$

Express each function as a trigonometric function of x. See Example 5.

69. $\tan^2 2x$ **70.** $\cos^2 2x$ **71.** $\cos 3x$ **72.** $\sin 4x$

73. $\tan 3x$ **74.** $\cos 4x$ **75.** $\tan 4x$ **76.** $\sin 5x$

Use a fundamental identity to simplify each expression.

77. $\sin^2 2x + \cos^2 2x$ **78.** $1 + \tan^2 4\alpha$ **79.** $\cot^2 3r + 1$ **80.** $\sin^2 11\alpha + \cos^2 11\alpha$

If an object is dropped in a vacuum, then the distance, d, that the object falls in t seconds is given by

$$d = \frac{1}{2}gt^2,$$

where g is the acceleration due to gravity. At any particular point on the Earth's surface, the value of g is a constant, roughly 978 cm per sec². A more exact value of g at any point on the Earth's surface is given by

$$g = 978.0524(1 + .005297 \sin^2 \phi - .0000059 \sin^2 2\phi) - .000094h$$

in cm per sec², where ϕ is the latitude of the point and h is the altitude of the point in feet. Find g, rounding to the nearest thousandth, given the following.

81. $\phi = 47° \, 12', h = 387.0$ ft

82. $\phi = 68° \, 47', h = 1145$ ft

Use the formulas developed in the Connections box to rewrite each of the following as a sum or difference of trigonometric functions.

83. $\cos 45° \sin 25°$ **84.** $2 \sin 74° \cos 114°$ **85.** $3 \cos 5x \cos 3x$ **86.** $2 \sin 2x \sin 4x$

87. $\sin(-\theta) \sin(-3\theta)$ **88.** $4 \cos 8\alpha \sin(-4\alpha)$ **89.** $-8 \cos 4y \cos 5y$ **90.** $2 \sin 3k \sin 14k$

91. Refer to Example 6. Use an identity to determine values for a, c, and ω so that $W = a \cos(\omega t) + c$. Check your answer by graphing both expressions for W on the same coordinate axes.

92. Amperage is a measure of the amount of electricity that is moving through a circuit whereas voltage is a measure of the force pushing the electricity. The wattage W consumed by an electrical device can be determined by calculating the product of the amperage I and voltage V. (*Source:* Wilcox, G. and C. Hesselberth, *Electricity for Engineering Technology,* Allyn and Bacon, Inc., Boston, 1970.)

(a) A household circuit has a voltage of $V = 163 \sin (120\pi t)$ when an incandescent lightbulb is turned on with an amperage of $I = 1.23 \sin(120\pi t)$. Graph the wattage $W = VI$ consumed by the lightbulb over the interval $0 \le t \le .05$.

(b) Determine the maximum and minimum wattages used by the lightbulb.

(c) Use identities to determine values for a, c, and ω so that $W = a \cos(\omega t) + c$.

(d) Check your answer by graphing both expressions for W on the same coordinate axes.

(e) Use the graph to estimate the average wattage used by the light. How many watts do you think this incandescent lightbulb is rated for?

93. Refer to Exercise 92. Suppose that for an electric heater the voltage is given by $V = a \sin(2\pi\omega t)$ and the amperage by $I = b \sin(2\pi\omega t)$ where t is time in seconds.

(a) Find the period of the graph for the voltage.
(b) Show that the graph of the wattage $W = VI$ will have half the period of the voltage. Interpret this result.

5.6 Half-Angle Identities ▼▼▼

From the alternative forms of the identity for $\cos 2A$, we can derive three additional identities for $\sin A/2$, $\cos A/2$, and $\tan A/2$. These are known as **half-angle identities.**

To derive the identity for $\sin A/2$, start with the following double-angle identity for cosine.

$$\cos 2x = 1 - 2 \sin^2 x$$

Then solve for $\sin x$.

$$2 \sin^2 x = 1 - \cos 2x$$

$$\sin x = \pm \sqrt{\frac{1 - \cos 2x}{2}}$$

Now let $2x = A$, so that $x = A/2$, and substitute into this last expression.

$$\sin \frac{A}{2} = \pm \sqrt{\frac{1 - \cos A}{2}}$$

The $\pm$ sign in the identity above indicates that, in practice, the appropriate sign is chosen depending upon the quadrant of $A/2$. For example, if $A/2$ is a third-quadrant angle, we choose the negative sign since the sine function is negative there.

The identity for $\cos A/2$ is derived in a very similar way, starting with the double-angle identity $\cos 2x = 2 \cos^2 x - 1$. Solve for $\cos x$.

$$\cos 2x + 1 = 2 \cos^2 x$$

$$\cos x = \pm \sqrt{\frac{1 + \cos 2x}{2}}$$

Replacing x with $A/2$ gives

$$\cos \frac{A}{2} = \pm \sqrt{\frac{1 + \cos A}{2}}.$$

The $\pm$ sign is used as described earlier.

Finally, an identity for $\tan A/2$ comes from the half-angle identities for sine and cosine.

$$\tan \frac{A}{2} = \frac{\pm \sqrt{\dfrac{1 - \cos A}{2}}}{\pm \sqrt{\dfrac{1 + \cos A}{2}}}$$

or $\quad \tan \dfrac{A}{2} = \pm \sqrt{\dfrac{1 - \cos A}{1 + \cos A}} \qquad \pm$ chosen depending upon quadrant of $\frac{A}{2}$

An alternative identity for $\tan A/2$ can be derived using the fact that $\tan A/2 = (\sin A/2)/(\cos A/2)$.

$$\tan \frac{A}{2} = \frac{\sin \dfrac{A}{2}}{\cos \dfrac{A}{2}}$$

$$= \frac{2 \sin \dfrac{A}{2} \cos \dfrac{A}{2}}{2 \cos^2 \dfrac{A}{2}} \qquad \text{Multiply by } 2 \cos \tfrac{A}{2} \text{ in numerator and denominator.}$$

$$= \frac{\sin 2\left(\dfrac{A}{2}\right)}{1 + \cos 2\left(\dfrac{A}{2}\right)} \qquad \text{Use double-angle identities.}$$

$$\tan \frac{A}{2} = \frac{\sin A}{1 + \cos A}$$

From this identity for $\tan A/2$, we can also derive

$$\tan \frac{A}{2} = \frac{1 - \cos A}{\sin A}.$$

See Exercise 57. These last two identities for $\tan A/2$ do not require a sign choice, as required in the first one.

Half-Angle Identities

$$\cos \frac{A}{2} = \pm \sqrt{\frac{1 + \cos A}{2}} \qquad \sin \frac{A}{2} = \pm \sqrt{\frac{1 - \cos A}{2}}$$

$$\tan \frac{A}{2} = \pm \sqrt{\frac{1 - \cos A}{1 + \cos A}} \qquad \tan \frac{A}{2} = \frac{\sin A}{1 + \cos A}$$

$$\tan \frac{A}{2} = \frac{1 - \cos A}{\sin A}$$

NOTE As mentioned earlier, the plus or minus sign is selected according to the quadrant in which $A/2$ terminates. For example, if A represents an angle of $324°$, then $A/2 = 162°$, which lies in quadrant II. In quadrant II, $\cos A/2$ and $\tan A/2$ are negative, while $\sin A/2$ is positive.

EXAMPLE 1
Using a half-angle identity to find an exact value

Find the exact value of $\cos 15°$ using the half-angle identity for cosine.

$$\cos 15° = \cos \frac{1}{2}(30°)$$

$$= \sqrt{\frac{1 + \cos 30°}{2}} \qquad \text{Choose the positive square root.}$$

$$= \sqrt{\frac{1 + \dfrac{\sqrt{3}}{2}}{2}}$$

$$= \sqrt{\frac{\left(1 + \dfrac{\sqrt{3}}{2}\right) \cdot 2}{2 \cdot 2}}$$

$$= \frac{\sqrt{2 + \sqrt{3}}}{2} \quad \blacktriangleright$$

NOTE Compare the value of $\cos 15°$ obtained in Example 1 to the value obtained in Example 1 of Section 5.3, where we used the identity for the cosine of the difference of two angles. Although the expressions look completely different, they are indeed equal, as suggested by a calculator approximation for both, .96592583.

EXAMPLE 2
Using a half-angle identity to find an exact value

Find the exact value of $\tan 22.5°$ using the identity $\tan A/2 = \dfrac{\sin A}{1 + \cos A}$.

Since $22.5° = (1/2)(45°)$, replacing A with $45°$ gives

$$\tan 22.5° = \tan \frac{45°}{2} = \frac{\sin 45°}{1 + \cos 45°} = \frac{\dfrac{\sqrt{2}}{2}}{1 + \dfrac{\sqrt{2}}{2}}.$$

Now multiply numerator and denominator by 2. Then rationalize the denominator.

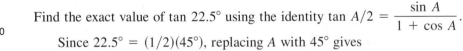

$$\tan 22.5° = \frac{\sqrt{2}}{2 + \sqrt{2}} = \frac{\sqrt{2}}{2 + \sqrt{2}} \cdot \frac{2 - \sqrt{2}}{2 - \sqrt{2}}$$

$$= \frac{2\sqrt{2} - 2}{2} = \sqrt{2} - 1 \quad \blacktriangleright$$

EXAMPLE 3
Finding functions of $A/2$ given information about A

Given $\cos s = 2/3$, with $3\pi/2 < s < 2\pi$, find $\cos s/2$, $\sin s/2$, and $\tan s/2$.

Since

$$\frac{3\pi}{2} < s < 2\pi,$$

dividing through by 2 gives

$$\frac{3\pi}{4} < \frac{s}{2} < \pi,$$

showing that $s/2$ terminates in quadrant II. In this quadrant the value of $\cos s/2$ is negative and the value of $\sin s/2$ is positive. Use the appropriate half-angle identities to get

$$\sin \frac{s}{2} = \sqrt{\frac{1 - \frac{2}{3}}{2}} = \sqrt{\frac{1}{6}} = \frac{\sqrt{6}}{6};$$

and

$$\cos \frac{s}{2} = -\sqrt{\frac{1 + \frac{2}{3}}{2}} = -\sqrt{\frac{5}{6}} = -\frac{\sqrt{30}}{6}.$$

Also,

$$\tan \frac{s}{2} = \frac{\sin \frac{s}{2}}{\cos \frac{s}{2}} = \frac{\frac{\sqrt{6}}{6}}{-\frac{\sqrt{30}}{6}} = -\frac{\sqrt{5}}{5}.$$

Notice that it is not necessary to use a half-angle identity for $\tan s/2$ once we find $\sin s/2$ and $\cos s/2$. However, using this identity would provide an excellent check. ▶

EXAMPLE 4
Simplifying expressions using the half-angle identities

Simplify each of the following using half-angle identities.

(a) $\pm\sqrt{\dfrac{1 + \cos 12x}{2}}$

Start with the identity for $\cos A/2$,

$$\cos \frac{A}{2} = \pm\sqrt{\frac{1 + \cos A}{2}},$$

and replace A with $12x$ to get

$$\pm\sqrt{\frac{1 + \cos 12x}{2}} = \cos \frac{12x}{2} = \cos 6x.$$

(b) $\dfrac{1 - \cos 5\alpha}{\sin 5\alpha}$

Use the third identity for $\tan A/2$ given earlier to get

$$\frac{1 - \cos 5\alpha}{\sin 5\alpha} = \tan \frac{5\alpha}{2}. \quad ▶$$

EXAMPLE 5
Verifying an identity

Verify the identity

$$\left(\sin \frac{x}{2} + \cos \frac{x}{2} \right)^2 = 1 + \sin x.$$

Work on the left.

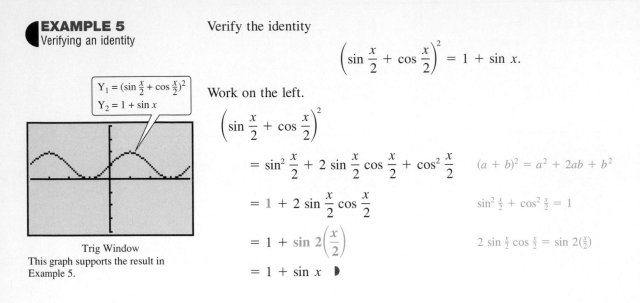

$Y_1 = (\sin \frac{x}{2} + \cos \frac{x}{2})^2$
$Y_2 = 1 + \sin x$

Trig Window
This graph supports the result in
Example 5.

$$\left(\sin \frac{x}{2} + \cos \frac{x}{2} \right)^2$$

$$= \sin^2 \frac{x}{2} + 2 \sin \frac{x}{2} \cos \frac{x}{2} + \cos^2 \frac{x}{2} \qquad (a + b)^2 = a^2 + 2ab + b^2$$

$$= 1 + 2 \sin \frac{x}{2} \cos \frac{x}{2} \qquad\qquad \sin^2 \frac{x}{2} + \cos^2 \frac{x}{2} = 1$$

$$= 1 + \sin 2\left(\frac{x}{2} \right) \qquad\qquad 2 \sin \frac{x}{2} \cos \frac{x}{2} = \sin 2(\frac{x}{2})$$

$$= 1 + \sin x \quad \blacktriangleright$$

5.6 Exercises ▼▼▼▼▼▼▼▼▼▼▼▼▼▼▼▼▼▼▼▼▼▼▼▼▼▼▼▼▼▼▼▼▼▼▼▼▼▼

In Exercises 1 and 2, the given graphing calculator screen is obtained for a particular stored value of X. What will the screen display for the value of the expression in the final line of the display?

1.
```
cos X
          .9682458366
sin X
                   .25
tan (X/2)
```

2.
```
cos X
                -.75
sin X
          .6614378278
tan (X/2)
```

Determine whether the positive or negative square root should be selected.

3. $\sin 195° = \pm \sqrt{\dfrac{1 - \cos 390°}{2}}$

4. $\cos 58° = \pm \sqrt{\dfrac{1 + \cos 116°}{2}}$

5. $\tan 225° = \pm \sqrt{\dfrac{1 - \cos 450°}{1 + \cos 450°}}$

6. $\sin(-10°) = \pm \sqrt{\dfrac{1 - \cos(-20°)}{2}}$

Use the half-angle identities of this section to find the exact value. See Examples 1 and 2.

7. $\sin 15°$ **8.** $\tan 15°$ **9.** $\cos \dfrac{\pi}{8}$ **10.** $\tan\left(-\dfrac{\pi}{8}\right)$ **11.** $\tan 67.5°$ **12.** $\cos 67.5°$

13. $\sin 67.5°$ **14.** $\sin 195°$ **15.** $\cos 195°$ **16.** $\tan 195°$ **17.** $\cos 165°$ **18.** $\sin 165°$

19. Explain how you could use an identity of this section to find the exact value of sin 7.5°. (*Hint:* 7.5 = (1/2)(1/2)(30).)

20. The identity

$$\tan \frac{A}{2} = \pm \sqrt{\frac{1 - \cos A}{1 + \cos A}}$$

can be used to find $\tan 22.5° = \sqrt{3 - 2\sqrt{2}}$, and the identity

$$\tan \frac{A}{2} = \frac{\sin A}{1 + \cos A}$$

can be used to get $\tan 22.5° = \sqrt{2} - 1$. Show that these answers are the same, without using a calculator. (*Hint:* If $a > 0$ and $b > 0$ and $a^2 = b^2$, then $a = b$.)

Find each of the following. See Example 3.

21. $\cos \theta/2$, given $\cos \theta = 1/4$, with $0 < \theta < \pi/2$

22. $\sin \theta/2$, given $\cos \theta = -5/8$, with $\pi/2 < \theta < \pi$

23. $\tan \theta/2$, given $\sin \theta = 3/5$, with $90° < \theta < 180°$

24. $\cos \theta/2$, given $\sin \theta = -1/5$, with $180° < \theta < 270°$

25. $\sin \alpha/2$, given $\tan \alpha = 2$, with $0 < \alpha < \pi/2$

26. $\cos \alpha/2$, given $\cot \alpha = -3$, with $\pi/2 < \alpha < \pi$

27. $\tan \beta/2$, given $\tan \beta = \sqrt{7}/3$, with $180° < \beta < 270°$

28. $\cot \beta/2$, given $\tan \beta = -\sqrt{5}/2$, with $90° < \beta < 180°$

29. $\sin \theta$, given $\cos 2\theta = 3/5$ and θ terminates in quadrant I

30. $\cos \theta$, given $\cos 2\theta = 1/2$ and θ terminates in quadrant II

31. $\cos x$, given $\cos 2x = -5/12$ and $\pi/2 < x < \pi$

32. $\sin x$, given $\cos 2x = 2/3$ and $\pi < x < 3\pi/2$

Use an identity to write each expression as a single trigonometric function. See Example 4.

33. $\sqrt{\dfrac{1 - \cos 40°}{2}}$

34. $\sqrt{\dfrac{1 + \cos 76°}{2}}$

35. $\sqrt{\dfrac{1 - \cos 147°}{1 + \cos 147°}}$

36. $\sqrt{\dfrac{1 + \cos 165°}{1 - \cos 165°}}$

37. $\dfrac{1 - \cos 59.74°}{\sin 59.74°}$

38. $\dfrac{\sin 158.2°}{1 + \cos 158.2°}$

39. $\pm\sqrt{\dfrac{1 + \cos 18x}{2}}$

40. $\pm\sqrt{\dfrac{1 + \cos 20\alpha}{2}}$

41. $\pm\sqrt{\dfrac{1 - \cos 8\theta}{1 + \cos 8\theta}}$

42. $\pm\sqrt{\dfrac{1 - \cos 5A}{1 + \cos 5A}}$

43. $\pm\sqrt{\dfrac{1 + \cos x/4}{2}}$

44. $\pm\sqrt{\dfrac{1 - \cos 3\theta/5}{2}}$

🖩 *In Exercises 45–48, graph each expression and use the graph to conjecture an identity. Then verify your conjecture as in Example 5.*

45. $\dfrac{\sin x}{1 + \cos x}$

46. $\dfrac{1 - \cos x}{\sin x}$

47. $\dfrac{\tan \dfrac{x}{2} + \cot \dfrac{x}{2}}{\cot \dfrac{x}{2} - \tan \dfrac{x}{2}}$

48. $1 - 8 \sin^2 \dfrac{x}{2} \cos^2 \dfrac{x}{2}$

Verify that the equation is an identity. See Example 5.

49. $\sec^2 \dfrac{x}{2} = \dfrac{2}{1 + \cos x}$

50. $\cot^2 \dfrac{x}{2} = \dfrac{(1 + \cos x)^2}{\sin^2 x}$

51. $\sin^2 \dfrac{x}{2} = \dfrac{\tan x - \sin x}{2 \tan x}$

52. $\dfrac{\sin 2x}{2 \sin x} = \cos^2 \dfrac{x}{2} - \sin^2 \dfrac{x}{2}$

53. $\dfrac{2}{1 + \cos x} - \tan^2 \dfrac{x}{2} = 1$

54. $\tan \dfrac{\gamma}{2} = \csc \gamma - \cot \gamma$

55. $1 - \tan^2 \dfrac{\theta}{2} = \dfrac{2 \cos \theta}{1 + \cos \theta}$

56. $\cos x = \dfrac{1 - \tan^2 \dfrac{x}{2}}{1 + \tan^2 \dfrac{x}{2}}$

57. In the text we derived the identity

$$\tan \frac{A}{2} = \frac{\sin A}{1 + \cos A}.$$

Multiply both the numerator and denominator of the right side by $1 - \cos A$ to obtain the equivalent form

$$\tan \frac{A}{2} = \frac{1 - \cos A}{\sin A}.$$

An airplane flying faster than sound sends out sound waves that form a cone, as shown in the figure. The cone intersects the ground to form a hyperbola. As this hyperbola passes over a particular point on the ground, a sonic boom is heard at that point. If α is the angle at the vertex of the cone, then

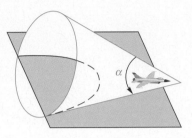

$$\sin \frac{\alpha}{2} = \frac{1}{m},$$

where m is the Mach number for the speed of the plane. (We assume $m > 1$.) The Mach number is the ratio of the speed of the plane and the speed of sound. Thus, a speed of Mach 1.4 means that the plane is flying at 1.4 times the speed of sound. Find α or m, as indicated.

58. $m = \dfrac{3}{2}$ **59.** $m = \dfrac{5}{4}$ **60.** $m = 2$ **61.** $m = \dfrac{5}{2}$ **62.** $\alpha = 30°$ **63.** $\alpha = 60°$

▼▼▼▼▼▼▼▼▼▼▼▼▼ **DISCOVERING CONNECTIONS** (Exercises 64–71) ▼▼▼▼▼▼▼▼▼▼▼▼▼

These exercises use results from plane geometry, instead of the half-angle formulas, to obtain exact values of the trigonometric functions of 15°. Start with a right triangle having a 60° angle at A and a 30° angle at B. Let the hypotenuse of this triangle have length 2. Extend side BC and draw a semicircle with diameter along BC extended, center at B, and radius AB. Draw segment AE. (See the figure.) Since any angle inscribed in a semicircle is a right angle, triangle AED is a right triangle.

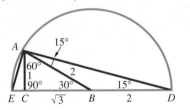

64. Why does $AB = BD$? Conclude that triangle ABD is isosceles.

65. Why does angle ABD have measure 150°?

66. Why do angles DAB and ADB both have measures of 15°?

67. What is the length of DC?

68. Use the Pythagorean theorem to show that the length of AD is $\sqrt{6} + \sqrt{2}$. (*Note:* $(\sqrt{6} + \sqrt{2})^2 = 8 + 4\sqrt{3}$.)

69. Use angle ADB of triangle ADE to find $\cos 15°$.

70. Show that AE has length $\sqrt{6} - \sqrt{2}$ and find $\sin 15°$.

71. Use triangle ACE and find $\tan 15°$.

Chapter 5   Summary ▼▼▼▼▼▼▼▼▼▼▼▼▼▼▼▼▼▼▼▼▼▼▼▼▼▼▼▼▼▼▼▼▼

SECTION	KEY IDEAS
5.1 Fundamental Identities	
	Reciprocal Identities $$\cot \theta = \frac{1}{\tan \theta} \qquad \sec \theta = \frac{1}{\cos \theta} \qquad \csc \theta = \frac{1}{\sin \theta}$$ **Quotient Identities** $$\tan \theta = \frac{\sin \theta}{\cos \theta} \qquad \cot \theta = \frac{\cos \theta}{\sin \theta}$$ **Pythagorean Identities** $$\sin^2 \theta + \cos^2 \theta = 1 \quad \tan^2 \theta + 1 = \sec^2 \theta \quad 1 + \cot^2 \theta = \csc^2 \theta$$ **Negative-Angle Identities** $$\sin(-\theta) = -\sin \theta \qquad \cos(-\theta) = \cos \theta \qquad \tan(-\theta) = -\tan \theta$$
5.3 Sum and Difference Identities for Cosine **5.4 Sum and Difference Identities for Sine and Tangent**	
	Cofunction Identities $$\cos(90° - \theta) = \sin \theta \qquad \cot(90° - \theta) = \tan \theta$$ $$\sin(90° - \theta) = \cos \theta \qquad \sec(90° - \theta) = \csc \theta$$ $$\tan(90° - \theta) = \cot \theta \qquad \csc(90° - \theta) = \sec \theta$$ **Sum and Difference Identities** $$\cos(A - B) = \cos A \cos B + \sin A \sin B$$ $$\cos(A + B) = \cos A \cos B - \sin A \sin B$$ $$\sin(A + B) = \sin A \cos B + \cos A \sin B$$ $$\sin(A - B) = \sin A \cos B - \cos A \sin B$$ $$\tan(A + B) = \frac{\tan A + \tan B}{1 - \tan A \tan B}$$ $$\tan(A - B) = \frac{\tan A - \tan B}{1 + \tan A \tan B}$$
5.5 Double-Angle Identities	
	Double-Angle Identities $$\cos 2A = \cos^2 A - \sin^2 A$$ $$\cos 2A = 1 - 2\sin^2 A$$ $$\cos 2A = 2\cos^2 A - 1$$ $$\sin 2A = 2 \sin A \cos A$$ $$\tan 2A = \frac{2 \tan A}{1 - \tan^2 A}$$

SECTION	KEY IDEAS
5.6 Half-Angle Identities	
	Half-Angle Identities $$\sin \frac{A}{2} = \pm\sqrt{\frac{1 - \cos A}{2}} \qquad \tan \frac{A}{2} = \frac{1 - \cos A}{\sin A}$$ $$\cos \frac{A}{2} = \pm\sqrt{\frac{1 + \cos A}{2}} \qquad \tan \frac{A}{2} = \frac{\sin A}{1 + \cos A}$$ $$\tan \frac{A}{2} = \pm\sqrt{\frac{1 - \cos A}{1 + \cos A}}$$ (The sign is chosen based on the quadrant of $A/2$.)

Chapter 5   Review Exercises ▼▼▼▼▼▼▼▼▼▼▼▼▼▼▼▼▼▼▼▼▼▼▼▼▼▼▼▼

1. Give all the trigonometric functions that satisfy the condition $f(-x) = -f(x)$.

2. Give all the trigonometric functions that satisfy the condition $f(-x) = f(x)$.

In Exercises 3 and 4, the given graphing calculator screen is obtained for a particular stored value of X. What will the screen display for the value of the expression in the final line of the display?

3.
```
sin X
             .6
X>(π/2)
             1
X<π
             1
sin (2X)
```

4.
```
sin X
             .8
X>0
             1
X<(π/2)
             1
tan (X/2)
```

5. Use the trigonometric identities to find the remaining five trigonometric functions of x, given that $\cos x = 3/5$ and x is in quadrant IV.

6. Given $\tan x = -5/4$, where $\pi/2 < x < \pi$, use the trigonometric identities to find the other trigonometric functions of x.

7. Find the exact value of $\sin x$, $\cos x$, and $\tan x$, for $x = \pi/12$, using
 (a) difference identities;   **(b)** half-angle identities.

8. Find the exact values of the six trigonometric functions of $165°$.

In Exercises 9–17, for each item in Column I, give the letter of the item in Column II that completes an identity.

Column I	Column II
9. $\cos 210°$	**(a)** $\sin(-35°)$
10. $\sin 35°$	**(b)** $\cos 55°$
11. $\tan(-35°)$	**(c)** $\sqrt{\dfrac{1 + \cos 150°}{2}}$

12. $-\sin 35°$ **(d)** $2 \sin 150° \cos 150°$

13. $\cos 35°$ **(e)** $\cos 150° \cos 60° - \sin 150° \sin 60°$

14. $\cos 75°$ **(f)** $\cot(-35°)$

15. $\sin 75°$ **(g)** $\cos^2 150° - \sin^2 150°$

16. $\sin 300°$ **(h)** $\sin 15° \cos 60° + \cos 15° \sin 60°$

17. $\cos 300°$ **(i)** $\cos(-35°)$

              (j) $\cot 125°$

In Exercises 18–24, for each item in Column I, give the letter of the item in Column II that completes an identity.

Column I   Column II

18. $\sec x$ **(a)** $\dfrac{1}{\sin x}$

19. $\csc x$ **(b)** $\dfrac{1}{\cos x}$

20. $\tan x$ **(c)** $\dfrac{\sin x}{\cos x}$

21. $\cot x$ **(d)** $\dfrac{1}{\cot^2 x}$

22. $\sin^2 x$ **(e)** $\dfrac{1}{\cos^2 x}$

23. $\tan^2 x + 1$ **(f)** $\dfrac{\cos x}{\sin x}$

24. $\tan^2 x$ **(g)** $\dfrac{1}{\sin^2 x}$

              (h) $1 - \cos^2 x$

Use identities to write each expression in terms of $\sin \theta$ *and* $\cos \theta$*, and simplify.*

25. $\sec^2 \theta - \tan^2 \theta$ **26.** $\dfrac{\cot \theta}{\sec \theta}$ **27.** $\tan^2 \theta(1 + \cot^2 \theta)$

28. $\csc \theta + \cot \theta$ **29.** $\csc^2 \theta + \sec^2 \theta$ **30.** $\tan \theta - \sec \theta \csc \theta$

For each of the following find $\sin(x + y)$*,* $\cos(x - y)$*,* $\tan(x + y)$*, and the quadrant of* $x + y$*.*

31. $\sin x = -1/4$, $\cos y = -4/5$, x and y in quadrant III

32. $\sin y = -2/3$, $\cos x = -1/5$, x in quadrant II, y in quadrant III

33. $\sin x = 1/10$, $\cos y = 4/5$, x in quadrant I, y in quadrant IV

34. $\cos x = 2/9$, $\sin y = -1/2$, x in quadrant IV, y in quadrant III

Find sine and cosine of each of the following.

35. θ, given $\cos 2\theta = -3/4$, $90° < 2\theta < 180°$ **36.** B, given $\cos 2B = 1/8$, B in quadrant IV

37. $2x$, given $\tan x = 3$, $\sin x < 0$ **38.** $2y$, given $\sec y = -5/3$, $\sin y > 0$

Find each of the following.

39. $\cos \theta/2$, given $\cos \theta = -1/2$, with $90° < \theta < 180°$

40. $\sin A/2$, given $\cos A = -3/4$, with $90° < A < 180°$

41. $\tan x$, given $\tan 2x = 2$, $\pi < x < 3\pi/2$

42. $\sin y$, given $\cos 2y = -1/3$, $\pi/2 < y < \pi$

In Exercises 43–48, graph each expression and use the graph to conjecture an identity. Then verify your conjecture.

43. $-\dfrac{\sin 2x + \sin x}{\cos 2x - \cos x}$

44. $\dfrac{1 - \cos 2x}{\sin 2x}$

45. $\dfrac{\sin x}{1 - \cos x}$

46. $\dfrac{\cos x \sin 2x}{1 + \cos 2x}$

47. $\dfrac{2(\sin x - \sin^3 x)}{\cos x}$

48. $\csc x - \cot x$

Verify that the equation is an identity.

49. $\sin^2 x - \sin^2 y = \cos^2 y - \cos^2 x$

50. $2 \cos^3 x - \cos x = \dfrac{\cos^2 x - \sin^2 x}{\sec x}$

51. $\dfrac{\sin^2 x}{2 - 2 \cos x} = \cos^2 \dfrac{x}{2}$

52. $\dfrac{\sin 2x}{\sin x} = \dfrac{2}{\sec x}$

53. $2 \cos A - \sec A = \cos A - \dfrac{\tan A}{\csc A}$

54. $\dfrac{2 \tan B}{\sin 2B} = \sec^2 B$

55. $1 + \tan^2 \alpha = 2 \tan \alpha \csc 2\alpha$

56. $-\dfrac{\sin(A - B)}{\sin(A + B)} = \dfrac{\cot A - \cot B}{\cot A + \cot B}$

57. $2 \cos(A + B) \sin(A + B) = \sin 2A \cos 2B + \sin 2B \cos 2A$

58. $\dfrac{2 \cot x}{\tan 2x} = \csc^2 x - 2$

59. $\tan \theta \sin 2\theta = 2 - 2 \cos^2 \theta$

60. $\csc A \sin 2A - \sec A = \cos 2A \sec A$

61. $2 \tan x \csc 2x - \tan^2 x = 1$

62. $2 \cos^2 \theta - 1 = \dfrac{1 - \tan^2 \theta}{1 + \tan^2 \theta}$

63. $\tan \theta \cos^2 \theta = \dfrac{2 \tan \theta \cos^2 \theta - \tan \theta}{1 - \tan^2 \theta}$

64. $-\cot \dfrac{x}{2} = \dfrac{\sin 2x + \sin x}{\cos 2x - \cos x}$

65. $2 \cos^3 x - \cos x = \dfrac{\cos^2 x - \sin^2 x}{\sec x}$

66. $\sin^3 \theta = \sin \theta - \cos^2 \theta \sin \theta$

67. $\cos^4 \theta = \dfrac{3}{8} + \dfrac{1}{2} \cos 2\theta + \dfrac{1}{8} \cos 4\theta$

68. $\tan \dfrac{7}{2}x = \dfrac{2 \tan \dfrac{7}{4}x}{1 - \tan^2 \dfrac{7}{4}x}$

69. $\sec^2 \alpha - 1 = \dfrac{\sec 2\alpha - 1}{\sec 2\alpha + 1}$

70. $\dfrac{\sin 3t + \sin 2t}{\sin 3t - \sin 2t} = \dfrac{\tan \dfrac{5t}{2}}{\tan \dfrac{t}{2}}$

71. $\tan 4\theta = \dfrac{2 \tan 2\theta}{2 - \sec^2 2\theta}$

72. $2 \cos^2 \dfrac{x}{2} \tan x = \tan x + \sin x$

73. $\tan\left(\dfrac{x}{2} + \dfrac{\pi}{4}\right) = \sec x + \tan x$

74. $\dfrac{1}{2} \cot \dfrac{x}{2} - \dfrac{1}{2} \tan \dfrac{x}{2} = \cot x$

6

Inverse Trigonometric Functions and Trigonometric Equations

Music is both art and science. During the Greek and Roman eras, music played an important role in philosophy and science. Although Pythagoras is usually associated with the Pythagorean theorem, in 500 B.C. he also discovered the mathematical relationships between lengths of strings and musical intervals. This discovery of the mathematical ratios that govern pitch and motion was the beginning of the science of musical sound. In the Middle Ages, music was studied together with arithmetic, geometry, and astronomy as part of the liberal arts curriculum. Later in 1862 the psychologist and scientist Hermann von Helmholtz published a classic work that opened a new direction for music using mathematics and technology. This direction has played its biggest role in the recording and reproduction of music. Max Mathews first created complex musical sounds using a computer in 1957. Since then computers have created new sounds that have been possible only through a technical knowledge of music.

When musicians tune instruments, they are able to compare like tones and accurately determine whether their pitches are the same frequency simply by listening, even though these tones vibrate hundreds or thousands of times per second. Some radios and telephones have small speakers that cannot vibrate slower than 200 times per second—yet 35 keys on a piano have frequencies below 200 and all of them can be clearly heard on these speakers.

Sources: Benade, Arthur, *Fundamentals of Musical Acoustics,* Oxford University Press, New York, 1976.

Pierce, John, *The Science of Musical Sound,* Scientific American Books, 1992.

How can we explain these phenomena? What is the advantage of having larger speakers for a stereo if small speakers are capable of reproducing the lower tones?

Explanations of musical phenomena like these require a mathematical understanding of sound. Music is made up of sound waves that cause rapid increases and decreases in air pressure on a person's eardrum. Sound often involves periodic motion through the air. As we shall see in this chapter, this periodic motion can be modeled using trigonometric functions. Using trigonometric equations and graphs, important aspects of music can be analyzed. Knowledge about waves and vibrations is necessary in order to create different musical sounds. This knowledge should continue to provide new sounds that will appeal to large and diverse audiences.

6.1 Inverse Trigonometric Functions ▼▼▼

In this chapter we learn how to solve trigonometric equations. To do this we will need to "work backwards"—that is, to find the angle or number given its trigonometric function value. For example, given $\sin x = .5$, one solution is $x = 30°$. This section begins with a review of the basic concepts of inverse functions, a topic usually studied in intermediate or college algebra courses. Recall from Section 1.1 that for a function f, every element x in the domain corresponds to one and only one element y, or $f(x)$, in the range. This means that if the point (a, b) lies on the graph of f, then there is no other point on the graph that has a as a first coordinate. However, there may be other points having b as a second coordinate, since the definition of a function allows range elements to be used more than once.

If a function is defined so that each range element is used only once, then it is called a **one-to-one function.** For example, the function $f(x) = x^3$ is a one-to-one function, because every real number has exactly one real cube root. On the other hand, $f(x) = x^2$ is not one-to-one, because, for example, $f(2) = 4$ and $f(-2) = 4$. We can find two domain elements, 2 and -2, that correspond to the range element 4.

There is a graphical test, called the horizontal line test, that will help determine whether a function is one-to-one.

> ### Horizontal Line Test
>
> Any horizontal line will intersect the graph of a one-to-one function in at most one point. If it is possible to draw a horizontal line that intersects the graph of a function in more than one point, then the function is not one-to-one.

Figure 1(a) shows the graph of $f(x) = x^3$. Since any horizontal line will intersect the graph in exactly one point, it passes the horizontal line test. Figure 1(b) shows the graph of $f(x) = x^2$. Since many horizontal lines will intersect the graph in two points, it is not a one-to-one function.

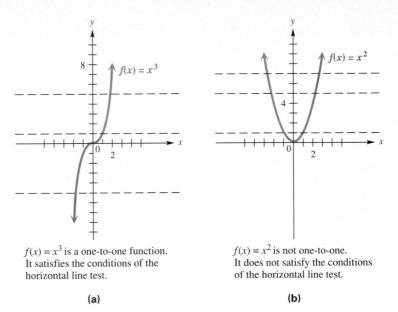

$f(x) = x^3$ is a one-to-one function. It satisfies the conditions of the horizontal line test.

(a)

$f(x) = x^2$ is not one-to-one. It does not satisfy the conditions of the horizontal line test.

(b)

FIGURE 1

By interchanging the components of the ordered pairs of a one-to-one function f, we obtain a set of ordered pairs that satisfies the definition of function. This new function is called the **inverse function,** or **inverse,** of f, and is symbolized by f^{-1}.

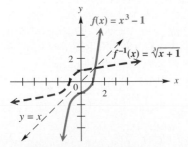

(b, a) is a reflection of (a, b) about the line $y = x$.

(a)

The graph of f^{-1} is a "mirror image" of the graph of f about the line $y = x$.

(b)

FIGURE 2

Inverse Function

The inverse function of the one-to-one function f is defined as

$$f^{-1} = \{(y, x) \mid (x, y) \text{ belongs to } f\}.$$

CAUTION In this context, the -1 is not an exponent. That is,

$$f^{-1}(x) \neq \frac{1}{f(x)}.$$

(f^{-1} names a function that has a special relationship with function f.)

Based on the definition of an inverse function, we can make the following general statements about inverses.

1. If the point (a, b) lies on the graph of the one-to-one function f, then the point (b, a) lies on the graph of f^{-1}.
2. The domain of f is equal to the range of f^{-1}, and the range of f is equal to the domain of f^{-1}.
3. For all x in the domain of f, $f^{-1}[f(x)] = x$, and for all x in the domain of f^{-1}, $f[f^{-1}(x)] = x$.
4. Because the point (b, a) is a reflection of the point (a, b) about the line $y = x$, the graph of f^{-1} is a reflection of the graph of f about this line. See Figure 2.

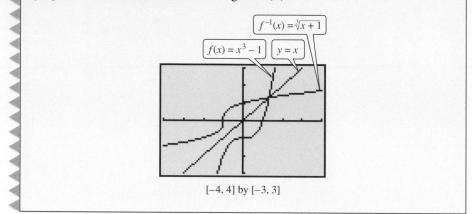

Some calculators have a DRAW feature that will draw the graph of the inverse of a given function. The figure shows a calculator-generated graph of the functions shown in Figure 2(b).

$f^{-1}(x) = \sqrt[3]{x+1}$

$f(x) = x^3 - 1$ $y = x$

[−4, 4] by [−3, 3]

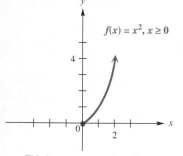

$f(x) = x^2, x \geq 0$

This is a one-to-one function.

FIGURE 3

As we shall see later in this section, there may be a reason for restricting the domain of a function that is not one-to-one so that it is. In such a restriction, we will also require that the range be unchanged. For example, we saw that $f(x) = x^2$, with its natural domain $(-\infty, \infty)$, is not one-to-one. However, if we restrict its domain to be the set of non-negative numbers $[0, \infty)$, we obtain a new function that is one-to-one, and has the same range as before the restriction, $[0, \infty)$. See Figure 3.

We could also have chosen to restrict the domain to $(-\infty, 0]$ so that a one-to-one function was obtained. For important functions, such as the trigonometric functions, such choices are usually made based on general agreement by mathematicians.

Now let us consider the function $y = \sin x$. From Figure 4 and the horizontal line test, it is clear that $y = \sin x$ is not a one-to-one function. By suitably restricting the domain of the sine function, however, a one-to-one function can be defined. It is common to restrict the domain of $y = \sin x$ to the interval $[-\pi/2, \pi/2]$, which gives the part of the graph shown in color in Figure 4. As Figure 4 shows, the range of $y = \sin x$ is $[-1, 1]$. Reflecting the graph of $y = \sin x$ on the restricted domain about the line $y = x$ gives the graph of the inverse function, shown in Figure 5. Some key points are labeled on the graph.

The roles of x and y are reversed in a pair of inverse functions. Therefore, the equation of the inverse of $y = \sin x$ is found by exchanging x and y to get $x = \sin y$. This equation then is solved for y by writing **$y = \sin^{-1} x$**, read "inverse sine of x." (Note that $\sin^{-1} x$ does not mean $1/\sin x$.) As Figure 5 shows, the domain of $y = \sin^{-1} x$ is $[-1, 1]$, while the restricted domain of $y = \sin x$, $[-\pi/2, \pi/2]$, is the range of $y = \sin^{-1} x$. An alternative notation for $\sin^{-1} x$ is **arcsin x.**

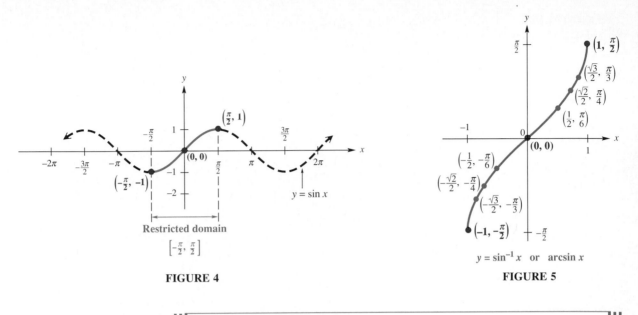

FIGURE 4

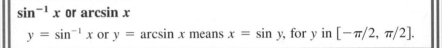

$y = \sin^{-1} x$ or   arcsin x

FIGURE 5

---

sin⁻¹ x or arcsin x

$y = \sin^{-1} x$ or $y = $ arcsin x means $x = \sin y$, for y in $[-\pi/2, \pi/2]$.

Thus, we may think of $y = \sin^{-1} x$ or $y = $ arcsin x as "y is the number in $[-\pi/2, \pi/2]$ whose sine is x." These two types of notation will be used in the rest of this book.

Graphing calculators use the notation $\sin^{-1} x$ (rather than arcsin x) for the inverse sine of x. A calculator-generated graph of $\sin^{-1} x$ is shown in the figure below.

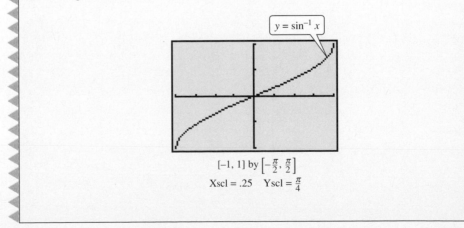

$y = \sin^{-1} x$

$[-1, 1]$ by $\left[-\frac{\pi}{2}, \frac{\pi}{2}\right]$

Xscl $= .25$ Yscl $= \frac{\pi}{4}$

EXAMPLE 1
Finding inverse sine values

Find y in each of the following.

(a) $y = \arcsin \dfrac{1}{2}$

The graph of the function $y = \arcsin x$ (Figure 5) shows that the point $(1/2, \pi/6)$ lies on the graph. Therefore, $\arcsin(1/2) = \pi/6$. Alternatively, we may think of $y = \arcsin(1/2)$ as

$$y \text{ is the number in } \left[-\frac{\pi}{2}, \frac{\pi}{2} \right] \text{ whose sine is } \frac{1}{2}.$$

Then we can rewrite the equation as $\sin y = 1/2$. Since $\sin \pi/6 = 1/2$ and $\pi/6$ is in the range of the arcsin function, $y = \pi/6$.

(b) $y = \sin^{-1}(-1)$

Writing the alternative equation, $\sin y = -1$, shows that $y = -\pi/2$. This can be verified by noticing that the point $(-1, -\pi/2)$ is on the graph of $y = \sin^{-1} x$. ▶

CAUTION   In Example 1(b), it is tempting to give the value of $\sin^{-1}(-1)$ as $3\pi/2$, since $\sin(3\pi/2) = -1$. Notice, however, that $3\pi/2$ is not in the range of the inverse sine function. Be certain (in dealing with *all* inverse trigonometric functions) that the number given for an inverse function value is in the range of the particular inverse function being considered.

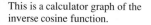

$[-1, 1]$ by $[0, \pi]$
Xscl = .5 Yscl = $\frac{\pi}{6}$

This is a calculator graph of the inverse cosine function.

The function $y = \cos^{-1} x$ (or $y = \arccos x$) is defined by restricting the domain of $y = \cos x$ to $[0, \pi]$. This domain becomes the range of $y = \cos^{-1} x$. The range of $y = \cos x$, the interval $[-1, 1]$, becomes the domain of $y = \cos^{-1} x$. The graphs of $y = \cos x$ with domain $[0, \pi]$ and $y = \cos^{-1} x$ are shown in Figure 6. Compare the key points on the two graphs.

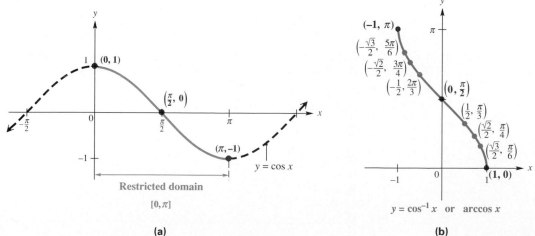

FIGURE 6

> **cos^{-1} x or arccos x**
>
> $y = \cos^{-1} x$ or $y = \arccos x$ means $x = \cos y$, for y in $[0, \pi]$.

◀EXAMPLE 2
Finding inverse cosine values

Find y in each of the following.

(a) $y = \arccos 1$

Since the point $(1, 0)$ lies on the graph of $y = \arccos x$, the value of y is 0. Alternatively, we may think of $y = \arccos 1$ as "y is the number in $[0, \pi]$ whose cosine is 1," or $\cos y = 1$. Then $y = 0$, since $\cos 0 = 1$ and 0 is in the range of the arccos function.

(b) $y = \cos^{-1}\left(-\dfrac{\sqrt{2}}{2}\right)$

We must find the value of y that satisfies $\cos y = -\sqrt{2}/2$, where y is in the interval $[0, \pi]$, the range of the function $y = \cos^{-1} x$. The only value for y that satisfies these conditions is $3\pi/4$. This can be verified from the graph in Figure 6(b). ▶

The inverse tangent function is defined below. Its graph is given in Figure 7(b).

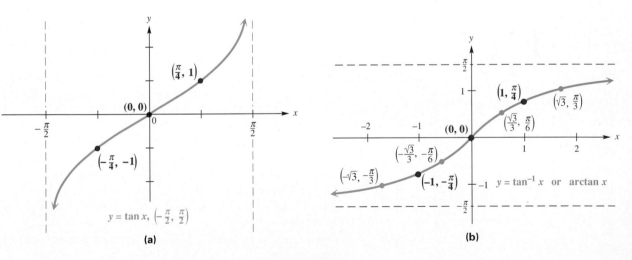

FIGURE 7

> **tan^{-1} x or arctan x**
>
> $y = \tan^{-1} x$ or $y = \arctan x$ means $x = \tan y$, for y in $(-\pi/2, \pi/2)$.

From the graph we can see that $\arctan 0 = 0$, $\arctan 1 = \pi/4$, $\arctan(-\sqrt{3}) = -\pi/3$, and so on.

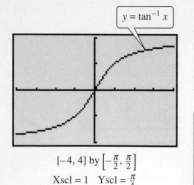

$y = \tan^{-1} x$

$[-4, 4]$ by $\left[-\frac{\pi}{2}, \frac{\pi}{2}\right]$

Xscl $= 1$ Yscl $= \frac{\pi}{4}$

This is a calculator graph of the inverse tangent function.

We have defined the inverse sine, cosine, and tangent functions with suitable restrictions on the domains. The other three inverse trigonometric functions are similarly defined.* The six inverse trigonometric functions with their domains and ranges are given in the table. This information, particularly the range for each function, should be memorized. (The graphs of the last three inverse trigonometric functions are left for the exercises.)

Inverse Trigonometric Functions

Function	Domain	Range	Quadrants of the Unit Circle Range Values Come from
$y = \sin^{-1} x$	$[-1, 1]$	$[-\pi/2, \pi/2]$	I and IV
$y = \cos^{-1} x$	$[-1, 1]$	$[0, \pi]$	I and II
$y = \tan^{-1} x$	$(-\infty, \infty)$	$(-\pi/2, \pi/2)$	I and IV
$y = \cot^{-1} x$	$(-\infty, \infty)$	$(0, \pi)$	I and II
$y = \sec^{-1} x$	$(-\infty, -1] \cup [1, \infty)$	$[0, \pi], y \neq \pi/2$	I and II
$y = \csc^{-1} x$	$(-\infty, -1] \cup [1, \infty)$	$[-\pi/2, \pi/2], y \neq 0$	I and IV

CONNECTIONS The sine function with the restricted domain $[-\pi/2, \pi/2]$ is sometimes differentiated from the sine function with its natural domain of $(-\infty, \infty)$ by the notation Sin x (with a capital S). Its inverse function is then denoted Sin$^{-1} x$. The other trigonometric functions with restricted domains and their inverses are also denoted with capital letters. We have chosen not to use this notation because graphing calculators use lower-case letters to denote the inverse trigonometric functions.

As we mention in a footnote, sometimes the inverse secant and inverse cosecant functions are defined with different ranges than we have given here. We have elected to use intervals that match their reciprocal functions, except for one missing point. In calculus, different ranges are considered more convenient: for sec$^{-1} x$, $[0, \pi/2) \cup [\pi, 3\pi/2)$; for csc$^{-1} x$, $[\pi/2, \pi) \cup [3\pi/2, 2\pi)$.

---

* sec^{-1} and csc^{-1} are sometimes defined differently.

The inverse trigonometric functions are formally defined with real number ranges. However, there are times when it may be convenient to find the degree-measured angles equivalent to these real number values. It is also often convenient to think in terms of the unit circle, and choose the inverse function values based on the quadrants given earlier in the table that summarizes these functions. The next example uses these ideas.

EXAMPLE 3
Finding inverse values
(degree-measured angles)

Find the *degree measure* of θ in each of the following.

(a) θ, if $\theta = \arctan 1$
  Here θ must be in $(-90°, 90°)$, but since $1 > 0$, θ must be in quadrant I. The alternative statement, $\tan \theta = 1$, leads to $\theta = 45°$.

(b) θ, if $\theta = \sec^{-1} 2$
  Write the equation as $\sec \theta = 2$. For $\sec^{-1} x$, θ is in quadrant I or II. Because 2 is positive, θ is in quadrant I and $\theta = 60°$, since $\sec 60° = 2$. Note that $60°$ (the degree equivalent of $\pi/3$) is in the range of the inverse secant function. ▶

The inverse trigonometric function keys on a calculator give results in the proper quadrant, for the $\sin^{-1}$, $\cos^{-1}$, and $\tan^{-1}$ functions, according to the definitions of these functions. For example, on a calculator, in degrees,

$$\sin^{-1} .5 = 30°, \qquad \sin^{-1}(-.5) = -30°,$$
$$\tan^{-1}(-1) = -45°, \qquad \text{and} \qquad \cos^{-1}(-.5) = 120°.$$

Similar results are found when the calculator is set for radian measure. This is not the case for $\cot^{-1}$. For example, since we can take the reciprocal of the inverse tangent to find $\cot^{-1}$, the calculator gives values of $\cot^{-1}$ with the same range as $\tan^{-1}$, $(-\pi/2, \pi/2)$, which is not the correct range for $\cot^{-1}$. For $\cot^{-1}$ the proper range must be considered and the results adjusted accordingly.

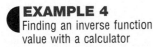

EXAMPLE 4
Finding an inverse function
value with a calculator

Find θ in degrees if $\theta = \text{arccot}(-.3541)$.
  A calculator gives $\tan^{-1}(1/-.3541) \approx -70.500946°$. The restriction on the range of arccot means that θ must be in quadrant II, and the absolute value of the angle obtained in the display is the first quadrant angle with the same cotangent. Therefore,

$$\theta = 180° - |-70.500946°| = 109.499054°.$$

Use a calculator to verify that $\cot 109.499054° \approx -.3541$. ▶

EXAMPLE 5
Finding function values without a calculator

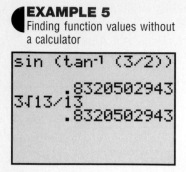

```
sin (tan⁻¹ (3/2))
           .8320502943
3√13/13
           .8320502943
```

This screen supports the result in Example 5(a).

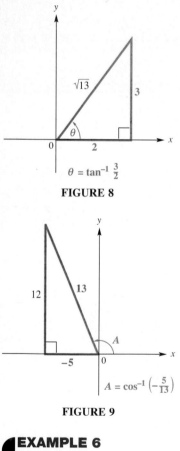

$\theta = \tan^{-1}\frac{3}{2}$

FIGURE 8

$A = \cos^{-1}\left(-\frac{5}{13}\right)$

FIGURE 9

Evaluate each of the following without a calculator.

(a) $\sin\left(\tan^{-1}\frac{3}{2}\right)$

Let

$$\theta = \tan^{-1}\frac{3}{2}, \text{ so that } \tan\theta = \frac{3}{2}.$$

Since $\tan^{-1}$ is defined only in quadrants I and IV and since $3/2$ is positive, θ is in quadrant I. Sketch θ in quadrant I, and label a triangle as shown in Figure 8. The hypotenuse is $\sqrt{13}$ and the value of sine is the ratio of the side opposite and the hypotenuse, so

$$\sin\left(\tan^{-1}\frac{3}{2}\right) = \sin\theta = \frac{3}{\sqrt{13}} = \frac{3\sqrt{13}}{13}.$$

To check this result on a calculator, enter $3/2$ as 1.5. Then find $\tan^{-1} 1.5$, and finally find $\sin(\tan^{-1} 1.5)$. Store this result and calculate $3\sqrt{13}/13$, which should agree with the result for $\sin(\tan^{-1} 1.5)$. Since the values are only approximations, this check does not *prove* that the result is correct, but it is highly suggestive that it is correct.

(b) $\tan\left(\cos^{-1}\left(-\frac{5}{13}\right)\right)$

Let $A = \cos^{-1}(-5/13)$. Then $\cos A = -5/13$. Since $\cos^{-1} x$ for a negative value of x is in quadrant II, sketch A in quadrant II, as shown in Figure 9.
From the triangle in Figure 9,

$$\tan\left(\cos^{-1}\left(-\frac{5}{13}\right)\right) = \tan A = -\frac{12}{5}.$$

(c) $\cos(\cos^{-1}(-.5))$
Recall, for inverse functions f and $g, f(g(x)) = x$. Since cosine and inverse cosine are inverse functions, $\cos(\cos^{-1}(-.5)) = -.5$.

(d) $\cos^{-1}\left(\cos\frac{5\pi}{4}\right) = \cos^{-1}\left(-\frac{\sqrt{2}}{2}\right) = \frac{3\pi}{4}.$ ▶

EXAMPLE 6
Finding function values using sum and double-angle formulas

Evaluate the following without using a calculator.

(a) $\cos\left(\arctan\sqrt{3} + \arcsin\frac{1}{3}\right)$
Let $A = \arctan\sqrt{3}$ and $B = \arcsin 1/3$ so that $\tan A = \sqrt{3}$ and $\sin B = 1/3$. Sketch both A and B in quadrant I, as shown in Figure 10.

Now use the identity for $\cos(A + B)$.

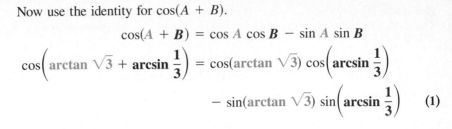

$$\cos(A + B) = \cos A \cos B - \sin A \sin B$$

$$\cos\left(\arctan \sqrt{3} + \arcsin \frac{1}{3}\right) = \cos(\arctan \sqrt{3}) \cos\left(\arcsin \frac{1}{3}\right)$$

$$- \sin(\arctan \sqrt{3}) \sin\left(\arcsin \frac{1}{3}\right) \quad (1)$$

From the sketch in Figure 10,

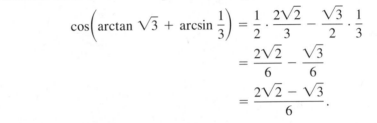

$$\cos(\arctan \sqrt{3}) = \cos A = \frac{1}{2}, \qquad \cos\left(\arcsin \frac{1}{3}\right) = \cos B = \frac{2\sqrt{2}}{3},$$

$$\sin(\arctan \sqrt{3}) = \sin A = \frac{\sqrt{3}}{2}, \qquad \sin\left(\arcsin \frac{1}{3}\right) = \sin B = \frac{1}{3}.$$

FIGURE 10

Substitute these values into equation (1) to get

$$\cos\left(\arctan \sqrt{3} + \arcsin \frac{1}{3}\right) = \frac{1}{2} \cdot \frac{2\sqrt{2}}{3} - \frac{\sqrt{3}}{2} \cdot \frac{1}{3}$$

$$= \frac{2\sqrt{2}}{6} - \frac{\sqrt{3}}{6}$$

$$= \frac{2\sqrt{2} - \sqrt{3}}{6}.$$

(b) $\tan\left(2 \arcsin \frac{2}{5}\right)$

Let $\arcsin(2/5) = B$. Then, from the identity for the tangent of the double angle,

$$\tan\left(2 \arcsin \frac{2}{5}\right) = \tan(2B)$$

$$= \frac{2 \tan B}{1 - \tan^2 B}.$$

Since $\arcsin (2/5) = B$, $\sin B = 2/5$. Sketch a triangle in quadrant I, find the length of the third side, and then find $\tan B$. From the triangle in Figure 11, $\tan B = 2/\sqrt{21}$, and

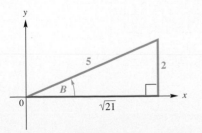

FIGURE 11

$$\tan\left(2 \arcsin \frac{2}{5}\right) = \frac{2\left(\dfrac{2}{\sqrt{21}}\right)}{1 - \left(\dfrac{2}{\sqrt{21}}\right)^2} = \frac{\dfrac{4}{\sqrt{21}}}{1 - \dfrac{4}{21}}$$

$$= \frac{\dfrac{4}{\sqrt{21}}}{\dfrac{17}{21}} = \frac{4\sqrt{21}}{17}. \quad \blacktriangleright$$

EXAMPLE 7
Writing a function value in terms of u

Write $\sin(\tan^{-1} u)$ as an expression in u.

Let $\theta = \tan^{-1} u$, so that $\tan \theta = u$. Here u may be positive or negative. Since $-\pi/2 < \tan^{-1} u < \pi/2$, sketch θ in quadrants I and IV and label two triangles as shown in Figure 12. Since sine is given by the ratio of the opposite side and the hypotenuse,

$$\sin(\tan^{-1} u) = \sin \theta = \frac{u}{\sqrt{u^2 + 1}}$$

$$= \frac{u\sqrt{u^2 + 1}}{u^2 + 1}.$$

The result is positive when u is positive and negative when u is negative. ▶

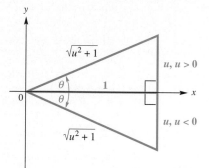

FIGURE 12

6.1 Exercises ▼▼▼▼▼▼▼▼▼▼▼▼▼▼▼▼▼▼▼▼▼▼▼▼▼▼▼▼▼▼▼▼▼▼▼▼

1. The accompanying screen contains the graph of $Y_1 = \sin^{-1} x$ along with the coordinates of a point on the graph. What is the exact value of Y?

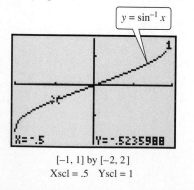

[−1, 1] by [−2, 2]
Xscl = .5   Yscl = 1

2. The accompanying screen contains the graph of $Y_1 = \cos^{-1} x$ along with the coordinates of a point on the graph. What is the exact value of Y?

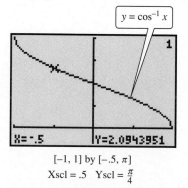

[−1, 1] by [−.5, π]
Xscl = .5   Yscl = $\frac{\pi}{4}$

3. Is $\sec^{-1} a$ calculated as $\cos^{-1}(1/a)$ or as $\dfrac{1}{\cos^{-1} a}$?

4. For positive values of a, $\cot^{-1} a$ is calculated as $\tan^{-1}(1/a)$. How is $\cot^{-1} a$ calculated for negative values of a?

Find the exact value of y in each of the following. Do not use a calculator. See Examples 1 and 2.

5. $y = \arcsin\left(-\dfrac{1}{2}\right)$
6. $y = \arccos\left(\dfrac{\sqrt{3}}{2}\right)$
7. $y = \tan^{-1} 1$
8. $y = \sin^{-1} 0$

9. $y = \cos^{-1}(-1)$
10. $y = \arctan(-1)$
11. $y = \sin^{-1}\left(-\dfrac{\sqrt{3}}{2}\right)$
12. $y = \cos^{-1}\left(\dfrac{1}{2}\right)$

13. $y = \arctan 0$
14. $y = \arcsin\left(-\dfrac{\sqrt{3}}{2}\right)$
15. $y = \arccos 0$
16. $y = \tan^{-1}(-1)$

17. $y = \sin^{-1}\left(\dfrac{\sqrt{2}}{2}\right)$
18. $y = \cos^{-1}\left(-\dfrac{1}{2}\right)$
19. $y = \arccos\left(-\dfrac{\sqrt{3}}{2}\right)$
20. $y = \arcsin\left(-\dfrac{\sqrt{2}}{2}\right)$

21. $y = \cot^{-1}(-1)$
22. $y = \sec^{-1}(-\sqrt{2})$
23. $y = \csc^{-1}(-2)$
24. $y = \text{arccot}(-\sqrt{3})$

25. $y = \text{arcsec}\left(\dfrac{2\sqrt{3}}{3}\right)$
26. $y = \csc^{-1}\sqrt{2}$
27. $y = \text{arccot}\left(\dfrac{\sqrt{3}}{3}\right)$
28. $y = \text{arcsec } 2$

Give the degree measure of θ. Do not use a calculator. See Example 3.

29. $\theta = \arctan(-1)$
30. $\theta = \arccos\left(-\dfrac{1}{2}\right)$
31. $\theta = \arcsin\left(-\dfrac{\sqrt{3}}{2}\right)$
32. $\theta = \arcsin\left(-\dfrac{\sqrt{2}}{2}\right)$

33. $\theta = \cot^{-1}\left(-\dfrac{\sqrt{3}}{3}\right)$
34. $\theta = \sec^{-1}(-2)$
35. $\theta = \csc^{-1}(-2)$
36. $\theta = \csc^{-1}(-1)$

Use a calculator to give the real number value. See Example 4.

37. $\theta = \arctan 1.1111111$
38. $\theta = \arcsin .81926439$
39. $\theta = \cot^{-1}(-.92170128)$
40. $\theta = \sec^{-1}(-1.2871684)$
41. $\theta = \arcsin .92837781$
42. $\theta = \arccos .44624593$

Use a calculator to give the value in decimal degrees. See Example 4.

43. $\theta = \sin^{-1}(-.13349122)$
44. $\theta = \cos^{-1}(-.13348816)$
45. $\theta = \arccos(-.39876459)$
46. $\theta = \arcsin .77900016$
47. $\theta = \csc^{-1} 1.9422833$
48. $\theta = \cot^{-1} 1.7670492$

Graph each function as defined in the text, and give the domain and range.

49. $y = \cot^{-1} x$
50. $y = \text{arccsc } x$
51. $y = \text{arcsec } x$

52. The following expressions were used by the mathematicians who computed the value of π to 100,000 decimal places. Use a calculator to verify that each is (approximately) correct.

(a) $\pi = 16 \tan^{-1} \dfrac{1}{5} - 4 \tan^{-1} \dfrac{1}{239}$

(b) $\pi = 24 \tan^{-1} \dfrac{1}{8} + 8 \tan^{-1} \dfrac{1}{57} + 4 \tan^{-1} \dfrac{1}{239}$

(c) $\pi = 48 \tan^{-1} \dfrac{1}{18} + 32 \tan^{-1} \dfrac{1}{57} - 20 \tan^{-1} \dfrac{1}{239}$

53. Explain why attempting to find $\sin^{-1} 1.003$ on your calculator will result in an error message.

54. Explain why you are able to find $\tan^{-1} 1.003$ on your calculator. Why is this situation different from the one described in Exercise 53?

55. Find sin 1.74 on your calculator (set for radians) and then find the inverse sine of that number. You get 1.401592654 instead of 1.74. What happened?

56. Recall that a graphing calculator will return a 1 for a true statement and a 0 for a false statement. Determine the possible values stored in X for which the screen will come up on a graphing calculator.

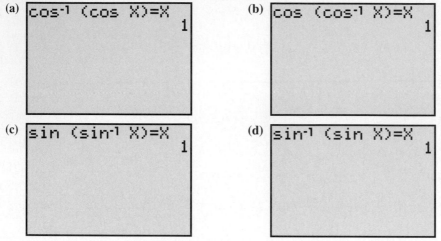

(a) `cos⁻¹ (cos X)=X`
    `1`

(b) `cos (cos⁻¹ X)=X`
    `1`

(c) `sin (sin⁻¹ X)=X`
    `1`

(d) `sin⁻¹ (sin X)=X`
    `1`

Give each value without using a calculator. See Examples 5 and 6.

57. $\tan\left(\arccos\dfrac{3}{4}\right)$

58. $\sin\left(\arccos\dfrac{1}{4}\right)$

59. $\cos(\tan^{-1}(-2))$

60. $\sec\left(\sin^{-1}\left(-\dfrac{1}{5}\right)\right)$

61. $\cot\left(\arcsin\left(-\dfrac{2}{3}\right)\right)$

62. $\cos\left(\arctan\dfrac{8}{3}\right)$

63. $\sec(\sec^{-1} 2)$

64. $\csc(\csc^{-1}\sqrt{2})$

65. $\arccos\left(\cos\dfrac{\pi}{4}\right)$

66. $\arctan\left(\tan\left(-\dfrac{\pi}{4}\right)\right)$

67. $\arcsin\left(\sin\dfrac{\pi}{3}\right)$

68. $\arccos(\cos 0)$

69. $\sin\left(2\tan^{-1}\dfrac{12}{5}\right)$

70. $\cos\left(2\sin^{-1}\dfrac{1}{4}\right)$

71. $\cos\left(2\arctan\dfrac{4}{3}\right)$

72. $\tan\left(2\cos^{-1}\dfrac{1}{4}\right)$

73. $\sin\left(2\cos^{-1}\dfrac{1}{5}\right)$

74. $\cos(2\arctan(-2))$

75. $\tan\left(2\arcsin\left(-\dfrac{3}{5}\right)\right)$

76. $\sin\left(2\arccos\dfrac{2}{9}\right)$

77. $\sin\left(\sin^{-1}\dfrac{1}{2}+\tan^{-1}(-3)\right)$

78. $\cos\left(\tan^{-1}\dfrac{5}{12}-\cot^{-1}\dfrac{4}{3}\right)$

79. $\cos\left(\arcsin\dfrac{3}{5}+\arccos\dfrac{5}{13}\right)$

80. $\tan\left(\arccos\dfrac{\sqrt{3}}{2}-\arcsin\left(-\dfrac{3}{5}\right)\right)$

Use a calculator to find each value. Give answers as real numbers.

81. $\cos(\tan^{-1}.5)$

82. $\sin(\cos^{-1}.25)$

83. $\tan(\arcsin .12251014)$

84. $\cot(\arccos .58236841)$

Write each of the following as an expression in u. See Example 7.

85. $\sin(\arccos u)$

86. $\tan(\arccos u)$

87. $\cot(\arcsin u)$

88. $\cos(\arcsin u)$

89. $\sin\left(\sec^{-1}\dfrac{u}{2}\right)$

90. $\cos\left(\tan^{-1}\dfrac{3}{u}\right)$

91. $\tan\left(\arcsin\dfrac{u}{\sqrt{u^2+2}}\right)$

92. $\cos\left(\arccos\dfrac{u}{\sqrt{u^2+5}}\right)$

93. Explain why $\sin^{-1}(1/2)\neq 5\pi/6$, despite the fact that $\sin 5\pi/6=1/2$.

94. A student observed, "$\mathrm{Cos}^{-1}(1/2)$ has two values: $\pi/3$ and $5\pi/3$." Comment on this observation.

95. Suppose an airplane flying faster than sound goes directly over you. Assume that the plane is flying level. At the instant you feel the sonic boom from the plane, the angle of elevation to the plane is given by

$$\alpha = 2 \arcsin \frac{1}{m},$$

where m is the Mach number of the plane's speed. (See the exercises at the end of Section 5.6.) Find α to the nearest degree for each of the following values of m.

(a) $m = 1.2$ (b) $m = 1.5$
(c) $m = 2$ (d) $m = 2.5$

96. A painting 1 m high and 3 m from the floor will cut off an angle θ to an observer, where

$$\theta = \tan^{-1}\left(\frac{x}{x^2 + 2}\right).$$

Assume that the observer is x m from the wall where the painting is displayed and that the eyes of the observer are 2 m above the ground. (See the figure.) Find the value of θ for the following values of x. Round to the nearest degree.

(a) 1   (b) 2   (c) 3
(d) Derive the formula given above. (*Hint:* Use the identity for tan $(\theta + \alpha)$. Use right triangles.)

(e) Graph the function for θ with a graphing calculator and determine the distance that maximizes the angle.

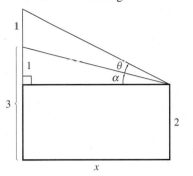

97. The following calculator trick will not work on all calculators. However, if you have a Texas Instruments or a Sharp scientific calculator, it will work. The calculator must be in the *degree* mode.

(a) Enter the year of your birth (all four digits).
(b) Subtract the number of years that have elapsed since 1980. For example, if it is 1997, subtract 17.
(c) Find the sine of the display.
(d) Find the inverse sine of the new display.

The result should be your age when you celebrate your birthday this year.

98. Explain why the procedure in Exercise 97 works as it does.

6.2 Trigonometric Equations ▼▼▼

In Chapter 5 we studied trigonometric equations that were identities. We now consider trigonometric equations that are **conditional;** that is, equations that are satisfied by some values but not others.

Conditional equations with trigonometric (or circular) functions can usually be solved by using algebraic methods and trigonometric identities. For example, suppose we wish to find the solutions of the equation

$$2 \sin \theta + 1 = 0$$

for all θ in the interval $[0°, 360°)$. We use the same method here as we would in solving the algebraic equation $2y + 1 = 0$. Subtract 1 from both sides, and divide by 2.

$$2 \sin \theta + 1 = 0$$
$$2 \sin \theta = -1$$
$$\sin \theta = -\frac{1}{2}$$

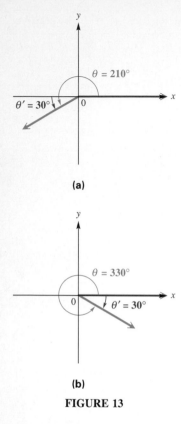

(a)

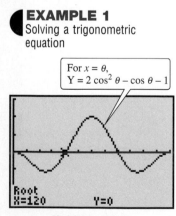

(b)

FIGURE 13

We know that θ must be in either quadrant III or IV, since the sine function is negative in these two quadrants. Furthermore, the reference angle must be 30°, since sin 30° = 1/2. The sketches in Figure 13 show the two possible values of θ, 210° and 330°.

CAUTION   One value that satisfies $\sin \theta = -1/2$ is $\sin^{-1}(-1/2)$, which in degrees is −30°. However, when solving an equation such as this, we must pay close attention to the *domain*, which in this case is [0°, 360°).

In some cases we are required to find *all* solutions of conditional trigonometric equations. All solutions of the equation $2 \sin \theta + 1 = 0$ would be written as

$$\theta = 210° + 360° \cdot n \qquad \text{or} \qquad \theta = 330° + 360° \cdot n,$$

where n is any integer. We add integer multiples of 360° to obtain all angles coterminal with 210° or 330°. If we had been required to solve this equation for real numbers (or angles in radians) in the interval [0, 2π), the two solutions would be $7\pi/6$ and $11\pi/6$, while all solutions would be written as

$$\theta = \frac{7\pi}{6} + 2n\pi \qquad \text{or} \qquad \theta = \frac{11\pi}{6} + 2n\pi,$$

where n is any integer.

In the examples in this section, we will find solutions in the intervals [0°, 360°) or [0, 2π). Remember that *all* solutions can be found using the methods described above.

EXAMPLE 1
Solving a trigonometric equation

For $x = \theta$,
$Y = 2 \cos^2 \theta - \cos \theta - 1$

[0, 360] by [−3, 3]
Xscl = 30   Yscl = 1

By graphing $Y = 2 \cos^2 \theta - \cos \theta - 1$ (where $x = \theta$), we can support the solutions in Example 1. The calculator must be in degree mode. The display shows that 120° is a solution; the others can be supported similarly.

Solve $2 \cos^2 \theta - \cos \theta - 1 = 0$ in the interval [0°, 360°).

We solve this equation by factoring. (It is quadratic in the term cos θ.)

$$2 \cos^2 \theta - \cos \theta - 1 = 0$$
$$(2 \cos \theta + 1)(\cos \theta - 1) = 0$$
$$2 \cos \theta + 1 = 0 \qquad \text{or} \qquad \cos \theta - 1 = 0$$
$$\cos \theta = -\frac{1}{2} \qquad\qquad\qquad \cos \theta = 1$$

In the first case, we have cos θ = −1/2, indicating that θ must be in either quadrant II or III, with a reference angle of 60°. Using a sketch similar to those in Figure 13 would indicate that two solutions are 120° and 240°. The second case, cos θ = 1, has the quadrantal angle 0° as its only solution in the interval. (We do not include 360° since it is not in the stated interval.) Therefore, the solutions of this equation are 0°, 120°, and 240°. Check these solutions by substituting them in the given equation.   ▶

EXAMPLE 2
Solving a trigonometric equation

Solve $\sin x \tan x = \sin x$ in the interval $[0°, 360°)$.

Subtract $\sin x$ from both sides, then factor on the left.

$$\sin x \tan x = \sin x$$

$$\sin x \tan x - \sin x = 0$$

$$\sin x(\tan x - 1) = 0$$

Now set each factor equal to 0.

$$\sin x = 0 \quad \text{or} \quad \tan x - 1 = 0$$

$$\tan x = 1$$

$$x = 0° \quad \text{or} \quad x = 180° \qquad x = 45° \quad \text{or} \quad x = 225°$$

Verify these solutions by substitution. ▶

CAUTION There are four solutions for Example 2. Trying to solve the equation by dividing both sides by $\sin x$ would give just $\tan x = 1$, which would give $x = 45°$ or $x = 225°$. The other two solutions would not appear. The missing solutions are the ones that make the divisor, $\sin x$, equal 0. For this reason, it is best to avoid dividing by a variable expression.

Recall from algebra that squaring both sides of an equation, such as $\sqrt{x + 4} = x + 2$, will yield all solutions, but may also give extraneous values. (In this equation, 0 is a solution, while -3 is extraneous. Verify this.) The same situation may occur when trigonometric equations are solved in this manner, as shown in the next example.

EXAMPLE 3
Solving a trigonometric equation

Solve $\tan x + \sqrt{3} = \sec x$ in the interval $[0, 2\pi)$.

Since the tangent and secant functions are related by the identity $1 + \tan^2 x = \sec^2 x$, one method of solving this equation is to square both sides, and express $\sec^2 x$ in terms of $\tan^2 x$.

$$\tan x + \sqrt{3} = \sec x$$

$$\tan^2 x + 2\sqrt{3} \tan x + 3 = \sec^2 x$$

$$\tan^2 x + 2\sqrt{3} \tan x + 3 = 1 + \tan^2 x$$

$$2\sqrt{3} \tan x = -2$$

$$\tan x = -\frac{1}{\sqrt{3}} = -\frac{\sqrt{3}}{3}$$

The possible solutions in the given interval are $5\pi/6$ and $11\pi/6$. Now check the possible solutions. Try $5\pi/6$ first.

Left side: $\quad \tan x + \sqrt{3} = \tan\dfrac{5\pi}{6} + \sqrt{3} = -\dfrac{\sqrt{3}}{3} + \sqrt{3} = \dfrac{2\sqrt{3}}{3}$

Right side: $\quad \sec x = \sec\dfrac{5\pi}{6} = \dfrac{-2\sqrt{3}}{3}$

The check shows that $5\pi/6$ is not a solution. Now check $11\pi/6$.

Left side: $\quad \tan\dfrac{11\pi}{6} + \sqrt{3} = -\dfrac{\sqrt{3}}{3} + \sqrt{3} = \dfrac{2\sqrt{3}}{3}$

Right side: $\quad \sec\dfrac{11\pi}{6} = \dfrac{2\sqrt{3}}{3}$

This solution satisfies the equation, so $11\pi/6$ is the only solution of the given equation. ◗

We have suggested checking the solutions of trigonometric equations by substitution. Another way to check solutions is to solve the equation by using a graphing calculator. For instance, in Example 3, let $Y_1 = \tan x + \sqrt{3}$ and $Y_2 = \sec x = 1/\cos x$. Graph both functions in the same window and use the capabilities of the calculator to find the intersection points. The figure shown at left below shows the graphs of Y_1 and Y_2 in the same window, with the value of the only intersection point at the bottom of the graph. The graph supports our results in Example 3. It is not clear from the graph whether there are one or two other solutions near $x = \pi/2$. To determine the number of solutions in the interval $[0, 2\pi)$, we graph $Y = \tan x + \sqrt{3} - \sec x$ and find the values of the x-intercepts. See the figure at right below. The only x-intercept in the given interval is approximately 5.7595865. Since $11\pi/6 \approx 5.7595865$, this supports our algebraic solution.

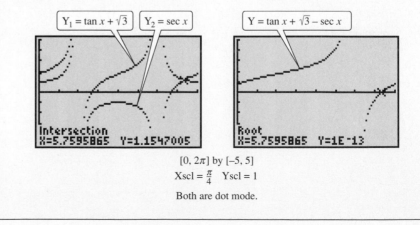

$[0, 2\pi]$ by $[-5, 5]$

$Xscl = \frac{\pi}{4} \quad Yscl = 1$

Both are dot mode.

In some cases trigonometric equations require a calculator to obtain approximate solutions, as in the next example.

Solving a trigonometric
equation using a calculator

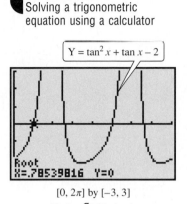

$Y = \tan^2 x + \tan x - 2$

Root
X=.78539816 Y=0

$[0, 2\pi]$ by $[-3, 3]$
$Xscl = \frac{\pi}{4}$ $Yscl = 1$

This graph supports the results in
Example 4. Here, the calculator is
in radian mode. The display indicates
that $\frac{\pi}{4} \approx .78539816$ is a solution. The
other three can be supported similarly.

Solve $\tan^2 x + \tan x - 2 = 0$ in the interval $[0, 2\pi)$.

Like Example 1, this equation is quadratic in form and may be solved for $\tan x$ by factoring.

$$\tan^2 x + \tan x - 2 = 0$$
$$(\tan x - 1)(\tan x + 2) = 0$$

Set each factor equal to 0.

$$\tan x - 1 = 0 \quad \text{or} \quad \tan x + 2 = 0$$
$$\tan x = 1 \quad \text{or} \quad \tan x = -2$$

The solutions for $\tan x = 1$ in the interval $[0, 2\pi)$ are $x = \pi/4$ or $5\pi/4$. To solve $\tan x = -2$ in that interval, use a calculator set in the *radian* mode. We find that $\tan^{-1}(-2) \approx -1.1071487$. This is a quadrant IV number, based on the range of the inverse tangent function. However, since we want solutions in the interval $[0, 2\pi)$, we must first add π to -1.1071487, and then add 2π:

$$x \approx -1.1071487 + \pi \approx 2.03444394$$
$$x \approx -1.1071487 + 2\pi \approx 5.1760366.$$

The solutions in the required interval are

$$\underbrace{\frac{\pi}{4}, \quad \frac{5\pi}{4},}_{\substack{\text{Exact} \\ \text{values}}} \quad \underbrace{2.0, \quad 5.2.}_{\substack{\text{Approximate} \\ \text{values to the} \\ \text{nearest tenth}}}$$

Note that these solutions are the x-intercepts of the graph of $y = \tan^2 x + \tan x - 2$. ▶

When a trigonometric equation that is quadratic in form cannot be factored, the quadratic formula can be used to solve the equation.

Solving a trigonometric
equation with the quadratic
formula

Solve $\cot^2 x + 3 \cot x = 1$ in $[0°, 360°)$.

Write the equation in quadratic form, with 0 on one side.

$$\cot^2 x + 3 \cot x - 1 = 0$$

Since this quadratic equation cannot be solved by factoring, use the quadratic formula, with $a = 1$, $b = 3$, $c = -1$, and $\cot x$ as the variable.

$$\cot x = \frac{-3 \pm \sqrt{9 + 4}}{2} = \frac{-3 \pm \sqrt{13}}{2} \approx \frac{-3 \pm 3.6055513}{2}$$

$$\cot x \approx .30277564 \quad \text{or} \quad \cot x \approx -3.3027756$$
$$x \approx 73.2°, 253.2°, 163.2°, 343.2°$$

The final answers were obtained using a calculator set in the degree mode. ▶

The methods for solving trigonometric equations illustrated in the examples can be summarized as follows.

Solving Trigonometric Equations Analytically

1. Begin by deciding whether the equation is linear or quadratic, so that the method of solution can be determined.
2. If only one trigonometric function is present, first solve the equation for that function.
3. If more than one trigonometric function is present, rearrange the equation so that one side equals 0. Then try to factor and set each factor equal to 0 to solve.
4. If Step 3 does not work, try using identities to change the form of the equation. It may be helpful to square both sides of the equation first. If this is done, check for extraneous solutions.
5. If the equation is quadratic in form, but not factorable, use the quadratic formula. Check for extraneous solutions.

The final example in this section is a continuation of the discussion in the chapter opener.

◀ EXAMPLE 6
Describing a musical tone by interpreting a graph

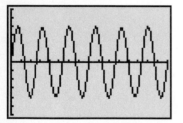

[0, .04] by [−.006, .006]
Xscl = .0025   Yscl = .001

FIGURE 14

 A basic component of music is a pure tone. The calculator-generated graph in Figure 14 shows the sinusoidal pressure P in pounds per square foot from a pure tone at time t seconds.

(a) The frequency of a pure tone is often measured in a unit called *hertz*. One hertz is equal to one cycle per second and is abbreviated as *Hz*. What is the frequency f in hertz of the pure tone shown in the graph?

From the graph we can see that there are 6 cycles in .04 sec. This is equivalent to $6/.04 = 150$ cycles per second. The pure tone has a frequency of $f = 150$ Hz.

(b) The time for the tone to produce one complete cycle is called the *period*. Approximate the period T of the pure tone in seconds.

Six periods cover a time of .04 second. One period would be equal to $T = .04/6 = 1/150$ or $.00\overline{6}$ sec.

(c) An equation for the graph is $y = .004 \sin(300\pi x)$. Use a calculator to estimate all solutions to the equation $y = .004$ on the interval $[0, .02]$.

If we reproduce the graph in Figure 14 on a calculator and graph the second function $y = .004$, we can determine that the approximate values of x at the points of intersection of the graphs are .0017, .0083, and .015. Verify these values with your own calculator. ▶

6.2 Exercises ▼▼▼▼▼▼▼▼▼▼▼▼▼▼▼▼▼▼▼▼▼▼▼▼▼▼▼▼▼▼▼▼▼▼▼

1. An equation of the form $\sin x - b = 0$ has either zero, one, or two solutions in the interval $[0, 2\pi)$. For what values of b will there be two solutions? One solution? No solutions? (*Hint:* Look at the graph of $y = \sin x$.)

2. Explain why an equation of the form $\tan x - b = 0$ always has exactly one solution in the interval $(-\pi/2, \pi/2)$ and two solutions in the interval $[0, 2\pi)$.

3. Suppose an equation of the form $\tan x - b = 0$ has the single solution $x = a$ in the interval $(-\pi/2, \pi/2)$. Give an expression for all solutions to the equation.

4. Suppose an equation of the form $\sin x - b = 0$ has the two solutions $x = a_1$ and $x = a_2$ in the interval $[0, 2\pi)$. Give expressions for all solutions to the equation.

Recall that a graphing calculator will return a 1 for a true statement and a 0 for a false statement. In Exercises 5 and 6, what possible values for X, in the interval $[0, 2\pi)$, will produce the accompanying graphing calculator screen?

5.
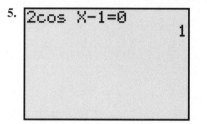

6.
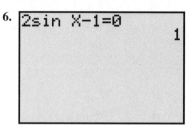

Solve each equation for solutions in the interval $[0, 2\pi)$ by first solving for the trigonometric function. See Examples 1–3.

7. $2 \cot x + 1 = -1$

8. $\sin x + 2 = 3$

9. $2 \sin x + 3 = 4$

10. $2 \sec x + 1 = \sec x + 3$

11. $\tan^2 x + 3 = 0$

12. $\sec^2 x + 2 = -1$

13. $(\cot x - 1)(\sqrt{3} \cot x + 1) = 0$

14. $(\csc x + 2)(\csc x - \sqrt{2}) = 0$

15. $\cos^2 x + 2 \cos x + 1 = 0$

16. $2 \cos^2 x - \sqrt{3} \cos x = 0$

17. $-2 \sin^2 x = 3 \sin x + 1$

18. $\tan^3 x = 3 \tan x$

19. $2 \cos^4 x = \cos^2 x$

20. $4(1 + \sin x)(1 - \sin x) = 3$

Determine all solutions of the following equations in radians.

21. $2 \sin^2 x - \sin x - 1 = 0$

22. $2 \cos^2 x + \cos x = 1$

23. $4 \cos^2 x - 1 = 0$

24. $2 \cos^2 x + 5 \cos x + 2 = 0$

25. $\cos^2 x + \cos x - 6 = 0$

26. $\sin^2 x - \sin x = 0$

Solve each equation for solutions in the interval $[0°, 360°)$. See Examples 1–3.

27. $(\cot \theta - \sqrt{3})(2 \sin \theta + \sqrt{3}) = 0$

28. $(\tan \theta - 1)(\cos \theta - 1) = 0$

29. $2 \sin \theta - 1 = \csc \theta$

30. $\tan \theta + 1 = \sqrt{3} + \sqrt{3} \cot \theta$

31. $\tan \theta - \cot \theta = 0$

32. $\cos^2 \theta = \sin^2 \theta + 1$

33. $\csc^2 \theta - 2 \cot \theta = 0$

34. $\sin^2 \theta \cos \theta = \cos \theta$

35. $2 \tan^2 \theta \sin \theta - \tan^2 \theta = 0$

36. $\sin^2 \theta \cos^2 \theta = 0$

37. $\sec^2 \theta \tan \theta = 2 \tan \theta$

38. $\cos^2 \theta - \sin^2 \theta = 0$

39. $\sin \theta + \cos \theta = 1$

40. $\sec \theta - \tan \theta = 1$

Solve the following equations for solutions in the interval $[0°, 360°)$*. Use a calculator and express approximate solutions to the nearest tenth of a degree. In Exercises 47–54, you will need to use the quadratic formula. See Examples 4 and 5.*

41. $3 \sin^2 x - \sin x = 2$

42. $\dfrac{2 \tan x}{3 - \tan^2 x} = 1$

43. $\sec^2 \theta = 2 \tan \theta + 4$

44. $5 \sec^2 \theta = 6 \sec \theta$

45. $3 \cot^2 \theta = \cot \theta$

46. $8 \cos \theta = \cot \theta$

47. $9 \sin^2 x - 6 \sin x = 1$

48. $4 \cos^2 x + 4 \cos x = 1$

49. $\tan^2 x + 4 \tan x + 2 = 0$

50. $3 \cot^2 x - 3 \cot x - 1 = 0$

51. $\sin^2 x - 2 \sin x + 3 = 0$

52. $2 \cos^2 x + 2 \cos x - 1 = 0$

53. $\cot x + 2 \csc x = 3$

54. $2 \sin x = 1 - 2 \cos x$

55. Refer to Example 1. The solutions in the interval $[0°, 360°)$ are $0°$, $120°$, and $240°$. See the discussion at the beginning of this section, and express *all* solutions of this equation (in degrees).

56. Refer to Example 4. The solutions in the interval $[0, 2\pi)$ are $\pi/4$, $5\pi/4$, 2.0, and 5.2. See the discussion at the beginning of this section and express *all* solutions of this equation (in real numbers).

57. What is wrong with the following solution for all x in the interval $[0, 2\pi)$ of the equation $\sin^2 x - \sin x = 0$?

$$\sin^2 x - \sin x = 0$$
$$\sin x - 1 = 0 \qquad \text{Divide by } \sin x.$$
$$\sin x = 1 \qquad \text{Add 1.}$$
$$x = \frac{\pi}{2}$$

58. What is wrong with the following solutions for all θ in the interval $[0°, 360°)$ of the equation $\tan^2 \theta - 1 = 0$?

$$\tan^2 \theta - 1 = 0$$
$$\tan^2 \theta = 1 \qquad \text{Add 1.}$$
$$\tan \theta = 1 \qquad \text{Take square root on each side.}$$
$$\theta = 45°, 225°$$

---

59. In an electric circuit, let V represent the electromotive force in volts at t seconds. Assume $V = \cos 2\pi t$. Find the smallest positive value of t where $0 \le t \le 1/2$ for each of the following values of V.
 (a) $V = 0$
 (b) $V = .5$
 (c) $V = .25$

60. A coil of wire rotating in a magnetic field induces a voltage given by

$$e = 20 \sin\left(\frac{\pi t}{4} - \frac{\pi}{2}\right),$$

where t is time in seconds. Find the smallest positive time to produce the following voltages.
 (a) 0 (b) $10\sqrt{3}$

61. The equation

$$.342D \cos \theta + h \cos^2 \theta = \frac{16D^2}{V_0^2}$$

is used in reconstructing accidents in which a vehicle vaults into the air after hitting an obstruction. V_0 is the velocity in feet per second of the vehicle when it hits, D is the distance (in feet) from the obstruction to the landing point, and h is the difference in height (in feet) between the landing point and the takeoff point. Angle θ is the takeoff angle, the angle between the horizontal and the path of the vehicle. Find θ to the nearest degree if $V_0 = 60$, $D = 80$, and $h = 2$.

62. If $0 < k < 1$, how many solutions does the equation $\sin^2 \theta = k$ have in the interval $[0°, 360°)$?

63. Refer to Example 6. Determine a simple relationship between f and T.

64. No musical instrument can generate a true pure tone. A pure tone has a unique, constant frequency and amplitude that sounds rather dull and uninteresting. The pressures caused by pure tones on the eardrum are sinusoidal. The change in pressure P in pounds per square foot on a person's eardrum from a pure tone at time t in seconds can be modeled using the equation $P = A \sin(2\pi ft + \phi)$. f is the frequency in cycles per second and ϕ is the phase angle. When P is positive there is an increase in pressure and the eardrum is pushed inward; when P is negative there is a decrease in pressure and the eardrum is pushed outward. (*Source:* Roederer, Juan,

Introduction to the Physics and Psychophysics of Music, The English Universities Press, Ltd., London, 1973.)

(a) Middle C has a frequency of 261.63 cycles per second. Graph this tone with $A = .004$ and $\phi = \pi/7$ in the window $[0, .005]$ by $[-.005, .005]$.

(b) Determine analytically the values of t for which $P = 0$ in $[0, .005]$ and support your answers graphically.

(c) Determine graphically when $P \leq 0$ on $[0, .005]$.

(d) Would an eardrum hearing this tone be vibrating outward or inward when $P \leq 0$?

A particle moves along a straight line. The distance of the particle from the origin at time t is given by

$$s(t) = \sin t + 2 \cos t.$$

Find a value of t that satisfies the given equation.

65. $s(t) = \dfrac{2 + \sqrt{3}}{2}$

66. $s(t) = \dfrac{3\sqrt{2}}{2}$

6.3 Trigonometric Equations with Multiple Angles ▼▼▼

In this section we discuss trigonometric equations that involve functions of half angles and multiples of angles.

◀**EXAMPLE 1**
Solving an equation with a double angle

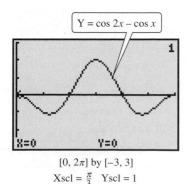

$Y = \cos 2x - \cos x$

$[0, 2\pi]$ by $[-3, 3]$
$Xscl = \frac{\pi}{3}$ $Yscl = 1$

This screen supports the solution 0 found in Example 1. By using $Xscl = \frac{\pi}{3}$ and observing that the graph intersects the x-axis at the second and fourth tick marks, we can feel fairly sure that $\frac{2\pi}{3}$ and $\frac{4\pi}{3}$ are also solutions, as found analytically.

Solve $\cos 2x = \cos x$ in the interval $[0, 2\pi)$.

First change $\cos 2x$ to a trigonometric function of x. Use the identity $\cos 2x = 2 \cos^2 x - 1$ so that the equation involves only the cosine of x. Then use the methods of the previous section.

$$\cos 2x = \cos x$$
$$2 \cos^2 x - 1 = \cos x$$
$$2 \cos^2 x - \cos x - 1 = 0$$
$$(2 \cos x + 1)(\cos x - 1) = 0$$

$$2 \cos x + 1 = 0 \quad \text{or} \quad \cos x - 1 = 0$$
$$\cos x = -\frac{1}{2} \quad \text{or} \quad \cos x = 1$$

In the required interval,

$$x = \frac{2\pi}{3} \quad \text{or} \quad \frac{4\pi}{3} \quad \text{or} \quad x = 0.$$

The solutions are 0, $2\pi/3$, and $4\pi/3$. ▶

$$\tan 3x + \sec 3x = 2$$

$$\tan 3x = 2 - \sec 3x \qquad \text{Subtract } \sec 3x.$$

$$\tan^2 3x = 4 - 4\sec 3x + \sec^2 3x \qquad \begin{array}{l}\text{Square both sides;} \\ (a-b)^2 = a^2 - 2ab + b^2\end{array}$$

$$\mathbf{sec^2\, 3x - 1} = 4 - 4\sec 3x + \sec^2 3x \qquad \text{Replace } \tan^2 3x \text{ with } \sec^2 3x - 1.$$

$$4\sec 3x = 5$$

$$\sec 3x = \frac{5}{4}$$

$$\frac{1}{\cos 3x} = \frac{5}{4} \qquad \sec \theta = \frac{1}{\cos \theta}$$

$$\cos 3x = \frac{4}{5} \qquad \text{Use reciprocals.}$$

Multiply the inequality $0 \le x < 2\pi$ by 3 to find the interval for $3x$: $[0, 6\pi)$. Using a calculator and knowing that cosine is positive in quadrants I and IV, we get

$$3x \approx .64350111,\ 5.6396842,\ 6.9266864,\ 11.922870,\ 13.209872,\ 18.206055.$$

Dividing by 3 gives

$$x \approx .21450037,\ 1.8798947,\ 2.3088955,\ 3.9742898,\ 4.4032906,\ 6.0686849.$$

Since both sides of the equation were squared, each of these proposed solutions must be checked. It can be verified by substitution in the given equation that the solutions are .21450037, 2.3088955, and 4.4032906. ▶

Comparing the intercepts of the graph of the equation $Y = \tan 3x + \sec 3x - 2$ in Example 4 with the solutions found analytically is a good way to verify which of the proposed solutions satisfy the equation. The calculator-generated graph in the figure below confirms the three solutions found above. The x-value shown in the display supports the solution 2.3088955.

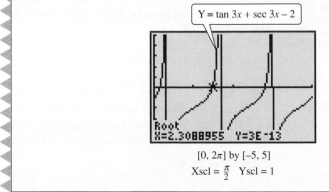

$$[0, 2\pi] \text{ by } [-5, 5]$$
$$\text{Xscl} = \frac{\pi}{2} \quad \text{Yscl} = 1$$

6.3 Exercises ▼▼▼▼▼▼▼▼▼▼▼▼▼▼▼▼▼▼▼▼▼▼▼▼▼▼▼▼▼▼▼▼▼

1. The accompanying screen shows the graphs of $Y_1 = 2 \sin x$ and $Y_2 = \cos(x/2)$ on the interval $[0, 2\pi)$ with Xscl = .5 and Yscl = 1. Estimate the solutions of the equation $2 \sin x = \cos(x/2)$ in the interval $[0, 2\pi)$.

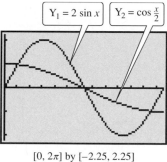

$[0, 2\pi]$ by $[-2.25, 2.25]$
Xscl = .5   Yscl = 1

2. The accompanying screen shows the graphs of $Y_1 = \sin 2x$ and $Y_2 = \cos^2 x$ on the interval $[0, 2\pi)$ with Xscl = .5 and Yscl = .5. The equation $\sin 2x = \cos^2 x$ has four solutions in the interval $[0, 2\pi)$. Two of the solutions can easily be found exactly. Find them. What are approximations of the other two solutions?

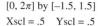

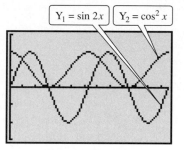

$[0, 2\pi]$ by $[-1.5, 1.5]$
Xscl = .5   Yscl = .5

Recall that a graphing calculator will return a 1 for a true statement and a 0 for a false statement. In Exercises 3 and 4, what possible values for X, in the interval $[0, 2\pi)$, will produce the accompanying graphing calculator screen?

3.
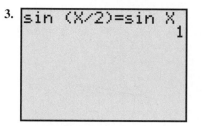

4.
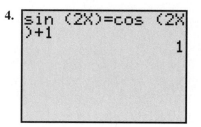

Solve the equation for solutions in the interval $[0, 2\pi)$. See Examples 1–4.

5. $\cos 2x = \dfrac{\sqrt{3}}{2}$

6. $\cos 2x = -\dfrac{1}{2}$

7. $\sin 3x = -1$

8. $\sin 3x = 0$

9. $3 \tan 3x = \sqrt{3}$

10. $\cot 3x = \sqrt{3}$

11. $\sqrt{2} \cos 2x = -1$

12. $2\sqrt{3} \sin 2x = \sqrt{3}$

13. $\sin \dfrac{x}{2} = \sqrt{2} - \sin \dfrac{x}{2}$

14. $\sin x = \sin 2x$

15. $\tan 4x = 0$

16. $\cos 2x - \cos x = 0$

17. $8 \sec^2 \dfrac{x}{2} = 4$

18. $\sin^2 \dfrac{x}{2} - 2 = 0$

19. $\sin \dfrac{x}{2} = \cos \dfrac{x}{2}$

20. $\sec \dfrac{x}{2} = \cos \dfrac{x}{2}$

21. $\cos 2x + \cos x = 0$

22. $\sin x \cos x = \dfrac{1}{4}$

Solve the equation for solutions in the interval $[0°, 360°)$. *When an approximation is appropriate, use a calculator as necessary to find solutions to the nearest tenth of a degree. See Examples 1–4.*

23. $\sqrt{2} \sin 3\theta - 1 = 0$

24. $-2 \cos 2\theta = \sqrt{3}$

25. $\cos \dfrac{\theta}{2} = 1$

26. $\sin \dfrac{\theta}{2} = 1$

27. $2\sqrt{3} \sin \dfrac{\theta}{2} = 3$

28. $2\sqrt{3} \cos \dfrac{\theta}{2} = -3$

29. $2 \sin \theta = 2 \cos 2\theta$

30. $\cos \theta - 1 = \cos 2\theta$

31. $1 - \sin \theta = \cos 2\theta$

32. $\sin 2\theta = 2 \cos^2 \theta$

33. $\csc^2 \dfrac{\theta}{2} = 2 \sec \theta$

34. $\cos \theta = \sin^2 \dfrac{\theta}{2}$

35. $2 - \sin 2\theta = 4 \sin 2\theta$

36. $4 \cos 2\theta = 8 \sin \theta \cos \theta$

37. $2 \cos^2 2\theta = 1 - \cos 2\theta$

---

38. In Section 6.2, we could easily find *all* solutions of a trigonometric equation after we found the solutions in the interval $[0, 2\pi)$. Why would this task be more complicated for the types of equations discussed in this section?

39. What is wrong with the following solution? Solve $2 \sin(1/2)x = -1$ in the interval $[0°, 360°)$.

$$2 \sin \frac{1}{2}x = -1$$

$$\sin \frac{1}{2}x = -\frac{1}{2}$$

$$\frac{1}{2}x = 210° \quad \text{or} \quad \frac{1}{2}x = 330°$$

$$x = 420° \qquad\qquad x = 660°$$

The solutions are $420°$ and $660°$.

40. What is wrong with the following solution? Solve $\tan 2\theta = 2$ in the interval $[0, 2\pi)$.

$$\tan 2\theta = 2$$

$$\frac{\tan 2\theta}{2} = \frac{2}{2}$$

$$\tan \theta = 1$$

$$\theta = \frac{\pi}{4} \quad \text{or} \quad \theta = \frac{5\pi}{4}$$

The solutions are $\pi/4$ and $5\pi/4$.

41. The seasonal variation in the length of daylight can be represented by a sine function. For example, the daily number of hours of daylight in New Orleans is given by

$$h = \frac{35}{3} + \frac{7}{3} \sin \frac{2\pi x}{365},$$

where x is the number of days after March 21 (disregarding leap year). (*Source:* Bushaw, Donald et al, *A Sourcebook of Applications of School Mathematics.* Copyright © 1980 by The Mathematical Association of America. Reprinted by permission. The material was prepared with the support of National Science Foundation Grant No. SED72-01123 A05. However, any opinions, findings, conclusions, or recommendations expressed herein are those of the authors and do not necessarily reflect the views of NSF.)

(a) On what date will there be about 14 hours of daylight?

(b) What date has the least number of hours of daylight?

(c) When will there be about 10 hours of daylight?

42. The British nautical mile is defined as the length of a minute of arc of a meridian. Since the Earth is flat at its poles, the nautical mile, in feet, is given by

$$L = 6077 - 31 \cos 2\theta,$$

where θ is the latitude in degrees. See the figure. (*Source:* Bushaw, Donald et al, *A Sourcebook of Applications of School Mathematics.* Copyright © 1980 by The Mathematical Association of America. Reprinted by permission. The material was prepared with the support of National Science Foundation Grant No. SED72-01123 A05. However, any opinions, findings, conclusions, or recommendations expressed herein are those of the authors and do not necessarily reflect the views of NSF.)

A nautical mile is the length on any of these meridians cut by a central angle of measure 1 minute.

(a) Find the latitude between 0° and 90° at which the nautical mile is 6074 feet.

(b) At what latitude between 0° and 180° is the nautical mile 6108 feet?

(c) In the United States the nautical mile is defined everywhere as 6080.2 feet. At what latitude between 0° and 90° does this agree with the British nautical mile?

The study of alternating electric current requires the solutions of equations of the form $i = I_{max} \sin 2\pi ft$, for time t in seconds, where i is instantaneous current in amperes, I_{max} is maximum current in amperes, and f is the number of cycles per second. (Source: Hannon, Ralph H., Basic Technical Mathematics with Calculus, W. B. Saunders Co., Philadelphia, 1978, pp. 300–302.) Find the smallest positive value of t, given the following data.

43. $i = 40, I_{max} = 100, f = 60$

44. $i = 50, I_{max} = 100, f = 120$

45. $i = I_{max}, f = 60$

46. $i = \frac{1}{2}I_{max}, f = 60$

47. A piano string can vibrate at more than one frequency when it is struck. It produces a complex wave that can mathematically be modeled by a sum of several pure tones. If a piano key with a frequency of f_1 is played, then the corresponding string will not only vibrate at f_1 but it will also vibrate at the higher frequencies of $2f_1, 3f_1, 4f_1, \ldots, nf_1$. f_1 is called the **fundamental frequency** of the string and higher frequencies are called the **upper harmonics.** The human ear will hear the sum of these frequencies as one complex tone. (*Source:* Roederer, Juan, *Introduction to the Physics*

and Psychophysics of Music, The English Universities Press, Ltd., London, 1973.)

(a) Suppose that the A key above middle C is played. Its fundamental frequency is $f_1 = 440$ Hz and its associated pressure is expressed as $P_1 = .002 \sin 880\pi t$. The string will also vibrate at 880, 1320, 1760, ... hertz. The corresponding pressures of these upper harmonics are

$$P_2 = \frac{.002}{2} \sin 1760\pi t, \quad P_3 = \frac{.002}{3} \sin 2640\pi t,$$

$$P_4 = \frac{.002}{4} \sin 3520\pi t, \text{ and } P_5 = \frac{.002}{5} \sin 4400\pi t.$$

Graph each of the following expressions for P in the window $[0, .01]$ by $[-.005, .005]$.

(i) $P = P_1$

(ii) $P = P_1 + P_2$

(iii) $P = P_1 + P_2 + P_3$

(iv) $P = P_1 + P_2 + P_3 + P_4$

(v) $P = P_1 + P_2 + P_3 + P_4 + P_5$

(b) Describe the final graph of P.

(c) What is the maximum pressure of $P = P_1 + P_2 + P_3 + P_4 + P_5$? When does this maximum occur on $[0, .01]$?

48. If a string with a fundamental frequency of 110 hertz is plucked in the middle, it will vibrate at the odd harmonics of 110, 330, 550, ... hertz but not at the even harmonics of 220, 440, 660, ... hertz. The resulting pressure P caused by the string can be approximated using the equation

$$P = .003 \sin 220\pi t + \frac{.003}{3} \sin 660\pi t$$

$$+ \frac{.003}{5} \sin 1100\pi t + \frac{.003}{7} \sin 1540\pi t.$$

(*Sources:* Benade, Arthur, *Fundamentals of Musical Acoustics,* Oxford University Press, New York, 1976; Roederer, Juan, *Introduction to the Physics and Psychophysics of Music,* The English Universities Press Ltd., London, 1973.)

(a) Graph P in the window $[0, .03]$ by $[-.005, .005]$.

(b) Use the graph to describe the shape of the sound wave that is produced.

(c) See Exercise 64 in Section 6.2. At lower frequencies, the inner ear will hear a tone only when the eardrum is moving outward. Determine the times on the interval $[0, .03]$ when this will occur.

6.4   Equations with Inverse Trigonometric Functions ▼▼▼

Section 6.1 introduced the inverse trigonometric functions. Recall, for example, that $x = \sin y$, with $-\pi/2 \le y \le \pi/2$, means the same thing as $y = \arcsin x$ or $y = \sin^{-1} x$. Sometimes the solution of a trigonometric equation with more than one variable requires inverse trigonometric functions, as shown in the following examples.

EXAMPLE 1
Solving an equation for a variable using inverse notation

Solve $y = 3 \cos 2x$ for x.

We want $\cos 2x$ alone on one side of the equation so we can solve for $2x$ and then for x. First, divide both sides of the equation by 3.

$$y = 3 \cos 2x$$

$$\frac{y}{3} = \cos 2x$$

Now write the statement in the alternative form

$$2x = \arccos \frac{y}{3}.$$

Finally, multiply both sides by 1/2.

$$x = \frac{1}{2} \arccos \frac{y}{3} \quad \blacktriangleright$$

The next examples show how to solve equations involving inverse trigonometric functions.

EXAMPLE 2
Solving an equation involving an inverse trigonometric function

Solve $2 \arcsin x = \pi$.

First solve for $\arcsin x$.

$$2 \arcsin x = \pi$$

$$\arcsin x = \frac{\pi}{2} \qquad \text{Divide by 2.}$$

Use the definition of $\arcsin x$ to get

$$x = \sin \frac{\pi}{2}$$

or $\qquad\qquad\qquad x = 1.$

Verify that the solution satisfies the given equation.   ▶

EXAMPLE 3
Solving an equation involving inverse trigonometric functions

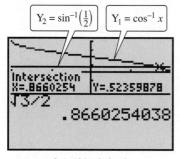

[-1, 1] by [-2, π]
Xscl = .5   Yscl = $\frac{\pi}{4}$

The exact solution found in Example 3, $\sqrt{3}/2$, is supported graphically in this screen. Notice that the approximation for X corresponds to that of $\sqrt{3}/2$.

Solve $\cos^{-1} x = \sin^{-1}(1/2)$.

Let $\sin^{-1}(1/2) = u$. Then $\sin u = 1/2$ and the equation becomes

$$\cos^{-1} x = u,$$

for u in quadrant I. This can be written as

$$\cos u = x.$$

Sketch a triangle and label it using the facts that u is in quadrant I and $\sin u = 1/2$. See Figure 15. Since $x = \cos u$,

$$x = \frac{\sqrt{3}}{2}.$$

Check the solution either by substitution or graphically. ▶

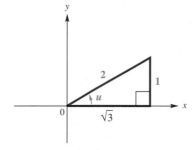

FIGURE 15

Some equations with inverse trigonometric functions require the use of identities to solve.

EXAMPLE 4
Solving an inverse trigonometric equation using an identity

Solve $\arcsin x - \arccos x = \pi/6$.

Begin by adding $\arccos x$ to both sides of the equation so that one inverse function is alone on one side of the equation.

$$\arcsin x - \arccos x = \frac{\pi}{6}$$

$$\arcsin x = \arccos x + \frac{\pi}{6} \qquad (1)$$

Use the definition of arcsin to write this statement as

$$\sin\left(\arccos x + \frac{\pi}{6}\right) = x.$$

Let $u = \arccos x$, so $0 \le u \le \pi$ by definition. Then

$$\sin\left(u + \frac{\pi}{6}\right) = x. \qquad (2)$$

Using the identity for $\sin(A + B)$,

$$\sin\left(u + \frac{\pi}{6}\right) = \sin u \cos \frac{\pi}{6} + \cos u \sin \frac{\pi}{6}.$$

Substitute this result into equation (2) to get

$$\sin u \cos \frac{\pi}{6} + \cos u \sin \frac{\pi}{6} = x. \qquad \text{(3)}$$

From equation (1) and by the definition of the arcsin function,

$$-\frac{\pi}{2} \le \operatorname{arccos} x + \frac{\pi}{6} \le \frac{\pi}{2}.$$

Subtract $\pi/6$ from each expression to get

$$-\frac{2\pi}{3} \le \operatorname{arccos} x \le \frac{\pi}{3}.$$

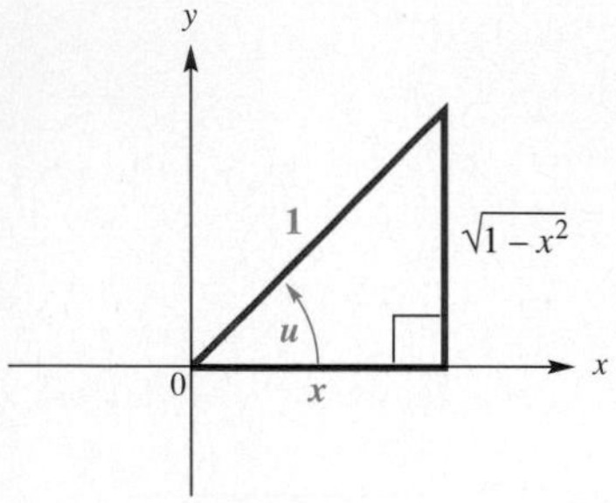

FIGURE 16

Since $0 \le \operatorname{arccos} x \le \pi$, it therefore follows that here we must have $0 \le \operatorname{arccos} x \le \pi/3$. Thus $x > 0$, and we can sketch the triangle in Figure 16. From this triangle we find that $\sin u = \sqrt{1 - x^2}$. Now substitute into equation (3) using $\sin u = \sqrt{1 - x^2}$, $\sin \pi/6 = 1/2$, $\cos \pi/6 = \sqrt{3}/2$, and $\cos u = x$.

$$(\sqrt{1 - x^2})\frac{\sqrt{3}}{2} + x \cdot \frac{1}{2} = x$$

$$(\sqrt{1 - x^2})\sqrt{3} + x = 2x$$

$$(\sqrt{3})\sqrt{1 - x^2} = x$$

Squaring both sides gives

$$3(1 - x^2) = x^2$$

$$3 - 3x^2 = x^2$$

$$3 = 4x^2$$

$$x = \sqrt{\frac{3}{4}} \qquad \text{Choose the positive square root because } x > 0.$$

$$= \frac{\sqrt{3}}{2}.$$

To check, replace x with $\sqrt{3}/2$ in the original equation:

$$\arcsin \frac{\sqrt{3}}{2} - \operatorname{arccos} \frac{\sqrt{3}}{2} = \frac{\pi}{3} - \frac{\pi}{6} = \frac{\pi}{6},$$

as required. The solution is $\sqrt{3}/2$. ▶

EXAMPLE 5
Determining the
pressure produced by
a sum of tones*

When two sources located at different positions produce the same pure tone, the human ear will often hear one sound that is equal to the sum of the individual tones. Since the sources are at different locations, they will have different phase angles ϕ. If two speakers located at different positions produce pure tones $P_1 = A_1 \sin(2\pi f + \phi_1)$ and $P_2 = A_2 \sin(2\pi f + \phi_2)$, where $-\pi/4 \le \phi_1, \phi_2 \le \pi/4$, then the resulting tone heard by a listener can be written as $P = A \sin(2\pi f + \phi)$, where

$$A = \sqrt{(A_1 \cos \phi_1 + A_2 \cos \phi_2)^2 + (A_1 \sin \phi_1 + A_2 \sin \phi_2)^2}$$

$$\text{and } \phi = \arctan\left(\frac{A_1 \sin \phi_1 + A_2 \sin \phi_2}{A_1 \cos \phi_1 + A_2 \cos \phi_2}\right).$$

(a) Calculate A and ϕ if $A_1 = .0012$, $\phi_1 = .052$, $A_2 = .004$, and $\phi_2 = .61$. Also find an expression for $P = A \sin(2\pi ft + \phi)$ if $f = 220$.

$$\begin{aligned} A &= \sqrt{(A_1 \cos \phi_1 + A_2 \cos \phi_2)^2 + (A_1 \sin \phi_1 + A_2 \sin \phi_2)^2} \\ &= \sqrt{(.0012 \cos .052 + .004 \cos .61)^2 + (.0012 \sin .052 + .004 \sin .61)^2} \\ &\approx .00506 \end{aligned}$$

$$\begin{aligned} \phi &= \arctan\left[\frac{A_1 \sin \phi_1 + A_2 \sin \phi_2}{A_1 \cos \phi_1 + A_2 \cos \phi_2}\right] \\ &= \arctan\left[\frac{.0012 \sin .052 + .004 \sin .61}{.0012 \cos .052 + .004 \cos .61}\right] \\ &\approx .484 \end{aligned}$$

Thus, $P = .00506 \sin(440\pi t + .484)$.

(b) Graph $Y_1 = P$ and $Y_2 = P_1 + P_2$ on the same coordinate axes on the interval $[0, .01]$. Are the two graphs the same?

As shown in Figure 17, the graphs of $P = .00506 \sin(440\pi t + .484)$ and $P_1 + P_2 = .0012 \sin(440\pi t + .052) + .004 \sin(440\pi t + .61)$ are the same. ▶

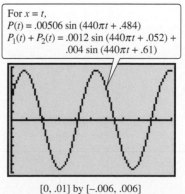

For $x = t$,
$P(t) = .00506 \sin (440\pi t + .484)$
$P_1(t) + P_2(t) = .0012 \sin (440\pi t + .052) + .004 \sin (440\pi t + .61)$

[0, .01] by [−.006, .006]
Xscl = .001   Yscl = .001

FIGURE 17

Source: Fletcher, N. and T. Rossing, *The Physics of Musical Instruments,* Springer-Verlag, 1991.

6.4 Exercises ▼▼▼▼▼▼▼▼▼▼▼▼▼▼▼▼▼▼▼▼▼▼▼▼▼▼▼▼▼▼▼▼▼▼▼▼▼▼

1. The accompanying screen shows the graphs of $Y_1 =$ arcsin x and $Y_2 =$ arccos x in the window $[-1, 1]$ by $[-\pi/2, \pi]$, with Xscl $= .5$ and Yscl $= \pi/6$. What are the exact coordinates of the point of intersection of the two graphs? What is the solution of arcsin $x -$ arccos $x = 0$?

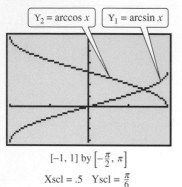

$[-1, 1]$ by $\left[-\frac{\pi}{2}, \pi\right]$
Xscl $= .5$ Yscl $= \frac{\pi}{6}$

2. The accompanying screen shows the graphs of $Y_1 = \cos^{-1} 2x$ and $Y_2 = (5\pi/3) - \cos^{-1} x$ in the window $[-1, 0]$ by $[0, 2\pi]$, with Xscl $= .5$ and Yscl $= \pi/3$. What are the exact coordinates of the point of intersection of the two graphs? What is the solution of $\cos^{-1} 2x + \cos^{-1} x = 5\pi/3$?

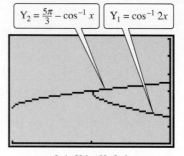

$[-1, 0]$ by $[0, 2\pi]$
Xscl $= .5$ Yscl $= \frac{\pi}{3}$

Recall that a graphing calculator will return a 1 for a true statement and a 0 for a false statement. In Exercises 3 and 4, what possible values for X, in the interval $[0, 1]$, will produce the accompanying graphing calculator screen?

3.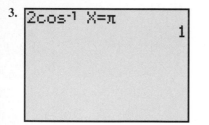

4.
```
4tan-1 X=π
                    1
```

Solve the equation for x. See Example 1.

5. $y = 5 \cos x$

6. $4y = \sin x$

7. $2y = \cot 3x$

8. $6y = \dfrac{1}{2} \sec x$

9. $y = 3 \tan 2x$

10. $y = 3 \sin \dfrac{x}{2}$

11. $y = 6 \cos \dfrac{x}{4}$

12. $y = -\sin \dfrac{x}{3}$

13. $y = -2 \cos 5x$

14. $y = 3 \cot 5x$

15. $y = \cos(x + 3)$

16. $y = \tan(2x - 1)$

17. $y = \sin x - 2$

18. $y = \cot x + 1$

19. $y = 2 \sin x - 4$

20. $y = 4 + 3 \cos x$

21. Refer to Exercise 17. A student attempting to solve this problem wrote as the first step

$$y = \sin(x - 2),$$

inserting parentheses as shown. Explain why this is incorrect.

22. Explain why the equation

$$\sin^{-1} x = \cos^{-1} 2$$

cannot have a solution. (No work needs to be shown here.)

Solve the equation. See Examples 2 and 3.

23. $\dfrac{4}{3} \cos^{-1} \dfrac{y}{4} = \pi$

24. $4\pi + 4 \tan^{-1} y = \pi$

25. $2 \arccos\left(\dfrac{y - \pi}{3}\right) = 2\pi$

26. $\arccos\left(y - \dfrac{\pi}{3}\right) = \dfrac{\pi}{6}$

27. $\arcsin x = \arctan \dfrac{3}{4}$

28. $\arctan x = \arccos \dfrac{5}{13}$

29. $\cos^{-1} x = \sin^{-1} \dfrac{3}{5}$

30. $\cot^{-1} x = \tan^{-1} \dfrac{4}{3}$

Solve the equation. See Example 4.

31. $\sin^{-1} x - \tan^{-1} 1 = -\dfrac{\pi}{4}$

32. $\sin^{-1} x + \tan^{-1} \sqrt{3} = \dfrac{2\pi}{3}$

33. $\arccos x + 2 \arcsin \dfrac{\sqrt{3}}{2} = \pi$

34. $\arccos x + 2 \arcsin \dfrac{\sqrt{3}}{2} = \dfrac{\pi}{3}$

35. $\arcsin 2x + \arccos x = \dfrac{\pi}{6}$

36. $\arcsin 2x + \arcsin x = \dfrac{\pi}{2}$

37. $\cos^{-1} x + \tan^{-1} x = \dfrac{\pi}{2}$

38. $\sin^{-1} x + \tan^{-1} x = 0$

39. Solve $d = 550 + 450 \cos \dfrac{\pi}{50} t$ for t in terms of d.

40. Solve $d = 40 + 60 \cos \dfrac{\pi}{6}(t - 2)$ for t in terms of d.

---

41. In the study of alternating electric current, instantaneous voltage is given by

$$e = E_{max} \sin 2\pi ft,$$

where f is the number of cycles per second, E_{max} is the maximum voltage, and t is time in seconds.
(a) Solve the equation for t.
(b) Find the smallest positive value of t if $E_{max} = 12$, $e = 5$, and $f = 100$. Use a calculator.

42. When a large-view camera is used to take a picture of an object that is not parallel to the film, the lens board

should be tilted so that the planes containing the subject, the lens board, and the film intersect in a line (see the figure). This gives the best "depth of field." (*Source:* Bushaw, Donald et al., *A Sourcebook of Applications of School Mathematics.* Copyright © 1980 by The Mathematical Association of America. Reprinted by permission. The material was prepared with the support of National Science Foundation Grant No. SED72-01123 A05. However, any opinions, findings, conclusions, or recommendations expressed herein are those of the authors and do not necessarily reflect the views of NSF.)

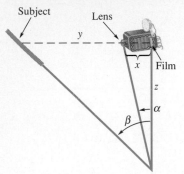

(a) Write two equations, one relating α, x, and z, and the other relating β, x, y, and z.
(b) Eliminate z from the equations in part (a) to get one equation relating α, β, x, and y.
(c) Solve the equation from part (b) for α.
(d) Solve the equation from part (b) for β.

43. In many computer languages, such as BASIC and FORTRAN, only the arctan function is available. To use the other inverse trigonometric functions, it is necessary to express them in terms of arctangent. This can be done as follows.
(a) Let $u = \arcsin x$. Solve the equation for x in terms of u.

(b) Use the result of part (a) to label the three sides of the triangle in the figure in terms of x.

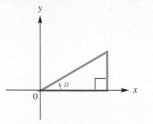

(c) Use the triangle from part (b) to write an equation for $\tan u$ in terms of x.
(d) Solve the equation from part (c) for u.

44. In the exercises for Section 4.1 we found the equation

$$y = \frac{1}{3} \sin \frac{4\pi t}{3},$$

where t is time (in seconds) and y is the angle formed by a rhythmically moving arm.
(a) Solve the equation for t.
(b) At what time(s) does the arm form an angle of .3 radian?

45. Repeat Example 5 with $A_1 = .0025$, $\phi_1 = \pi/7$, $A_2 = .001$, $\phi_2 = \pi/6$, and $f = 300$.

While visiting a museum, Patricia Quinlan views a painting that is 3 ft high and hanging 6 ft above the ground. See the figure. Assume Patricia's eyes are 5 ft above the ground, and let x be the distance from the spot where Patricia is standing to the wall displaying the painting.

46. Show that θ, the viewing angle subtended by the painting, is given by

$$\theta = \tan^{-1} \frac{4}{x} - \tan^{-1} \frac{1}{x}.$$

47. Find the value of x for each value of θ.
(a) $\theta = \dfrac{\pi}{6}$ (b) $\theta = \dfrac{\pi}{8}$

48. Find the value of θ for each value of x.
(a) $x = 4$ (b) $x = 3$

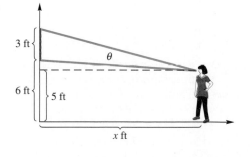

Chapter 6 Summary ▼▼▼▼▼▼▼▼▼▼▼▼▼▼▼▼▼▼▼▼▼▼▼▼▼▼▼▼▼▼▼

SECTION	KEY IDEAS
6.1 Inverse Trigonometric Functions	

Inverse Trigonometric Functions

Function	Domain	Range	Quadrants of the Unit Circle Range Values Come from
$y = \sin^{-1} x$	$[-1, 1]$	$\left[-\dfrac{\pi}{2}, \dfrac{\pi}{2}\right]$	I and IV
$y = \cos^{-1} x$	$[-1, 1]$	$[0, \pi]$	I and II
$y = \tan^{-1} x$	$(-\infty, \infty)$	$\left(-\dfrac{\pi}{2}, \dfrac{\pi}{2}\right)$	I and IV
$y = \cot^{-1} x$	$(-\infty, \infty)$	$(0, \pi)$	I and II
$y = \sec^{-1} x$	$(-\infty, -1] \cup [1, \infty)$	$[0, \pi], y \neq \pi/2$	I and II
$y = \csc^{-1} x$	$(-\infty, -1] \cup [1, \infty)$	$\left[-\dfrac{\pi}{2}, \dfrac{\pi}{2}\right], y \neq 0$	I and IV

$y = \sin^{-1} x$ or $\arcsin x$

$y = \cos^{-1} x$ or $\arccos x$

$y = \tan^{-1} x$ or $\arctan x$

SECTION	KEY IDEAS
6.2 Trigonometric Equations	
	Solving Trigonometric Equations
	1. Begin by deciding whether the equation is linear or quadratic, so that the method of solution can be determined.
	2. If only one trigonometric function is present, first solve the equation for that function.
	3. If more than one trigonometric function is present, rearrange the equation so that one side equals 0. Then try to factor and set each factor equal to 0 to solve.
	4. If Step 3 does not work, try using identities to change the form of the equation. It may be helpful to square both sides of the equation first. If this is done, check for extraneous solutions.
	5. If the equation is quadratic in form, but not factorable, use the quadratic formula. Check for extraneous solutions.

Chapter 6   Review Exercises ▼▼▼▼▼▼▼▼▼▼▼▼▼▼▼▼▼▼▼▼▼▼▼▼▼▼▼▼

1. Explain why cos(arccos x) always equals x, but arccos(cos x) may not equal x.

2. With a graphing calculator, sketch the graph of $\cos^{-1}(\cos x)$ in the window $[0, 4\pi]$ by $[0, 8]$ and explain its shape.

3. With a graphing calculator, sketch the graph of $\sin^{-1}(\sin x)$ in the window $[0, 4\pi]$ by $[-4, 4]$ and explain its shape.

4. Discuss how the accompanying screen supports the analytic work in Example 5(b) of Section 6.1. Does it matter whether the calculator is in radian mode or degree mode?

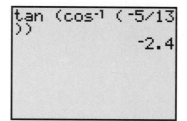

Give the exact real number value of y. Do not use a calculator.

5. $y = \sin^{-1}\left(\dfrac{\sqrt{2}}{2}\right)$

6. $y = \arccos\left(-\dfrac{1}{2}\right)$

7. $y = \tan^{-1}(-\sqrt{3})$

8. $y = \arcsin(-1)$

9. $y = \cos^{-1}\left(-\dfrac{\sqrt{2}}{2}\right)$

10. $y = \arctan\left(\dfrac{\sqrt{3}}{3}\right)$

11. $y = \sec^{-1}(-2)$

12. $y = \text{arccsc}\left(\dfrac{2\sqrt{3}}{3}\right)$

13. $y = \text{arccot}(-1)$

Give the degree measure of θ. Do not use a calculator.

14. $\theta = \arccos\left(\dfrac{1}{2}\right)$

15. $\theta = \arcsin\left(-\dfrac{\sqrt{3}}{2}\right)$

16. $\theta = \tan^{-1} 0$

Use a calculator to give the measure of θ.

17. $\theta = \arctan 1.7804675$

18. $\theta = \sin^{-1}(-.66045320)$

19. $\theta = \cos^{-1} .80396577$

20. $\theta = \cot^{-1} 4.5046388$

21. $\theta = \text{arcsec } 3.4723155$

22. $\theta = \csc^{-1} 7.4890096$

23. Explain why $\sin^{-1} 3$ cannot be defined.

24. $\text{Arcsin}(\sin 5\pi/6) \neq 5\pi/6$. Explain why this is so.

25. What is the domain of the arccot function?

26. What is the range of the arcsec function as defined in this text?

Find each of the following without using a calculator.

27. $\sin\left(\sin^{-1} \dfrac{1}{2}\right)$

28. $\tan\left(\tan^{-1} \dfrac{2}{3}\right)$

29. $\cos(\arccos(-1))$

30. $\sin\left(\arcsin\left(-\dfrac{\sqrt{3}}{2}\right)\right)$

31. $\arccos\left(\cos \dfrac{3\pi}{4}\right)$

32. $\text{arcsec}(\sec \pi)$

33. $\tan^{-1}\left(\tan \dfrac{\pi}{4}\right)$

34. $\cos^{-1}(\cos 0)$

35. $\sin\left(\arccos \dfrac{3}{4}\right)$

36. $\cos(\arctan 3)$

37. $\cos(\csc^{-1}(-2))$

38. $\sec\left(2 \sin^{-1}\left(-\dfrac{1}{3}\right)\right)$

39. $\tan\left(\arcsin \dfrac{3}{5} + \arccos \dfrac{5}{7}\right)$

Write each of the following as an expression in u.

40. $\sin(\tan^{-1} u)$

41. $\cos\left(\arctan \dfrac{u}{\sqrt{1 - u^2}}\right)$

42. $\tan\left(\text{arcsec } \dfrac{\sqrt{u^2 + 1}}{u}\right)$

Graph each of the following, and give the domain and range.

43. $y = \sin^{-1} x$

44. $y = \cos^{-1} x$

45. $y = \text{arccot } x$

Solve the equation for solutions in the interval $[0, 2\pi)$. Use a calculator in Exercises 47 and 48.

46. $\sin^2 x = 1$

47. $2 \tan x - 1 = 0$

48. $3 \sin^2 x - 5 \sin x + 2 = 0$

49. $\tan x = \cot x$

50. $\sec^4 2x = 4$

51. $\tan^2 2x - 1 = 0$

52. $\sec \dfrac{x}{2} = \cos \dfrac{x}{2}$

53. $\cos 2x + \cos x = 0$

54. $4 \sin x \cos x = \sqrt{3}$

Solve the equation for solutions in the interval $[0°, 360°)$. When appropriate, use a calculator and express solutions to the nearest tenth of a degree.

55. $\sin^2 \theta + 3 \sin \theta + 2 = 0$

56. $2 \tan^2 \theta = \tan \theta + 1$

57. $\sin 2\theta = \cos 2\theta + 1$

58. $2 \sin 2\theta = 1$

59. $3 \cos^2 \theta + 2 \cos \theta - 1 = 0$

60. $5 \cot^2 \theta - \cot \theta - 2 = 0$

61. $\sin 2\theta + \sin 4\theta = 0$

62. $\cos \theta - \cos 2\theta = 2 \cos \theta$

Solve the equation for x.

63. $4y = 2 \sin x$

64. $y = 3 \cos \dfrac{x}{2}$

65. $2y = \tan(3x + 2)$

66. $5y = 4 \sin x - 3$

67. $\dfrac{4}{3} \arctan \dfrac{x}{2} = \pi$

68. $\arccos x = \arcsin \dfrac{2}{7}$

69. $\arccos x + \arctan 1 = \dfrac{11\pi}{12}$

70. $\operatorname{arccot} x = \arcsin\left(\dfrac{-\sqrt{2}}{2}\right) + \dfrac{3\pi}{4}$

---

71. Recall Snell's law from Exercises 43–46 of Section 2.3:

$$\frac{c_1}{c_2} = \frac{\sin \theta_1}{\sin \theta_2},$$

where c_1 is the speed of light in one medium, c_2 is the speed of light in a second medium, and θ_1 and θ_2 are the angles shown in the figure. Suppose a light is shining up through water into the air as in the figure.

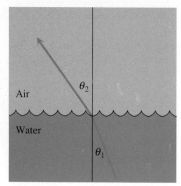

As θ_1 increases, θ_2 approaches 90°, at which point no light will emerge from the water. Assume the ratio c_1/c_2 in this case is .752. For what value of θ_1 does $\theta_2 = 90°$? This value of θ_1 is called the *critical angle* for water.

72. Refer to Exercise 71. What happens when θ_1 is greater than the critical angle?

73. The function $y = \sec^{-1} x$ is not found on graphing calculators. However, with some models it can be graphed as

$$y = \frac{\pi}{2} - ((x > 0) - (x < 0))$$

$$\times \left(\frac{\pi}{2} - \tan^{-1}\left(\sqrt{x^2 - 1}\right)\right).$$

(This formula appears as Y_1 in the screen below.) Use the formula to obtain the graph of $y = \sec^{-1} x$ in the window $[-10, 10]$ by $[0, \pi]$.

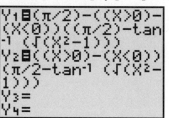

74. The function $y = \csc^{-1} x$ is not found on graphing calculators. However, with some models it can be graphed as

$$y = ((x > 0) - (x < 0))$$

$$\times \left(\frac{\pi}{2} - \tan^{-1}\left(\sqrt{x^2 - 1}\right)\right)$$

(This formula appears as Y_2 in the screen above.) Use the formula to obtain the graph of $y = \csc^{-1} x$ in the window $[-10, 10]$ by $[-\pi/2, \pi/2]$.

75. Musicians sometimes tune instruments by playing the same tone on two different instruments and listening for a phenomenon known as **beats.** Beats occur when two tones vary in frequency by only a few hertz. When the two instruments are in tune the beats disappear. The ear hears beats because the pressure slowly rises and falls as a result of this slight variation in the frequency. This phenomenon can be seen using a graphing calculator. (*Source:* Pierce, John, *The Science of Musical Sound,* Scientific American Books, 1992.)

(a) Consider two tones with frequencies of 220 and 223 hertz and pressures $P_1 = .005 \sin 440\pi t$ and $P_2 = .005 \sin 446\pi t$, respectively. Graph the pressure $P = P_1 + P_2$ felt by an eardrum over the one-second interval $[.15, 1.15]$. How many beats are there in one second?

(b) Repeat part (a) with frequencies of 220 and 216.

(c) Determine a simple way to find the number of beats per second if the frequency of each tone is given.

 76. Small speakers like those found in older radios and telephones often cannot vibrate slower than 200 hertz—yet 35 keys on a piano have frequencies below 200 hertz. When a musical instrument creates a tone of 110 hertz it also creates tones at 220, 330, 440, 550, 660, . . . hertz. A small speaker cannot reproduce the 110-hertz vibration but it can reproduce the higher frequencies, which are called the upper harmonics. The low tones can still be heard because the speaker produces **difference tones** of the upper harmonics. The difference between consecutive frequencies is 110 hertz and this difference tone will be heard by a listener. We can see this phenomenon using a graphing calculator. (*Source:* Benade, Arthur, *Fundamentals of Musical Acoustics,* Oxford University Press, New York, 1976.)

(a) Graph the upper harmonics represented by the pressure

$$P = \frac{1}{2}\sin[2\pi(220)t] + \frac{1}{3}\sin[2\pi(330)t]$$
$$+ \frac{1}{4}\sin[2\pi(440)t]$$

in the window $[0, .03]$ by $[-2, 2]$.

(b) Estimate all t-coordinates where P is maximum.

(c) What does a person hear in addition to the frequencies of 220, 330, and 440 hertz?

(d) Graph the pressure produced by a speaker that can vibrate at 110 Hz and above.

7

Applications of Trigonometry and Vectors

Aerial photography first began in 1858 when French photographer Gaspard Tournachon took pictures of Paris from a hot-air balloon that had a makeshift darkroom. Since then aerial photography has been used in a variety of applications including surveying, road design, weather forecasting, military surveillance, topographic maps, and even archaeology. The first archaeological aerial photographs were taken of Stonehenge in 1906. By searching these photographs for unusual soil and marks caused by structures lying below the ground, Stonehenge Avenue was discovered. Today, hot-air balloons have been replaced by airplanes, helicopters, and satellites.

In aerial photography a series of photographs are usually taken with sufficient overlap to allow for the stereoscopic vision necessary to obtain accurate ground measurements. These photographs are often used to construct a map that gives both the coordinates and elevations of important features located on the ground. The perspective of these photographs can be affected if the airplane is not perfectly horizontal, the ground below is not level, or the camera is tilted. For example, how much distance will an aerial photograph cover if it is taken at an altitude of 5000 feet with a camera having a 12-inch focal length and a 35°-tilt to the vertical? If the distance between two objects in a photograph is 6 inches, what other information do we need to know in order to determine the actual ground distance between these objects? These questions are answered later in this chapter.

Being able to determine the measurements of the sides and angles in a triangle is essential to solving applications involving aerial photography. In this chapter we will learn techniques to solve many applications using

Sources: Brooks, R. and Dieter Johannes, *Phytoarchaeology*, Dioscorides Press, Portland, Oregon, 1990.

Moffitt, F., *Photogrammetry*, International Textbook Company, Scranton, Pennsylvania, 1967.

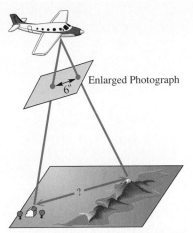

Enlarged Photograph

trigonometry. These trigonometric applications will allow us to interpret aerial photographs, determine the distance to the moon, calculate the area of complicated plots of land, and find the velocity of a distant star.

Until now, our applied work with trigonometry has been limited to right triangles. However, the concepts developed in the earlier chapters can be extended so that our work can apply to *all* triangles. Every triangle has three sides and three angles. In this chapter we show that if any three of the six measures of a triangle (provided at least one measure is a side) are known, then the other three measures can be found. This process is called solving a triangle. Later in the chapter this knowledge is used to solve problems involving vectors.

7.1 Oblique Triangles and the Law of Sines ▼▼▼

The following axioms from geometry allow us to prove that two triangles are congruent (that is, their corresponding sides and angles are equal).

Congruence Axioms	
Side-Angle-Side (SAS)	If two sides and the included angle of one triangle are equal, respectively, to two sides and the included angle of a second triangle, then the triangles are congruent.
Angle-Side-Angle (ASA)	If two angles and the included side of one triangle are equal, respectively, to two angles and the included side of a second triangle, then the triangles are congruent.
Side-Side-Side (SSS)	If three sides of one triangle are equal, respectively, to three sides of a second triangle, then the triangles are congruent.

Throughout this chapter keep in mind that whenever any of the groups of data described above are given, the triangle is uniquely determined; that is, all other data in the triangle are given by one and only one set of measures. We will continue to label triangles as we did earlier with right triangles: side *a* opposite angle *A*, side *b* opposite angle *B*, and side *c* opposite angle *C*.

A triangle that is not a right triangle is called an **oblique triangle.** The measures of the three sides and the three angles of a triangle can be found if at least one side and any other two measures are known. There are four possible cases.

Solving Oblique Triangles

1. One side and two angles are known.
2. Two sides and one angle not included between the two sides are known. This case may lead to more than one triangle.
3. Two sides and the angle included between the two sides are known.
4. Three sides are known.

NOTE If we know three angles of a triangle, we cannot find unique side lengths, since AAA assures us only of similarity, not congruence. For example, there are infinitely many triangles *ABC* with $A = 35°$, $B = 65°$, and $C = 80°$.

The first two cases require the *law of sines,* which is discussed in this section and the next. The last two cases require the *law of cosines,* which is discussed in Section 7.3.

To derive the law of sines, we start with an oblique triangle, such as the acute triangle in Figure 1(a) or the obtuse triangle in Figure 1(b). (Recall: These terms were defined in Section 1.3.) The following discussion applies to both triangles. First, construct the perpendicular from *B* to side *AC*. Let *h* be the length of this perpendicular. Then *c* is the hypotenuse of right triangle *ADB*, and *a* is the hypotenuse of right triangle *BDC*. By results from Chapter 2,

in triangle *ADB*, $\qquad \sin A = \dfrac{h}{c}$ or $h = c \sin A$,

in triangle *BDC*, $\qquad \sin C = \dfrac{h}{a}$ or $h = a \sin C$.

Since $h = c \sin A$ and $h = a \sin C$,

$$a \sin C = c \sin A,$$

or, upon dividing both sides by $\sin A \sin C$,

$$\frac{a}{\sin A} = \frac{c}{\sin C}.$$

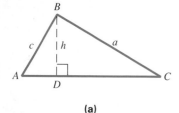

(a)

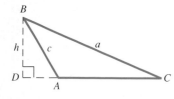

(b)

FIGURE 1

In a similar way, by constructing the perpendiculars from other vertices, it can be shown that

$$\frac{a}{\sin A} = \frac{b}{\sin B} \quad \text{and} \quad \frac{b}{\sin B} = \frac{c}{\sin C}.$$

This discussion proves the following theorem.

Law of Sines

In any triangle ABC, with sides a, b, and c,

$$\frac{a}{\sin A} = \frac{b}{\sin B}, \qquad \frac{a}{\sin A} = \frac{c}{\sin C}, \qquad \text{and} \qquad \frac{b}{\sin B} = \frac{c}{\sin C}.$$

This can be written in compact form as

$$\frac{a}{\sin A} = \frac{b}{\sin B} = \frac{c}{\sin C}.$$

Sometimes an alternative form of the law of sines,

$$\frac{\sin A}{a} = \frac{\sin B}{b} = \frac{\sin C}{c},$$

is more convenient to use.

If two angles and the side opposite one of the angles are known, the law of sines can be used directly to solve for the side opposite the other known angle. The triangle can then be solved completely, as shown in the first example.

◀EXAMPLE 1
Using the law of sines to solve a triangle

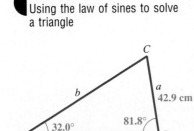

FIGURE 2

Solve triangle ABC if $A = 32.0°$, $B = 81.8°$, and $a = 42.9$ centimeters. See Figure 2.

Start by drawing a triangle, roughly to scale, and labeling the given parts as in Figure 2. Since the values of A, B, and a are known, use the part of the law of sines that involves these variables.

$$\frac{a}{\sin A} = \frac{b}{\sin B}$$

Substituting the known values gives

$$\frac{42.9}{\sin 32.0°} = \frac{b}{\sin 81.8°}.$$

Multiply both sides of the equation by $\sin 81.8°$.

$$b = \frac{42.9 \sin 81.8°}{\sin 32.0°}$$

When using a calculator to find b, keep intermediate answers in the calculator until the final result is found. Then round to the proper number of significant digits. In this case, find sin 81.8°, and then multiply that number by 42.9. Keep the result in the calculator while you find sin 32.0°, and then divide. Since the given information is accurate to three significant digits, round the value of b to get

$$b = \mathbf{80.1} \text{ centimeters.}$$

Find C from the fact that the sum of the angles of any triangle is 180°.

$$A + B + C = 180°$$
$$C = 180° - A - B$$
$$C = 180° - 32.0° - 81.8°$$
$$C = \mathbf{66.2°}$$

Now use the law of sines again to find c. (Why does the Pythagorean theorem not apply?)

$$\frac{a}{\sin A} = \frac{c}{\sin C}$$

$$\frac{42.9}{\sin 32.0°} = \frac{c}{\sin \mathbf{66.2°}}$$

$$c = \frac{42.9 \sin 66.2°}{\sin 32.0°}$$

$$c = \mathbf{74.1} \text{ centimeters} \quad \blacktriangleright$$

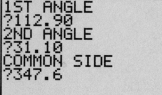

The triangle in Example 2 is completely solved above, using a program. Programs such as this one and the ones illustrated later in this chapter are available from user's groups and the manufacturer.

CAUTION   In applications of oblique triangles, such as the one in Example 1, a correctly labeled sketch is essential in order to set up the correct equation.

⬚ The law of sines is a good example of a formula that can be programmed into a graphing calculator. See your owner's manual for guidelines to programming.

◀**EXAMPLE 2**
Using the law of sines in an application

Tri Nguyen wishes to measure the distance across the Big Muddy River. See Figure 3. He finds that $C = 112.90°$, $A = 31.10°$, and $b = 347.6$ feet. Find the required distance.

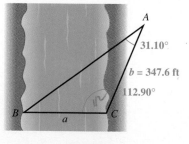

FIGURE 3

To use the law of sines, one side and the angle opposite it must be known. Since the only side whose length is given is b, angle B must be found before the law of sines can be used.

$$B = 180° - A - C$$
$$= 180° - 31.10° - 112.90° = 36.00°$$

Now use the form of the law of sines involving A, B, and b to find a.

$$\frac{a}{\sin A} = \frac{b}{\sin B}$$

$$\frac{a}{\sin 31.10°} = \frac{347.6}{\sin 36.00°} \qquad \text{Substitute.}$$

$$a = \frac{347.6 \sin 31.10°}{\sin 36.00°} \qquad \text{Multiply by } \sin 31.10°.$$

$$a = \textbf{305.5 feet} \qquad \text{Use a calculator.} \quad \blacktriangleright$$

The next example involves the use of bearing, first discussed in Section 2.5.

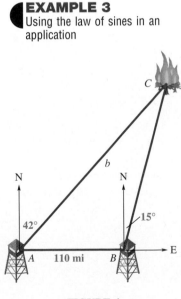

◀EXAMPLE 3
Using the law of sines in an application

Two tracking stations are on an east-west line 110 miles apart. A large forest fire is located on a bearing of N 42° E from the western station and a bearing of N 15° E from the eastern station. How far is the fire from the western station?

Figure 4 shows the two stations at points A and B and the fire at point C. Angle $BAC = 90° - 42° = 48°$, the obtuse angle at B equals $90° + 15° = 105°$, and the third angle, C, equals $180° - 105° - 48° = 27°$. Using the law of sines to find side b gives

$$\frac{b}{\sin 105°} = \frac{110}{\sin 27°}$$
$$b = 234,$$

or 230 miles (to two significant digits). ▶

FIGURE 4

AREA   The method used to derive the law of sines can also be used to derive a useful formula to find the area of a triangle. A familiar formula for the area of a triangle is $K = (1/2)bh$, where K represents the area, b the base, and h the height. This formula cannot always be used easily, since in practice h is often unknown. To find a more useful formula, refer to acute triangle ABC in Figure 5(a) or obtuse triangle ABC in Figure 5(b).

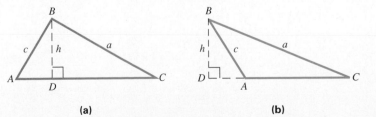

(a)

(b)

FIGURE 5

A perpendicular has been drawn from B to the base of the triangle (or the extension of the base). This perpendicular forms two right triangles. Using triangle ABD,

$$\sin A = \frac{h}{c},$$

or
$$h = c \sin A.$$

Substituting into the formula $K = (1/2)bh$,

$$K = \frac{1}{2}b(c \sin A)$$

or
$$K = \frac{1}{2}bc \sin A.$$

Any other pair of sides and the angle between them could have been used, as stated in the next theorem.

Area of a Triangle

In any triangle ABC, the area K is given by any of the following formulas:

$$K = \frac{1}{2}bc \sin A, \qquad K = \frac{1}{2}ab \sin C, \qquad K = \frac{1}{2}ac \sin B.$$

That is, the area is given by half the product of the lengths of two sides and the sine of the angle included between them.

EXAMPLE 4
Finding the area of a triangle
using $K = (1/2)ab \sin C$

Find the area of triangle ABC if $A = 24° \, 40'$, $b = 27.3$ centimeters, and $C = 52° \, 40'$.

Before we can use the formula given above, we must use the law of sines to find either a or c. Since the sum of the measures of the angles of any triangle is 180°,

$$B = 180° - 24° \, 40' - 52° \, 40' = 102° \, 40'.$$

Now use the form of the law of sines that relates a, b, A, and B to find a.

$$\frac{a}{\sin A} = \frac{b}{\sin B}$$

$$\frac{a}{\sin 24° \, 40'} = \frac{27.3}{\sin 102° \, 40'}$$

Solve for a to verify that $a = 11.7$ centimeters. Now find the area.

$$K = \frac{1}{2}ab \sin C = \frac{1}{2}(11.7)(27.3) \sin 52° \, 40' = 127$$

The area of triangle ABC is 127 square centimeters to three significant digits. ▶

NOTE Whenever possible, it is a good idea to use given values in solving triangles or finding areas rather than values obtained in intermediate steps. This avoids possible rounding errors.

CONNECTIONS As mentioned in the Chapter Opener, aerial photography has become important in many situations. Sometimes it is helpful to use coordinates of ordered pairs to determine distances on the ground. Suppose we assign coordinates as shown in the figure.

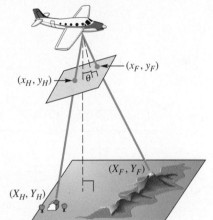

If an object's photographic coordinates are (x, y), then its ground coordinates (X, Y) in feet can be computed using the following formulas.

$$X = \frac{(a - h)x}{f \sec \theta - y \sin \theta}, \qquad Y = \frac{(a - h)y \cos \theta}{f \sec \theta - y \sin \theta}$$

Here, f is the focal length of the camera in inches, a is the altitude in feet of the airplane, and h is the elevation in feet of the object. Suppose that a house has photographic coordinates of $(x_H, y_H) = (.9, 3.5)$ with an elevation of 150 feet, while a nearby forest fire has photographic coordinates $(x_F, y_F) = (2.1, -2.4)$ and is at an elevation of 690 feet. If the photograph was taken at 7400 feet by a camera with a focal length of 6 inches and a tilt angle of $\theta = 4.1°$, we can use these formulas to find the distance on the ground in feet between the house and the forest fire. (*Source:* Moffitt, F., *Photogrammetry,* International Textbook Company, Scranton, Pennsylvania, 1967.)

FOR DISCUSSION OR WRITING
1. Use the formulas to find the ground coordinates of the house and the fire.
2. Use the distance formula given in Section 1.1 to find the required distance on the ground to the nearest tenth of a foot.

7.1 Exercises ▼▼▼▼▼▼▼▼▼▼▼▼▼▼▼▼▼▼▼▼▼▼▼▼▼▼▼▼▼▼▼▼▼▼▼▼▼

In Exercises 1 and 2, solve the equation for x.

1. $\dfrac{6}{\sin 30°} = \dfrac{x}{\sin 45°}$

2. $\dfrac{\sqrt{2}}{\sin 45°} = \dfrac{x}{\sin 60°}$

In Exercises 3 and 4, find the length of side a. Do not use a calculator.

3.

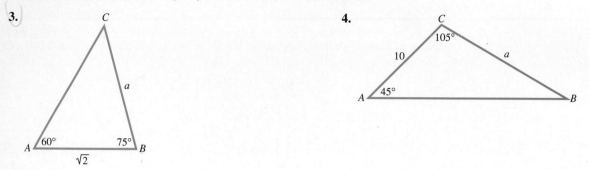

4.

In Exercises 5–20, determine the remaining sides and angles. See Example 1.

5.

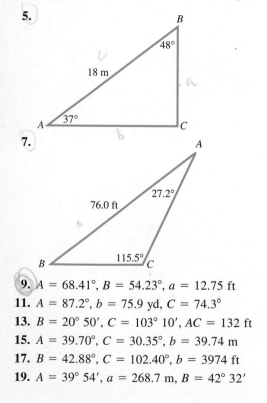

6.

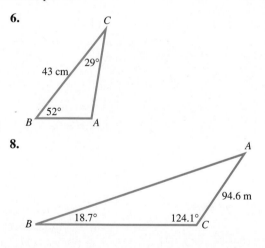

7.

8.

9. $A = 68.41°$, $B = 54.23°$, $a = 12.75$ ft

10. $C = 74.08°$, $B = 69.38°$, $c = 45.38$ m

11. $A = 87.2°$, $b = 75.9$ yd, $C = 74.3°$

12. $B = 38°\ 40'$, $a = 19.7$ cm, $C = 91°\ 40'$

13. $B = 20°\ 50'$, $C = 103°\ 10'$, $AC = 132$ ft

14. $A = 35.3°$, $B = 52.8°$, $AC = 675$ ft

15. $A = 39.70°$, $C = 30.35°$, $b = 39.74$ m

16. $C = 71.83°$, $B = 42.57°$, $a = 2.614$ cm

17. $B = 42.88°$, $C = 102.40°$, $b = 3974$ ft

18. $A = 18.75°$, $B = 51.53°$, $c = 2798$ yd

19. $A = 39°\ 54'$, $a = 268.7$ m, $B = 42°\ 32'$

20. $C = 79°\ 18'$, $c = 39.81$ mm, $A = 32°\ 57'$

21. Explain why the law of sines cannot be used to solve a triangle if we are given the lengths of the three sides of a triangle.

22. In Example 1, we ask the question, "Why does the Pythagorean theorem not apply?" Answer this question.

23. State the law of sines in your own words.

24. Kathleen Burk, a perceptive trigonometry student, makes the statement, "If we know *any* two angles and one side of a triangle, then the triangle is uniquely determined." Is this a valid statement? Explain, referring to the congruence axioms given in this section.

25. Can the law of sines be written as $a/b = \sin A/\sin B$? Explain.

26. If *a* is twice as long as *b*, is *A* necessarily twice as large as *B*?

Solve each of the following problems. See Examples 2 and 3.

27. To find the distance *AB* across a river, a distance *BC* = 354 m is laid off on one side of the river. It is found that *B* = 112° 10′ and *C* = 15° 20′. Find *AB*.

28. To determine the distance *RS* across a deep canyon, Joanna lays off a distance *TR* = 582 yd. She then finds that *T* = 32° 50′ and *R* = 102° 20′. Find *RS*.

29. Radio direction finders are placed at points *A* and *B*, which are 3.46 mi apart on an east-west line, with *A* west of *B*. From *A* the bearing of a certain radio transmitter is 47.7°, and from *B* the bearing is 302.5°. Find the distance of the transmitter from *A*.

30. A ship is sailing due north. At a certain point the bearing of a lighthouse 12.5 km distant is N 38.8° E. Later on, the captain notices that the bearing of the lighthouse has become S 44.2° E. How far did the ship travel between the two observations of the lighthouse?

31. A folding chair is to have a seat 12.0 inches deep with angles as shown in the figure. How far down from the seat should the crossing legs be joined? (Find *x* in the figure.)

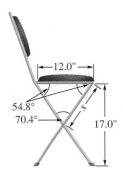

10.8

32. Mark notices that the bearing of a tree on the opposite bank of a river flowing north is 115.45°. Lisa is on the same bank as Mark, but 428.3 m away. She notices that the bearing of the tree is 45.47°. The two banks are parallel. What is the distance across the river?

33. Three gears are arranged as shown in the figure. Find angle θ.

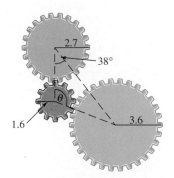

34. Three atoms with atomic radii of 2.0, 3.0, and 4.5 are arranged as in the figure. Find the distance between the centers of atoms A and C.

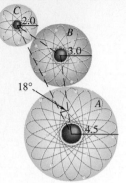

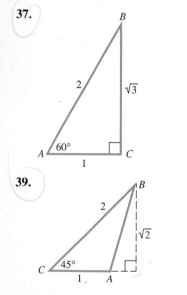

35. The bearing of a lighthouse from a ship was found to be N 37° E. After the ship sailed 2.5 miles due south,

the new bearing was N 25° E. Find the distance between the ship and the lighthouse at each location.

36. A balloonist is directly above a straight road 1.5 miles long that joins two villages. She finds that the town closer to her is at an angle of depression of 35° and the farther town is at an angle of depression of 31°. How high above the ground is the balloon? See the figure.

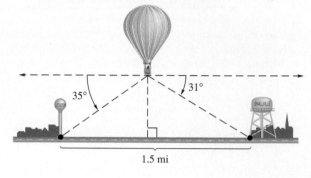

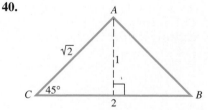

1.5 mi

Find the area of the triangle using the formula $K = (1/2)bh$ and then verify that the formula $K = (1/2)ab \sin C$ gives the same result.

37.

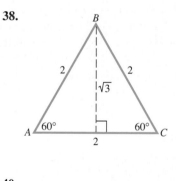

38.

39.

40.

Find the area of the triangle. See Example 4.

41. $A = 42.5°$, $b = 13.6$ m, $c = 10.1$ m

42. $C = 72.2°$, $b = 43.8$ ft, $a = 35.1$ ft

43. $B = 124.5°$, $a = 30.4$ cm, $c = 28.4$ cm

44. $C = 142.7°$, $a = 21.9$ km, $b = 24.6$ km

45. $A = 56.80°$, $b = 32.67$ in, $c = 52.89$ in

46. $A = 34.97°$, $b = 35.29$ m, $c = 28.67$ m

47. A painter is going to apply a special coating to a triangular metal plate on a new building. Two sides measure 16.1 m and 15.2 m. She knows that the angle between these sides is 125°. What is the area of the surface she plans to cover with the coating?

48. A real estate agent wants to find the area of a triangular lot. A surveyor takes measurements and finds that two sides are 52.1 m and 21.3 m, and the angle between them is 42.2°. What is the area of the triangular lot?

▼▼▼▼▼▼▼▼▼▼▼▼▼ **DISCOVERING CONNECTIONS** (Exercises 49–53) ▼▼▼▼▼▼▼▼▼▼▼▼▼

In any triangle, the longest side is opposite the largest angle. This result from geometry can be proven with trigonometry. Let us first prove it for acute triangles. (The result will be proven for obtuse triangles in 7.3 Exercises.)

49. Is the graph of the function $y = \sin x$ increasing or decreasing over the interval $0 \le x \le \pi/2$?

50. Suppose angle A is the largest angle of an acute triangle and let B be an angle smaller than A. Explain why $\dfrac{\sin B}{\sin A} < 1$.

51. Solve for b in the first form of the law of sines.

52. Use the result in Exercise 50 to show that $b < a$.

53. Use the result proved in Exercises 49–52 to explain why no triangle ABC satisfies $A = 83°$, $a = 14$, $b = 20$.

54. 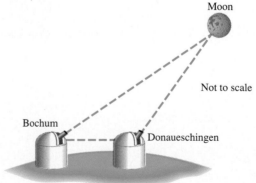 Since the moon is a relatively close celestial object, its distance can be measured directly using trigonometry. To find this distance, two different photographs of the moon are taken at precisely the same time in two different locations with a known distance between them. The moon will have a different angle of elevation at each location. On April 29, 1976, at 11:35 A.M., the lunar angles of elevation during a partial solar eclipse at Bochum in upper Germany and at Donaueschingen in lower Germany were measured as 52.6997° and 52.7430°, respectively. The two cities are 398 kilometers apart. Calculate the distance to the moon from Bochum on this day and compare it with the actual value of 406,000 km. Disregard the curvature of the Earth in this calculation. (*Source:* Scholosser, W., T. Schmidt-Kaler, and E. Milone, *Challenges of Astronomy,* Springer-Verlag, New York, 1991.)

55. The distance covered by an aerial photograph is determined by both the focal length of the camera and the tilt of the camera from the perpendicular to the ground. Although the tilt is usually small, both archaeological and Canadian photographs often use larger tilts. A camera lens with a 12-inch focal length will have an angular coverage of 60°. If an aerial photograph is taken with this camera tilted $\theta = 35°$ at an altitude of 5000 feet, calculate the ground distance d in miles that will be shown in this photograph. (*Sources:* Brooks, R. and Dieter Johannes, *Phytoarchaeology,* Dioscorides Press, Portland, Oregon, 1990; Moffitt, F., *Photogrammetry,* International Textbook Company, Scranton, Pennsylvania, 1967.)

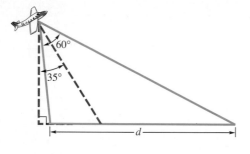

56. Refer to the previous exercise. A camera lens with a 6-inch focal length has an angular coverage of 86°. Suppose an aerial photograph is taken vertically with no tilt at an altitude of 3500 feet over ground with an increasing slope of 5° as shown in the figure. Calculate the ground distance CB that would appear in the resulting photograph. (*Source:* Moffitt, F., *Photogrammetry,* International Textbook Company, Scranton, Pennsylvania, 1967.)

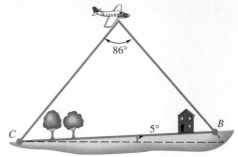

57. Repeat the previous exercise if the camera lens has an 8.25-inch focal length with an angular coverage of 72°. Why do cameras used in aerial photography usually have shorter focal lengths?

7.2 The Ambiguous Case of the Law of Sines ▼▼▼

The law of sines can be used when given two angles and the side opposite one of these angles. Also, if two angles and the included side are known, then the third angle can be found by using the fact that the sum of the angles of a triangle is 180°, and then the law of sines can be applied. However, if we are given the lengths of two sides and the angle opposite one of them, it is possible that 0, 1, or 2 such triangles exist. (Recall that there is no "SSA" congruence theorem.)

To illustrate these facts, suppose that the measure of acute angle A of triangle ABC, the length of side a, and the length of side b are given. Draw angle A having a terminal side of length b. Now draw a side of length a opposite angle A. The following chart shows that there might be more than one possible outcome. This situation is called the **ambiguous case of the law of sines.**

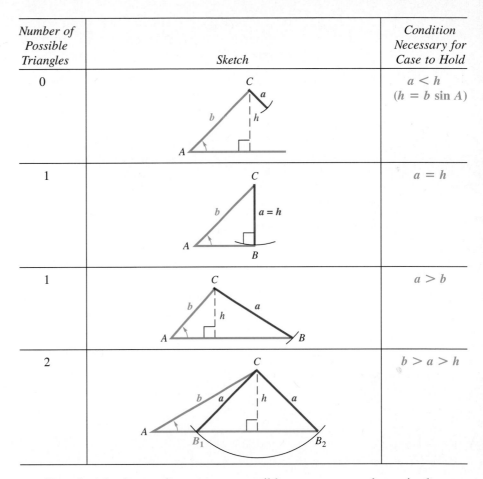

Number of Possible Triangles	Sketch	Condition Necessary for Case to Hold
0		$a < h$ $(h = b \sin A)$
1		$a = h$
1		$a > b$
2		$b > a > h$

If angle A is obtuse, there are two possible outcomes, as shown in the next chart.

Number of Possible Triangles	Sketch	Condition Necessary for Case to Hold
0		$a \le b$
1		$a > b$

Applying the law of sines to the values of a, b, and A and some basic properties of geometry and trigonometry will allow us to determine which case applies. The following facts should be kept in mind.

1. For any angle θ of a triangle, $0 < \sin \theta \leq 1$. If $\sin \theta = 1$, then $\theta = 90°$ and the triangle is a right triangle.
2. $\sin \theta = \sin(180° - \theta)$ (That is, supplementary angles have the same sine value.)
3. The smallest angle is opposite the shortest side, the largest angle is opposite the longest side, and the middle-valued angle is opposite the medium side (assuming the triangle is scalene).

> 📉🖩 You may wish to write a calculator program for the ambiguous case of the law of sines, or you may get one from a user's group or the manufacturer of your calculator.

EXAMPLE 1
Solving a triangle using the law of sines (no such triangle)

Solve the triangle ABC if $B = 55° \, 40'$, $b = 8.94$ meters, and $a = 25.1$ meters.
Since we are given B, b, and a, use the law of sines to find A.

$$\frac{\sin A}{a} = \frac{\sin B}{b}$$

Substitute the given values.

$$\frac{\sin A}{25.1} = \frac{\sin 55° \, 40'}{8.94}$$

$$\sin A = \frac{25.1 \sin 55° \, 40'}{8.94}$$

$$\sin A = 2.3184379$$

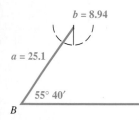

$b = 8.94$

$a = 25.1$

$55° \, 40'$

B

FIGURE 6

Since $\sin A$ cannot be greater than 1, there can be no such angle A and thus no triangle with the given information. An attempt to sketch such a triangle leads to the situation seen in Figure 6. ▶

EXAMPLE 2
Solving a triangle using the law of sines (two triangles)

Solve triangle ABC if $A = 55.3°$, $a = 22.8$ feet, and $b = 24.9$ feet.
To begin, use the law of sines to find angle B.

$$\frac{a}{\sin A} = \frac{b}{\sin B}$$

$$\frac{22.8}{\sin 55.3°} = \frac{24.9}{\sin B}$$

$$\sin B = \frac{24.9 \sin 55.3°}{22.8}$$

$$\sin B = .8978678$$

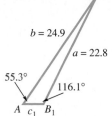

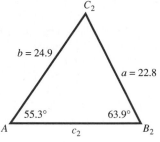

FIGURE 7

Since $\sin B = .8978678$, to the nearest tenth we have one value of B as

$$B = 63.9°.$$

Supplementary angles have the same sine value, so another *possible* value of B is

$$B = 180° - 63.9° = 116.1°.$$

To see if $B = 116.1°$ is a valid possibility, simply add $116.1°$ to the measure of the given value of A, $55.3°$. Since $116.1° + 55.3° = 171.4°$, and this sum is less than $180°$ (the sum of the angles of a triangle), we know that it is a valid angle measure for this triangle.

To keep track of these two different values of B, let

$$B_1 = 116.1° \quad \text{and} \quad B_2 = 63.9°.$$

Now separately solve triangles $AB_1 C_1$ and $AB_2 C_2$ shown in Figure 7. Let us begin with $AB_1 C_1$. Find C_1 first.

$$C_1 = 180° - A - B_1 = 8.6°.$$

Now, use the law of sines to find c_1.

$$\frac{a}{\sin A} = \frac{c_1}{\sin C_1}$$

$$\frac{22.8}{\sin 55.3°} = \frac{c_1}{\sin 8.6°}$$

$$c_1 = \frac{22.8 \sin 8.6°}{\sin 55.3°}$$

$$c_1 = 4.15 \text{ feet}$$

To solve triangle $AB_2 C_2$, first find C_2.

$$C_2 = 180° - A - B_2 = 60.8°$$

By the law of sines,

$$\frac{22.8}{\sin 55.3°} = \frac{c_2}{\sin 60.8°}$$

$$c_2 = \frac{22.8 \sin 60.8°}{\sin 55.3°}$$

$$c_2 = 24.2 \text{ feet.} \quad \blacktriangleright$$

CAUTION   When solving a triangle using the type of data given in Example 2, do not forget to find the possible obtuse angle. The inverse sine function of the calculator will not give it directly. As we shall see in the next example, it is possible that the obtuse angle will not be a valid measure.

EXAMPLE 3
Solving a triangle using the law of sines (one triangle)

Solve triangle ABC given $A = 43.5°$, $a = 10.7$ inches, and $b = 7.2$ inches.

To find angle B use the law of sines.

$$\frac{\sin B}{7.2} = \frac{\sin 43.5°}{10.7}$$

$$\sin B = \frac{7.2 \sin 43.5°}{10.7} = .46319186$$

The inverse sine function of the calculator gives us

$$B = 27.6°$$

as the acute angle. The other possible value of B is $180° - 27.6° = 152.4°$. However, when we add this possible obtuse angle to the given angle $A = 43.5°$, we get $152.4° + 43.5° = 195.9°$, which is greater than $180°$. So there can be only one triangle. (Notice that this is the third case listed in the chart at the beginning of this section.) Then angle $C = 180° - 27.6° - 43.5° = 108.9°$, and side c can be found with the law of sines.

$$\frac{c}{\sin 108.9°} = \frac{10.7}{\sin 43.5°}$$

$$c = \frac{10.7 \sin 108.9°}{\sin 43.5°}$$

$$c = 14.7 \text{ inches} \quad \blacktriangleright$$

EXAMPLE 4
Analyzing data involving an obtuse angle

Without using the law of sines, explain why the data

$$A = 104°, \ a = 26.8 \text{ meters}, \ b = 31.3 \text{ meters}$$

cannot be valid for a triangle ABC.

Since A is an obtuse angle, the largest side of the triangle must be a, the side opposite A. However, we are given $b > a$, which is impossible if A is obtuse. Therefore, no such triangle ABC exists. $\quad \blacktriangleright$

7.2 Exercises ▼▼▼▼▼▼▼▼▼▼▼▼▼▼▼▼▼▼▼▼▼▼▼▼▼▼▼▼▼▼▼▼▼▼▼▼

In Exercises 1 and 2, solve the equation for θ, where θ is the measure of an angle in a triangle. Give all possible values of θ.

1. $\dfrac{\sqrt{2}}{\sin 30°} = \dfrac{2}{\sin \theta°}$

2. $\dfrac{\sqrt{6}}{\sin 45°} = \dfrac{3}{\sin \theta°}$

3. In the figure shown below, a line of length h is to be drawn from the point $(3, 4)$ to the positive x-axis in order to form a triangle. For what value(s) of h can you draw the following?
(a) Two triangles   **(b)** Exactly one triangle
(c) No triangle

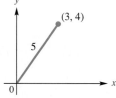

4. In the figure shown below, a line of length h is to be drawn from the point $(-3, 4)$ to the positive x-axis in order to form a triangle. For what value(s) of h can you draw the following?
(a) Two triangles   **(b)** Exactly one triangle
(c) No triangle

Determine the number of triangles possible with the given parts. See Examples 1–4.

5. $a = 50,\ b = 26,\ A = 95°$

6. $b = 60,\ a = 82,\ B = 100°$

7. $a = 31,\ b = 26,\ B = 48°$

8. $a = 35,\ b = 30,\ A = 40°$

9. $a = 50,\ b = 61,\ A = 58°$

10. $B = 54°,\ c = 28,\ b = 23$

In Exercises 11 and 12, find angle B. Do not use a calculator.

11.

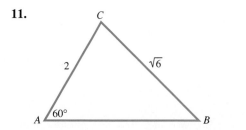

12.

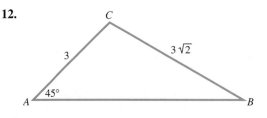

Find the unknown angles in triangle ABC for each triangle that exists. See Examples 1–3.

13. $A = 29.7°,\ b = 41.5$ ft, $a = 27.2$ ft

14. $B = 48.2°,\ a = 890$ cm, $b = 697$ cm

15. $C = 41° 20',\ b = 25.9$ m, $c = 38.4$ m

16. $B = 48° 50',\ a = 3850$ in, $b = 4730$ in

17. $B = 74.3°,\ a = 859$ m, $b = 783$ m

18. $C = 82.2°,\ a = 10.9$ km, $c = 7.62$ km

19. $A = 142.13°,\ b = 5.432$ ft, $a = 7.297$ ft

20. $B = 113.72°,\ a = 189.6$ yd, $b = 243.8$ yd

Solve each triangle that exists. See Examples 1–3.

21. $A = 42.5°,\ a = 15.6$ ft, $b = 8.14$ ft

22. $C = 52.3°,\ a = 32.5$ yd, $c = 59.8$ yd

23. $B = 72.2°,\ b = 78.3$ m, $c = 145$ m

24. $C = 68.5°,\ c = 258$ cm, $b = 386$ cm

25. $A = 38° 40',\ a = 9.72$ km, $b = 11.8$ km

26. $C = 29° 50',\ a = 8.61$ m, $c = 5.21$ m

27. $A = 96.80°,\ b = 3.589$ ft, $a = 5.818$ ft

28. $C = 88.70°,\ b = 56.87$ yd, $c = 112.4$ yd

29. $B = 39.68°,\ a = 29.81$ m, $b = 23.76$ m

30. $A = 51.20°,\ c = 7986$ cm, $a = 7208$ cm

31. Apply the law of sines to the following: $a = \sqrt{5}$, $c = 2\sqrt{5}$, $A = 30°$. What is the value of sin C? What is the measure of C? Based on its angle measures, what kind of triangle is triangle ABC?

32. In your own words, explain the condition that must exist to determine that there is no triangle satisfying the given values of a, b, and B, once the value of sin B is found.

33. Without using the law of sines, explain why no triangle ABC exists satisfying $A = 103° 20'$, $a = 14.6$ ft, $b = 20.4$ ft.

34. Apply the law of sines to the data given in Example 4. Describe in your own words what happens when you try to find the measure of angle B using a calculator.

35. A surveyor reported the following data about a piece of property: "The property is triangular in shape, with dimensions as shown in the figure." Use the law of sines to see whether such a piece of property could exist.

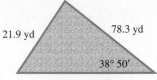

Can such a triangle exist?

36. The surveyor tries again: "A second triangular piece of property has dimensions as shown." This time it turns out that the surveyor did not consider every possible case. Use the law of sines to show why.

Use the law of sines to prove that the statement is true for any triangle ABC, with corresponding sides a, b, and c.

37. $\dfrac{a + b}{b} = \dfrac{\sin A + \sin B}{\sin B}$

38. $\dfrac{a - b}{a + b} = \dfrac{\sin A - \sin B}{\sin A + \sin B}$

7.3 The Law of Cosines ▼▼▼

As mentioned in Section 7.1, if we are given two sides and the included angle or three sides of a triangle, a unique triangle is formed. These are the SAS and SSS cases, respectively. In these cases, however, we cannot begin the solution of the triangle by using the law of sines because we are not given a side and the angle opposite it. Both cases require the use of the law of cosines, introduced in this section.

It will be helpful to remember the following property of triangles when applying the law of cosines.

> In any triangle, the sum of the lengths of any two sides must be greater than the length of the remaining side.

For example, it would be impossible to construct a triangle with sides of lengths 3, 4, and 10. See Figure 8.

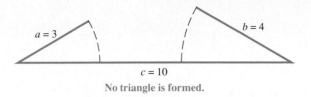

c = 10

No triangle is formed.

FIGURE 8

To derive the law of cosines, let *ABC* be any oblique triangle. Choose a coordinate system so that vertex *B* is at the origin and side *BC* is along the positive *x*-axis. See Figure 9.

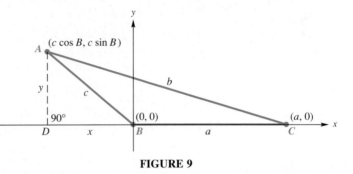

FIGURE 9

Let (x, y) be the coordinates of vertex *A* of the triangle. Verify that for angle *B*, whether obtuse or acute,

$$\sin B = \frac{y}{c} \quad \text{and} \quad \cos B = \frac{x}{c}.$$

(Here *x* is negative if *B* is obtuse.) From these results

$$y = c \sin B \quad \text{and} \quad x = c \cos B,$$

so that the coordinates of point *A* become

$$(c \cos B, c \sin B).$$

Point *C* has coordinates $(a, 0)$, and *AC* has length *b*. By the distance formula,

$$b = \sqrt{(c \cos B - a)^2 + (c \sin B)^2}.$$

Squaring both sides and simplifying gives

$$
\begin{aligned}
b^2 &= (c \cos B - a)^2 + (c \sin B)^2 \\
&= c^2 \cos^2 B - 2ac \cos B + a^2 + c^2 \sin^2 B \\
&= a^2 + c^2(\cos^2 B + \sin^2 B) - 2ac \cos B \\
&= a^2 + c^2(1) - 2ac \cos B \\
&= a^2 + c^2 - 2ac \cos B.
\end{aligned}
$$

This result is one form of the law of cosines. In the work above, we could just as easily have placed A or C at the origin. This would have given the same result, but with the variables rearranged. These various forms of the law of cosines are summarized in the following theorem.

Law of Cosines

In any triangle ABC, with sides a, b, and c,

$$a^2 = b^2 + c^2 - 2bc \cos A$$
$$b^2 = a^2 + c^2 - 2ac \cos B$$
$$c^2 = a^2 + b^2 - 2ab \cos C.$$

The law of cosines says that the square of a side of a triangle is equal to the sum of the squares of the other two sides, minus twice the product of those two sides and the cosine of the angle included between them.

NOTE If we let $C = 90°$ in the third form of the law of cosines given above, we have $\cos C = \cos 90° = 0$, and the formula becomes

$$c^2 = a^2 + b^2,$$

the familiar equation of the Pythagorean theorem. Thus, the Pythagorean theorem is a special case of the law of cosines.

The first example shows how the law of cosines can be used to solve an applied problem.

◀EXAMPLE 1
Using the law of cosines in an application

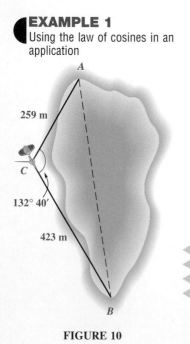

A surveyor wishes to find the distance between two inaccessible points A and B on opposite sides of a lake. While standing at point C, she finds that $AC = 259$ meters, $BC = 423$ meters, and angle ACB measures $132° \, 40'$. Find the distance AB. See Figure 10.

The law of cosines can be used here, since we know the lengths of two sides of the triangle and the measure of the included angle.

$$AB^2 = 259^2 + 423^2 - 2(259)(423) \cos 132° \, 40'$$
$$AB^2 = 394{,}510.6 \qquad \text{Use a calculator.}$$
$$AB \approx 628 \qquad \text{Take the square root and round to 3 significant digits.}$$

The distance between the points is approximately 628 meters. ▶

▢ The law of cosines is another useful formula to program into a graphing calculator. For the SSS case, only one of the three forms needs to be programmed. Be sure to consider the SAS case also.

259 m

C

132° 40′

423 m

A

B

FIGURE 10

EXAMPLE 2
Using the law of cosines to solve a triangle (SAS)

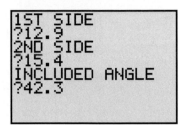

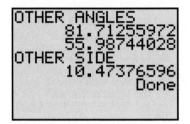

A program can be used to solve the triangle in Example 2. (There are slight discrepancies due to roundoff error.)

Solve triangle ABC if $A = 42.3°$, $b = 12.9$ meters, and $c = 15.4$ meters. See Figure 11.

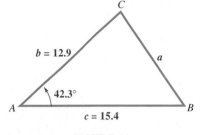

FIGURE 11

Start by finding a with the law of cosines.

$$a^2 = b^2 + c^2 - 2bc \cos A$$
$$a^2 = 12.9^2 + 15.4^2 - 2(12.9)(15.4) \cos 42.3°$$
$$a^2 = 109.7$$
$$a = 10.5 \text{ meters}$$

We now must find the measures of angles B and C. There are several approaches that can be used at this point. Let us use the law of sines to find one of these angles. Of the two remaining angles, B must be the smaller since it is opposite the shorter of the two sides b and c. Therefore, it cannot be obtuse, and we will avoid any ambiguity when we find its sine.

$$\frac{\sin 42.3°}{10.5} = \frac{\sin B}{12.9}$$

$$\sin B = \frac{12.9 \sin 42.3°}{10.5}$$

$$B = 55.8° \qquad \text{Use the inverse sine function of a calculator.}$$

The easiest way to find C is to subtract the measures of A and B from $180°$.

$$C = 180° - A - B = 81.9°. \quad \blacktriangleright$$

CAUTION Had we chosen to use the law of sines to find C rather than B in Example 2, we would not have known whether C equals $81.9°$ or its supplement, $98.1°$.

EXAMPLE 3
Using the law of cosines to solve a triangle (SSS)

Solve triangle ABC if $a = 9.47$ feet, $b = 15.9$ feet, and $c = 21.1$ feet.
    We are given the lengths of three sides of the triangle, so we may use the law of cosines to solve for any angle of the triangle. Let us solve for C, the largest angle, using the law of cosines. We will be able to tell if C is obtuse if $\cos C < 0$. Use the form of the law of cosines that involves C.

$$c^2 = a^2 + b^2 - 2ab \cos C,$$

or

$$\cos C = \frac{a^2 + b^2 - c^2}{2ab}.$$

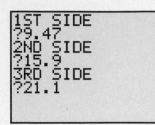

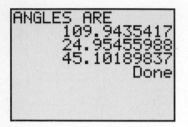

A program can be used to solve the triangle in Example 3.

Inserting the given values leads to

$$\cos C = \frac{(9.47)^2 + (15.9)^2 - (21.1)^2}{2(9.47)(15.9)}$$

$$\cos C = -.34109402. \qquad \text{Use a calculator.}$$

Using the inverse cosine function of the calculator, we get the obtuse angle C.

$$C = 109.9°$$

We can use either the law of sines or the law of cosines to find $B = 45.1°$. (Verify this.) Since $A = 180° - B - C$,

$$A = 25.0°. \quad \blacktriangleright$$

As shown in this section and the previous one, four possible cases can occur when solving an oblique triangle. These cases are summarized in the following chart, along with a suggested procedure for solving in each case. There are other procedures that work, but we give the one that is most efficient. In all four cases, it is assumed that the given information actually produces a triangle.

Case	*Suggested Procedure for Solving*
One side and two angles are known. (SAA or ASA)	**1.** Find the remaining angle using the angle sum formula ($A + B + C = 180°$). **2.** Find the remaining sides using the law of sines.
Two sides and one angle (not included between the two sides) are known. (SSA)	*Be aware of the ambiguous case; there may be two triangles.* **1.** Find an angle using the law of sines. **2.** Find the remaining angle using the angle sum formula. **3.** Find the remaining side using the law of sines. *If two triangles exist, repeat Steps 2 and 3.*
Two sides and the included angle are known. (SAS)	**1.** Find the third side using the law of cosines. **2.** Find the smaller of the two remaining angles using the law of sines. **3.** Find the remaining angle using the angle sum formula.
Three sides are known. (SSS)	**1.** Find the largest angle using the law of cosines. **2.** Find either remaining angle using the law of sines. **3.** Find the remaining angle using the angle sum formula.

AREA The law of cosines can be used to derive a formula for the area of a triangle when only the lengths of the three sides are known. This formula is known as Heron's formula, named after the Greek mathematician Heron of Alexandria, who lived around A.D. 75. It is found in his work *Metrica*.

Heron's Area Formula

If a triangle has sides of lengths a, b, and c, and if the **semi-perimeter** is

$$s = \frac{1}{2}(a + b + c),$$

then the area of the triangle is

$$K = \sqrt{s(s - a)(s - b)(s - c)}.$$

A proof of Heron's formula is suggested in Exercises 55–60.

EXAMPLE 4
Finding an area using Heron's formula

Find the area of the triangle having sides of lengths $a = 29.7$ feet, $b = 42.3$ feet, and $c = 38.4$ feet.

To use Heron's area formula, first find s.

$$s = \frac{1}{2}(a + b + c)$$

$$s = \frac{1}{2}(29.7 + 42.3 + 38.4)$$

$$= 55.2$$

The area is

$$K = \sqrt{s(s - a)(s - b)(s - c)}$$
$$K = \sqrt{55.2(55.2 - 29.7)(55.2 - 42.3)(55.2 - 38.4)}$$
$$= \sqrt{55.2(25.5)(12.9)(16.8)}$$
$$\approx 552 \text{ square feet.} \quad \blacktriangleright$$

```
ENTER SIDES
A
?29.7
B
?42.3
C
?38.4
```

```
AREA =
      552.3179085
           Done
```

A program for Heron's formula can be used to find the area of a triangle. The screens above support the result of Example 4.

CONNECTIONS We have introduced two new formulas for the area of a triangle in this chapter. You should now be able to find the area K of a triangle using one of three formulas:

(a) $K = (1/2)bh$

(b) $K = (1/2)ab \sin C$ (or $K = (1/2)ac \sin B$ or $K = (1/2)bc \sin A$)

(c) $K = \sqrt{s(s - a)(s - b)(s - c)}$.

Another area formula can be used when the coordinates of the vertices of the triangle are given. If the vertices are the ordered pairs (x_1, y_1), (x_2, y_2), and (x_3, y_3), then

$$K = \frac{1}{2} \left| (x_1 y_2 - y_1 x_2 + x_2 y_3 - y_2 x_3 + x_3 y_1 - y_3 x_1) \right|.$$

FOR DISCUSSION OR WRITING

Consider triangle PQR with vertices $P(2, 5)$, $Q(-1, 3)$, and $R(4, 0)$.

1. Find the area of the triangle using the new formula just introduced.
2. Find the area of the triangle using (c) above. Use the distance formula to find the lengths of the three sides.
3. Find the area of the triangle using (b) above. First use the law of cosines to find the measure of an angle.

7.3 Exercises ▼▼▼▼▼▼▼▼▼▼▼▼▼▼▼▼▼▼▼▼▼▼▼▼▼▼▼▼▼▼▼▼▼▼▼▼▼▼

1. In your own words, describe the types of problems that require the law of sines and those that require the law of cosines.

2. Can the law of cosines be used to solve any triangle for which two angles and a side are known? Explain your answer.

In Exercises 3 and 4, find the length of the remaining side. Do not use a calculator.

3.

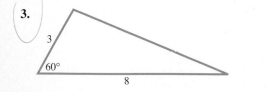

4.

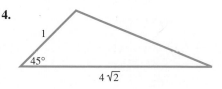

In Exercises 5 and 6, find the value of θ. Do not use a calculator.

5.

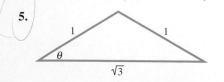

6.

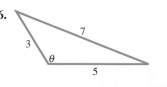

Solve the triangle. See Example 2.

7. $C = 28.3°$, $b = 5.71$ in, $a = 4.21$ in

8. $A = 41.4°$, $b = 2.78$ yd, $c = 3.92$ yd

9. $C = 45.6°$, $b = 8.94$ m, $a = 7.23$ m

10. $A = 67.3°$, $b = 37.9$ km, $c = 40.8$ km

11. $A = 80° \, 40'$, $b = 143$ cm, $c = 89.6$ cm

12. $C = 72° \, 40'$, $a = 327$ ft, $b = 251$ ft

13. $B = 74.80°$, $a = 8.919$ in, $c = 6.427$ in

14. $C = 59.70°$, $a = 3.725$ mi, $b = 4.698$ mi

15. $A = 112.8°$, $b = 6.28$ m, $c = 12.2$ m

16. $B - 168.2°$, $a = 15.1$ cm, $c - 19.2$ cm

Find all the angles in each triangle. See Example 3.

17. $a = 3.0$ ft, $b = 5.0$ ft, $c = 6.0$ ft

18. $a = 4.0$ ft, $b = 5.0$ ft, $c = 8.0$ ft

19. $a = 9.3$ cm, $b = 5.7$ cm, $c = 8.2$ cm

20. $a = 28$ ft, $b = 47$ ft, $c = 58$ ft

21. $a = 42.9$ m, $b = 37.6$ m, $c = 62.7$ m

22. $a = 187$ yd, $b = 214$ yd, $c = 325$ yd

23. $AB = 1240$ ft, $AC = 876$ ft, $BC = 918$ ft

24. $AB = 298$ m, $AC = 421$ m, $BC = 324$ m

Find the measure of the angle θ to two decimal places.

25.

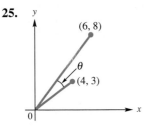

26.

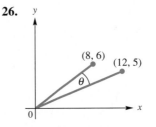

Find the area of the triangle using the formula $K = (1/2)bh$, and then verify that Heron's formula gives the same result.

27.

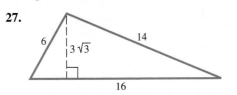

28.

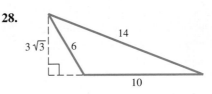

Find the area of the triangle. See Example 4.

29. $a = 12$ m, $b = 16$ m, $c = 25$ m

30. $a = 22$ in, $b = 45$ in, $c = 31$ in

31. $a = 154$ cm, $b = 179$ cm, $c = 183$ cm

32. $a = 25.4$ yd, $b = 38.2$ yd, $c = 19.8$ yd

33. $a = 76.3$ ft, $b = 109$ ft, $c = 98.8$ ft

34. $a = 15.89$ in, $b = 21.74$ in, $c = 10.92$ in

Solve the following problems.

35. A painter needs to cover a triangular region 75 m by 68 m by 85 m. A can of paint covers 75 sq m of area. How may cans (to the next higher number of cans) will be needed?

36. How many cans of paint would be needed in Exercise 35 if the region were 8.2 m by 9.4 m by 3.8 m?

37. Find the area of the Bermuda Triangle if the sides of the triangle have the approximate lengths 850 miles, 925 miles, and 1300 miles.

38. Find the area of a triangle in a rectangular coordinate plane whose vertices are $(0, 0)$, $(3, 4)$, and $(-8, 6)$ using Heron's area formula.

Solve the problem, using the law of sines or the law of cosines. See Example 1.

39. Points A and B are on opposite sides of Lake Yankee. From a third point, C, the angle between the lines of sight to A and B is $46.3°$. If AC is 350 m long and BC is 286 m long, find AB.

40. The sides of a parallelogram are 4.0 cm and 6.0 cm. One angle is $58°$ while another is $122°$. Find the lengths of the diagonals of the parallelogram.

41. Airports A and B are 450 km apart, on an east-west line. Tom flies in a northeast direction from A to airport C. From C he flies 359 km on a bearing of $128°$ $40'$ to B. How far is C from A?

42. Two ships leave a harbor together, traveling on courses that have an angle of $135°$ $40'$ between them. If they each travel 402 mi, how far apart are they?

43. The layout for a child's playhouse in her backyard shows the dimensions given in the figure. Find x.

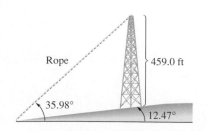

44. A hill slopes at an angle of $12.47°$ with the horizontal. From the base of the hill, the angle of elevation of a 459.0 ft tower at the top of the hill is $35.98°$. How much rope would be required to reach from the top of the tower to the bottom of the hill?

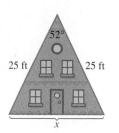

45. A crane with a counterweight is shown in the figure. Find the horizontal distance between points A and B.

46. A weight is supported by cables attached to both ends of a balance beam, as shown in the figure. What angles are formed between the beam and the cables?

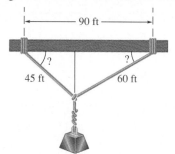

47. A satellite traveling in a circular orbit 1600 km above Earth is due to pass directly over a tracking station at noon. See the figure. Assume that the satellite takes 2 hr to make an orbit and that the radius of the Earth is 6400 km. Find the distance between the satellite and the tracking station at 12:03 P.M. (*Source:* Kastner, Bernice, Ph.D., *Spacemathematics*, National Aeronautics and Space Administration (NASA), 1985.)

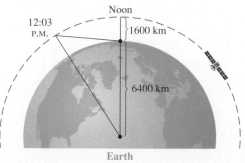

48. Two factories blow their whistles at exactly 5:00. A man hears the two blasts at 3 seconds and 6 seconds after 5:00, respectively. The angle between his lines of sight to the two factories is 42.2°. If sound travels 344 m per sec, how far apart are the factories?

49. A parallelogram has sides of lengths 25.9 cm and 32.5 cm. The longer diagonal has a length of 57.8 cm.

Find the measure of the angle opposite the diagonal.

50. A person in a plane flying a straight course observes a mountain at a bearing 24.1° to the right of its course. At that time the plane is 7.92 km from the mountain. A short time later, the bearing to the mountain becomes 32.7°. How far is the airplane from the mountain when the second bearing is taken?

To help predict eruptions from the volcano Mauna Loa on the island of Hawaii, scientists keep track of the volcano's movement by using a "super triangle" with vertices on the three volcanoes shown on the map at the right. (For example, in a recent year, Mauna Loa moved 6 inches, a result of increasing internal pressure.) Refer to the map to work Exercises 51 and 52.

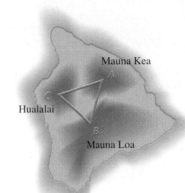

51. $AB = 22.47928$ mi, $AC = 28.14276$ mi, $A = 58.56989°$; find BC

52. $AB = 22.47928$ mi, $BC = 25.24983$ mi, $A = 58.56989°$; find B

53. Refer to Figure 8. If you attempt to find any angle of a triangle using the values $a = 3$, $b = 4$, and $c = 10$ with the law of cosines, what happens?

54. A familiar saying is "The shortest distance between two points is a straight line." Explain how this relates to the geometric property that states that the sum of the lengths of any two sides of a triangle must be greater than the remaining side.

Use the fact that $\cos A = \dfrac{b^2 + c^2 - a^2}{2bc}$ *to show that each of the following is true.*

55. $1 + \cos A = \dfrac{(b + c + a)(b + c - a)}{2bc}$

56. $1 - \cos A = \dfrac{(a - b + c)(a + b - c)}{2bc}$

57. $\cos \dfrac{A}{2} = \sqrt{\dfrac{s(s - a)}{bc}}$ $\left(Hint: \cos \dfrac{A}{2} = \sqrt{\dfrac{1 + \cos A}{2}} \right)$

58. $\sin \dfrac{A}{2} = \sqrt{\dfrac{(s - b)(s - c)}{bc}}$ $\left(Hint: \sin \dfrac{A}{2} = \sqrt{\dfrac{1 - \cos A}{2}} \right)$

59. The area of a triangle having sides b and c and angle A is given by $(1/2)bc \sin A$. Show that this result can be written as

$$\sqrt{\frac{1}{2}bc(1 + \cos A) \cdot \frac{1}{2}bc(1 - \cos A)}.$$

60. Use the results of Exercises 55–59 to prove Heron's area formula.

61. Let a and b be the equal sides of an isosceles triangle. Prove that $c^2 = 2a^2(1 - \cos C)$.

62. Let point D on side AB of triangle ABC be such that CD bisects angle C. Show that $AD/DB = b/a$.

63. In addition to the law of sines and the law of cosines, there is a **law of tangents.** In any triangle *ABC*,

$$\frac{\tan \frac{1}{2}(A - B)}{\tan \frac{1}{2}(A + B)} = \frac{a - b}{a + b}.$$

Verify this law for the triangle *ABC* with $a = 2$, $b = 2\sqrt{3}$, $A = 30°$, and $B = 60°$.

64. Prove the law of tangents by referring to Exercise 38 in Section 7.2 and applying identities found in the Connections box in Section 5.5.

▼▼▼▼▼▼▼▼▼▼▼▼▼ **DISCOVERING CONNECTIONS** (Exercises 65–68) ▼▼▼▼▼▼▼▼▼▼▼▼▼

In any triangle, the longest side is opposite the largest angle. This result from geometry was proven for acute triangles in 7.1 Exercises. Let's now prove it for obtuse triangles.

65. Suppose angle *A* is the largest angle of an obtuse triangle. Explain why cos *A* is negative.

66. Consider the law of cosines expression for *a*, and show that $a^2 > b^2 + c^2$.

67. Use the result of Exercise 66 to show that $a > b$ and $a > c$.

68. Use the result of Exercise 67 to explain why no triangle *ABC* satisfies $A = 103°$, $a = 25$, $c = 30$.

7.4   Introduction to Vectors ▼▼▼

We have seen that the measures of all six parts of a triangle can be found, given at least one side and any two other measures. This section and the next show applications of this work to *vectors.* In this section the basic ideas of vectors are presented, and in the next section the law of sines and the law of cosines are applied to vector problems.

Many quantities in mathematics involve magnitudes, such as 45 pounds or 60 miles per hour. These quantities are called **scalars.** Other quantities, called **vector quantities,** involve both magnitude and direction. Typical vector quantities are velocity, acceleration, and force.

A vector quantity is often represented with a directed line segment, which is called a **vector.** The length of the vector represents the **magnitude** of the vector quantity. The direction of the vector, indicated with an arrowhead, represents the direction of the quantity. For example, the vector in Figure 12 represents a force of 10 pounds applied at an angle of 30° from the horizontal.

The symbol for a vector is often printed in boldface type. When writing vectors by hand, it is customary to use an arrow over the letter or letters. Thus **OP** and $\overrightarrow{OP}$ both represent vector **OP**. Vectors may be named with either one lowercase or uppercase letter, or two uppercase letters. When two letters are used, the first indicates the **initial point** and the second indicates the **terminal point** of the vector. Knowing these points gives the direction of the vector. For example, vectors **OP** and **PO** in Figure 13 are not the same vector. They have the same magnitude, but they have opposite directions. The magnitude of vector **OP** is written $|\mathbf{OP}|$.

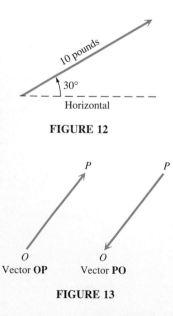

FIGURE 12

Vector **OP**   Vector **PO**

FIGURE 13

Two vectors are *equal* if and only if they both have the same direction and the same magnitude. In Figure 14 vectors **A** and **B** are equal, as are vectors **C** and **D.** As Figure 14 shows, equal vectors need not coincide, but they must be parallel. Vectors **A** and **E** are unequal because they do not have the same direction, while **A** ≠ **F** because they have different magnitudes, as indicated by their different lengths.

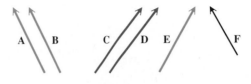

FIGURE 14

To find the *sum* of two vectors **A** and **B,** written **A** + **B,** we place the initial point of vector **B** at the terminal point of vector **A,** as shown in Figure 15. The vector with the same initial point as **A** and the same terminal point as **B** is the sum **A** + **B.** The sum of two vectors is also a vector.

Another way to find the sum of two vectors is to use the **parallelogram rule.** Place vectors **A** and **B** so that their initial points coincide. Then complete a parallelogram that has **A** and **B** as two adjacent sides. The diagonal of the parallelogram with the same initial point as **A** and **B** is the same vector sum **A** + **B** found by the definition. See Figure 16.

The vector sum **A** + **B** is the **resultant** of vectors **A** and **B.** Each of the vectors **A** and **B** is a **component** of vector **A** + **B.** In many practical applications, such as surveying, it is necessary to break a vector into its **vertical** and **horizontal components.** These components are two vectors, one vertical and one horizontal, whose resultant is the original vector. As shown in Figure 17, vector **OR** is the vertical component and vector **OS** is the horizontal component of **OP.**

For every vector **v** there is a vector −**v** with the same magnitude as **v** but opposite direction. Vector −**v** is the **opposite** of **v.** See Figure 18. The sum of **v** and −**v** has magnitude 0 and is a **zero vector.** As with real numbers, to *subtract* vector **B** from vector **A,** we find the vector sum **A** + (−**B**). See Figure 19.

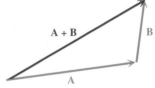

FIGURE 15

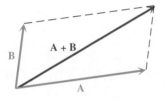

FIGURE 16

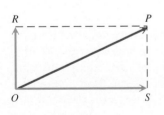

FIGURE 17

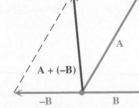

FIGURE 18

FIGURE 19

The **scalar product** of a real number (or scalar) k and a vector **u** is the vector $k\mathbf{u}$, which has magnitude $|k|$ times the magnitude of **u**. As shown in Figure 20, $k\mathbf{u}$ has the same direction as **u** if $k > 0$, and the opposite direction if $k < 0$.

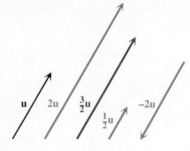

FIGURE 20

EXAMPLE 1
Finding magnitudes of vertical and horizontal components

Vector **w** has magnitude 25.0 and is inclined at an angle of 40° from the horizontal. Find the magnitudes of the horizontal and vertical components of the vector.

In Figure 21, the vertical component is labeled **v** and the horizontal component is labeled **u.** Vectors **u, v,** and **w** form a right triangle. In this right triangle,

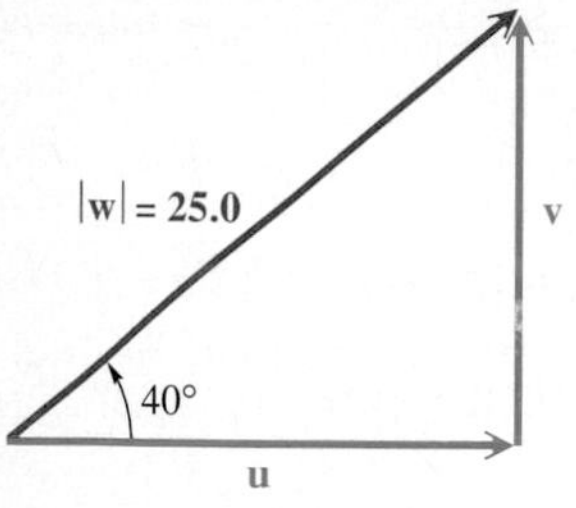

$$\sin 40° = \frac{|\mathbf{v}|}{|\mathbf{w}|} = \frac{|\mathbf{v}|}{25.0},$$

and

$$|\mathbf{v}| = 25.0 \sin 40° = 16.1.$$

In the same way,

$$\cos 40° = \frac{|\mathbf{u}|}{25.0},$$

with

$$|\mathbf{u}| = 25.0 \cos 40° = 19.2. \quad \blacktriangleright$$

FIGURE 21

$|w| = 25.0$

$40°$

u

v

It is helpful to review some of the properties of parallelograms when studying vectors. A parallelogram is a quadrilateral whose opposite sides are parallel. The opposite sides and opposite angles of a parallelogram are equal, and adjacent angles of a parallelogram are supplementary. The diagonals of a parallelogram bisect each other, but do not necessarily bisect the angles of the parallelogram.

Some of these properties are used in the following example.

EXAMPLE 2
Finding the magnitude of the resultant of two vectors in an application

Two forces of 15 newtons and 22 newtons (a newton is a unit of force used in physics) act at a point in the plane. If the angle between the forces is 100°, find the magnitude of the resultant force.

As shown in Figure 22, a parallelogram that has the forces as adjacent sides can be formed. The angles of the parallelogram adjacent to angle P each measure

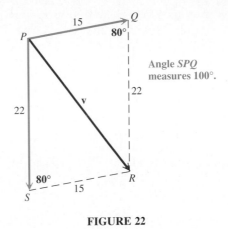

FIGURE 22

80°, since adjacent angles of a parallelogram are supplementary. (Angle *SPQ* measures 100°.) Opposite sides of the parallelogram are equal in length. The resultant force divides the parallelogram into two triangles. Use the law of cosines to get

$$|\mathbf{v}|^2 = 15^2 + 22^2 - 2(15)(22) \cos 80°$$
$$|\mathbf{v}| = 24. \quad \blacktriangleright$$

CONNECTIONS It is often useful to place a vector in the coordinate plane with its initial point at the origin and its endpoint at the point (a, b). This orientation gives us a one-to-one correspondence between points in the plane and the set of all vectors. If its endpoint is at (a, b), vector **u** is also denoted vector $\langle a, b \rangle$. The angle θ from the positive *x*-axis to the vector is called the **direction angle** of **u.** (See the figure.) From earlier results, if $\mathbf{u} = \langle a, b \rangle$, we get the following relationships.

(a) $|\mathbf{u}| = \sqrt{a^2 + b^2}$

(b) $a = |\mathbf{u}| \cos \theta$

(c) $b = |\mathbf{u}| \sin \theta$

(d) $\tan \theta = b/a$

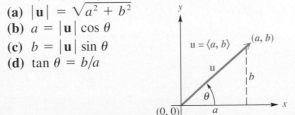

FOR DISCUSSION OR WRITING

1. Find the magnitude and direction angle rounded to the nearest tenth for $\mathbf{u} = \langle 3, -2 \rangle$.
2. Use the definitions of a and b given above to find $\mathbf{u} = \langle a, b \rangle$ if **u** has magnitude 5 and direction angle 60°.
3. For real numbers a, b, c, d, and k, $\langle a, b \rangle + \langle c, d \rangle = \langle a + c, b + d \rangle$ and $k\langle a, b \rangle = \langle ka, kb \rangle$. Let $\mathbf{u} = \langle -2, 1 \rangle$ and $\mathbf{v} = \langle 4, 3 \rangle$. Find the following.
   (a) $\mathbf{u} + \mathbf{v}$ **(b)** $-2\mathbf{u}$ **(c)** $4\mathbf{u} - 3\mathbf{v}$

7.4 Exercises ▼▼

1. In your own words, write a few sentences describing how a vector differs from a scalar.

2. Is a scalar product a vector or a scalar? Explain.

Exercises 3–6 refer to the vectors at the right.

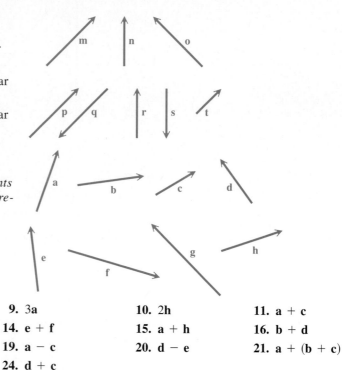

3. Name all pairs of vectors that appear to be equal.

4. Name all pairs of vectors that are opposites.

5. Name all pairs of vectors where the first is a scalar multiple of the other, with the scalar positive.

6. Name all pairs of vectors where the first is a scalar multiple of the other, with the scalar negative.

Exercises 7–24 refer to the vectors at the right. Draw a sketch to represent each vector. For example, find **a** + **e** *by placing* **a** *and* **e** *so that their initial points coincide. Then use the parallelogram rule to find the resultant, as shown in the figure below.*

7. −**b**	**8.** −**g**	**9.** 3**a**	**10.** 2**h**	**11.** **a** + **c**
12. **a** + **b**	**13.** **h** + **g**	**14.** **e** + **f**	**15.** **a** + **h**	**16.** **b** + **d**
17. **h** + **d**	**18.** **a** + **f**	**19.** **a** − **c**	**20.** **d** − **e**	**21.** **a** + (**b** + **c**)
22. (**a** + **b**) + **c**	**23.** **c** + **d**	**24.** **d** + **c**		

25. From the results of Exercises 21 and 22, do you think vector addition is associative?

26. From the results of Exercises 23 and 24, do you think vector addition is commutative?

For each pair of vectors **u** *and* **w** *with angle θ between them, sketch the resultant.*

27. $|\mathbf{u}| = 12$, $|\mathbf{w}| = 20$, $\theta = 27°$
28. $|\mathbf{u}| = 8$, $|\mathbf{w}| = 12$, $\theta = 20°$
29. $|\mathbf{u}| = 20$, $|\mathbf{w}| = 30$, $\theta = 30°$
30. $|\mathbf{u}| = 27$, $|\mathbf{w}| = 50$, $\theta = 12°$
31. $|\mathbf{u}| = 50$, $|\mathbf{w}| = 70$, $\theta = 40°$

For each of the following, vector **v** *has the given magnitude and direction. Find the magnitudes of the horizontal and vertical components of* **v**, *if α is the inclination of* **v** *from the horizontal. See Example 1.*

32. $\alpha = 20°$, $|\mathbf{v}| = 50$
33. $\alpha = 38°$, $|\mathbf{v}| = 12$
34. $\alpha = 70°$, $|\mathbf{v}| = 150$
35. $\alpha = 50°$, $|\mathbf{v}| = 26$
36. $\alpha = 35° \, 50'$, $|\mathbf{v}| = 47.8$
37. $\alpha = 27° \, 30'$, $|\mathbf{v}| = 15.4$
38. $\alpha = 128.5°$, $|\mathbf{v}| = 198$
39. $\alpha = 146.3°$, $|\mathbf{v}| = 238$

State a condition on **a** *and* **b** *that implies each of the following equations or inequalities.*

40. $|\mathbf{a} + \mathbf{b}| = 0$
41. $|\mathbf{a} + \mathbf{b}| = |\mathbf{a}| + |\mathbf{b}|$
42. $|\mathbf{a} + \mathbf{b}| = |\mathbf{a} - \mathbf{b}|$
43. $|\mathbf{a} + \mathbf{b}| > |\mathbf{a} - \mathbf{b}|$

44. Explain why the sum of two nonzero vectors can be a zero vector.

Two forces act at a point in the plane. The angle betweeen the two forces is given. Find the magnitude of the resultant force. See Example 2.

45. Forces of 250 and 450 newtons, forming an angle of 85°

46. Forces of 19 and 32 newtons, forming an angle of 118°

47. Forces of 17.9 and 25.8 lb, forming an angle of 105.5°

48. Forces of 75.6 and 98.2 lb, forming an angle of 82° 50′

49. Forces of 116 and 139 lb, forming an angle of 140° 50′

50. Forces of 37.8 and 53.7 lb, forming an angle of 68.5°

Find the magnitude and direction angle for **u** *rounded to the nearest tenth. See the Connections box.*

51. $\mathbf{u} = \langle 5, 12 \rangle$ **52.** $\mathbf{u} = \langle 6, -8 \rangle$ **53.** $\mathbf{u} = \langle -3, 4 \rangle$ **54.** $\mathbf{u} = \langle -4, 5 \rangle$

Write **u** *in the form* $\langle a, b \rangle$. *See the Connections box.*

55. $|\mathbf{u}| = 6$, direction angle of $\mathbf{u} = 30°$ **56.** $|\mathbf{u}| = 8$, direction angle of $\mathbf{u} = 45°$

57. $|\mathbf{u}| = 4$, direction angle of $\mathbf{u} = 120°$ **58.** $|\mathbf{u}| = 10$, direction angle of $\mathbf{u} = 150°$

Let $\mathbf{u} = \langle a_1, b_1 \rangle$, $\mathbf{v} = \langle a_2, b_2 \rangle$, $\mathbf{w} = \langle a_3, b_3 \rangle$, *and* $\mathbf{0} = \langle 0, 0 \rangle$. *Let k be any real number. Prove each statement.*

59. $\mathbf{u} + \mathbf{v} = \mathbf{v} + \mathbf{u}$ **60.** $\mathbf{u} + (\mathbf{v} + \mathbf{w}) = (\mathbf{u} + \mathbf{v}) + \mathbf{w}$ **61.** $-1(\mathbf{u}) = -\mathbf{u}$

62. $k(\mathbf{u} + \mathbf{v}) = k\mathbf{u} + k\mathbf{v}$ **63.** $\mathbf{u} + \mathbf{0} = \mathbf{u}$ **64.** $\mathbf{u} + (-\mathbf{u}) = \mathbf{0}$

If we define the unit vectors **i** *and* **j** *as follows,*

$$\mathbf{i} = \langle 1, 0 \rangle, \qquad \mathbf{j} = \langle 0, 1 \rangle,$$

then the vector $\mathbf{v} = \langle a, b \rangle$ *may be written as a linear combination of* **i** *and* **j**:

$$\mathbf{v} = a\mathbf{i} + b\mathbf{j}.$$

Let $\mathbf{u} = a_1\mathbf{i} + b_1\mathbf{j}$ *and* $\mathbf{v} = a_2\mathbf{i} + b_2\mathbf{j}$. *The dot product, or inner product, of* **u** *and* **v,** *written* $\mathbf{u} \cdot \mathbf{v}$, *is defined as*

$$\mathbf{u} \cdot \mathbf{v} = a_1 a_2 + b_1 b_2.$$

Find $\mathbf{u} \cdot \mathbf{v}$ *for each of the following pairs of vectors.*

65. $\mathbf{u} = 6\mathbf{i} - 2\mathbf{j}$ and $\mathbf{v} = -3\mathbf{i} + 2\mathbf{j}$ **66.** $\mathbf{u} = 3\mathbf{i} + 2\mathbf{j}$ and $\mathbf{v} = -3\mathbf{i} + 7\mathbf{j}$

67. $\mathbf{u} = \langle -6, 8 \rangle$ and $\mathbf{v} = \langle 3, -4 \rangle$ **68.** $\mathbf{u} = \langle 0, -2 \rangle$ and $\mathbf{v} = \langle -2, 6 \rangle$

69. Let α be the angle between the vectors **u** and **v**, where $0° \le \alpha \le 180°$. Show that $\mathbf{u} \cdot \mathbf{v} = |\mathbf{u}| \cdot |\mathbf{v}| \cdot \cos \alpha$.

Use the result of Exercise 69 to find the angle between each of the following pairs of vectors.

70. $\mathbf{u} = \langle -2, 5 \rangle$ and $\mathbf{v} = \langle 3, -4 \rangle$ **71.** $\mathbf{u} = \langle 1, 8 \rangle$ and $\mathbf{v} = \langle 2, -5 \rangle$

72. $\mathbf{u} = \langle -6, -2 \rangle$ and $\mathbf{v} = \langle 3, -1 \rangle$

Prove the following properties of the dot product. Assume that **u**, **v**, *and* **w** *are vectors, and k is a nonzero real number.*

73. $\mathbf{u} \cdot \mathbf{v} = \mathbf{v} \cdot \mathbf{u}$

74. $\mathbf{u} \cdot \mathbf{u} = |\mathbf{u}|^2$

75. $\mathbf{u} \cdot (\mathbf{v} + \mathbf{w}) = \mathbf{u} \cdot \mathbf{v} + \mathbf{u} \cdot \mathbf{w}$

76. $(k \cdot \mathbf{u}) \cdot \mathbf{v} = k \cdot (\mathbf{u} \cdot \mathbf{v})$

77. If **u** and **v** are not **0**, and if $\mathbf{u} \cdot \mathbf{v} = 0$, then **u** and **v** are perpendicular.

78. If **u** and **v** are perpendicular, then $\mathbf{u} \cdot \mathbf{v} = 0$.

7.5 Applications of Vectors ▼▼▼

The previous section covered methods for finding the resultant of two vectors. In many applications it is necessary to find a vector that will counterbalance the resultant. This opposite vector is called the *equilibrant:* the **equilibrant** of vector **u** is the vector **−u**.

EXAMPLE 1
Finding the magnitude and direction of an equilibrant

Find the magnitude of the equilibrant of forces of 48 newtons and 60 newtons acting on a point *A*, if the angle between the forces is 50°. Then find the angle between the equilibrant and the 48-newton force.

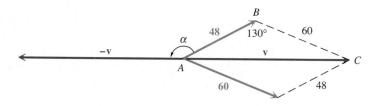

FIGURE 23

In Figure 23, the equilibrant is **−v.** The magnitude of **v,** and hence of **−v,** is found by using triangle *ABC* and the law of cosines:

$$|\mathbf{v}|^2 = 48^2 + 60^2 - 2(48)(60) \cos 130°$$
$$= 9606.5$$
$$|\mathbf{v}| = 98 \text{ newtons}$$

to two significant digits.

The required angle, labeled α in Figure 23, can be found by subtracting angle *CAB* from 180°. Use the law of sines to find angle *CAB*.

$$\frac{98}{\sin 130°} = \frac{60}{\sin CAB}$$
$$\sin CAB = .46900680$$
$$CAB = 28°$$

Finally,

$$\alpha = 180° - 28°$$
$$= 152°. \quad \blacktriangleright$$

EXAMPLE 2
Pulling a weight up an incline

Use vectors to solve the following incline problems.

(a) Find the force required to pull a 50-pound weight up a ramp inclined at 20° to the horizontal.

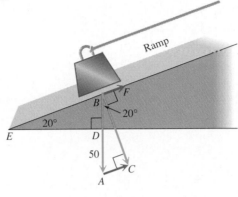

FIGURE 24

In Figure 24, the vertical 50-pound force **BA** represents the force of gravity. Its components are **BC** and **AC.** The component **BC** represents the force with which the body pushes against the ramp. The vector **BF** represents a force that would pull the body up the ramp. Since vectors **BF** and **AC** are equal, $|\mathbf{AC}|$ gives the magnitude of the required force.

Vectors **BF** and **AC** are parallel, so angle *EBD* equals angle *A*. Since angle *BDE* and angle *C* are right angles, triangles *CBA* and *DEB* have two corresponding angles equal and so are similar triangles. Therefore, angle *ABC* equals angle *E*, which is 20°. From right triangle *ABC*,

$$\sin 20° = \frac{|\mathbf{AC}|}{50}$$
$$|\mathbf{AC}| = 50 \sin 20°$$
$$|\mathbf{AC}| = 17.$$

To the nearest pound, a 17-pound force will be required to pull the weight up the ramp.

(b) A force of 16 pounds is required to hold a 40-pound lawn mower on an incline. What angle does the incline make with the horizontal?

Figure 25 illustrates the situation. Consider right triangle *ABC*. Angle $B = $ angle θ, the magnitude of vector **BA** represents the weight of the mower, and vector **AC** equals vector **BE,** which represents the force required to hold the mower on the incline. From the figure,

$$\sin B = \frac{16}{40}$$
$$= .4$$
$$B \approx 23.5782°.$$

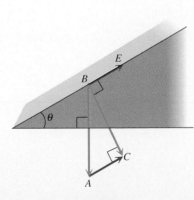

FIGURE 25

Therefore, the hill makes an angle of about 24° with the horizontal. ▶

Problems involving bearing (defined in Section 2.5) can also be worked with vectors, as shown in the next example.

EXAMPLE 3
Applying vectors to a navigation problem

A ship leaves port on a bearing of 28° and travels 8.2 miles. The ship then turns due east and travels 4.3 miles. How far is the ship from port? What is its bearing from port?

In Figure 26, vectors **PA** and **AE** represent the ship's path. The magnitude and bearing of the resultant **PE** can be found as follows. Triangle *PNA* is a right triangle, so angle *NAP* = 90° − 28° = 62°. Then angle *PAE* = 180° − 62° = 118°. Use the law of cosines to find |**PE**|, the magnitude of vector **PE**.

$$|\mathbf{PE}|^2 = 8.2^2 + 4.3^2 - 2(8.2)(4.3) \cos 118°$$
$$|\mathbf{PE}|^2 = 118.84$$

Therefore, $\quad |\mathbf{PE}| \approx 10.9,$

or 11 miles, rounded to two significant digits.

To find the bearing of the ship from port, first find angle *APE*. Use the law of sines, along with the value of |**PE**| before rounding.

$$\frac{\sin APE}{4.3} = \frac{\sin 118°}{10.9}$$

$$\sin APE = \frac{4.3 \sin 118°}{10.9}$$

$$\text{angle } APE \approx 20.4°$$

After rounding, angle *APE* is 20°, and the ship is 11 miles from port on a bearing of 28° + 20° = 48°. ◗

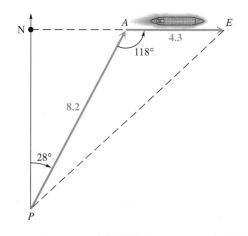

FIGURE 26

In air navigation, the **airspeed** of a plane is its speed relative to the air, while the **groundspeed** is its speed relative to the ground. Because of wind, these two speeds are usually different. The groundspeed of the plane is represented by the vector sum of the airspeed and windspeed vectors. See Figure 27.

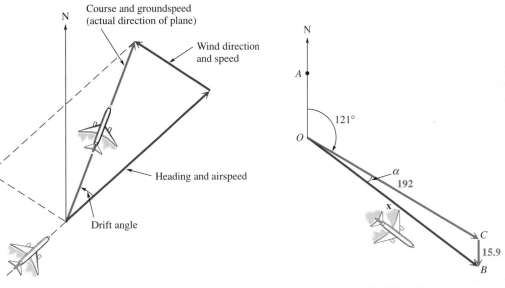

FIGURE 27   **FIGURE 28**

EXAMPLE 4
Applying vectors to a navigation problem

A plane with an airspeed of 192 miles per hour is headed on a bearing of 121°. A north wind is blowing (from north to south) at 15.9 miles per hour. Find the groundspeed and the actual bearing of the plane.

In Figure 28 the groundspeed is represented by $|\mathbf{x}|$. We must find angle α to determine the bearing, which will be $121° + \alpha$. From Figure 28, angle BCO equals angle AOC, which equals 121°. Find $|\mathbf{x}|$ by the law of cosines.

$$|\mathbf{x}|^2 = 192^2 + 15.9^2 - 2(192)(15.9)\cos 121°$$
$$|\mathbf{x}|^2 = 40{,}261$$

Therefore, $|\mathbf{x}| \approx 200.7,$

or 201 miles per hour. Now find α by using the law of sines. As before, use the value of $|\mathbf{x}|$ before rounding.

$$\frac{\sin \alpha}{15.9} = \frac{\sin 121°}{200.7}$$
$$\sin \alpha \approx .0679$$
$$\alpha \approx 3.89°$$

After rounding, α is 3.9°. The groundspeed is about 201 miles per hour, on a bearing of 125°, to three significant digits. ▶

CONNECTIONS When "reading" an electrocardiogram, a cardiologist measures the heights and the directions of certain peaks that appear. Also, depending on where the electrodes are attached to the patient, a certain angle is associated with the reading. If we call the measures of the peaks **a** and **b** and the angle θ, then **a, b,** and θ are related as shown in the figure. The cardiologist needs to know the length and direction of vector **v** in the figure. The following equations give these values.

$$|\mathbf{v}| = \frac{\sqrt{a^2 + b^2 - 2ab\cos\theta}}{\sin\theta} \qquad \alpha = \cos^{-1}\frac{a}{|\mathbf{v}|}$$

(Do not assume that line L is perpendicular to **b.**)

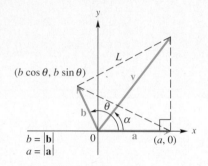

FOR DISCUSSION OR WRITING

1. As shown in Exercises 80–83 in Section 5.4, the slope of a line equals the tangent of its angle of inclination. Use this fact and the information given in the figure to write the equation of line L.
2. Use the result from Item 1 to find the coordinates of the endpoint of **v.**
3. Use the distance formula to find the magnitude of **v.** Rewrite the answer so that all trigonometric functions are expressed in terms of sine and cosine. This result should be the first formula given above.
4. What line in the figure corresponds to the quantity under the radical in the numerator of the expression for the magnitude of **v?**
5. Explain how to get the formula given above for α.

7.5 Exercises ▼▼▼▼▼▼▼▼▼▼▼▼▼▼▼▼▼▼▼▼▼▼▼▼▼▼▼▼▼▼▼▼▼▼▼▼▼▼

Solve the problem. See Examples 1–4.

1. Two forces of 692 newtons and 423 newtons act at a point. The resultant force is 786 newtons. Find the angle between the forces.

2. Two forces of 128 lb and 253 lb act at a point. The equilibrant is 320 lb. Find the angle between the forces.

3. A force of 25 lb is required to push an 80-lb crate up a hill. What angle does the hill make with the horizontal?

4. Find the force required to keep a 3000-lb car parked on a hill that makes an angle of 15° with the horizontal.

5. To build the pyramids in Egypt, it is believed that giant causeways were built to transport the building materials to the site. One such causeway is said to have been 3000 ft long, with a slope of about 2.3°. How much force would be required to pull a 60-ton monolith along this causeway?

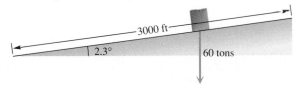

6. A force of 500 lb is required to pull a boat up a ramp inclined at 18° with the horizontal. How much does the boat weigh?

7. Two tugboats are pulling a disabled speedboat into port with forces of 1240 lb and 1480 lb. The angle between these forces is 28.2°. Find the direction and magnitude of the equilibrant.

8. Two people are carrying a box. One person exerts a force of 150 lb at an angle of 62.4° with the horizontal. The other person exerts a force of 114 lb at an angle of 54.9°. Find the weight of the box.

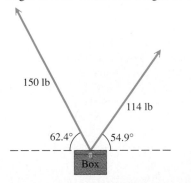

9. A crate is supported by two ropes. One rope makes an angle of 46° 20' with the horizontal and has a tension of 89.6 lb on it. The other rope is horizontal. Find the weight of the crate and the tension in the horizontal rope.

10. Three forces acting at a point are in equilibrium. The forces are 980 lb, 760 lb, and 1220 lb. Find the angles between the directions of the forces. (*Hint:* Arrange the forces to form the sides of a triangle.)

11. A force of 176 lb makes an angle of 78° 50' with a second force. The resultant of the two forces makes an angle of 41° 10' with the first force. Find the magnitudes of the second force and of the resultant.

12. A force of 28.7 lb makes an angle of 42° 10' with a second force. The resultant of the two forces makes an angle of 32° 40' with the first force. Find the magnitudes of the second force and of the resultant.

13. A plane flies 650 mph on a bearing of 175.3°. A 25-mph wind, from a direction of 266.6°, blows against the plane. Find the resulting course of the plane.

14. A pilot wants to fly on a bearing of 74.9°. By flying due east, he finds that a 42-mph wind, blowing from the south, puts him on course. Find the airspeed and the groundspeed.

15. Starting at point *A*, a ship sails 18.5 km on a bearing of 189°, then turns and sails 47.8 km on a bearing of 317°. Find the distance of the ship from point *A*.

16. Two towns 21 mi apart are separated by a dense forest. See the figure. To travel from town *A* to town *B*, a person must go 17 mi on a bearing of 325°, then turn and continue for 9 mi to reach town *B*. Find the bearing of *B* from *A*.

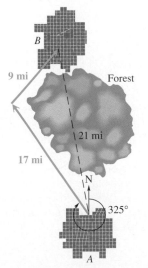

17. An airline route from San Francisco to Honolulu is on a bearing of 233°. A jet flying at 450 mph on that bearing flies into a wind blowing at 39 mph from a direction of 114°. Find the resulting bearing and groundspeed of the plane.

18. A pilot is flying at 168 mph. She wants her flight path to be on a bearing of 57° 40'. A wind is blowing from the south at 27.1 mph. Find the bearing the pilot should fly, and find the plane's groundspeed.

19. What bearing and airspeed are required for a plane to fly 400 mi due north in 2.5 hr if the wind is blowing from a direction of 328° at 11 mph?

20. A plane is headed due south with an airspeed of 192 mph. A wind from a direction of 78° is blowing at 23 mph. Find the groundspeed and resulting bearing of the plane.

21. An airplane is headed on a bearing of 174° at an airspeed of 240 km per hr. A 30 km per hr wind is blowing from a direction of 245°. Find the groundspeed and resulting bearing of the plane.

22. A ship sailing due east in the North Atlantic has been warned to change course to avoid a group of icebergs. The captain turns and sails on a bearing of 62° for a while, then changes course again to a bearing of 115° until the ship reaches its original course. See the figure. How much farther did the ship have to travel to avoid the icebergs?

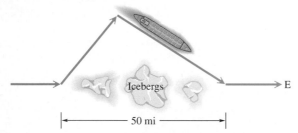

23. The aircraft carrier *Tallahassee* is traveling at sea on a steady course with a bearing of 30° at 32 mph. Patrol planes on the carrier have enough fuel for 2.6 hr of flight when traveling at a speed of 520 mph. One of the pilots takes off on a bearing of 338° and then turns and heads in a straight line, so as to be able to catch the carrier and land on the deck at the exact instant that his fuel runs out. If the pilot left at 2 P.M., at what time did he turn to head for the carrier?

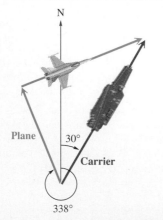

24. A car going around a banked curve is subject to the forces shown in the figure. If the radius of the curve is 100 ft, what value of θ to the nearest degree would allow an automobile to travel around the curve at a speed of 40 ft per sec without depending on friction?

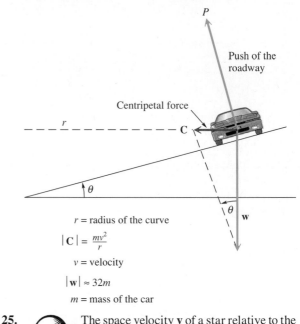

r = radius of the curve

$$|\mathbf{C}| = \frac{mv^2}{r}$$

v = velocity

$$|\mathbf{w}| \approx 32m$$

m = mass of the car

25. 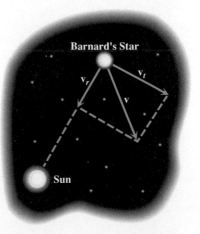 The space velocity $\mathbf{v}$ of a star relative to the sun can be expressed as the resultant vector of two perpendicular vectors—the radial velocity $\mathbf{v}_r$ and the tangential velocity $\mathbf{v}_t$ where $\mathbf{v} = \mathbf{v}_r + \mathbf{v}_t$. Refer to the figure. If a star is located near the sun and its space velocity is large, then its motion across the sky will also be large. Barnard's Star is a relatively close star with a distance

Not to scale

of 35 trillion miles from the sun. It moves across the sky through an angle 10.34″ per year, which is the largest motion of any known star. Its radial velocity is $v_r = 67$ mi/sec toward the sun. (*Sources:* Zelik, M., S. Gregory, and E. Smith, *Introductory Astronomy and Astrophysics,* Saunders College Publishing, 1992;

Acker, A. and C. Jaschek, *Astronomical Methods and Calculations,* John Wiley & Sons, 1986.)

(a) Approximate the tangential velocity v_t of Barnard's Star. (*Hint:* Use the Length of Arc Formula: $s = r\theta$.)

(b) Compute the magnitude of **v.**

Chapter 7   Summary ▼▼▼▼▼▼▼▼▼▼▼▼▼▼▼▼▼▼▼▼▼▼▼▼▼▼▼▼▼▼▼▼▼▼▼▼▼

SECTION	KEY IDEAS
7.1 Oblique Triangles and the Law of Sines	
	Law of Sines In any triangle *ABC,* with sides *a*, *b*, and *c*, $$\frac{a}{\sin A} = \frac{b}{\sin B}$$ $$\frac{a}{\sin A} = \frac{c}{\sin C}$$ $$\frac{b}{\sin B} = \frac{c}{\sin C}.$$ **Area of a Triangle** The area of a triangle is given by half the product of the lengths of two sides and the sine of the angle between the two sides. $$K = \frac{1}{2}bc \sin A, \qquad K = \frac{1}{2}ab \sin C, \qquad K = \frac{1}{2}ac \sin B$$
7.3 The Law of Cosines	
	Law of Cosines In any triangle *ABC,* with sides *a*, *b*, and *c*, $$a^2 = b^2 + c^2 - 2bc \cos A$$ $$b^2 = a^2 + c^2 - 2ac \cos B$$ $$c^2 = a^2 + b^2 - 2ab \cos C.$$ **Heron's Area Formula** If a triangle has sides of lengths *a*, *b*, and *c*, and if the semiperimeter is $$s = \frac{1}{2}(a + b + c),$$ then the area of the triangle is $$K = \sqrt{s(s - a)(s - b)(s - c)}.$$

SECTION	KEY IDEAS
7.4 Introduction to Vectors	

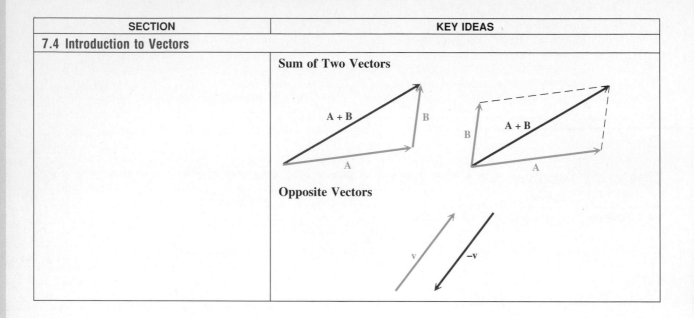

Sum of Two Vectors

Opposite Vectors

Chapter 7 Review Exercises ▼▼▼▼▼▼▼▼▼▼▼▼▼▼▼▼▼▼▼▼▼▼▼▼▼▼▼▼▼▼▼▼▼

Use the law of sines to find the indicated part of each triangle ABC.

1. $C = 74.2°$, $c = 96.3$ m, $B = 39.5°$; find b

2. $A = 129.7°$, $a = 127$ ft, $b = 69.8$ ft; find B

3. $C = 51.3°$, $c = 68.3$ m, $b = 58.2$ m; find B

4. $a = 165$ m, $A = 100.2°$, $B = 25.0°$; find b

5. $B = 39° 50'$, $b = 268$ m, $a = 340$ m; find A

6. $C = 79° 20'$, $c = 97.4$ mm, $a = 75.3$ mm; find A

7. If we are given a, A, and C in a triangle ABC, does the possibility of the ambiguous case exist? If not, explain why.

8. Can triangle ABC exist if $a = 4.7$, $b = 2.3$, and $c = 7.0$? If not, explain why. Answer this question without using trigonometry.

9. Given $a = 10$ and $B = 30°$, determine the values of b for which A has
(a) Exactly one value **(b)** Two values **(c)** No value.

10. Given $a = 10$ and $B = 150°$, determine the values of b for which A has
(a) Exactly one value **(b)** Two values **(c)** No value.

Use the law of cosines to find the indicated part of each triangle ABC.

11. $a = 86.14$ in, $b = 253.2$ in, $c = 241.9$ in; find A

12. $B = 120.7°$, $a = 127$ ft, $c = 69.8$ ft; find b

13. $A = 51° 20'$, $c = 68.3$ m, $b = 58.2$ m; find a

14. $a = 14.8$ m, $b = 19.7$ m, $c = 31.8$ m; find B

15. $A = 46° 10'$, $b = 184$ cm, $c = 192$ cm; find a

16. $a = 7.5$ ft, $b = 12.0$ ft, $c = 6.9$ ft; find C

Solve each triangle ABC having the given information.

17. $A = 25.2°$, $a = 6.92$ yd, $b = 4.82$ yd

18. $A = 61.7°$, $a = 78.9$ m, $b = 86.4$ m

19. $a = 27.6$ cm, $b = 19.8$ cm, $C = 42° 30'$

20. $a = 94.6$ yd, $b = 123$ yd, $c = 109$ yd

Find the area of each triangle ABC with the given information.

21. $b = 840.6$ m, $c = 715.9$ m, $A = 149.3°$

22. $a = 6.90$ ft, $b = 10.2$ ft, $C = 35° 10'$

23. $a = .913$ km, $b = .816$ km, $c = .582$ km

24. $a = 43$ m, $b = 32$ m, $c = 51$ m

The following identities involve all six parts of a triangle, ABC, and are thus useful for checking answers.

Newton's formula $\dfrac{a + b}{c} = \dfrac{\cos \frac{1}{2}(A - B)}{\sin \frac{1}{2}C}$

Mollweide's formula $\dfrac{a - b}{c} = \dfrac{\sin \frac{1}{2}(A - B)}{\cos \frac{1}{2}C}$

A diagram shows triangle ABC with a right angle at C, angle $A = 30°$, angle $B = 60°$, side $b = 7\sqrt{3}$, side $c = 14$, and side $a = 7$.

25. Apply Newton's formula to the triangle shown in the figure to verify the accuracy of the information.

26. Apply Mollweide's formula to the triangle shown in the figure to verify the accuracy of the information.

Solve the problem.

27. Raoul plans to paint a triangular wall in his A-frame cabin. Two sides measure 7 m each, and the third side measures 6 m. How much paint will he need to buy if a can of paint covers 7.5 sq m?

28. A lot has the shape of a quadrilateral. See the figure. What is its area?

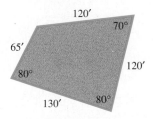

29. A tree leans at an angle of 8.0° from the vertical. See the figure. From a point 7.0 m from the bottom of the tree, the angle of elevation to the top of the tree is 68°. How tall is the tree?

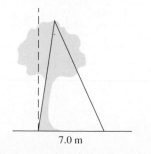

7.0 m

30. A hill makes an angle of 14.3° with the horizontal. From the base of the hill, the angle of elevation to the top of a tree on top of the hill is 27.2°. The distance along the hill from the base to the tree is 212 ft. Find the height of the tree.

31. A ship is sailing east. At one point, the bearing of a submerged rock is 45° 20′. After sailing 15.2 mi, the bearing of the rock has become 308° 40′. Find the distance of the ship from the rock at the latter point.

32. From an airplane flying over the ocean, the angle of depression to a submarine lying just under the surface is 24° 10′. At the same moment the angle of depression from the airplane to a battleship is 17° 30′. See the figure. The distance from the airplane to the battleship is 5120 ft. Find the distance between the battleship and the submarine. (Assume the airplane, submarine, and battleship are in a vertical plane.)

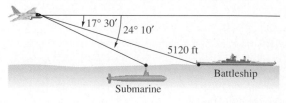

33. Two boats leave a dock together. Each travels in a straight line. The angle between their courses measures 54° 10′. One boat travels 36.2 km per hr, and the other travels 45.6 km per hr. How far apart will they be after 3 hr?

34. Find the lengths of both diagonals of a parallelogram with adjacent sides of 12 cm and 15 cm if the angle between these sides is 33°.

35. To measure the distance through a mountain for a proposed tunnel, a point C is chosen that can be reached from each end of the tunnel. See the figure.

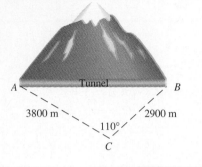

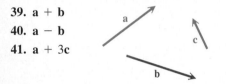

If $AC = 3800$ m, $BC = 2900$ m, and angle $C = 110°$, find the length of the tunnel.

36. A baseball diamond is a square, 90.0 ft on a side, with home plate and the three bases as vertices. The pitcher's rubber is located 60.5 ft from home plate. Find the distance from the pitcher's rubber to each of the bases.

37. The Vietnam Veterans' Memorial in Washington, D.C., is in the shape of an unenclosed isosceles triangle (that is, V-shaped) with equal sides of length 246.75 feet and the angle between these sides measuring 125° 12′. Find the distance between the ends of the two equal sides.

38. If angle C of a triangle ABC measures 90°, what does the law of cosines $c^2 = a^2 + b^2 - 2ab \cos C$ become?

In Exercises 39–41, use the vectors pictured here. Find the following.

39. a + b

40. a − b

41. a + 3c

42. True or false: Opposite angles of a parallelogram are congruent.

43. True or false: A diagonal of a parallelogram must bisect two angles of the parallelogram.

Find the horizontal and vertical components of each vector, where α is the inclination of the vector from the horizontal.

44. $\alpha = 45°$, magnitude 50

45. $\alpha = 75.0°$, magnitude 69.2

46. $\alpha = 154° 20′$, magnitude 964

Given two forces and the angle between them, find the magnitude of the resultant force.

47. Forces of 15 and 23 lb, forming an angle of 87°

48. Forces of 142 and 215 newtons, forming an angle of 112°

49. Forces of 85.2 and 69.4 newtons, forming an angle of 58° 20′

50. Forces of 475 and 586 lb, forming an angle of 78° 20′

*Find the magnitude and direction angle for **u** rounded to the nearest tenth. See the Connections box in Section 7.4.*

51. u = ⟨21, −20⟩

52. u = ⟨−9, 12⟩

*Write **u** in the form ⟨a, b⟩. See the Connections box in Section 7.4.*

53. |**u**| = 6, direction angle of **u** = 60°

54. |**u**| = 8, direction angle of **u** = 120°

Solve the problem.

55. One rope pulls a barge directly east with a force of 100 newtons. Another rope pulls the barge to the northeast with a force of 200 newtons. Find the resultant force acting on the barge and the angle between the resultant and the first rope.

56. Paula and Steve are pulling their daughter Jessie on a sled. Steve pulls with a force of 18 lb at an angle of 10°. Paula pulls with a force of 12 lb at an angle of 15°. What is the weight of Jessie and the sled? See the figure. (*Hint:* Find the resultant force.)

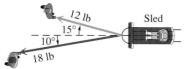

57. A 186-lb force just keeps a 2800-lb car from rolling down a hill. What angle does the hill make with the horizontal?

58. A plane has an airspeed of 520 mph. The pilot wishes to fly on a bearing of 310°. A wind of 37 mph is blowing from a bearing of 212°. What direction should the pilot fly, and what will be her actual speed?

59. A boat travels 15 km per hr in still water. The boat is traveling across a large river, on a bearing of 130°. The current in the river, coming from the west, has a speed of 7 km per hr. Find the resulting speed of the boat and its resulting direction of travel.

60. A long-distance swimmer starts out swimming a steady 3.2 mph due north. A 5.1-mph current is flowing on a bearing of 12°. What is the swimmer's resulting bearing and speed?

61. In order to obtain accurate aerial photographs, ground control must determine the coordinates of **control points** located on the ground that can be identified in the

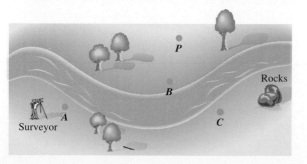

photographs. Using these known control points, the orientation and scale of each photograph can be determined. Then, unknown positions and distances can easily be determined. Before an aerial photograph is taken for highway design, horizontal control points must be found and the distance between them calculated. The figure shows three consecutive control points *A*, *B*, and *C*. A surveyor measures a baseline distance of 92.13 feet from *B* to an arbitrary point *P*. Angles *BAP* and *BCP* are found to be 2°22′47″ and 5°13′11″, respectively. Then, angles *APB* and *CPB* are determined to be 63°4′25″ and 74°19′49″, respectively. Determine the distance between the control points *A* and *B* and between *B* and *C*. (*Source:* Moffitt, F., *Photogrammetry,* International Textbook Company, Scranton, Pennsylvania, 1967.)

62. 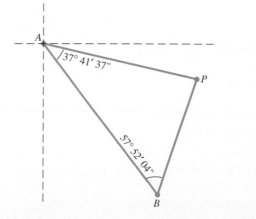 In order to find the coordinates of control points for aerial photographs, ground control must first locate basic control monuments established by the U.S. Coast and Geodetic Survey and the U.S. Geological Survey. These monuments have published *x*- and *y*-coordinates called **state plane coordinates.** Using these monuments and common surveying techniques, the coordinates of the control points can be determined. Two basic control monuments *A* and *B* have coordinates in feet of $x_A = 2{,}101{,}345.1$, $y_A = 998{,}764.3$ and $x_B = 2{,}131{,}667.8$, $y_B = 923{,}541.7$. The location of an unknown control point *P* is to be determined. If angles *PAB* and *PBA* are measured as 37°41′37″ and 57°52′04″, respectively, discuss the steps you would take to determine the state plane coordinates of control point *P*. (*Source:* Moffitt, F., *Photogrammetry,* International Textbook Company, Scranton, Pennsylvania, 1967.)

8

Complex Numbers and Polar Equations

The concept of counting began when people matched objects in a one-to-one correspondence with parts of the body such as fingers and joints. The Bugilia of British New Guinea used their left hand, wrist, elbow, shoulder, left and right breasts to count to ten. Gradually over time, the natural numbers were developed and used by most cultures. However, the number zero was not as easily accepted. The early Greeks never recognized zero as a number and Roman numerals had no zero symbol. After all, why would one need a number to represent nothing? The Greeks represented irrational numbers like $\sqrt{2}$ as lengths of line segments in geometry but avoided the problem of what irrational numbers were and how to represent them. Finally in 1872 Georg Cantor and Richard Dedekind published papers that played important roles in our present-day understanding of irrational numbers. In A.D. 628, Hindu mathematicians introduced the concept of negative numbers and arithmetic operations on them. Negative numbers were more difficult for mathematicians to accept than irrational numbers. Even as late as the 16th and 17th centuries, many mathematicians did not fully understand negative numbers. Negative numbers seemed to have no physical significance. This may also be true for many people today—until they open their first checking accounts. Then, negative numbers take on a whole new reality! One of the earliest encounters with the square root of a negative number was in A.D. 50 by Heron of Alexandria when he derived the expression $\sqrt{81 - 144}$. Taking square roots of negative numbers led to the need for the complex numbers. As one might imagine, if 16th- and 17th-century mathematicians felt uneasy about negative numbers, they felt even more uneasy about taking square roots of negative numbers. The famous mathematician René Descartes rejected complex numbers and coined the term "imaginary" numbers.

Today, complex numbers are readily accepted and play an important role in many new and exciting fields of applied mathematics and technology. Complex numbers are no more imaginary than are negative numbers and zero. Their development has been necessary in order to solve new problems like the design of airplane wings, ships, electrical circuits, noise control, and fractals.

During the past 20 years, computer graphics and complex numbers have made it possible to produce many beautiful fractals. Benoit B. Mandelbrot in 1975 first used the term *fractal*. Largely because of his efforts, fractal geometry has become a new field of study. At its basic level, a fractal is a unique and enchanting geometric figure with an endless self-similarity property. A fractal image repeats itself infinitely with ever-decreasing dimensions. If you look at smaller and smaller portions of the figure, you will continue to see the whole—much like looking into two parallel mirrors that are facing each other. Although most current applications for fractals are related to creating fascinating images and pictures, they have a tremendous potential for applied science. An example of a fractal is shown in the figure. It is an amazing graphical solution to a difficult problem first presented by Sir Arthur Cayley in 1879. The fractal is called *Newton's basins of attraction for the cube roots of unity*. Like many other fractals, it was created with the aid of high-resolution computer graphics and complex numbers.

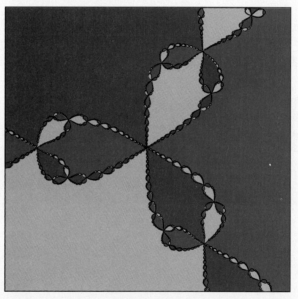

Source: Kincaid, D. and W. Cheney, *Numerical Analysis,*
Brooks/Cole Publishing Company, Pacific Grove, California, 1991.

In order to better understand how fractals like this are created, our current set of numbers must be expanded to include the complex numbers. Through acceptance of the importance and necessity of complex numbers, a new window of understanding about modern mathematics and science is opened.

Sources: Crownover, R., *Introduction to Fractals and Chaos,* Jones and Bartlett Publishers, Boston, 1995.

Kline, M., *Mathematics: The Loss of Certainty,* Oxford University Press, New York, 1980.

Lauwerier, H., *Fractals,* Princeton University Press, Princeton, New Jersey, 1991.

National Council of Teachers of Mathematics, *Historical Topics for the Mathematics Classroom,* Thirty-first Yearbook, Washington, D.C., 1969.

In this chapter we show the connection between complex numbers and vectors in the plane. We introduce a new notation for complex numbers involving the sine and cosine functions.

8.1 Operations on Complex Numbers ▼▼▼

The complex number system involves a number that is new to us, the imaginary unit i, defined as follows.

Definition of i

$$i = \sqrt{-1} \qquad \text{or} \qquad i^2 = -1$$

The letters j and k are also used to represent $\sqrt{-1}$ in calculus and some applications (electronics, for example).

A complex number is a number that has the form $a + bi$, where a and b are real numbers. The form $a + bi$ is called the **standard,** or **rectangular form** of the complex number. The real number a is called the **real part,** and the real number b is called the **imaginary part.*** If $b \neq 0$, $a + bi$ is also called an **imaginary number.** Each real number is a complex number, since a real number a may be thought of as the complex number $a + 0i$. The set of real numbers is a subset of the set of complex numbers. See Figure 1.

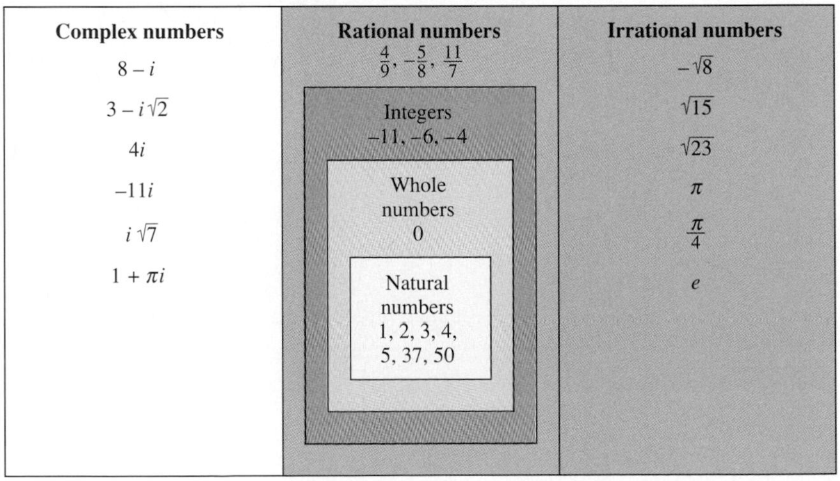

Complex numbers	Rational numbers $\frac{4}{9}, -\frac{5}{8}, \frac{11}{7}$	Irrational numbers
$8 - i$		$-\sqrt{8}$
$3 - i\sqrt{2}$	Integers $-11, -6, -4$	$\sqrt{15}$
$4i$		$\sqrt{23}$
$-11i$	Whole numbers 0	π
$i\sqrt{7}$		$\frac{\pi}{4}$
$1 + \pi i$	Natural numbers $1, 2, 3, 4, 5, 37, 50$	e

Real numbers are shaded.

FIGURE 1

---

* Some texts define bi as the imaginary part.

NOTE   The form $a + ib$ is often used for symbols such as $i\sqrt{5}$, since $\sqrt{5}i$ could be too easily mistaken for $\sqrt{5i}$.

The square root of a negative number can be written as the product of a real number and i, using the definition of $\sqrt{-a}$ which follows.

Definition of $\sqrt{-a}$

For positive real numbers a,

$$\sqrt{-a} = i\sqrt{a}.$$

For example, $\sqrt{-16} = i\sqrt{16} = 4i$, and $\sqrt{-75} = i\sqrt{75} = 5i\sqrt{3}$.

The first example shows how complex numbers may occur as solutions of quadratic equations.

EXAMPLE 1

Solving quadratic equations with complex solutions

Solve each equation for its complex solutions.

(a) $x^2 = -9$

Take the square root on both sides, remembering that we must find both roots, indicated by the $\pm$ sign.

$$x^2 = -9$$
$$x = \pm\sqrt{-9}$$
$$x = \pm i\sqrt{9} \qquad \sqrt{-a} = i\sqrt{a}$$
$$x = \pm 3i \qquad \sqrt{9} = 3$$

(b) $9x^2 + 5 = 6x$

Write the equation in standard form, $9x^2 - 6x + 5 = 0$. We will use the quadratic formula, with $a = 9$, $b = -6$, and $c = 5$.

$$x = \frac{-b \pm \sqrt{b^2 - 4ac}}{2a} \qquad \text{Quadratic formula}$$

$$= \frac{-(-6) \pm \sqrt{(-6)^2 - 4(9)(5)}}{2(9)} \qquad a = 9, b = -6, c = 5$$

$$= \frac{6 \pm \sqrt{-144}}{18}$$

$$= \frac{6 \pm 12i}{18} \qquad \sqrt{-144} = 12i$$

$$= \frac{6(1 \pm 2i)}{18} \qquad \text{Factor.}$$

$$= \frac{1 \pm 2i}{3} \qquad \text{Lowest terms}$$

The solutions may be written in standard form as

$$\frac{1}{3} \pm \frac{2}{3}i. \quad \blacktriangleright$$

NOTE   In the quadratic formula, the expression $b^2 - 4ac$ is called the *discriminant*. If a, b, and c are real numbers, where $a \neq 0$, the quadratic equation $ax^2 + bx + c = 0$ has two real solutions if the discriminant is positive, one real solution if it is 0, and two nonreal, complex solutions if the discriminant is negative. This last case was seen in Example 1.

When working with negative radicands, it is very important to use the definition $\sqrt{-a} = i\sqrt{a}$ before using any of the other rules for radicals. In particular, the rule $\sqrt{c} \cdot \sqrt{d} = \sqrt{cd}$ is valid only when c and d are not both negative. For example, multiplying $\sqrt{-2} \cdot \sqrt{-32}$ to get $\sqrt{64}$ (which is incorrect) gives 8, but

$$\sqrt{-2} \cdot \sqrt{-32} = i\sqrt{2} \cdot i\sqrt{32}$$
$$= i^2\sqrt{64}$$
$$= (-1)8$$
$$= -8,$$

which is the correct result.

◀ EXAMPLE 2
Simplifying products and quotients with negative radicands

Express each product or quotient as a real number, or a product of a real number and i.

(a) $\sqrt{-7} \cdot \sqrt{-7} = i\sqrt{7} \cdot i\sqrt{7}$
$\qquad = i^2 \cdot (\sqrt{7})^2$
$\qquad = (-1) \cdot 7 = -7$

(b) $\sqrt{-6} \cdot \sqrt{-10} = i\sqrt{6} \cdot i\sqrt{10} = i^2 \cdot \sqrt{6 \cdot 10} = -1 \cdot 2\sqrt{15} = -2\sqrt{15}$

(c) $\dfrac{\sqrt{-20}}{\sqrt{-2}} = \dfrac{i\sqrt{20}}{i\sqrt{2}} = \sqrt{\dfrac{20}{2}} = \sqrt{10}$

(d) $\dfrac{\sqrt{-48}}{\sqrt{24}} = \dfrac{i\sqrt{48}}{\sqrt{24}} = i\sqrt{2}$ ▶

Addition and subtraction of complex numbers is defined in a manner similar to these operations on binomials.

Sum and Difference of Complex Numbers

For complex numbers $a + bi$ and $c + di$,

$$(a + bi) + (c + di) = (a + c) + (b + d)i$$

and $\qquad (a + bi) - (c + di) = (a - c) + (b - d)i$

To add complex numbers, add their real parts and add their imaginary parts. Subtraction is accomplished in a similar manner.

EXAMPLE 3
Adding and subtracting
complex numbers

Add or subtract as indicated.

(a) $(3 - 4i) + (-2 + 6i) = [3 + (-2)] + [-4 + 6]i = 1 + 2i$

(b) $(-9 + 7i) + (3 - 15i) = -6 - 8i$

(c) $(-4 + 3i) - (6 - 7i) = (-4 - 6) + [3 - (-7)]i = -10 + 10i$

(d) $(12 - 5i) - (8 - 3i) = 4 - 2i$ ▶

The *product* of two complex numbers can be found by multiplying as if the numbers were binomials and using the fact that $i^2 = -1$, as follows.

$$(a + bi)(c + di) = ac + adi + bic + bidi$$
$$= ac + adi + bci + bdi^2$$
$$= ac + (ad + bc)i + bd(-1)$$
$$= (ac - bd) + (ad + bc)i$$

Thus, the product of the complex numbers $a + bi$ and $c + di$ is defined as follows.

Product of Complex Numbers

For complex numbers $a + bi$ and $c + di$,

$$(a + bi)(c + di) = (ac - bd) + (ad + bc)i.$$

The formal definition is rarely used when multiplying complex numbers. It is usually easier just to multiply as with binomials, using the FOIL method.

EXAMPLE 4
Multiplying complex
numbers

Find each product.

(a) $(5 - 4i)(7 - 2i) = 5(7) + 5(-2i) - 4i(7) - 4i(-2i)$
$$= 35 - 10i - 28i + 8i^2$$
$$= 35 - 38i + 8(-1) \qquad \text{Replace } i^2 \text{ with } -1.$$
$$= 27 - 38i \qquad \qquad \text{Combine terms.}$$

(b) $(3 - i)(3 + i) = 9 + 3i - 3i - i^2$
$$= 9 - (-1) = 10$$ ▶

The factors in Example 4(b) are called *conjugates*. The **conjugate** of the complex number $a + bi$ is the complex number $a - bi$. Notice that the product of a pair of conjugates is the difference of squares, so Example 4(b) could have been written as $(3 - i)(3 + i) = 3^2 - i^2 = 9 - (-1) = 10$. The product of conjugates is always a real number. Specifically, the product is the sum of the squares of the real and imaginary parts.

$$(a + bi)(a - bi) = a^2 + b^2$$

Recall that the product of the conjugates $\sqrt{a} + \sqrt{b}$ and $\sqrt{a} - \sqrt{b}$ is also always a real number. We use this fact to rationalize denominators with radicals. Conjugates are used in the same way in division of complex numbers.

> **Rule for the Quotient of Complex Numbers**
>
> To divide complex numbers, multiply both the numerator and the denominator (divisor) by the conjugate of the denominator.

EXAMPLE 5
Dividing complex numbers

Find each quotient.

(a) $\dfrac{3 + 2i}{5 - i}$

Multiply the numerator and denominator by $5 + i$, the conjugate of $5 - i$.

$$\frac{3 + 2i}{5 - i} = \frac{(3 + 2i)(5 + i)}{(5 - i)(5 + i)}$$

$$= \frac{15 + 3i + 10i + 2i^2}{26} \qquad (5 - i)(5 + i) = 5^2 + 1^2 = 26$$

$$= \frac{13 + 13i}{26} = \frac{1}{2} + \frac{1}{2}i$$

To check this answer, show that

$$(5 - i)\left(\frac{1}{2} + \frac{1}{2}i\right) = 3 + 2i.$$

(b) $\dfrac{3}{i} = \dfrac{3(-i)}{i(-i)}$ $-i$ is the conjugate of i.

$$= \frac{-3i}{-i^2}$$

$$= \frac{-3i}{1} \qquad -i^2 = -(-1) = 1$$

$$= -3i \quad \blacktriangleright$$

The fact that i^2 is equal to -1 can be used to find higher powers of i.

$$i^0 = 1 \qquad\qquad\qquad i^4 = i^2 \cdot i^2 = (-1)(-1) = 1$$
$$i^1 = i \qquad\qquad\qquad i^5 = i \cdot i^4 = i \cdot 1 = i$$
$$i^2 = -1 \qquad\qquad\quad i^6 = i^2 \cdot i^4 = (-1) \cdot 1 = -1$$
$$i^3 = i \cdot i^2 = i(-1) = -i \qquad i^7 = i^3 \cdot i^4 = (-i) \cdot 1 = -i$$

As these examples show, the powers of i rotate through the four numbers 1, i, -1, and $-i$. Larger powers of i can be simplified by using the fact that $i^4 = 1$. For example $i^{75} = (i^4)^{18} \cdot i^3 = 1^{18} \cdot i^3 = 1 \cdot i^3 = i^3 = -i$.

EXAMPLE 6
Simplifying powers of i

Find each power of i.

(a) $i^{12} = (i^4)^3 = 1^3 = 1$

(b) $i^{39} = i^{36} \cdot i^3$

$= (i^4)^9 \cdot i^3$

$= 1^9 \cdot (-i)$

$= -i$

(c) $i^{-2} = \dfrac{1}{i^2} = \dfrac{1}{-1} = -1$ ▶

8.1 Exercises ▼▼▼▼▼▼▼▼▼▼▼▼▼▼▼▼▼▼▼▼▼▼▼▼▼▼▼▼▼▼▼▼▼▼▼

1. Explain why a real number must be a complex number, but a complex number need not be a real number.

2. If the complex number $a + bi$ is real, then what can be said about the value of b?

3. We know that $\sqrt{-25} = 5i$. What are the solutions to the equation $x^2 = -25$?

4. Is 0 a complex number?

Simplify each of the following. See Example 2.

5. $\sqrt{-4}$

6. $\sqrt{-49}$

7. $\sqrt{-\dfrac{25}{9}}$

8. $\sqrt{-\dfrac{1}{81}}$

9. $\sqrt{-150}$

10. $\sqrt{-180}$

11. $\sqrt{-80}$

12. $\sqrt{-72}$

13. $\sqrt{-3} \cdot \sqrt{-3}$

14. $\sqrt{-2} \cdot \sqrt{-2}$

15. $\sqrt{-5} \cdot \sqrt{-6}$

16. $\sqrt{-27} \cdot \sqrt{-3}$

17. $\dfrac{\sqrt{-12}}{\sqrt{-8}}$

18. $\dfrac{\sqrt{-15}}{\sqrt{-3}}$

19. $\dfrac{\sqrt{-24}}{\sqrt{72}}$

20. $\dfrac{\sqrt{-27}}{\sqrt{9}}$

Solve the quadratic equation and express all complex solutions in terms of i. See Example 1.

21. $x^2 = -16$

22. $y^2 = -36$

23. $z^2 + 12 = 0$

24. $w^2 + 48 = 0$

25. $3x^2 + 4x + 2 = 0$

26. $2k^2 + 3k = -2$

27. $m^2 - 6m + 14 = 0$

28. $p^2 + 4p + 11 = 0$

29. $4z^2 = 4z - 7$

30. $9a^2 + 7 = 6a$

31. $m^2 + 1 = -m$

32. $y^2 = 2y - 2$

33. Explain how to determine whether a quadratic equation of the form $ax^2 + bx + c = 0$ has nonreal, complex solutions without actually solving the equation.

34. Explain how to determine whether a quadratic equation of the form $ax^2 + bx + c = 0$ has nonreal, complex solutions by looking at the graph of $y = ax^2 + bx + c$.

35. What is wrong with the following?

$$\sqrt{-5} \cdot \sqrt{-5} = \sqrt{-5(-5)}$$
$$= \sqrt{25}$$
$$= 5$$

After explaining what is wrong, evaluate the expression correctly.

36. Discuss the similarity between rationalizing the denominator in a fraction of the form $\dfrac{a + \sqrt{b}}{c + \sqrt{d}}$ and dividing two complex numbers.

Perform the operation and express all results in standard form. See Examples 3–5.

37. $(2 - 5i) + (3 + 2i)$ **38.** $(5 - i) + (3 + 4i)$ **39.** $(-2 + 3i) - (3 + i)$

40. $(4 + 6i) - (-2 - i)$ **41.** $(1 - i) - (5 - 2i)$ **42.** $(-2 + 6i) - (-3 - 8i)$

43. $(2 + i)(3 - 2i)$ **44.** $(-2 + 3i)(4 - 2i)$ **45.** $(2 + 4i)(-1 + 3i)$

46. $(1 + 3i)(2 - 5i)$ **47.** $(2 - i)(2 + i)$ **48.** $(5 + 4i)(5 - 4i)$

49. $\dfrac{5}{i}$ **50.** $\dfrac{-4}{i}$ **51.** $\dfrac{4 - 3i}{4 + 3i}$ **52.** $\dfrac{5 + 6i}{5 - 6i}$ **53.** $\dfrac{4 + i}{6 + 2i}$ **54.** $\dfrac{3 - 2i}{5 + 3i}$

55. A student makes the following statement. "I can simplify a large positive power of i by dividing the exponent by 4, and looking at the remainder. If the remainder is 0, it simplifies to 1, if the remainder is 1, it simplifies to i, if the remainder is 2, it simplifies to -1, and if the remainder is 3, it simplifies to $-i$." Explain why this statement is true.

56. Using the procedure of Exercise 55, why don't we have to consider getting a remainder of 4?

Simplify each power of i. See Example 6.

57. i^{12} **58.** i^{9} **59.** i^{18} **60.** i^{99} **61.** i^{-3} **62.** i^{-5} **63.** i^{-10} **64.** i^{-40}

65. Evaluate $1 + i + i^2 + i^3$.   **66.** Evaluate $1 + i + i^2 + \cdots + i^{100}$.

Two complex numbers $a + bi$ and $c + di$ are equal if and only if $a = c$ and $b = d$. Use this definition of equality of complex numbers to solve each of the following equations for x and y.

67. $2x + yi = 4 - 3i$ **68.** $x + 3yi = 5 + 2i$ **69.** $7 - 2yi = 14x - 30i$

70. $-5 + yi = x + 6i$ **71.** $x + yi = (2 + 3i)(4 - 2i)$ **72.** $x + yi = (5 - 7i)(1 + i)$

73. Explain why the method of dividing complex numbers (that is, multiplying both the numerator and the denominator by the conjugate of the denominator) works. That is, what property justifies this process?

74. What complex numbers are their own conjugates?

75. Show that the sum of a complex number and its conjugate is a real number.

76. Show that the product of a complex number and its conjugate is a real number.

77. Show that $\dfrac{\sqrt{2}}{2} + \dfrac{\sqrt{2}}{2}i$ is a square root of i.   **78.** Show that $-\dfrac{\sqrt{3}}{2} + \dfrac{1}{2}i$ is a cube root of i.

79. **Impedance** is a measure of the opposition to the flow of alternating electrical current found in common electrical outlets. It consists of two parts called the resistance and reactance. Resistance occurs when a light bulb is turned on, while reactance is produced when electricity passes through a coil of wire like that found in electric motors. Impedance Z in ohms (Ω) can be expressed as a complex number where the real part represents

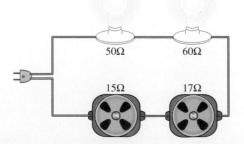

the resistance and the imaginary part represents the reactance. For example, if the resistive part is 3 ohms and the reactive part is 4 ohms, then the impedance could be described by the complex number $Z = 3 + 4i$. In the series circuit shown in the figure, the total impedance will be the sum of the individual impedances. (*Source:* Wilcox, G. and C. Hesselberth, *Electricity for Engineering Technology,* Allyn and Bacon, Inc., Boston, 1970.)

(a) The circuit contains two light bulbs and two electric motors. Assuming that the light bulbs are pure resistive and the motors are pure reactive, find the total impedance in this circuit and express it in the form $Z = a + bi$.

(b) The phase angle θ measures the phase difference between the voltage and the current in an electrical circuit. θ can be determined by the equation $\tan \theta = b/a$. Find θ for this circuit.

 In work with alternating current, complex numbers are used to describe current, I, voltage, E, and impedance, Z (the opposition to current). These three quantities are related by the equation $E = IZ$. Thus, if any two of these quantities are known, the third can be found. In each of the following problems, solve the equation $E = IZ$ for the missing variable.

80. $I = 8 + 6i$, $Z = 6 + 3i$

81. $I = 10 + 6i$, $Z = 8 + 5i$

82. $I = 7 + 5i$, $E = 28 + 54i$

83. $E = 35 + 55i$, $Z = 6 + 4i$

8.2 Trigonometric Form of Complex Numbers ▼▼▼

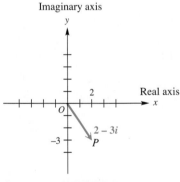

FIGURE 2

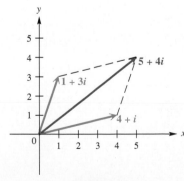

FIGURE 3

Unlike real numbers, complex numbers cannot be ordered. One way to organize and illustrate them is by using a graph. To graph a complex number such as $2 - 3i$, the familiar coordinate system must be modified. We do this by calling the horizontal axis the **real axis** and the vertical axis the **imaginary axis.** Then complex numbers can be graphed in this **complex plane,** as shown in Figure 2 for the complex number $2 - 3i$.

Each nonzero complex number graphed in this way determines a unique directed line segment, the segment from the origin to the point representing the complex number. Recall from Chapter 7 that such directed line segments (like **OP** of Figure 2) are called vectors.

The previous section showed how to find the sum of two complex numbers, such as $4 + i$ and $1 + 3i$.

$$(4 + i) + (1 + 3i) = 5 + 4i$$

Graphically, the sum of two complex numbers is represented by the vector that is the resultant of the vectors corresponding to the two numbers. The vectors representing the complex numbers $4 + i$ and $1 + 3i$ and the resultant vector that represents their sum, $5 + 4i$, are shown in Figure 3.

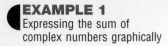

EXAMPLE 1
Expressing the sum of
complex numbers graphically

Find the sum of $6 - 2i$ and $-4 - 3i$. Graph both complex numbers and their resultant.

The sum is found by adding the two numbers.

$$(6 - 2i) + (-4 - 3i) = 2 - 5i$$

The graphs are shown in Figure 4. ▶

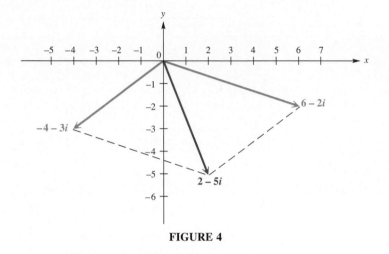

FIGURE 4

CONNECTIONS In Section 7.4 we saw that the vector **u** with its initial point at the origin and its endpoint at (a, b) could be designated $\langle a, b \rangle$. We then showed how to add and subtract vectors using this new notation. Now we see that the complex number $a + bi$ corresponds to the vector **u** described above. Thus, we have

$$\mathbf{u} = \langle a, b \rangle = a + bi$$

as three ways to designate a complex number or vector.

We can use addition of vectors in the form $\langle a, b \rangle$ to find the sum in Example 1.

$$(6 - 2i) + (-4 - 3i) = \langle 6, -2 \rangle + \langle -4, -3 \rangle$$
$$= \langle 2, -5 \rangle$$
$$= 2 - 5i$$

FOR DISCUSSION OR WRITING
1. Find $(6 - 2i) - (-4 - 3i)$ using vectors in the form $\langle a, b \rangle$. Then graph the given complex numbers and the difference.
2. Describe a general method for finding the difference of two vectors.

Figure 5 shows the complex number $x + yi$ that corresponds to a vector **OP** with direction angle θ and magnitude r. The following relationships among r, θ, x, and y can be verified from Figure 5.

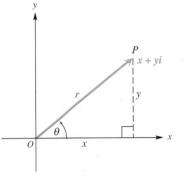

FIGURE 5

Relationships among x, y, r, and θ

$$x = r \cos \theta \qquad r = \sqrt{x^2 + y^2}$$

$$y = r \sin \theta \qquad \tan \theta = \frac{y}{x}, \text{ if } x \neq 0$$

Substituting $x = r \cos \theta$ and $y = r \sin \theta$ from the relationships given above into $x + yi$ gives

$$x + yi = r \cos \theta + (r \sin \theta)i$$
$$= r(\cos \theta + i \sin \theta).$$

Trigonometric or Polar Form of a Complex Number

The expression

$$r(\cos \theta + i \sin \theta)$$

is called the **trigonometric form** or **polar form** of the complex number $x + yi$. The expression $\cos \theta + i \sin \theta$ is sometimes abbreviated cis θ. Using this notation,

$$r(\cos \theta + i \sin \theta) \text{ is written as } r \text{ cis } \theta.$$

The number r is called the **modulus** or **absolute value** of $x + yi$, while θ is the **argument** of $x + yi$. In this section we will choose the value of θ in the interval $[0°, 360°)$. However, angles coterminal with such angles are also possible; that is, the argument for a particular complex number is not unique.

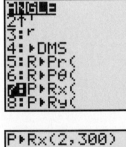

Choices 7 and 8 in the top screen show how a graphing calculator can convert from trigonometric (*polar*) form to rectangular (*x-y*) form. The screen at the bottom supports the result of Example 2. The calculator is in degree mode.

◖**EXAMPLE 2**
Converting from trigonometric form to standard form

Express $2(\cos 300° + i \sin 300°)$ in standard form.
Since $\cos 300° = 1/2$ and $\sin 300° = -\sqrt{3}/2$,

$$2(\cos 300° + i \sin 300°) = 2\left(\frac{1}{2} - i\frac{\sqrt{3}}{2}\right) = 1 - i\sqrt{3}. \quad ▶$$

NOTE   In the examples of this section, we will write arguments using degree measure. Arguments may also be written with radian measure.

In order to convert from standard form to trigonometric form, the following procedure is used.

> **Steps for Converting from Standard to Trigonometric Form**
>
> **1.** Sketch a graph of the number in the complex plane.
> **2.** Find r by using the equation $r = \sqrt{x^2 + y^2}$.
> **3.** Find θ by using the equation $\tan \theta = y/x$, $x \neq 0$, choosing the quadrant indicated in Step 1.

CAUTION   Errors often occur in Step 3 described above. Be sure that the correct quadrant for θ is chosen by referring to the graph sketched in Step 1.

EXAMPLE 3
Converting from standard form to trigonometric form

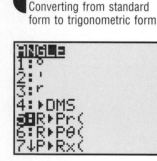

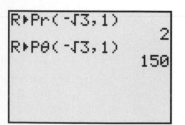

Choices 5 and 6 in the top screen show how a graphing calculator can convert from rectangular (*x-y*) form to trigonometric (*polar*) form. The screen at the bottom supports the result of Example 3(a). The calculator is in degree mode.

Write the following complex numbers in trigonometric form.

(a) $-\sqrt{3} + i$

Start by sketching the graph of $-\sqrt{3} + i$ in the complex plane, as shown in Figure 6. Next, find r. Since $x = -\sqrt{3}$ and $y = 1$,

$$r = \sqrt{x^2 + y^2} = \sqrt{(-\sqrt{3})^2 + 1^2} = \sqrt{3 + 1} = 2.$$

Then find θ.

$$\tan \theta = \frac{y}{x} = \frac{1}{-\sqrt{3}} = -\frac{\sqrt{3}}{3}$$

Since $\tan \theta = -\sqrt{3}/3$, the reference angle for θ is $30°$. From the sketch we see that θ is in quadrant II, so $\theta = 180° - 30° = 150°$. Therefore, in trigonometric form,

$$-\sqrt{3} + i = 2(\cos 150° + i \sin 150°)$$
$$= 2 \operatorname{cis} 150°.$$

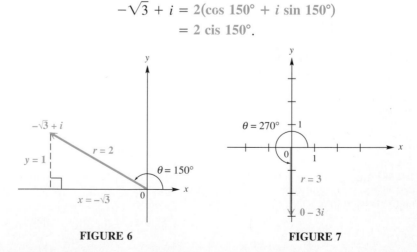

FIGURE 6   **FIGURE 7**

(b) $-3i$

The sketch of $-3i$ is shown in Figure 7. Since $-3i = 0 - 3i$, we have $x = 0$ and $y = -3$. Find r as follows.

$$r = \sqrt{0^2 + (-3)^2} = \sqrt{0 + 9} = \sqrt{9} = 3$$

We cannot find θ by using $\tan \theta = y/x$, since $x = 0$. In a case like this, refer to the graph and determine the argument directly from the sketch. A value for θ here is $270°$. In trigonometric form,

$$-3i = 3(\cos 270° + i \sin 270°)$$
$$= 3 \text{ cis } 270°. \quad \blacktriangleright$$

NOTE   In Examples 2 and 3 we gave answers in both forms: $r(\cos \theta + i \sin \theta)$ and r cis θ. These forms will be used interchangeably throughout the rest of this chapter.

In the next example we use a calculator to convert between trigonometric and standard forms.

◄ EXAMPLE 4
Converting between trigonometric and standard forms using a calculator

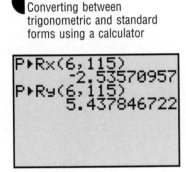

The result of Example 4(a) is supported in the screen above. The calculator is in degree mode.

(a) Write the complex number $6(\cos 115° + i \sin 115°)$ in standard form.

Since $115°$ does not have a special angle as a reference angle, we cannot find exact values for $\cos 115°$ and $\sin 115°$. Use a calculator set in the degree mode to find $\cos 115° \approx -.4226182617$ and $\sin 115° \approx .906307787$. Therefore, in standard form,

$$6(\cos 115° + i \sin 115°) \approx 6(-.4226182617 + .906307787i)$$
$$= -2.53570957 + 5.437846722i.$$

(b) Write $5 - 4i$ in trigonometric form.

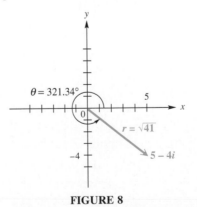

FIGURE 8

A sketch of $5 - 4i$ shows that θ must be in quadrant IV. See Figure 8. Here $r = \sqrt{5^2 + (-4)^2} = \sqrt{41}$ and $\tan \theta = -4/5$. Use a calculator to find that one measure of θ is $-38.66°$. In order to express θ in the interval $[0, 360°)$, we find that $\theta = 360° - 38.66° = 321.34°$. Use these results to get

$$5 - 4i = \sqrt{41} \text{ cis } 321.34°. \quad \blacktriangleright$$

The final example discusses a fractal, a relatively new application of complex numbers.

EXAMPLE 5
Deciding whether a complex number is in the Julia set

The fractal called the **Julia set** is shown in Figure 9. It can be created by graphing a special set of complex numbers. To determine if a complex number $z = a + bi$ is in this Julia set, perform the following sequence of calculations. Repeatedly compute the values of $z^2 - 1$, $(z^2 - 1)^2 - 1$, $[(z^2 - 1)^2 - 1]^2 - 1$, ... If the moduli of any of the resulting complex numbers exceeds 2, then the complex number z is not in the Julia set. Otherwise z is part of this set and the point (a, b) should be shaded in the graph. Determine whether the following numbers belong to the Julia set.*

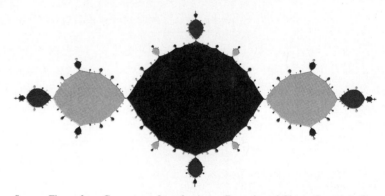

Source: Figure from Crownover: *Introduction to Fractals and Chaos*. Copyright © 1995 Boston: Jones and Bartlett Publisher. Reprinted with permission.

FIGURE 9

(a) $z = 0 + 0i$

Since $z = 0 + 0i = 0$, $z^2 - 1 = 0^2 - 1 = -1$,
$$(z^2 - 1)^2 - 1 = (-1)^2 - 1 = 0,$$
$$[(z^2 - 1)^2 - 1]^2 - 1 = 0^2 - 1 = -1,$$

and so on. We see that the calculations repeat as $0, -1, 0, -1$, and so on. The moduli are either 0 or 1, which do not exceed 2, so $0 + 0i$ is in the Julia set and the point $(0, 0)$ is part of the graph.

(b) $z = 1 + 1i$

We have $z^2 - 1 = (1 + i)^2 - 1 = (1 + 2i + i^2) - 1 = -1 + 2i$. The modulus is $\sqrt{(-1)^2 + 2^2} = \sqrt{5}$. Since $\sqrt{5}$ is greater than 2, $1 + 1i$ is not in the Julia set and $(1, 1)$ is not part of the graph. ▶

---

*Crownover, R., *Introduction to Fractals and Chaos,* Jones and Bartlett Publishers, Boston, 1995.

8.2 Exercises ▼▼▼▼▼▼▼▼▼▼▼▼▼▼▼▼▼▼▼▼▼▼▼▼▼▼▼▼▼▼▼▼▼▼▼

1. The modulus of a complex number represents the _____ of the vector representing it in the complex plane.

2. Describe geometrically the argument of a complex number.

Graph each complex number. See Example 1.

3. $-2 + 3i$ 4. $-4 + 5i$ 5. $8 - 5i$ 6. $6 - 5i$ 7. $2 - 2i\sqrt{3}$
8. $4\sqrt{2} + 4i\sqrt{2}$ 9. $-4i$ 10. $3i$ 11. -8 12. 2

13. What must be true in order for a complex number to also be a real number?

14. If a real number is graphed in the complex plane, on what axis does the vector lie?

Find the resultant of each pair of complex numbers. See Example 1.

15. $4 - 3i, -1 + 2i$ 16. $2 + 3i, -4 - i$ 17. $5 - 6i, -2 + 3i$ 18. $7 - 3i, -4 + 3i$
19. $-3, 3i$ 20. $6, -2i$ 21. $2 + 6i, -2i$ 22. $4 - 2i, 5$
23. $7 + 6i, 3i$ 24. $-5 - 8i, -1$

Write the complex number in standard form. See Example 2.

25. $2(\cos 45° + i \sin 45°)$ 26. $4(\cos 60° + i \sin 60°)$ 27. $10(\cos 90° + i \sin 90°)$
28. $8(\cos 270° + i \sin 270°)$ 29. $4(\cos 240° + i \sin 240°)$ 30. $2(\cos 330° + i \sin 330°)$
31. cis 30°   32. 3 cis 150°   33. 5 cis 300°
34. 6 cis 135°   35. $\sqrt{2}$ cis 180°   36. $\sqrt{3}$ cis 315°

Write the complex number in trigonometric form $r(\cos \theta + i \sin \theta)$, with θ in the interval $[0°, 360°)$. See Example 3.

37. $3 - 3i$ 38. $-2 + 2i\sqrt{3}$ 39. $-3 - 3i\sqrt{3}$ 40. $1 + i\sqrt{3}$ 41. $\sqrt{3} - i$
42. $4\sqrt{3} + 4i$ 43. $-5 - 5i$ 44. $-\sqrt{2} + i\sqrt{2}$ 45. $2 + 2i$ 46. $-\sqrt{3} + i$
47. -4 48. $5i$ 49. $-2i$ 50. 7

Perform the conversion, using a calculator as necessary. See Example 4.

	Standard Form	Trigonometric Form
51.	$2 + 3i$	_____
52.	_____	$(\cos 35° + i \sin 35°)$
53.	_____	$3(\cos 250° + i \sin 250°)$
54.	$-4 + i$	_____
55.	$12i$	_____
56.	_____	3 cis 180°
57.	$3 + 5i$	_____
58.	_____	cis 110.5°

The complex number z, where $z = x + yi$, can be graphed in the plane as (x, y). Describe the graphs of all complex numbers z satisfying the conditions in Exercises 59–62.

59. The modulus of z is 1.   60. The real and imaginary parts of z are equal.

61. The real part of z is 1.   62. The imaginary part of z is 1.

In Exercises 63 and 64, suppose $z = r(\cos \theta + i \sin \theta)$.

63. Use vectors in order to show that the conjugate of $z = r(\cos(360° - \theta) + i \sin(360° - \theta)) = r(\cos \theta - i \sin \theta)$.

64. Use vectors to show that $-z = r(\cos(\theta + \pi) + i \sin(\theta + \pi))$.

65. Use vectors to show that the sum of a complex number and its conjugate is a real number.

In Exercises 66–68 identify which geometric condition (A, B, or C) implies the situation.

A. The corresponding vectors have opposite directions.
B. The terminal points of the vectors corresponding to $a + bi$ and $c + di$ lie on a horizontal line.
C. The corresponding vectors have the same direction.

66. The difference between two nonreal complex numbers $a + bi$ and $c + di$ is a real number.

67. The modulus of the sum of two complex numbers $a + bi$ and $c + di$ is equal to the sum of their moduli.

68. The modulus of the difference of two complex numbers $a + bi$ and $c + di$ is equal to the sum of their moduli.

69. Refer to Example 5. Is $z = -.2i$ in the Julia set?

70. Refer to Example 5. The graph of the Julia set in Figure 9 appears to be symmetric with respect to both the x-axis and y-axis. Complete the following to show that this is true.

(a) Show that complex conjugates have the same modulus.
(b) Compute $z_1^2 - 1$ and $z_2^2 - 1$ where $z_1 = a + bi$ and $z_2 = a - bi$.
(c) Discuss why if (a, b) is in the Julia set then so is $(a, -b)$.
(d) Conclude that the graph of the Julia set must be symmetric with respect to the x-axis.
(e) Using a similar argument, show that the Julia set must also be symmetric with respect to the y-axis.

8.3 Product and Quotient Theorems ▼▼▼

The product of the two complex numbers $1 + i\sqrt{3}$ and $-2\sqrt{3} + 2i$ can be found by the method shown in Section 8.1.

$$(1 + i\sqrt{3})(-2\sqrt{3} + 2i) = -2\sqrt{3} + 2i - 2i(3) + 2i^2\sqrt{3}$$
$$= -2\sqrt{3} + 2i - 6i - 2\sqrt{3}$$
$$= -4\sqrt{3} - 4i$$

This same product also can be found by first converting the complex numbers $1 + i\sqrt{3}$ and $-2\sqrt{3} + 2i$ to trigonometric form. Using the method explained in the previous section,

$$1 + i\sqrt{3} = 2(\cos 60° + i \sin 60°)$$
and
$$-2\sqrt{3} + 2i = 4(\cos 150° + i \sin 150°).$$

If the trigonometric forms are now multiplied together and if the trigonometric identities for the cosine and the sine of the sum of two angles are used, the result is

$$[2(\cos 60° + i \sin 60°)][4(\cos 150° + i \sin 150°)]$$
$$= 2 \cdot 4(\cos 60° \cdot \cos 150° + i \sin 60° \cdot \cos 150°$$
$$+ i \cos 60° \cdot \sin 150° + i^2 \sin 60° \cdot \sin 150°)$$
$$= 8[(\cos 60° \cdot \cos 150° - \sin 60° \cdot \sin 150°)$$
$$+ i(\sin 60° \cdot \cos 150° + \cos 60° \cdot \sin 150°)]$$
$$= 8[\cos(60° + 150°) + i \sin(60° + 150°)]$$
$$= 8(\cos 210° + i \sin 210°).$$

The modulus of the product, 8, is equal to the product of the moduli of the factors, $2 \cdot 4$, while the argument of the product, 210°, is the sum of the arguments of the factors, $60° + 150°$.

As we would expect, the product obtained upon multiplying by the first method is the standard form of the product obtained upon multiplying by the second method.

$$8(\cos 210° + i \sin 210°) = 8\left(-\frac{\sqrt{3}}{2} - \frac{1}{2}i\right)$$
$$= -4\sqrt{3} - 4i$$

Generalizing, the product of the two complex numbers, $r_1(\cos \theta_1 + i \sin \theta_1)$ and $r_2(\cos \theta_2 + i \sin \theta_2)$, is

$$[r_1(\cos \theta_1 + i \sin \theta_1)] \cdot [r_2(\cos \theta_2 + i \sin \theta_2)]$$
$$= r_1 r_2(\cos \theta_1 \cos \theta_2 + i \sin \theta_1 \cos \theta_2 + i \cos \theta_1 \sin \theta_2 + i^2 \sin \theta_1 \sin \theta_2)$$
$$= r_1 r_2[(\cos \theta_1 \cos \theta_2 - \sin \theta_1 \sin \theta_2) + i(\sin \theta_1 \cos \theta_2 + \cos \theta_1 \sin \theta_2)]$$
$$= r_1 r_2[\cos(\theta_1 + \theta_2) + i \sin(\theta_1 + \theta_2)].$$

This work is summarized in the following *product theorem.*

Product Theorem

If $r_1(\cos \theta_1 + i \sin \theta_1)$ and $r_2(\cos \theta_2 + i \sin \theta_2)$ are any two complex numbers, then

$$[r_1(\cos \theta_1 + i \sin \theta_1)] \cdot [r_2(\cos \theta_2 + i \sin \theta_2)]$$
$$= r_1 r_2[\cos(\theta_1 + \theta_2) + i \sin(\theta_1 + \theta_2)].$$

In compact form, this is written

$$(r_1 \operatorname{cis} \theta_1)(r_2 \operatorname{cis} \theta_2) = r_1 r_2 \operatorname{cis}(\theta_1 + \theta_2).$$

EXAMPLE 1
Using the product theorem

Find the product of $3(\cos 45° + i \sin 45°)$ and $2(\cos 135° + i \sin 135°)$.

    Using the product theorem,

$$[3(\cos 45° + i \sin 45°)][2(\cos 135° + i \sin 135°)]$$
$$= 3 \cdot 2[\cos(45° + 135°) + i \sin(45° + 135°)]$$
$$= 6(\cos 180° + i \sin 180°),$$

which can be expressed as $6(-1 + i \cdot 0) = 6(-1) = -6$. The two complex numbers in this example are complex factors of -6. ▶

    Using the method shown in Section 8.1, in standard form the quotient of the complex numbers $1 + i\sqrt{3}$ and $-2\sqrt{3} + 2i$ is

$$\frac{1 + i\sqrt{3}}{-2\sqrt{3} + 2i} = \frac{(1 + i\sqrt{3})(-2\sqrt{3} - 2i)}{(-2\sqrt{3} + 2i)(-2\sqrt{3} - 2i)}$$
$$= \frac{-2\sqrt{3} - 2i - 6i - 2i^2\sqrt{3}}{12 - 4i^2}$$
$$= \frac{-8i}{16} = -\frac{1}{2}i.$$

Writing $1 + i\sqrt{3}$, $-2\sqrt{3} + 2i$, and $-\frac{1}{2}i$ in trigonometric form gives

$$1 + i\sqrt{3} = 2(\cos 60° + i \sin 60°)$$
$$-2\sqrt{3} + 2i = 4(\cos 150° + i \sin 150°)$$
$$-\frac{1}{2}i = \frac{1}{2}(\cos(-90°) + i \sin(-90°)).$$

The modulus of the quotient, $1/2$, is the quotient of the two moduli, 2 and 4. The argument of the quotient, $-90°$, is the difference of the two arguments, $60° - 150° = -90°$. It would be easier to find the quotient of these two complex numbers in trigonometric form than in standard form. Generalizing from this example leads to another theorem. The proof is similar to the proof of the product theorem, after the numerator and denominator are multiplied by the conjugate of the denominator.

Quotient Theorem

If $r_1(\cos \theta_1 + i \sin \theta_1)$ and $r_2(\cos \theta_2 + i \sin \theta_2)$ are complex numbers, where $r_2(\cos \theta_2 + i \sin \theta_2) \neq 0$, then

$$\frac{r_1(\cos \theta_1 + i \sin \theta_1)}{r_2(\cos \theta_2 + i \sin \theta_2)} = \frac{r_1}{r_2}[\cos(\theta_1 - \theta_2) + i \sin(\theta_1 - \theta_2)].$$

In compact form, this is written

$$\frac{r_1 \operatorname{cis} \theta_1}{r_2 \operatorname{cis} \theta_2} = \frac{r_1}{r_2} \operatorname{cis}(\theta_1 - \theta_2).$$

EXAMPLE 2
Using the quotient theorem

Find the quotient

$$\frac{10 \text{ cis}(-60°)}{5 \text{ cis } 150°}.$$

Write the result in standard form.
By the quotient theorem,

$$\frac{10 \text{ cis}(-60°)}{5 \text{ cis } 150°} = \frac{10}{5} \text{ cis}(-60° - 150°) \qquad \text{Quotient theorem}$$

$$= 2 \text{ cis}(-210°) \qquad \text{Subtract.}$$

$$= 2[\cos(-210°) + i \sin(-210°)]$$

$$= 2\left[-\frac{\sqrt{3}}{2} + i\left(\frac{1}{2}\right)\right] \qquad \cos(-210°) = -\frac{\sqrt{3}}{2};$$
$$\qquad\qquad\qquad\qquad\qquad\qquad \sin(-210°) = \frac{1}{2}$$

$$= -\sqrt{3} + i \qquad \text{Standard form} \blacktriangleright$$

EXAMPLE 3
Using the product and quotient
theorems with a calculator

Use a calculator to find the following. Write the results in standard form.

(a) $(9.3 \text{ cis } 125.2°)(2.7 \text{ cis } 49.8°)$
By the product theorem,

$$(9.3 \text{ cis } 125.2°)(2.7 \text{ cis } 49.8°) = (9.3)(2.7) \text{ cis}(125.2° + 49.8°)$$

$$= 25.11 \text{ cis } 175°$$

$$= 25.11(\cos 175° + i \sin 175°)$$

$$= 25.11(-.99619470 + i(.08715574))$$

$$= -25.014449 + 2.1884807i.$$

(b) $$\dfrac{10.42\left(\cos \dfrac{3\pi}{4} + i \sin \dfrac{3\pi}{4}\right)}{5.21\left(\cos \dfrac{\pi}{5} + i \sin \dfrac{\pi}{5}\right)}$$

Use the quotient theorem.

$$\frac{10.42\left(\cos \dfrac{3\pi}{4} + i \sin \dfrac{3\pi}{4}\right)}{5.21\left(\cos \dfrac{\pi}{5} + i \sin \dfrac{\pi}{5}\right)} = \frac{10.42}{5.21}\left[\cos\left(\frac{3\pi}{4} - \frac{\pi}{5}\right) + i \sin\left(\frac{3\pi}{4} - \frac{\pi}{5}\right)\right]$$

$$= 2\left(\cos \frac{11\pi}{20} + i \sin \frac{11\pi}{20}\right)$$

$$= -.31286893 + 1.9753767i \blacktriangleright$$

8.3 Exercises ▼▼

1. Figure 3 of Section 8.2 shows a geometric method for finding the sum of two complex numbers. Suppose you are given a similar figure containing two blue arrows corresponding to complex numbers (without the actual complex numbers given) and are asked to draw the red arrow corresponding to their product. Explain how you would proceed if you had a protractor and a ruler.

2. Answer the problem in Exercise 1 for the quotient of two complex numbers.

Find each product. Write each product in standard form. See Example 1.

3. $[3(\cos 60° + i \sin 60°)][2(\cos 90° + i \sin 90°)]$

4. $[4(\cos 30° + i \sin 30°)][5(\cos 120° + i \sin 120°)]$

5. $[2(\cos 45° + i \sin 45°)][2(\cos 225° + i \sin 225°)]$

6. $[8(\cos 300° + i \sin 300°)][5(\cos 120° + i \sin 120°)]$

7. $[4(\cos 60° + i \sin 60°)][6(\cos 330° + i \sin 330°)]$

8. $[8(\cos 210° + i \sin 210°)][2(\cos 330° + i \sin 330°)]$

9. $[5 \text{ cis } 90°][3 \text{ cis } 45°]$

10. $[6 \text{ cis } 120°][5 \text{ cis}(-30°)]$

11. $[\sqrt{3} \text{ cis } 45°][\sqrt{3} \text{ cis } 225°]$

12. $[\sqrt{2} \text{ cis } 300°][\sqrt{2} \text{ cis } 270°]$

Find each quotient. Write each quotient in standard form. In Exercises 19–24, first convert the numerator and the denominator to trigonometric form. See Example 2.

13. $\dfrac{4(\cos 120° + i \sin 120°)}{2(\cos 150° + i \sin 150°)}$

14. $\dfrac{10(\cos 225° + i \sin 225°)}{5(\cos 45° + i \sin 45°)}$

15. $\dfrac{16(\cos 300° + i \sin 300°)}{8(\cos 60° + i \sin 60°)}$

16. $\dfrac{24(\cos 150° + i \sin 150°)}{2(\cos 30° + i \sin 30°)}$

17. $\dfrac{3 \text{ cis } 305°}{9 \text{ cis } 65°}$

18. $\dfrac{12 \text{ cis } 293°}{6 \text{ cis } 23°}$

19. $\dfrac{8}{\sqrt{3} + i}$

20. $\dfrac{2i}{-1 - i\sqrt{3}}$

21. $\dfrac{-i}{1 + i}$

22. $\dfrac{1}{2 - 2i}$

23. $\dfrac{2\sqrt{6} - 2i\sqrt{2}}{\sqrt{2} - i\sqrt{6}}$

24. $\dfrac{4 + 4i}{2 - 2i}$

Use a calculator to perform the indicated operations. Give answers in standard form. See Example 3.

25. $[2.5(\cos 35° + i \sin 35°)][3.0(\cos 50° + i \sin 50°)]$

26. $[4.6(\cos 12° + i \sin 12°)][2.0(\cos 13° + i \sin 13°)]$

27. $(12 \text{ cis } 18.5°)(3 \text{ cis } 12.5°)$

28. $(4 \text{ cis } 19.25°)(7 \text{ cis } 41.75°)$

29. $\dfrac{45(\cos 127° + i \sin 127°)}{22.5(\cos 43° + i \sin 43°)}$

30. $\dfrac{30(\cos 130° + i \sin 130°)}{10(\cos 21° + i \sin 21°)}$

31. $\left[2 \text{ cis } \dfrac{5\pi}{9} \right]^2$

32. $\left[24.3 \left(\cos \dfrac{7\pi}{12} + i \sin \dfrac{7\pi}{12} \right) \right]^2$

▼▼▼▼▼▼▼▼▼▼▼▼▼▼ **DISCOVERING CONNECTIONS** (Exercises 33–39) ▼▼▼▼▼▼▼▼▼▼▼▼▼▼

Consider the complex numbers

$$w = -1 + i \quad and \quad z = -1 - i.$$

33. Multiply w and z using their standard forms and the "FOIL" method from Section 8.1. Leave the product in standard form.

34. Find the trigonometric forms of w and z.

35. Multiply w and z using their trigonometric forms and the method described in this section.

36. Use the result of Exercise 35 to find the standard form of wz. How does this compare to your result in Exercise 33?

37. Find the quotient w/z using their standard forms and multiplying both the numerator and the denominator by the conjugate of the denominator. Leave the quotient in standard form.

38. Use the trigonometric forms of w and z, found in Exercise 34, to divide w by z using the method described in this section.

39. Use the result of Exercise 38 to find the standard form of w/z. How does this compare to your result in Exercise 37?

---

40. Show that $1/z = (1/r)(\cos \theta - i \sin \theta)$, where $z = r(\cos \theta + i \sin \theta)$.

41. Without actually performing the operations, state why the products

$$[2(\cos 45° + i \sin 45°)] \cdot [5(\cos 90° + i \sin 90°)]$$

and

$$[2(\cos(-315°) + i \sin(-315°))] \cdot [5(\cos(-270°) + i \sin(-270°))]$$

are the same.

42. Notice that $(r \operatorname{cis} \theta)^2 = (r \operatorname{cis} \theta)(r \operatorname{cis} \theta) = r^2 \operatorname{cis}(\theta + \theta) = r^2 \operatorname{cis} 2\theta$. State in your own words how we can square a complex number in trigonometric form. (In the next section, we will develop this idea more fully.)

---

43. The alternating current in an electric inductor is

$$I = \frac{E}{Z}$$

amperes, where E is the voltage and $Z = R + X_L i$ is the impedance. If $E = 8(\cos 20° + i \sin 20°)$, $R = 6$, and $X_L = 3$, find the current. Give the answer in standard form.

44. The current I in a circuit with voltage E, resistance R, capacitive reactance X_c, and inductive reactance X_L is

$$I = \frac{E}{R + (X_L - X_c)i}.$$

Find I if $E = 12(\cos 25° + i \sin 25°)$, $R = 3$, $X_L = 4$, and $X_c = 6$. Give the answer in standard form.

45. Refer to Exercise 79 of Section 8.1. In the parallel electrical circuit shown in the figure here, the impedance Z can be calculated using the equation

$$Z = \frac{1}{\dfrac{1}{Z_1} + \dfrac{1}{Z_2}}, \text{ where } Z_1 \text{ and } Z_2 \text{ are the impedances}$$

for the branches of the circuit.

(a) If $Z_1 = 50 + 25i$ and $Z_2 = 60 + 20i$, calculate Z.

(b) Determine the phase angle θ.

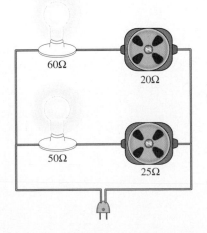

8.4 Powers and Roots of Complex Numbers ▼▼▼

In Section 8.3 we studied the product and quotient theorems for complex numbers in trigonometric form. Because raising a number to a positive integer power is a repeated application of the product rule, it would seem likely that a theorem for finding powers of complex numbers exists. This is indeed the case. For example, the square of the complex number $r(\cos \theta + i \sin \theta)$ is

$$[r(\cos \theta + i \sin \theta)]^2 = [r(\cos \theta + i \sin \theta)][r(\cos \theta + i \sin \theta)]$$
$$= r \cdot r[\cos(\theta + \theta) + i \sin(\theta + \theta)]$$
$$= r^2(\cos 2\theta + i \sin 2\theta).$$

In the same way,

$$[r(\cos \theta + i \sin \theta)]^3 = r^3(\cos 3\theta + i \sin 3\theta).$$

These results suggest the plausibility of the following theorem for positive integer values of n. Although the following theorem is stated and can be proved for all n, we will use it only for positive integer values of n and their reciprocals.

De Moivre's Theorem

If $r(\cos \theta + i \sin \theta)$ is a complex number and if n is any real number, then

$$[r(\cos \theta + i \sin \theta)]^n = r^n(\cos n\theta + i \sin n\theta).$$

In compact form, this is written

$$[r \operatorname{cis} \theta]^n = r^n(\operatorname{cis} n\theta).$$

This theorem is named after the French expatriate friend of Isaac Newton, Abraham De Moivre (1667–1754), although he never explicitly stated it.

EXAMPLE 1
Applying De Moivre's theorem (finding a power of a complex number)

Find $(1 + i\sqrt{3})^8$ and express the result in standard form.

To use De Moivre's theorem, first convert $1 + i\sqrt{3}$ into trigonometric form using the methods of Section 8.2.

$$1 + i\sqrt{3} = 2(\cos 60° + i \sin 60°)$$

Now apply De Moivre's theorem.

$$(1 + i\sqrt{3})^8 = [2(\cos 60° + i \sin 60°)]^8$$
$$= 2^8[\cos (8 \cdot 60°) + i \sin (8 \cdot 60°)]$$
$$= 256(\cos 480° + i \sin 480°)$$
$$= 256(\cos 120° + i \sin 120°) \qquad \text{480° and 120° are coterminal.}$$
$$= 256\left(-\frac{1}{2} + i\frac{\sqrt{3}}{2}\right) \qquad \cos 120° = -\frac{1}{2};$$
$$\qquad\qquad\qquad\qquad\qquad \sin 120° = \frac{\sqrt{3}}{2}$$
$$= -128 + 128i\sqrt{3} \qquad \text{Standard form} \blacktriangleright$$

In algebra it is shown that every nonzero complex number has exactly n distinct complex nth roots. De Moivre's theorem can be extended to find all nth roots of a complex number. An nth root of a complex number is defined as follows.

---

nth Root

For a positive integer n, the complex number $a + bi$ is an **nth root** of the complex number $x + yi$ if

$$(a + bi)^n = x + yi.$$

To find the cube roots of the complex number $8(\cos 135° + i \sin 135°)$, for example, look for a complex number, say $r(\cos \alpha + i \sin \alpha)$, that will satisfy

$$[r(\cos \alpha + i \sin \alpha)]^3 = 8(\cos 135° + i \sin 135°).$$

By De Moivre's theorem, this equation becomes

$$r^3(\cos 3\alpha + i \sin 3\alpha) = 8(\cos 135° + i \sin 135°).$$

One way to satisfy this equation is to set $r^3 = 8$ and also $\cos 3\alpha + i \sin 3\alpha = \cos 135° + i \sin 135°$. The first of these conditions implies that $r = 2$, and the second implies that

$$\cos 3\alpha = \cos 135° \quad \text{and} \quad \sin 3\alpha = \sin 135°.$$

For these equations to be satisfied, 3α must represent an angle that is coterminal with $135°$. Therefore, we must have

$$3\alpha = 135° + 360° \cdot k, \quad k \text{ any integer,}$$

or

$$\alpha = \frac{135° + 360° \cdot k}{3}, \quad k \text{ any integer.}$$

Now let k take on the integer values 0, 1, and 2.

If $k = 0$,
$$\alpha = \frac{135° + 0°}{3} = 45°.$$

If $k = 1$,
$$\alpha = \frac{135° + 360°}{3} = \frac{495°}{3} = 165°.$$

If $k = 2$,
$$\alpha = \frac{135° + 720°}{3} = \frac{855°}{3} = 285°.$$

In the same way, $\alpha = 405°$ when $k = 3$. But note that $\sin 405° = \sin 45°$ and $\cos 405° = \cos 45°$. If $k = 4$, $\alpha = 525°$ which has the same sine and cosine values as $165°$. To continue with larger values of k would just be repeating solutions already found. Therefore, all of the cube roots (three of them) can be found by letting $k = 0$, 1, and 2.

When $k = 0$, the root is

$$2(\cos 45° + i \sin 45°).$$

When $k = 1$, the root is

$$2(\cos 165° + i \sin 165°).$$

When $k = 2$, the root is

$$2(\cos 285° + i \sin 285°).$$

In conclusion, we see that $2(\cos 45° + i \sin 45°)$, $2(\cos 165° + i \sin 165°)$, and $2(\cos 285° + i \sin 285°)$ are the three cube roots of $8(\cos 135° + i \sin 135°)$.

Notice that the formula for α in the discussion above can be written in an alternative form as

$$\alpha = \frac{135°}{3} + \frac{360° \cdot k}{3} = 45° + 120° \cdot k,$$

for $k = 0$, 1, and 2, which is easier to use.

Generalizing the work above leads to the following theorem.

---

nth Root Theorem

If n is any positive integer and r is a positive real number, then the complex number $r(\cos \theta + i \sin \theta)$ has exactly n distinct nth roots, given by

$$\sqrt[n]{r}(\cos \alpha + i \sin \alpha) \qquad \text{or} \qquad \sqrt[n]{r} \text{ cis } \alpha,$$

where

$$\alpha = \frac{\theta + 360° \cdot k}{n} \qquad \text{or} \qquad \alpha = \frac{\theta}{n} + \frac{360° \cdot k}{n},$$

$k = 0, 1, 2, \ldots, n - 1.$

EXAMPLE 2
Finding roots of a complex number

Find all fourth roots of $-8 + 8i\sqrt{3}$. Write the roots in standard form.

First write $-8 + 8i\sqrt{3}$ in trigonometric form as

$$-8 + 8i\sqrt{3} = 16 \text{ cis } 120°.$$

Here $r = 16$ and $\theta = 120°$. The fourth roots of this number have modulus $\sqrt[4]{16} = 2$ and arguments given as follows. Using the alternative formula for α,

$$\alpha = \frac{120°}{4} + \frac{360° \cdot k}{4} = 30° + 90° \cdot k.$$

If $k = 0$, $\alpha = 30° + 90° \cdot 0 = 30°.$

If $k = 1$, $\alpha = 30° + 90° \cdot 1 = 120°.$

If $k = 2$, $\alpha = 30° + 90° \cdot 2 = 210°.$

If $k = 3$, $\alpha = 30° + 90° \cdot 3 = 300°.$

Using these angles, the fourth roots are

$$2 \text{ cis } 30°,$$
$$2 \text{ cis } 120°,$$
$$2 \text{ cis } 210°,$$

and
$$2 \text{ cis } 300°.$$

These four roots can be written in standard form as $\sqrt{3} + i$, $-1 + i\sqrt{3}$, $-\sqrt{3} - i$, and $1 - i\sqrt{3}$. The graphs of these roots are all on a circle that has center at the origin and radius 2, as shown in Figure 10. Notice that the roots are equally spaced about the circle 90° apart. ▶

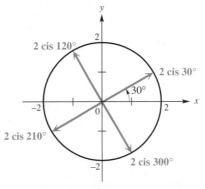

FIGURE 10

◀ **EXAMPLE 3**
Solving an equation by finding complex roots

Find all complex number solutions of $x^5 - 1 = 0$.
Write the equation as

$$x^5 - 1 = 0$$
$$x^5 = 1.$$

While there is only one real number solution, 1, there are five complex number solutions. To find these solutions, first write 1 in trigonometric form as

$$1 = 1 + 0i = 1(\cos 0° + i \sin 0°).$$

The modulus of the fifth roots is $\sqrt[5]{1} = 1$, and the arguments are given by

$$0° + 72° \cdot k, \qquad k = 0, 1, 2, 3, \text{ or } 4.$$

By using these arguments, the fifth roots are

$$1(\cos 0° + i \sin 0°), \qquad k = 0$$
$$1(\cos 72° + i \sin 72°), \qquad k = 1$$
$$1(\cos 144° + i \sin 144°), \qquad k = 2$$
$$1(\cos 216° + i \sin 216°), \qquad k = 3$$

and
$$1(\cos 288° + i \sin 288°). \qquad k = 4$$

The first of these roots equals 1; the others cannot easily be expressed in standard form. The five fifth roots all lie on a unit circle and are equally spaced around it every 72°, as shown in Figure 11. ▶

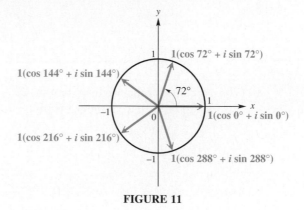

FIGURE 11

8.4 Exercises ▼▼▼▼▼▼▼▼▼▼▼▼▼▼▼▼▼▼▼▼▼▼▼▼▼▼▼▼▼▼▼▼▼

Find each power. Write each answer in standard form. See Example 1.

1. $[3(\cos 30° + i \sin 30°)]^3$

2. $[2(\cos 135° + i \sin 135°)]^4$

3. $(\cos 45° + i \sin 45°)^8$

4. $[2(\cos 120° + i \sin 120°)]^3$

5. $[3 \text{ cis } 100°]^3$

6. $[3 \text{ cis } 40°]^3$

7. $(\sqrt{3} + i)^5$

8. $(2\sqrt{2} - 2i\sqrt{2})^6$

9. $(2 - 2i\sqrt{3})^4$

10. $\left(\dfrac{\sqrt{2}}{2} - \dfrac{\sqrt{2}}{2}i\right)^8$

11. $(-2 - 2i)^5$

12. $(-1 + i)^7$

Find and graph all cube roots of each complex number. Leave answers in trigonometric form. See Example 2.

13. $(\cos 0° + i \sin 0°)$

14. $(\cos 90° + i \sin 90°)$

15. $8 \text{ cis } 60°$

16. $27 \text{ cis } 300°$

17. $-8i$

18. $27i$

19. -64

20. 27

21. $1 + i\sqrt{3}$

22. $2 - 2i\sqrt{3}$

23. $-2\sqrt{3} + 2i$

24. $\sqrt{3} - i$

Find and graph all the specified roots of 1.

25. Second (square)

26. Fourth

27. Sixth

28. Eighth

Find and graph all the specified roots of i.

29. Second (square)

30. Fourth

31. Explain why a real number must have a real *n*th root if *n* is odd.

32. How many complex 64th roots does 1 have? How many are real? How many are not?

33. True or false: Every real number must have two real square roots.

34. True or false: Some real numbers have three real cube roots.

35. Explain why a real number can have only one real cube root.

36. Explain why the n nth roots of 1 are equally spaced around the unit circle.

37. Refer to Figure 11. A regular pentagon can be created by joining the tips of the arrows. Explain how you can use this principle to create a regular octagon.

38. Show that if z is an nth root of 1, then so is $1/z$.

Find all solutions of each equation. Leave answers in trigonometric form. See Example 3.

39. $x^3 - 1 = 0$ **40.** $x^3 + 1 = 0$ **41.** $x^3 + i = 0$ **42.** $x^4 + i = 0$

43. $x^3 - 8 = 0$ **44.** $x^3 + 27 = 0$ **45.** $x^4 + 1 = 0$ **46.** $x^4 + 16 = 0$

47. $x^4 - i = 0$ **48.** $x^5 - i = 0$ **49.** $x^3 - (4 + 4i\sqrt{3}) = 0$ **50.** $x^4 - (8 + 8i\sqrt{3}) = 0$

Use a calculator to find all solutions of each equation in standard form.

51. $x^3 + 4 - 5i = 0$ **52.** $x^5 + 2 + 3i = 0$

53. Solve the equation $x^3 - 1 = 0$ by factoring the left side as the difference of two cubes and setting each factor equal to zero. Apply the quadratic formula as needed. Then compare your solutions to those of Exercise 39.

54. Solve the equation $x^3 + 27 = 0$ by factoring the left side as the sum of two cubes and setting each factor equal to zero. Apply the quadratic formula as needed. Then compare your solutions to those of Exercise 44.

55. One of the three cube roots of a complex number is $2 + 2\sqrt{3}i$. Determine the standard form of its other two cube roots.

▼▼▼▼▼▼▼▼▼▼▼▼▼ **DISCOVERING CONNECTIONS** (Exercises 56–58) ▼▼▼▼▼▼▼▼▼▼▼▼▼

56. De Moivre's theorem states that $(\cos \theta + i \sin \theta)^2 = $ _____ .

57. Expand the left side of the equation in Exercise 56 as a binomial and collect terms to write the left side in the form $a + bi$.

58. Use the equality principle stated in the directions to Exercises 67–72 of Section 8.1 to obtain the double-angle formulas for the cosine and sine.

59. Set your graphing calculator to degree and parametric modes, and set the window and functions as shown below. (*Note:* 72 is 360/5.) Graph to see the pentagon in the graphing calculator screen below whose corners are the five 5th roots of one in the complex plane. Trace to find the coordinates of these points and thereby determine the five 5th roots of one. Compare your answers with those found in Example 3.

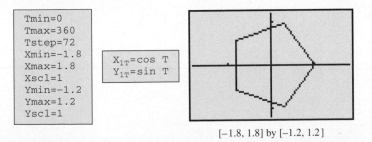

```
Tmin=0
Tmax=360
Tstep=72
Xmin=-1.8        X₁ᴛ=cos T
Xmax=1.8         Y₁ᴛ=sin T
Xscl=1
Ymin=-1.2
Ymax=1.2
Yscl=1
```

[−1.8, 1.8] by [−1.2, 1.2]

60. Use the method of Exercise 59 to find the first three of the ten 10th roots of one.

61. 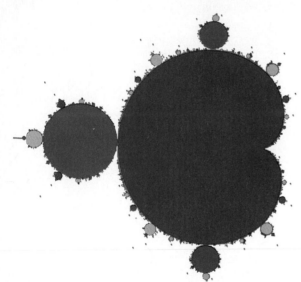 The fractal called the **Mandelbrot set** is shown in the figure. To determine if a complex number $z = a + bi$ is in this set, perform the following sequence of calculations. Repeatedly compute

$$z, \; z^2 + z, \; (z^2 + z)^2 + z, \; [(z^2 + z)^2 + z]^2 + z, \; \ldots$$

In a manner analogous to the Julia set, the complex number z does not belong to the Mandelbrot set if any of the resulting moduli exceed 2. Otherwise z is in the set and the point (a, b) should be shaded in the graph. Determine whether or not the following numbers belong to the Mandelbrot set. (*Source:* Lauwerier, H., *Fractals,* Princeton University Press, Princeton, New Jersey, 1991.)

(a) $z = 0 + 0i$ **(b)** $z = 1 - 1i$ **(c)** $z = -.5i$

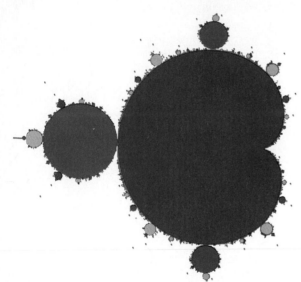

Source: Figure from Crownover: *Introduction to Fractals and Chaos.* Copyright © 1995. Boston: Jones and Bartlett Publisher. Reprinted with permission.

62. The fractal shown in the figure is the solution to Cayley's problem of determining the basins of attraction for the cube roots of unity. The three cube roots of unity are

$$w_1 = 1, \; w_2 = -\frac{1}{2} + \frac{\sqrt{3}}{2}i, \text{ and } w_3 = -\frac{1}{2} - \frac{\sqrt{3}}{2}i.$$

This fractal can be generated by repeatedly evaluating the function $f(z) = \dfrac{2z^3 + 1}{3z^2}$ where z is a complex number. One begins by picking $z_1 = a + bi$ and then successively computing $z_2 = f(z_1)$, $z_3 = f(z_2)$, $z_4 = f(z_3)$, ... If the resulting values of $f(z)$ approach w_1, color the pixel at (a, b) red. If it approaches w_2, color it blue, and if it approaches w_3, color it yellow. If this process continues for a large number of different z_1, the fractal in the figure will appear. Determine the appropriate color of the pixel for each value of z_1. (*Source:* Crownover, R., *Introduction to Fractals and Chaos,* Jones and Bartlett Publishers, Boston, 1995.)

(a) $z_1 = i$ **(b)** $z_1 = 2 + i$ **(c)** $z_1 = -1 - i$

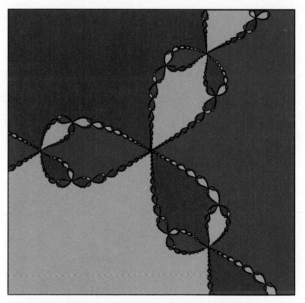

Source: Kincaid, D. and W. Cheney, *Numerical Analysis,* Brooks/Cole Publishing Company, Pacific Grove, California, 1991.

8.5 Polar Equations ▼▼▼

Pole

Polar axis

FIGURE 12

Throughout this text we have been using the Cartesian coordinate system to graph equations. Another coordinate system that is particularly useful for graphing many relations is the **polar coordinate system.** The system is based on a point, called the **pole,** and a ray, called the **polar axis.** The polar axis is usually drawn in the direction of the positive *x*-axis, as shown in Figure 12.

In Figure 13 the pole has been placed at the origin of a Cartesian coordinate system, so that the polar axis coincides with the positive *x*-axis. Point *P* has coordinates (*x*, *y*) in the Cartesian coordinate system. Point *P* can also be located by giving the directed angle θ from the positive *x*-axis to ray *OP* and the directed distance *r* from the pole to point *P*. The ordered pair (*r*, θ) gives the **polar coordinates** of point *P*.

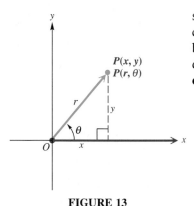

FIGURE 13

◀ **EXAMPLE 1**
Graphing points with polar coordinates

Plot each point, given its polar coordinates.

(a) *P*(2, 30°)

In this case, *r* = 2 and θ = 30°, so the point *P* is located 2 units from the origin in the positive direction on a ray making a 30° angle with the polar axis, as shown in Figure 14.

(b) *Q*(−4, 120°)

Since *r* is negative, *Q* is 4 units in the negative direction from the pole on an extension of the 120° ray. See Figure 15.

(c) *R*(5, −45°)

Point *R* is shown in Figure 16. Since θ is negative, the angle is measured in the clockwise direction. ▶

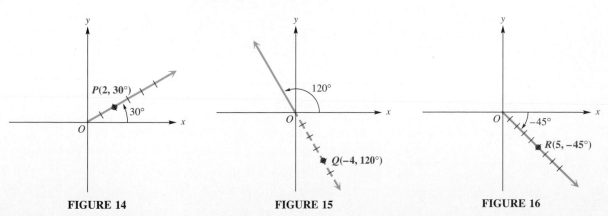

FIGURE 14   **FIGURE 15**   **FIGURE 16**

One important difference between Cartesian coordinates and polar coordinates is that while a given point in the plane can have only one pair of Cartesian coordinates, this same point can have an infinite number of pairs of polar coordinates. For example, (2, 30°) locates the same point as (2, 390°) or (2, −330°) or (−2, 210°).

EXAMPLE 2
Giving alternative forms of a pair of polar coordinates

Give three other pairs of polar coordinates for the point $P(3, 140°)$.

Three pairs that could be used for the point are $(3, -220°)$, $(-3, 320°)$, and $(-3, -40°)$. See Figure 17. ▶

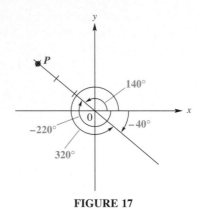

FIGURE 17

An equation like $r = 3 \sin \theta$ where r and θ are the variables, is a **polar equation.** (Equations in x and y are called **rectangular** or **Cartesian equations.**) The simplest equation for many useful curves turns out to be a polar equation.

Graphing a polar equation is much the same as graphing a Cartesian equation. Find some representative ordered pairs, (r, θ), satisfying the equation, and then sketch the graph.

EXAMPLE 3
Graphing a polar equation (cardioid)

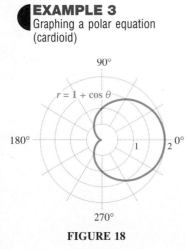

FIGURE 18

Graph $r = 1 + \cos \theta$.

To graph this equation, find some ordered pairs (as in the table) and then connect the points in order—from (2, 0°) to (1.9, 30°) to (1.7, 45°) and so on. The graph is shown in Figure 18. This curve is called a **cardioid** because of its heart shape.

θ	0°	30°	45°	60°	90°	120°	135°	150°	180°	270°	315°
$\cos \theta$	1	.9	.7	.5	0	−.5	−.7	−.9	−1	0	.7
$r = 1 + \cos \theta$	2	1.9	1.7	1.5	1	.5	.3	.1	0	1	1.7

Once the pattern of values of r becomes clear, it is not necessary to find more ordered pairs. That is why we stopped with the ordered pair (1.7, 315°) in the table above. From the pattern, the pair (1.9, 330°) also would satisfy the relation. ▶

A graphing calculator can be used to graph an equation in the form $r = f(\theta)$. Refer to your owner's manual to see how your model handles polar graphs. As always it is necessary to set the window appropriately and choose the correct angle mode (radians or degrees). You will need to decide on maximum and minimum values of θ. Keep in mind the periods of the functions, so that the entire set of function values are generated. For example, to graph $r = 1 + \cos \theta$, which we graphed by hand in Example 3, we use degree mode. We might choose the intervals $[0, 360°]$, $[-3, 3]$, and $[-3, 3]$ for θ, x, and y, respectively, using the results of our work in Example 3 to determine these choices. The calculator-generated graph and the corresponding window is shown below. (The second window is a continuation of the first one.)

You can experiment with your calculator by graphing the curves shown in other examples.

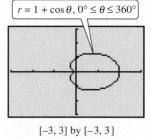

$r = 1 + \cos \theta, 0° \leq \theta \leq 360°$

$[-3, 3]$ by $[-3, 3]$

EXAMPLE 4
Graphing a polar equation
(lemniscate)

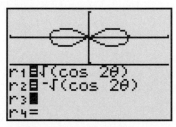

r1=√(cos 2θ)
r2=-√(cos 2θ)
r3=
r4=

$[-2, 2]$ by $[-1, 1]$

To graph $r^2 = \cos 2\theta$ with a graphing calculator, define r_1 as $\sqrt{\cos 2\theta}$ and r_2 as $-\sqrt{\cos 2\theta}$. Compare to the graph in Figure 19.

Graph $r^2 = \cos 2\theta$.

First complete a table of ordered pairs as shown, and then sketch the graph, as in Figure 19. The point $(-1, 0°)$, with r negative, may be plotted as $(1, 180°)$. Also, $(-.7, 30°)$ may be plotted as $(.7, 210°)$, and so on. This curve is called a **lemniscate.**

θ	0°	30°	45°	135°	150°	180°
2θ	0°	60°	90°	270°	300°	360°
$\cos 2\theta$	1	.5	0	0	.5	1
$r = \pm\sqrt{\cos 2\theta}$	±1	±.7	0	0	±.7	±1

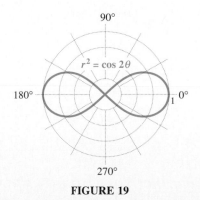

$r^2 = \cos 2\theta$

FIGURE 19

Values of θ for $45° < \theta < 135°$ are not included in the table because the corresponding values of cos 2θ are negative (quadrants II and III) and so do not have real square roots. Values of θ larger than $180°$ give 2θ larger than $360°$, and would repeat the points already found. ▶

EXAMPLE 5
Graphing a polar equation
(rose)

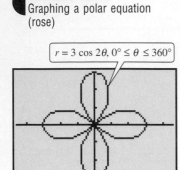

$r = 3 \cos 2\theta,\ 0° \leq \theta \leq 360°$

[−4.6, 4.6] by [−3, 3]

Compare this calculator-generated graph to the one in Figure 20.

Graph $r = 3 \cos 2\theta$.

Because of the 2θ, the graph requires a large number of points. A few ordered pairs are given below. You should complete the table similarly through the first $360°$.

θ	0°	15°	30°	45°	60°	75°	90°
2θ	0°	30°	60°	90°	120°	150°	180°
$\cos 2\theta$	1	.9	.5	0	−.5	−.9	−1
r	3	2.7	1.5	0	−1.5	−2.7	−3

Plotting these points in order gives the graph, called a **four-leaved rose.** Notice in Figure 20 how the graph is developed with a continuous curve, beginning with the upper half of the right horizontal leaf and ending with the lower half of that leaf. As the graph is traced, the curve goes through the pole four times. ▶

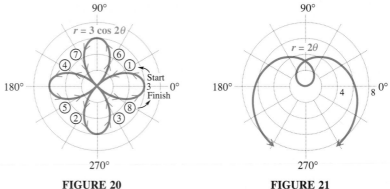

FIGURE 20   **FIGURE 21**

The graph in Figure 20 is one of a family of curves called **roses.** The graphs of $r = \sin n\theta$ and $r = \cos n\theta$ are roses, with n petals if n is odd, and $2n$ petals if n is even.

EXAMPLE 6
Graphing a polar equation
(spiral of Archimedes)

Graph $r = 2\theta$ (θ measured in radians).

Some ordered pairs are shown below. Since $r = 2\theta$, rather than a trigonometric function of θ, it is also necessary to consider negative values of θ. The radian measures have been rounded for simplicity.

θ (degrees)	−180	−90	−45	0	30	60	90	180	270	360
θ (radians)	−3.1	−1.6	−.8	0	.5	1	1.6	3.1	4.7	6.3
$r = 2\theta$	−6.2	−3.2	−1.6	0	1	2	3.2	6.2	9.4	12.6

Figure 21 shows this graph, called a **spiral of Archimedes.** ▶

It is quite tedious to graph the spiral of Archimedes using traditional methods. Much more of the graph is shown in the calculator-generated graph below.

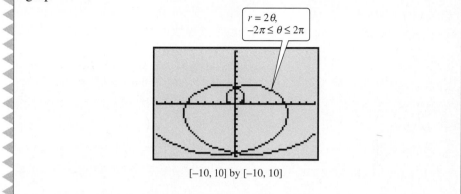

$$r = 2\theta, \quad -2\pi \le \theta \le 2\pi$$

[−10, 10] by [−10, 10]

Sometimes an equation given in polar form is easier to graph in Cartesian form. To convert a polar equation to a Cartesian equation, we use the following relationships, which were introduced in Section 8.2. See triangle *POQ* in Figure 22.

FIGURE 22

Converting Between Polar and Rectangular Coordinates

$$x = r \cos \theta \qquad r = \sqrt{x^2 + y^2}$$

$$y = r \sin \theta \qquad \tan \theta = \frac{y}{x}, \text{ if } x \ne 0$$

◀EXAMPLE 7
Converting a polar equation to a Cartesian equation

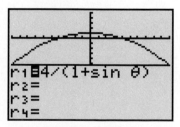

[−10, 10] by [−10, 10]
Xscl = 1   Yscl = 2

This screen shows a calculator-generated graph of the equation in Example 7, using the polar graphing mode. See Exercise 35 for more on this graph and its relationship to that of $x^2 = -8(y - 2)$.

Convert the equation

$$r = \frac{4}{1 + \sin \theta}$$

to Cartesian coordinates, and graph.

Multiply both sides of the equation by the denominator on the right, to clear the fraction.

$$r = \frac{4}{1 + \sin \theta}$$

$$r + r \sin \theta = 4$$

Now substitute $\sqrt{x^2 + y^2}$ for r and y for $r \sin \theta$.

$$\sqrt{x^2 + y^2} + y = 4$$

$$\sqrt{x^2 + y^2} = 4 - y$$

Square both sides to eliminate the radical.

$$x^2 + y^2 = (4 - y)^2$$
$$x^2 + y^2 = 16 - 8y + y^2$$
$$x^2 = -8y + 16$$
$$x^2 = -8(y - 2)$$

The final equation represents a parabola and can be graphed using rectangular coordinates. See Figure 23. ◗

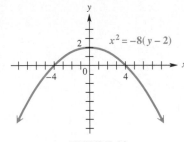

FIGURE 23

> ◗ **EXAMPLE 8**
> Converting a Cartesian
> equation to a polar equation

Convert the equation $3x + 2y = 4$ to a polar equation.

Use $x = r \cos \theta$ and $y = r \sin \theta$ to get

$$3x + 2y = 4$$
$$3r \cos \theta + 2r \sin \theta = 4.$$

Now solve for r. First factor out r on the left.

$$r(3 \cos \theta + 2 \sin \theta) = 4$$

$$r = \frac{4}{3 \cos \theta + 2 \sin \theta}$$

The polar equation of the line $3x + 2y = 4$ is

$$r = \frac{4}{3 \cos \theta + 2 \sin \theta}. \quad ◗$$

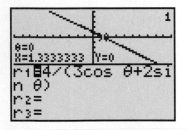

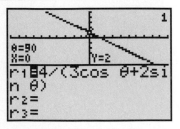

[−10, 10] by [−10, 10]
Xscl = 1   Yscl = 2

The screens here indicate that the
x- and y-intercepts of the line are
$\frac{4}{3}$ and 2, respectively, further
supporting the result of Example 8.
The rectangular form is $3x + 2y = 4$.

8.5 Exercises ▼▼▼▼▼▼▼▼▼▼▼▼▼▼▼▼▼▼▼▼▼▼▼▼▼▼▼▼▼▼▼▼▼▼▼▼

1. Explain why the ordered pairs (r, θ) and $(r, \theta + 360°)$ have the same graph.

2. Explain why, if $r > 0$, the ordered pairs (r, θ) and $(-r, \theta + 180°)$ have the same graph.

Plot each point, given its polar coordinates. Give two other pairs of polar coordinates for each point. See Examples 1 and 2.

3. $(1, 45°)$ 　　　　4. $(3, 120°)$ 　　　　5. $(-2, 135°)$ 　　　　6. $(-4, 27°)$ 　　　　7. $(5, -60°)$

8. $(2, -45°)$ 　　　　9. $(-3, -210°)$ 　　　　10. $(-1, -120°)$ 　　　　11. $(3, 300°)$ 　　　　12. $(4, 270°)$

13. If a point lies on an axis in the Cartesian plane, then what kind of angle must θ be if (r, θ) represents the point in polar coordinates?

14. What will the graph of $r = k$ be, for $k > 0$?

Graph the equation for θ in $[0°, 360°)$. See Examples 3–6.

15. $r = 2 + 2 \cos \theta$ （cardioid）

16. $r = 2(4 + 3 \cos \theta)$ （cardioid）

17. $r = 3 + \cos \theta$ （limaçon）

18. $r = 2 - \cos \theta$ （limaçon）

19. $r = 4 \cos 2\theta$ （four-leaved rose）

20. $r = 3 \cos 5\theta$ （five-leaved rose）

21. $r^2 = 4 \cos 2\theta$ （lemniscate）

22. $r^2 = 4 \sin 2\theta$ （lemniscate）

23. $r = 4(1 - \cos \theta)$ （cardioid）

24. $r = 3(2 - \cos \theta)$ （cardioid）

25. $r = 2 \sin \theta \tan \theta$ （cissoid）

26. $r = \dfrac{\cos 2\theta}{\cos \theta}$ （cissoid with a loop）

In Exercises 27–30 identify the geometric symmetry (A, B, or C) that the graph will possess.

A. Symmetry with respect to the origin

B. Symmetry with respect to the y-axis

C. Symmetry with respect to the x-axis

27. Whenever (r, θ) is on the graph, then so is $(-r, -\theta)$.

28. Whenever (r, θ) is on the graph, then so is $(-r, \theta)$.

29. Whenever (r, θ) is on the graph, then so is $(r, -\theta)$.

30. Whenever (r, θ) is on the graph, then so is $(r, \pi - \theta)$.

31. Graph the equations $r = 4 \sin \theta$ and $r = 4 \cos \theta$. How are they the same and how do they differ? Describe the graphs of $r = a \sin \theta$ and $r = a \cos \theta$ for positive a.

32. Graph the equations $r = 4(1 + \cos \theta)$, $r = 4(1 - \cos \theta)$, $r = 4(1 + \sin \theta)$, and $r = 4(1 - \sin \theta)$. How are they the same and how do they differ? Describe the graphs of $r = a(1 + \cos \theta)$, $r = a(1 - \cos \theta)$, $r = a(1 + \sin \theta)$, and $r = a(1 - \sin \theta)$ for positive a.

In Exercises 33 and 34, find the greatest value of $|r|$ of any point on the graph. Also, find all values of θ for which $r = 0$.

33. $r = 4 \cos 2\theta,\ 0° \leq \theta < 360°$

34. $r = 5 \sin 3\theta,\ 0° \leq \theta < 180°$

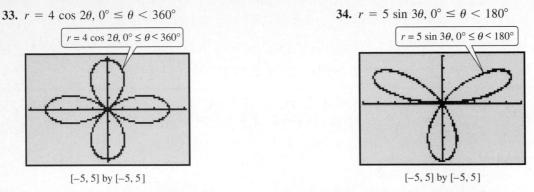

$r = 4 \cos 2\theta,\ 0° \leq \theta < 360°$

$[-5, 5]$ by $[-5, 5]$

$r = 5 \sin 3\theta,\ 0° \leq \theta < 180°$

$[-5, 5]$ by $[-5, 5]$

35. The screen displays below indicate the same point. Verify analytically that the polar coordinates shown in the left screen and the rectangular coordinates shown in the right screen are equivalent.

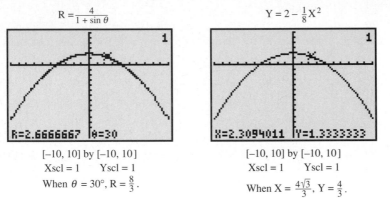

$R = \dfrac{4}{1 + \sin \theta}$

$Y = 2 - \dfrac{1}{8}X^2$

R=2.6666667   θ=30

X=2.3094011   Y=1.3333333

$[-10, 10]$ by $[-10, 10]$

Xscl = 1   Yscl = 1

When $\theta = 30°$, $R = \dfrac{8}{3}$.

$[-10, 10]$ by $[-10, 10]$

Xscl = 1   Yscl = 1

When $X = \dfrac{4\sqrt{3}}{3}$, $Y = \dfrac{4}{3}$.

Find the polar coordinates of the points of intersection of the given curves for the specified interval of θ.

36. $r = 4 \sin \theta,\ r = 1 + 2 \sin \theta;\quad 0 \leq \theta < 2\pi$

37. $r = 2 + \sin \theta,\ r = 2 + \cos \theta;\quad 0 \leq \theta < 2\pi$

38. $r = \sin 2\theta,\ r = \sqrt{2} \cos \theta;\quad 0 \leq \theta < \pi$

39. Explain the method used to plot a point (r, θ) in polar coordinates, if $r < 0$.

40. Refer to Example 8. Would you find it easier to graph the equation using the Cartesian form or the polar form? Why?

For each equation, find an equivalent equation in Cartesian coordinates, and sketch the graph. See Example 7.

41. $r = 2 \sin \theta$

42. $r = 2 \cos \theta$

43. $r = \dfrac{2}{1 - \cos \theta}$

44. $r = \dfrac{3}{1 - \sin \theta}$

45. $r + 2 \cos \theta = -2 \sin \theta$

46. $r = \dfrac{3}{4 \cos \theta - \sin \theta}$

47. $r = 2 \sec \theta$

48. $r = -5 \csc \theta$

49. $r(\cos \theta + \sin \theta) = 2$

50. $r(2 \cos \theta + \sin \theta) = 2$

For each equation, find an equivalent equation in polar coordinates. See Example 8.

51. $x + y = 4$

52. $2x - y = 5$

53. $x^2 + y^2 = 16$

54. $x^2 + y^2 = 9$

55. $y = 2$

56. $x = 4$

57. Graph $r = \theta$, a spiral of Archimedes. (See Example 6.) Use both positive and nonpositive values for θ.

58. Show that the distance between (r_1, θ_1) and (r_2, θ_2) is $\sqrt{r_1^2 + r_2^2 - 2r_1r_2 \cos(\theta_1 - \theta_2)}$.

In Exercises 59 and 60, write a polar equation of the line through the given points.

59. $(1, 0°), (2, 90°)$

60. $(2, 30°), (1, 90°)$

61. The polar equation $r = \dfrac{a(1 - e^2)}{1 + e \cos \theta}$ can be used to graph the orbits of the planets where a is the average distance in astronomical units from the sun and e is a constant called the eccentricity. The sun will be located at the pole. The table lists a and e for the planets. (*Sources:* Karttunen, H., P. Kröger, H. Oja, M. Putannen, and K. Donners (editors), *Fundamental Astronomy,* Springer-Verlag, 1994; Zeilik, M., S. Gregory, and E. Smith, *Introductory Astronomy and Astrophysics,* Saunders College Publishers, 1992.)

(a) Graph the orbits of the four planets closest to the sun on the same polar axis. Choose a viewing rectangle that results in a graph with nearly circular orbits.

(b) Plot the orbits of Earth, Jupiter, Uranus, and Pluto on the same polar axis. How does Earth's distance from the sun compare to these planets?

(c) Use graphing to determine whether or not Pluto is always the farthest planet from the sun.

Planet	a	e
Mercury	.39	.206
Venus	.78	.007
Earth	1.00	.017
Mars	1.52	.093
Jupiter	5.20	.048
Saturn	9.54	.056
Uranus	19.2	.047
Neptune	30.1	.009
Pluto	39.4	.249

8.6 Parametric Equations ▼▼▼

Throughout this text, we have graphed sets of ordered pairs of real numbers that corresponded to a function of the form $y = f(x)$ or $r = g(\theta)$. Another way to determine a set of ordered pairs involves two functions f and g defined by $x = f(t)$ and $y = g(t)$, where t is a real number in some interval I. Each value of t leads to a corresponding x-value and a corresponding y-value, and thus to an ordered pair (x, y).

> **Parametric Equations of a Plane Curve**
>
> A **plane curve** is a set of points (x, y) such that $x = f(t)$, $y = g(t)$, and f and g are both defined on an interval I. The equations $x = f(t)$ and $y = g(t)$ are **parametric equations** with **parameter t.**

EXAMPLE 1
Graphing a plane curve defined parametrically

Let $x = t^2$ and $y = 2t + 3$ for t in $[-3, 3]$. Graph the set of ordered pairs (x, y). Begin by making a table of values.

t	-3	-2	-1	0	1	2	3
x	9	4	1	0	1	4	9
y	-3	-1	1	3	5	7	9

Now graph the points (x, y) from the table of values and connect them with a smooth curve as in Figure 24. Since the domain of t is a closed interval, the graph has endpoints at $(9, -3)$ and $(9, 9)$. ▶

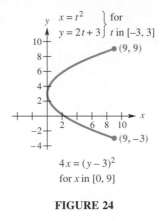

FIGURE 24

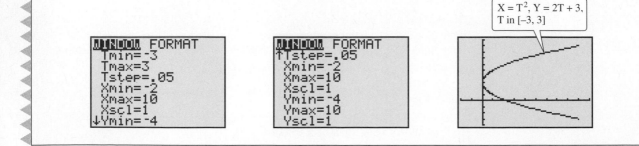

In addition to graphing rectangular and polar equations, graphing calculators are capable of graphing plane curves defined by parametric equations. The calculator must be set in parametric mode, and the window requires intervals for the parameter T, as well as for X and Y. The window and graph below are for the graph of $X = T^2$ and $Y = 2T + 3$, with T in $[-3, 3]$, as in Example 1.

Sometimes it is possible to eliminate the parameter from a pair of parametric equations to get a **rectangular equation,** an equation relating x and y.

EXAMPLE 2
Finding an equivalent rectangular equation

Find a rectangular equation for the plane curve defined as follows, and graph the curve.

$$x = t^2, \; y = 2t + 3, \text{ for } t \text{ in } [-3, 3]$$

This is the curve of Example 1. To eliminate the parameter t, solve either equation for t. Here, only the second equation, $y = 2t + 3$, leads to a unique solution for t, so choose it.

$$y = 2t + 3$$
$$2t = y - 3$$
$$t = \frac{y - 3}{2}$$

Now substitute the result in the first equation to get

$$x = t^2 = \left(\frac{y - 3}{2}\right)^2 = \frac{(y - 3)^2}{4}$$

or

$$4x = (y - 3)^2.$$

This is the equation of a horizontal parabola opening to the right, which agrees with the graph given in Figure 24. Because t is in $[-3, 3]$, x is in $[0, 9]$, and y is in $[-3, 9]$. The rectangular equation must be given with its restricted domain as

$$4x = (y - 3)^2, \quad \text{for } x \text{ in } [0, 9]. \; \blacktriangleright$$

Trigonometric functions are often used to define a plane curve parametrically.

EXAMPLE 3
Graphing a plane curve defined parametrically with trigonometric functions

Graph the plane curve defined by $x = 2 \sin t$, $y = 3 \cos t$, for t in $[0, 2\pi]$.

It is awkward to solve either equation for t. Instead, we use the fact that $\sin^2 t + \cos^2 t = 1$ to apply another approach. Square both sides of each equation; solve one for $\sin^2 t$, the other for $\cos^2 t$.

$$x = 2 \sin t \qquad y = 3 \cos t$$
$$x^2 = 4 \sin^2 t \qquad y^2 = 9 \cos^2 t$$
$$\frac{x^2}{4} = \sin^2 t \qquad \frac{y^2}{9} = \cos^2 t$$

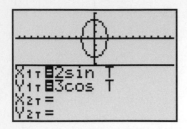

[−12.4, 12.4] by [−4, 4]

Compare this graph to the one in Figure 25. It is an ellipse with x-intercepts ± 2 and y-intercepts ± 3.

Now add corresponding sides of the two equations to get

$$\frac{x^2}{4} + \frac{y^2}{9} = \sin^2 t + \cos^2 t$$

$$\frac{x^2}{4} + \frac{y^2}{9} = 1,$$

the equation of an ellipse with vertical major axis, as shown in Figure 25. ▶

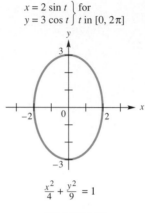

$\left. \begin{array}{l} x = 2 \sin t \\ y = 3 \cos t \end{array} \right\}$ for t in $[0, 2\pi]$

$$\frac{x^2}{4} + \frac{y^2}{9} = 1$$

FIGURE 25

Parametric representations of a curve are not unique. In fact, there are infinitely many parametric representations of a given curve. If the curve can be described by a rectangular equation $y = f(x)$, with domain X, then one simple parametric representation is

$$x = t, \ y = f(t), \ \text{for } t \text{ in } X.$$

EXAMPLE 4
Finding alternative parametric equation forms

Give three parametric representations for the parabola

$$y = (x - 2)^2 + 1.$$

The simplest choice is to let

$$x = t, \ y = (t - 2)^2 + 1, \quad \text{for } t \text{ in } (-\infty, \infty).$$

Another choice, that leads to a simpler equation for y is

$$x = t + 2, \ y = t^2 + 1, \quad \text{for } t \text{ in } (-\infty, \infty).$$

Sometimes trigonometric functions are desirable; one choice here might be

$$x = 2 + \tan t, \ y = \sec^2 t, \quad \text{for } t \text{ in } \left(-\frac{\pi}{2}, \frac{\pi}{2}\right). \quad ▶$$

An important application of parametric equations is to determine the path of a moving object whose position is given by the functions $x = f(t)$, $y = g(t)$, where t represents time. The parametric equations give the position of the object at any time t.

EXAMPLE 5
Examining parametric equations defining the position of an object in motion

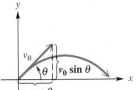

FIGURE 26

The motion of a projectile (neglecting air resistance) is given by

$$x = (v_0 \cos \theta)t, \; y = (v_0 \sin \theta)t - 16t^2, \quad \text{for } t \text{ in } [0, k],$$

where t is time in seconds, v_0 is the initial speed of the projectile in the direction θ with the horizontal, x and y are in feet, and k is a positive real number. See Figure 26.

Solving the first equation for t and substituting the result into the second equation gives (after simplification)

$$y = (\tan \theta)x - \frac{16}{v_0{}^2 \cos^2 \theta}x^2,$$

the equation of a vertical parabola opening downward, as shown in Figure 26. ▶

The path traced by a fixed point on the circumference of a circle rolling along a line is called a *cycloid*. See Figure 27. The **cycloid** is defined by

$$x = at - a \sin t, \; y = a - a \cos t, \quad \text{for } t \text{ in } (-\infty, \infty).$$

EXAMPLE 6
Graphing a cycloid

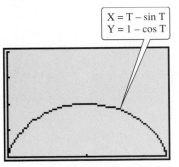

$[0, 2\pi]$ by $[0, 4]$
$\text{Xscl} = \pi \quad \text{Yscl} = 1$

This screen shows a cycloid with $a = 1$.

Graph the cycloid for t in $[0, 2\pi]$.

There is no simple way to find a rectangular equation for the cycloid from its parametric equations. Instead, begin with a table of values.

t	0	$\dfrac{\pi}{4}$	$\dfrac{\pi}{2}$	π	$\dfrac{3\pi}{2}$	2π
x	0	$.08a$	$.6a$	πa	$5.7a$	$2\pi a$
y	0	$.3a$	a	$2a$	a	0

Plotting the ordered pairs (x, y) from the table of values leads to the portion of the graph in Figure 27 from 0 to $2\pi a$. ▶

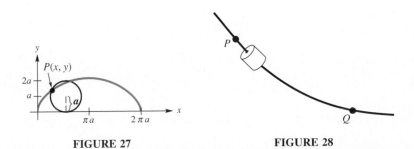

FIGURE 27   **FIGURE 28**

The cycloid has an interesting physical property. If a flexible cord or wire goes through points P and Q as in Figure 28, and a bead is allowed to slide without friction along this path from P to Q, the path that requires the shortest time takes the shape of the graph of an inverted cycloid.

8.6 Exercises ▼▼▼▼▼▼▼▼▼▼▼▼▼▼▼▼▼▼▼▼▼▼▼▼▼▼▼▼▼▼▼▼▼▼▼▼▼▼▼

1. The graphing calculator screen at the right below shows the graph of the parametric equations X = −2 + 6T, Y = 1 + 2T, T in [0, 1]. What is the first point graphed (corresponding to T = 0) and the last point graphed (corresponding to T = 1) as the curve is traced out? What point is plotted when T = .6?

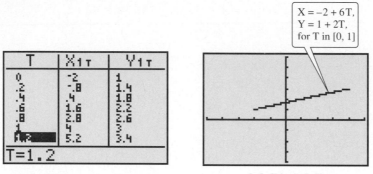

[−5, 5] by [−5, 5]

2. The parametric equations X = 4 − 6T, Y = 3 − 2T, T in [0, 1] have the same graph as in Exercise 1. What is the first point graphed (corresponding to T = 0) and the last point graphed (corresponding to T = 1) as the curve is traced out? What point is plotted when T = .6?

3. The parametric equations $x = \cos t$, $y = \sin t$, t in $[0, 2\pi]$ and the parametric equations $x = \cos t$, $y = -\sin t$, t in $[0, 2\pi]$ both have the unit circle as their graph. However, in one case the circle is traced out clockwise (as t moves from 0 to 2π) and in the other case the circle is traced out counterclockwise. For which equations is the circle traced out in the clockwise direction?

4. Consider the parametric equations $x = f(t)$, $y = g(t)$, t in $[a, b]$. How is the graph affected if the equation $x = f(t)$ is replaced by $x = c + f(t)$? How is the graph affected if the equation $y = g(t)$ is replaced by $y = d + g(t)$?

Use a table of values to graph each plane curve defined by the following parametric equations. Find a rectangular equation for each curve. See Examples 1 and 2.

5. $x = 2t$, $y = t + 1$, for t in $[-2, 3]$

6. $x = t + 2$, $y = t^2$, for t in $[-1, 1]$

7. $x = \sqrt{t}$, $y = 3t - 4$, for t in $[0, 4]$

8. $x = t^2$, $y = \sqrt{t}$, for t in $[0, 4]$

9. $x = t^3 + 1$, $y = t^3 - 1$, for t in $(-\infty, \infty)$

10. $x = 2t - 1$, $y = t^2 + 2$, for t in $(-\infty, \infty)$

11. $x = 2 \sin \theta$, $y = 2 \cos \theta$, for θ in $[0, 2\pi]$

12. $x = \sqrt{5} \sin \theta$, $y = \sqrt{3} \cos \theta$, for θ in $[0, 2\pi]$

13. $x = 3 \tan \theta$, $y = 2 \sec \theta$, for θ in $\left(-\dfrac{\pi}{2}, \dfrac{\pi}{2}\right)$

14. $x = \cot \theta$, $y = \csc \theta$, for θ in $(0, \pi)$

Find a rectangular equation for each curve defined as follows and graph the curve. See Examples 1 and 2.

15. $x = \sin \theta$, $y = \csc \theta$, for θ in $(0, \pi)$

16. $x = \tan \theta$, $y = \cot \theta$, for θ in $\left(0, \dfrac{\pi}{2}\right)$

17. $x = t$, $y = \sqrt{t^2 + 2}$, for t in $(-\infty, \infty)$

18. $x = \sqrt{t}$, $y = t^2 - 1$, for t in $[0, \infty)$

19. $x = 2 + \sin \theta$, $y = 1 + \cos \theta$, for θ in $[0, 2\pi]$

20. $x = 1 + 2 \sin \theta$, $y = 2 + 3 \cos \theta$, for θ in $[0, 2\pi]$

21. $x = t + 2$, $y = \dfrac{1}{t + 2}$, $t \neq -2$

22. $x = t - 3$, $y = \dfrac{2}{t - 3}$, $t \neq 3$

Graph each curve defined in Exercises 23 and 24. Assume the interval for t is all real numbers for which $x - f(t)$ and $y = g(t)$ are both defined. See Examples 2 and 3.

23. (a) $x = \sin t$, $y = \cos t$ **(b)** $x = t$, $y = \dfrac{\sqrt{4 - 4t^2}}{2}$

24. (a) $x = t + 2$, $y = t - 4$ **(b)** $x = t^2 + 2$, $y = t^2 - 4$

Graph each cycloid defined in Exercises 25 and 26 for θ in the given interval. See Example 6.

25. $x = \theta - \sin \theta$, $y = 1 - \cos \theta$, θ in $[0, 4\pi]$

26. $x = 2\theta - 2 \sin \theta$, $y = 2 - 2 \cos \theta$, θ in $[0, 8\pi]$

27. A projectile is fired with an initial velocity of 400 ft per sec at an angle of $45°$ with the horizontal. Find each of the following. See Example 5.
   (a) the time when it strikes the ground   **(b)** the range (horizontal distance covered)
   (c) the maximum altitude

28. Repeat Exercise 27 if the projectile is fired at 800 ft per sec at an angle of $30°$ with the horizontal.

29. Show that the rectangular equation for the curve describing the motion of a projectile defined by

$$x = (v_0 \cos \theta)t, \quad y = (v_0 \sin \theta)t - 16t^2, \text{ for } t \text{ in } [0, k],$$

is

$$y = (\tan \theta)x - \frac{16}{v_0^2 \cos^2 \theta}x^2.$$

30. Find the vertex of the parabola given by the rectangular equation of Exercise 29.

31. Give two parametric representations of the line through the point (x_1, y_1) with slope m.

32. Give two parametric representations of the parabola $y = a(x - h)^2 + k$.

33. Give a parametric representation of the hyperbola $(x^2/a^2) - (y^2/b^2) = 1$.

34. Give a parametric representation of the ellipse $(x^2/a^2) + (y^2/b^2) = 1$.

35. The spiral of Archimedes has polar equation $r = a\theta$, where $r^2 = x^2 + y^2$. Show that a parametric representation of the spiral of Archimedes is

$$x = a\theta \cos \theta, \quad y = a\theta \sin \theta, \quad \text{for } \theta \text{ in } (-\infty, \infty).$$

(θ in radians)

36. Show that the hyperbolic spiral $r\theta = a$, where $r^2 = x^2 + y^2$, is given parametrically by

$$x = \frac{a \cos \theta}{\theta}, \quad y = \frac{a \sin \theta}{\theta}, \quad \text{for } \theta \text{ in } (-\infty, 0) \cup (0, \infty).$$

(θ in radians)

Chapter 8   Summary ▼▼▼▼▼▼▼▼▼▼▼▼▼▼▼▼▼▼▼▼▼▼▼▼▼▼▼▼▼▼▼▼▼▼▼

SECTION	KEY IDEAS
8.1 Operations on Complex Numbers	
	Definition of i $$i = \sqrt{-1} \quad \text{or} \quad i^2 = -1$$ **Definition of $\sqrt{-a}$** For positive real numbers a, $$\sqrt{-a} = i\sqrt{a}.$$ **Conjugate** The conjugate of $a + bi$ is $a - bi$. **Operations on Complex Numbers** For complex numbers $a + bi$ and $c + di$: **Addition of Complex Numbers** $$(a + bi) + (c + di) = (a + c) + (b + d)i$$ **Subtraction of Complex Numbers** $$(a + bi) - (c + di) = (a - c) + (b - d)i.$$ **Product of Complex Numbers** To find the product $(a + bi)(c + di)$, use FOIL and the definition of i^2. **Quotient of Complex Numbers** To find the quotient $(a + bi)/(c + di)$, multiply both the numerator and the denominator by the conjugate of the denominator, $c - di$.
8.2 Trigonometric Form of Complex Numbers	
	Trigonometric Form of Complex Numbers If the complex number $x + yi$ corresponds to the vector with direction angle θ and magnitude r, then $$x = r \cos \theta \qquad r = \sqrt{x^2 + y^2}$$ $$y = r \sin \theta \qquad \tan \theta = \frac{y}{x}, \text{ if } x \neq 0$$ and $$r(\cos \theta + i \sin \theta) \quad \text{or} \quad r \text{ cis } \theta$$ is the trigonometric form (or polar form) of $x + yi$.
8.3 Product and Quotient Theorems	
	Product and Quotient Theorems For any two complex numbers $r_1(\cos \theta_1 + i \sin \theta_1)$ and $r_2(\cos \theta_2 + i \sin \theta_2)$, $$[r_1(\cos \theta_1 + i \sin \theta_1)] \cdot [r_2(\cos \theta_2 + i \sin \theta_2)]$$ $$= r_1 r_2 [\cos(\theta_1 + \theta_2) + i \sin(\theta_1 + \theta_2)]$$ and $$\frac{r_1(\cos \theta_1 + i \sin \theta_1)}{r_2(\cos \theta_2 + i \sin \theta_2)} = \frac{r_1}{r_2}[\cos(\theta_1 - \theta_2) + i \sin(\theta_1 - \theta_2)],$$ where $r_2 \text{ cis } \theta_2 \neq 0$.

SECTION	KEY IDEAS
8.4 Powers and Roots of Complex Numbers	
	De Moivre's Theorem $$[r(\cos \theta + i \sin \theta)]^n = r^n(\cos n\theta + i \sin n\theta)$$
	nth Root Theorem If n is any positive integer and r is a positive real number, then the nonzero complex number $r(\cos \theta + i \sin \theta)$ has exactly n distinct nth roots, given by $$\sqrt[n]{r}(\cos \alpha + i \sin \alpha),$$ where $$\alpha = \frac{\theta + 360°k}{n} \quad \text{or} \quad \alpha = \frac{\theta}{n} + \frac{360°k}{n},$$ $k = 0, 1, 2, \ldots, n - 1.$
8.5 Polar Equations	
	Polar coordinates determine a point by locating it θ degrees from the polar axis (the positive x-axis) and r units from the origin. Polar equations are graphed in the same way as Cartesian equations, by point plotting or with a graphing calculator.
8.6 Parametric Equations	
	Plane Curve A plane curve is a set of points (x, y) such that $x = f(t)$, $y = g(t)$, and f and g are both defined on an interval I. The equations $x = f(t)$ and $y = g(t)$ are parametric equations with parameter t.

Chapter 8   Review Exercises ▼▼▼▼▼▼▼▼▼▼▼▼▼▼▼▼▼▼▼▼▼▼▼▼▼▼▼▼▼▼▼▼

Write as a multiple of i.

1. $\sqrt{-9}$

2. $\sqrt{-12}$

Solve each quadratic equation.

3. $x^2 = -81$

4. $x(2x + 3) = -4$

Perform the indicated operation. Write answers in standard form.

5. $(1 - i) - (3 + 4i) + 2i$

6. $(2 - 5i) + (9 - 10i) - 3$

7. $(6 - 5i) + (2 + 7i) - (3 - 2i)$

8. $(4 - 2i) - (6 + 5i) - (3 - i)$

9. $(3 + 5i)(8 - i)$

10. $(4 - i)(5 + 2i)$

11. $(2 + 6i)^2$

12. $(6 - 3i)^2$

13. $(1 - i)^3$

14. $(2 + i)^3$

15. $\dfrac{6 + 2i}{3 - i}$

16. $\dfrac{2 - 5i}{1 + i}$

17. $\dfrac{2 + i}{1 - 5i}$

18. $\dfrac{3 + 2i}{i}$

19. i^{53}

20. i^{-41}

21. Evaluate $1 \cdot i \cdot i^2 \cdot i^3$.

22. Evaluate $1 \cdot i \cdot i^2 \cdot \ldots \cdot i^{100}$.

23. $[5(\cos 90° + i \sin 90°)][6(\cos 180° + i \sin 180°)]$ **24.** $[3 \operatorname{cis} 135°][2 \operatorname{cis} 105°]$

25. $\dfrac{2(\cos 60° + i \sin 60°)}{8(\cos 300° + i \sin 300°)}$ **26.** $\dfrac{4 \operatorname{cis} 270°}{2 \operatorname{cis} 90°}$ **27.** $(\sqrt{3} + i)^3$

28. $(2 - 2i)^5$ **29.** $(\cos 100° + i \sin 100°)^6$ **30.** $(\operatorname{cis} 20°)^3$

31. The vector representing a real number will lie on the _____-axis in the complex plane.

32. Explain the geometric similarity between the absolute value of a real number and the absolute value (or modulus) of a complex number. (*Hint:* Think in terms of distance.)

Graph each complex number.

33. $5i$ **34.** $-4 + 2i$ **35.** $3 - 3i\sqrt{3}$ **36.** $-5 + i\sqrt{3}$

Find and graph the resultant of each pair of complex numbers.

37. $7 + 3i$ and $-2 + i$ **38.** $2 - 4i$ and $5 + i$

Complete the chart in Exercises 39–46.

Standard Form	Trigonometric Form
39. $-2 + 2i$	_____
40. _____	$3(\cos 90° + i \sin 90°)$
41. _____	$2(\cos 225° + i \sin 225°)$
42. $-4 + 4i\sqrt{3}$	_____
43. $1 - i$	_____
44. _____	$4 \operatorname{cis} 240°$
45. $-4i$	_____
46. _____	$2 \operatorname{cis} 180°$

The complex number z, where $z = x + yi$, can be graphed in the plane as (x, y). Describe the graphs of all complex numbers z satisfying the conditions in Exercises 47 and 48.

47. The modulus of z is 2.

48. The imaginary part of z is the negative of the real part of z.

49. Give a geometric condition that implies that the square of a nonreal complex number $a + bi$ is a real number.

50. Give a geometric condition that implies that the square of a nonreal complex number is a complex number whose real part is 0.

51. Find the fifth roots of $-2 + 2i$.   **52.** Find the cube roots of $1 - i$.

53. How many real fifth roots does -32 have?   **54.** How many real sixth roots does -64 have?

Solve each equation.

55. $x^3 + 125 = 0$ **56.** $x^4 + 16 = 0$ **57.** $x^2 + i = 0$

Graph each polar equation for θ in $[0°, 360°)$.

58. $r = 4 \cos \theta$ **59.** $r = -1 + \cos \theta$ **60.** $r = 1 - \cos \theta$

61. $r = 2 \sin 4\theta$ **62.** $r = 3 \cos 3\theta$

Find an equivalent equation in rectangular coordinates.

63. $r = \dfrac{3}{1 + \cos \theta}$ **64.** $r = \dfrac{4}{2 \sin \theta - \cos \theta}$ **65.** $r = \sin \theta + \cos \theta$ **66.** $r = 2$

Find an equivalent equation in polar coordinates.

67. $y = x$ **68.** $y = x^2$ **69.** $x = y^2$

 In Exercises 70–72, find a polar equation having the given graph. (Note: The values of Xscl and Yscl are 1.)

70.   **71.**   **72.**

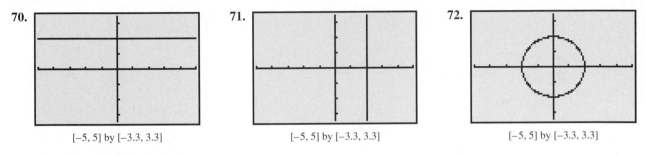

$[-5, 5]$ by $[-3.3, 3.3]$ $[-5, 5]$ by $[-3.3, 3.3]$ $[-5, 5]$ by $[-3.3, 3.3]$

Sketch the graph of each plane curve given by the parametric equations.

73. $x = 4t - 3, y = t^2$, for t in $[-3, 4]$ **74.** $x = t + \cos t, y = \sin t$, for t in $[0, 2\pi]$

Find a rectangular equation for each plane curve with the given parametric equations.

75. $x = 3t + 2, y = t - 1$, for t in $[-5, 5]$ **76.** $x = t^2 + 1, y = 2t^2 - 1$, for t in $[-2, 3]$

77. $x = \sqrt{t - 1}, y = \sqrt{t}$, for t in $[1, \infty)$ **78.** $x = t^2 + 5, y = \dfrac{1}{t^2 + 1}$, for t in $(-\infty, \infty)$

79. $x = 5 \tan t, y = 3 \sec t$, for t in $(-\pi/2, \pi/2)$ **80.** $x = \cos 2t, y = \sin t$, for t in $(-\pi, \pi)$

81. Show that the graph of $x = a + r \cos t, y = b + r \sin t, t$ in $[0, 2\pi]$ is a circle with center (a, b) and radius r.

82. Refer to Exercise 81. Find a pair of parametric equations whose graph is the circle with center $(3, 4)$ and containing the origin.

83. Follow the steps in Exercise 70 of Section 8.2 to show that the graph of the Mandelbrot set in Exercise 61 of Section 8.4 is symmetric with respect to the x-axis.

Exponential and Logarithmic Functions

In 1896 Swedish scientist Svante Arrhenius first predicted the greenhouse effect resulting from emissions of carbon dioxide by industrialized countries. In his classic calculation, he was able to estimate that a doubling of the carbon dioxide level in the atmosphere would raise the average global temperature by 7°F to 11°F. Since global warming would not be uniform, changes as small as 4.5°F in the average temperature could have drastic climatic effects, particularly on the central plains of North America. Sea levels could rise dramatically as a result of both thermal expansion and the melting of ice caps. The annual cost to the United States economy could reach $60 billion.

The burning of fossil fuels, deforestation, and changes in land use from 1850 to 1986 put approximately 312 billion tons of carbon into the atmosphere, mostly in the form of carbon dioxide. Burning of fossil fuels produces 5.4 billion tons of carbon each year which is absorbed by both the atmosphere and the oceans. A critical aspect of the accumulation of carbon dioxide in the atmosphere is that it is irreversible and its effect requires hundreds of years to disappear. In 1990 the International Panel on Climate Change (IPCC) reported that if current trends of burning of fossil fuel and deforestation continue, then future amounts of atmospheric carbon dioxide in parts per million (ppm) will increase as shown in the table.

Sources: Clime, W., *The Economics of Global Warming,* Institute for International Economics, Washington, D.C., 1992.

Kraljic, M. (Editor), *The Greenhouse Effect,* The H. W. Wilson Company, New York, 1992.

International Panel on Climate Change (IPCC), 1990.

Wuebbles, D. and J. Edmonds, *Primer of Greenhouse Gases,* Lewis Publishers, Inc., Chelsea, Michigan, 1991.

Year	Carbon Dioxide
1990	353
2000	375
2075	590
2175	1090
2275	2000

How can these data be used to predict when the amount of carbon dioxide will double? What will be the resulting global warming? How are carbon dioxide levels and global temperature increases related? These questions will be addressed in several sections of this chapter.

Since hard data on the greenhouse effect are lacking, mathematical models play a central role in analyzing the reality of the greenhouse effect and answering questions like these. In order for Svante Arrhenius to make his first calculation about global warming, he needed both logarithmic and exponential functions. These functions are central to many real applications found throughout science, business, and environmental forecasting. Using these functions we will be able to analyze the greenhouse effect, model population growth, and predict the time it takes for the planet Pluto to orbit the sun.

Many trigonometry courses include a chapter on exponential and logarithmic functions. Traditionally, these were taught with the trigonometric functions, because together they comprised the transcendental functions. This chapter is included for those who continue this approach.

9.1 Exponential Functions ▼▼▼

Recall from algebra the definition of a^r, where r is a rational number: if $r = m/n$, then for appropriate values of m and n,

$$a^{m/n} = (\sqrt[n]{a})^m.$$

For example,

$$16^{3/4} = (\sqrt[4]{16})^3 = 2^3 = 8,$$

$$27^{-1/3} = \frac{1}{27^{1/3}} = \frac{1}{\sqrt[3]{27}} = \frac{1}{3},$$

and

$$64^{-1/2} = \frac{1}{64^{1/2}} = \frac{1}{\sqrt{64}} = \frac{1}{8}.$$

In this section the definition of a^r is extended to include all real (not just rational) values of the exponent r. For example, the new symbol $2^{\sqrt{3}}$ might be evaluated by approximating the exponent $\sqrt{3}$ by the numbers 1.7, 1.73, 1.732, and so on. Since these decimals approach the value of $\sqrt{3}$ more and more closely, it seems reasonable that $2^{\sqrt{3}}$ should be approximated more and more closely by the numbers $2^{1.7}$, $2^{1.73}$, $2^{1.732}$, and so on. (Recall, for example, that $2^{1.7} = 2^{17/10} = \sqrt[10]{2^{17}}$.) In fact, this is exactly how $2^{\sqrt{3}}$ is defined (in a more advanced course). To show that this assumption is reasonable, Figure 1 gives the graphs of the function $f(x) = 2^x$ with three different domains.

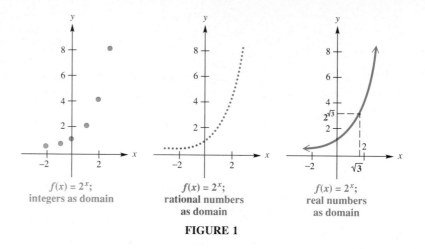

$f(x) = 2^x$;
integers as domain

$f(x) = 2^x$;
rational numbers
as domain

$f(x) = 2^x$;
real numbers
as domain

FIGURE 1

With this interpretation of real exponents, all rules and theorems for exponents are valid for real-number exponents as well as rational ones. In addition to the rules for exponents presented in algebra, several new properties are used in this chapter. For example, if $y = 2^x$, then each real value of x leads to exactly one value of y, and therefore, $y = 2^x$ defines a function. Furthermore,

$$\text{if } 3^x = 3^4, \quad \text{then} \quad x = 4,$$

and for $p > 0$,

$$\text{if } p^2 = 3^2, \quad \text{then} \quad p = 3.$$

Also,

$$4^2 < 4^3 \qquad \text{but} \qquad \left(\frac{1}{2}\right)^2 > \left(\frac{1}{2}\right)^3,$$

so that when $a > 1$, increasing the exponent on a leads to a *larger* number, but if $0 < a < 1$, increasing the exponent on a leads to a *smaller* number.

These properties are generalized below. Proofs of the properties are not given here, as they require more advanced mathematics.

Additional Properties of Exponents

For any real number $a > 0$, $a \neq 1$, and any real number x, the following statements are true:

(a) a^x **is a unique real number.**
(b) $a^b = a^c$ **if and only if** $b = c$.
(c) If $a > 1$ **and** $m < n$, **then** $a^m < a^n$.
(d) If $0 < a < 1$ **and** $m < n$, **then** $a^m > a^n$.

Properties (a) and (b) require $a > 0$ so that a^x is always defined. For example, $(-6)^x$ is not a real number if $x = 1/2$. This means that a^x will always be positive, since a must be positive. In part (a), $a \neq 1$ because $1^x = 1$ for every real-number value of x, so that each value of x leads to the *same* real number, 1. For Property (b) to hold, a must not equal 1 since, for example, $1^4 = 1^5$, even though $4 \neq 5$.

With most calculators, values of a^x are computed with either a key labeled x^y (or y^x or a^b) or with the key marked ^. In each case enter the base, then the appropriate exponentiation key, then the exponent.

GRAPHING EXPONENTIAL FUNCTIONS   As mentioned, the expression a^x satisfies all the properties of exponents from earlier work. We can now define a function $f(x) = a^x$ whose domain is the set of all real numbers (and not just the rationals).

Exponential Function

If $a > 0$ and $a \neq 1$, then

$$f(x) = a^x$$

defines the **exponential function** with base a.

NOTE   If $a = 1$, the function is the constant function $f(x) = 1$, and not an exponential function.

EXAMPLE 1
Evaluating an exponential expression

If $f(x) = 2^x$, find each of the following.

(a) $f(-1)$
Replace x with -1.

$$f(-1) = 2^{-1} = \frac{1}{2}$$

(b) $f(3) = 2^3 = 8$

(c) $f(5/2) = 2^{5/2} = (2^5)^{1/2} = 32^{1/2} = \sqrt{32} = 4\sqrt{2}$

(d) $f(4.92) \approx 30.2738447$ ▶

EXAMPLE 3
Graphing reflections and translations

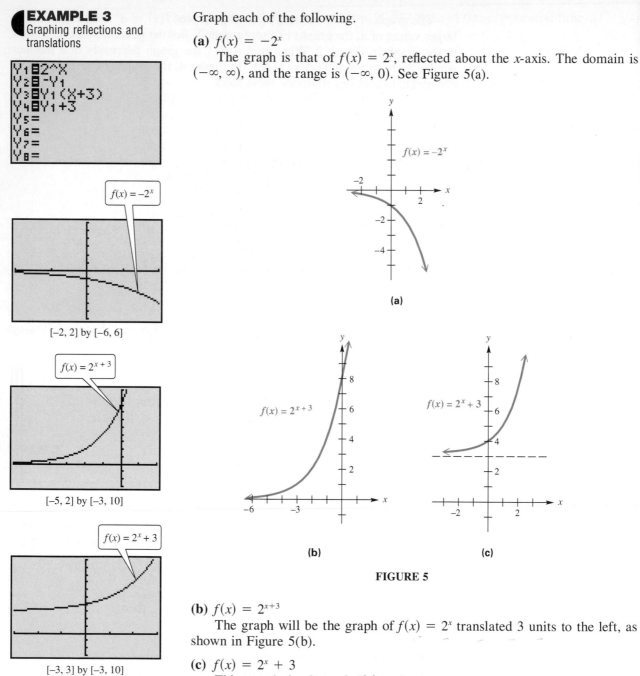

```
Y1∎2^X
Y2∎-Y1
Y3∎Y1(X+3)
Y4∎Y1+3
Y5=
Y6=
Y7=
Y8=
```

$f(x) = -2^x$

[−2, 2] by [−6, 6]

$f(x) = 2^{x+3}$

[−5, 2] by [−3, 10]

$f(x) = 2^x + 3$

[−3, 3] by [−3, 10]

The top screen shows how Y_2, Y_3, and Y_4 are defined in terms of Y_1, using function notation. Compare the graphs to those in Figure 5.

Graph each of the following.

(a) $f(x) = -2^x$

The graph is that of $f(x) = 2^x$, reflected about the x-axis. The domain is $(-\infty, \infty)$, and the range is $(-\infty, 0)$. See Figure 5(a).

$f(x) = -2^x$

(a)

$f(x) = 2^{x+3}$

(b)

$f(x) = 2^x + 3$

(c)

FIGURE 5

(b) $f(x) = 2^{x+3}$

The graph will be the graph of $f(x) = 2^x$ translated 3 units to the left, as shown in Figure 5(b).

(c) $f(x) = 2^x + 3$

This graph is that of $f(x) = 2^x$ translated 3 units upward. See Figure 5(c). ▶

EXAMPLE 4
Graphing a composite exponential function

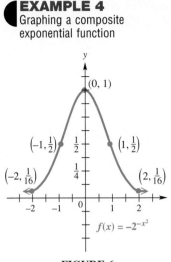

FIGURE 6

Graph $f(x) = 2^{-x^2}$.

Write $f(x) = 2^{-x^2}$ as $f(x) = 1/(2^{x^2})$ to find ordered pairs that belong to the function. Some ordered pairs are shown in the table.

x	-2	-1	0	1	2
y	$\frac{1}{16}$	$\frac{1}{2}$	1	$\frac{1}{2}$	$\frac{1}{16}$

As the table suggests, $0 < y \le 1$ for all values of x. The y-intercept is 1. The x-axis is a horizontal asymptote. Replacing x with $-x$ shows that the graph is symmetric with respect to the y-axis. Plotting the y-intercept and the points in the table, and drawing a smooth curve through them gives the graph in Figure 6. It is necessary to plot several points close to $(0, 1)$, to determine the correct shape of the graph there. This type of "bell-shaped" curve is important in statistics. ▶

EXPONENTIAL EQUATIONS Property (b) given at the beginning of this section is useful in solving equations, as shown by the next examples.

EXAMPLE 5
Using a property of exponents to solve an equation

Solve $\left(\dfrac{1}{3}\right)^x = 81$.

First, write $1/3$ as 3^{-1}, so that $(1/3)^x = (3^{-1})^x = 3^{-x}$. Since $81 = 3^4$,

$$\left(\frac{1}{3}\right)^x = 81$$

becomes

$$3^{-x} = 3^4.$$

By Property (b),

$$-x = 4, \quad \text{or} \quad x = -4.$$

The solution of the given equation is -4. ▶

Later in this chapter, we describe a more general method for solving exponential equations where the approach used in Example 5 is not possible. For instance, this method could not be used to solve an equation like $7^x = 12$, since it is not easy to express both sides as exponential expressions with the same base.

EXAMPLE 6
Using a property of exponents to solve an equation

Solve $81 = b^{4/3}$.

Begin by writing $b^{4/3}$ as $(\sqrt[3]{b})^4$.

$$81 = b^{4/3}$$
$$81 = (\sqrt[3]{b})^4 \qquad \text{Definition of rational exponent}$$
$$\pm 3 = \sqrt[3]{b} \qquad \text{Take fourth roots on both sides.}$$
$$\pm 27 = b \qquad \text{Cube both sides.}$$

Check both solutions in the original equation. Since both solutions check, the solutions are 27 and -27. ▶

COMPOUND INTEREST The formula for *compound interest* (interest paid on both principal and interest) is an important application of exponential functions. You may recall the formula for simple interest, $I = Prt$, where P is the principal (amount left at interest), r is the rate of interest expressed as a decimal, and t is time in years that the principal earns interest. Suppose $t = 1$ year. Then at the end of the year the amount has grown to

$$P + Pr = P(1 + r),$$

the original principal plus the interest. If this amount is left at the same interest rate for another year, the total amount becomes

$$[P(1 + r)] + [P(1 + r)]r = [P(1 + r)](1 + r)$$
$$= P(1 + r)^2.$$

After the third year, this will grow to

$$[P(1 + r)^2] + [P(1 + r)^2]r = [P(1 + r)^2](1 + r)$$
$$= P(1 + r)^3.$$

Continuing in this way produces the following formula for compound interest.

Compound Interest

If P dollars is deposited in an account paying an annual rate of interest r compounded (paid) m times per year, then after t years the account will contain A dollars, where

$$A = P\left(1 + \frac{r}{m}\right)^{tm}.$$

For example, let \$1000 be deposited in an account paying 8% per year compounded quarterly, or four times per year. After 10 years the account will contain

$$P\left(1 + \frac{r}{m}\right)^{tm} = 1000\left(1 + \frac{.08}{4}\right)^{10(4)}$$
$$= 1000(1 + .02)^{40}$$
$$= 1000(1.02)^{40}$$

dollars. The number $(1.02)^{40}$ can be found using a calculator. To five decimal places, $(1.02)^{40} = 2.20804$. The amount on deposit after 10 years is

$$1000(1.02)^{40} = 1000(2.20804) = 2208.04,$$

or \$2208.04.

In the formula for compound interest, A is sometimes called the **future value** and P the **present value.**

EXAMPLE 7
Finding present value

An accountant wants to buy a new computer in three years that will cost $20,000.

(a) How much should be deposited now, at 6% interest compounded annually, to give the required $20,000 in three years?

Since the money deposited should amount to $20,000 in three years, $20,000 is the future value of the money. To find the present value P of $20,000 (the amount to deposit now), use the compound interest formula with $A = 20,000$, $r = .06$, $m = 1$, and $t = 3$.

$$A = P\left(1 + \frac{r}{m}\right)^{tm}$$

$$20,000 = P\left(1 + \frac{.06}{1}\right)^{3(1)} = P(1.06)^3$$

$$\frac{20,000}{(1.06)^3} = P$$

$$P = 16,792.39$$

The accountant must deposit $16,792.39.

(b) If only $15,000 is available to deposit now, what annual interest rate is required for it to increase to $20,000 in three years?

Here $P = 15,000$, $A = 20,000$, $m = 1$, $t = 3$, and r is unknown. Substitute the known values into the compound interest formula and solve for r.

$$A = P\left(1 + \frac{r}{m}\right)^{tm}$$

$$20,000 = 15,000\left(1 + \frac{r}{1}\right)^3$$

$$\frac{4}{3} = (1 + r)^3 \qquad \text{Divide both sides by 15,000.}$$

$$\left(\frac{4}{3}\right)^{1/3} = 1 + r \qquad \text{Take the cube root on both sides.}$$

$$\left(\frac{4}{3}\right)^{1/3} - 1 = r \qquad \text{Subtract 1 on both sides.}$$

$$r \approx .10 \qquad \text{Use a calculator.}$$

An interest rate of 10% will produce enough interest to increase the $15,000 deposit to the $20,000 needed at the end of three years. ▶

Perhaps the single most useful base for an exponential function is the irrational number e. Base e exponential functions provide a good model for many natural, as well as economic, phenomena. The letter e was chosen to represent this number in honor of the Swiss mathematician Leonhard Euler (pronounced "oiler") (1707–1783). Applications of the exponential function with base e are given later in this section.

The number e comes up in a natural way when using the formula for compound interest. Suppose a lucky investment produces an annual interest rate of 100%, so that $r = 1.00$, or $r = 1$. Suppose also that only $1 can be deposited at this rate, and for only one year. Then $P = 1$ and $t = 1$. Substitute into the formula for compound interest:

$$P\left(1 + \frac{r}{m}\right)^{tm} = 1\left(1 + \frac{1}{m}\right)^{1(m)} = \left(1 + \frac{1}{m}\right)^{m}.$$

As interest is compounded more and more often, the value of this expression will increase. If interest is compounded annually, making $m = 1$, the total amount on deposit is

$$\left(1 + \frac{1}{m}\right)^{m} = \left(1 + \frac{1}{1}\right)^{1} = 2^1 = 2,$$

so an investment of $1 becomes $2 in one year. As interest is compounded more and more often, the value of this expression will increase.

A calculator with a y^x key was used to get the results in the table below. These results have been rounded when necessary to five decimal places. The table suggests that, as m increases, the value of $(1 + 1/m)^m$ gets closer and closer to some fixed number. It turns out that this is indeed the case. This fixed number is called e.

m	$\left(1 + \dfrac{1}{m}\right)^{m}$
1	2
2	2.25
5	2.48832
10	2.59374
25	2.66584
50	2.69159
100	2.70481
500	2.71557
1000	2.71692
10,000	2.71815
1,000,000	2.71828

Value of e

To nine decimal places,

$$e \approx 2.718281828.$$

NOTE Values of e^x can be found with a calculator that has a key marked e^x or by using a pair of keys marked INV and ln. See your instruction booklet for details or ask your instructor for assistance. The reason the second of these methods works will be apparent in the next section.

In Figure 7 the functions $y = 2^x$, $y = e^x$, and $y = 3^x$ are graphed for comparison.

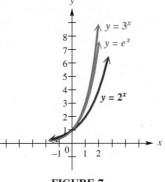

FIGURE 7

CONNECTIONS In calculus, it is shown that

$$e^x = 1 + x + \frac{x^2}{2 \cdot 1} + \frac{x^3}{3 \cdot 2 \cdot 1} + \frac{x^4}{4 \cdot 3 \cdot 2 \cdot 1} + \frac{x^5}{5 \cdot 4 \cdot 3 \cdot 2 \cdot 1} + \cdots .$$

By using more and more terms, a more and more accurate approximation may be obtained for e^x.

FOR DISCUSSION OR WRITING
1. Use the terms shown here and replace x with 1 to approximate $e^1 = e$ to three decimal places. Check your results with a calculator.
2. Use the terms shown here and replace x with $-.05$ to approximate $e^{-.05}$ to four decimal places. Check your results with a calculator.
3. Give the next term in the sum for e^x.

EXPONENTIAL GROWTH OR DECAY As mentioned above, the number e is important as the base of an exponential function because many practical applications require an exponential function with base e. For example, it can be shown that in situations involving growth or decay of a quantity, the amount or number present at time t often can be closely approximated by a function defined by

$$y = y_0 e^{kt},$$

where y_0 is the amount or number present at time $t = 0$ and k is a constant.

The next example, which refers to the problem stated at the beginning of this chapter, illustrates exponential growth.

EXAMPLE 8
Using data to determine an
exponential growth function

The International Panel on Climate Change (IPCC) in 1990 published
its finding that if current trends of burning fossil fuel and deforestation
continue, then future amounts of atmospheric carbon dioxide in parts
per million (ppm) will increase as shown in the table.*

Year	Carbon Dioxide
1990	353
2000	375
2075	590
2175	1090
2275	2000

(a) Plot the data. Do the carbon dioxide levels appear to grow exponentially?
We show a calculator-generated graph for the data in Figure 8(a). The data
do appear to have the shape of the graph of an increasing exponential function.

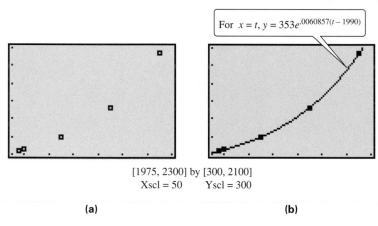

For $x = t$, $y = 353e^{.0060857(t - 1990)}$

[1975, 2300] by [300, 2100]
Xscl = 50   Yscl = 300

(a)   (b)

FIGURE 8

(b) The function defined by $y = 353e^{.0060857(t - 1990)}$ is a good model for the data.
A graph of the function in Figure 8(b) shows that it is very close to the data
points. From the graph, estimate when future levels of carbon dioxide will
double and triple over the preindustrial level of 280 ppm.
We graph $y = 2 \cdot 280 = 560$ and $y = 3 \cdot 280 = 840$ on the same coordinate axes as the function in Figure 9 and use the calculator to find the intersection points. The graph of the function intersects the horizontal lines at approx-

Source: International Panel on Climate Change (IPCC), 1990.

imately 2065.8 and 2132.5. Carbon dioxide levels will double by 2065 and triple by 2132. ◗

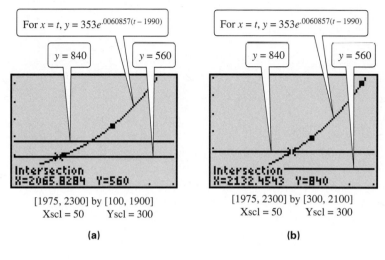

FIGURE 9

Further examples of exponential growth and decay are given in Section 9.4.

9.1 Exercises ▼▼▼

Decide whether the statements in Exercises 1–5 are true or false.

1. If $x = 2$, then $5^x = 5^2$.

2. If $x > 1$, then $x^3 > x^2$.

3. If $x < 0$, then $x^3 > x^2$.

4. The domain of $f(x) = .5^x$ is $[0, \infty)$.

5. The range of $f(x) = .5^x$ is $(0, \infty)$.

6. Graph f for each of the following. Compare the graphs to that of $f(x) = 2^x$. See Examples 2–4.
   (a) $f(x) = 2^x + 1$ **(b)** $f(x) = 2^x - 4$ **(c)** $f(x) = 2^{x+1}$ **(d)** $f(x) = 2^{x-4}$

7. Graph f for each of the following. See Examples 2–4.
   (a) $f(x) = 3^{-x} - 2$ **(b)** $f(x) = 3^{-x} + 4$ **(c)** $f(x) = 3^{-x-2}$ **(d)** $f(x) = 3^{-x+4}$

8. Explain how you could use the graph of $y = 4^x$ to graph $y = -4^x$.

Graph f for each of the following. See Examples 2–4.

9. $f(x) = 3^x$

10. $f(x) = 4^x$

11. $f(x) = \left(\dfrac{3}{2}\right)^x$

12. $f(x) = e^{-x}$

13. $f(x) = 2^{|x|}$

14. $f(x) = 2^{-|x|}$

In the given figure, the graphs of y = aˣ for a = 1.8, 2.3, 3.2, .4, .75, and .31 are given. They are identified by letter, but not necessarily in the same order as the values of a just given. Use your knowledge of how the exponential function behaves for various values of a to identify each lettered graph.

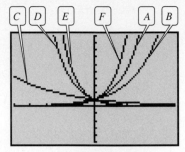

[−10, 10] by [−5, 10]

15. A   **16.** B   **17.** C

18. D   **19.** E   **20.** F

21. For $a > 1$, how does the value of $f(x) = a^x$ change as x increases? What if $0 < a < 1$?

22. What two points on the graph of $f(x) = a^x$ can be found with no computation?

23. A function of the form $f(x) = x^r$, where r is a constant, is called a *power function*. Discuss the differences between an exponential function and a power function.

24. If $f(x) = a^x$ and $f(3) = 27$, find the following values of $f(x)$.
  (a) $f(1)$ **(b)** $f(-1)$ **(c)** $f(2)$ **(d)** $f(0)$

Give an equation of the form f(x) = aˣ to define the exponential function whose graph contains the given point.

25. $(3, 8)$ **26.** $(-3, 64)$

Use properties of exponents to write each of the following in the form f(t) = kaᵗ, where k is a constant. (Hint: Recall 4ˣ⁺ʸ = 4ˣ · 4ʸ.)

27. $f(t) = 3^{2t+3}$ **28.** $f(t) = \left(\dfrac{1}{3}\right)^{1-2t}$

29. Explain why the exponential equation $3^x = 12$ cannot be solved by using the properties of exponentials given in this section.

Solve the equation. See Examples 5 and 6.

30. $4^x = 2$ **31.** $125^r = 5$ **32.** $\left(\dfrac{1}{2}\right)^k = 4$ **33.** $\left(\dfrac{2}{3}\right)^r = \dfrac{9}{4}$

34. $2^{3-y} = 8$ **35.** $5^{2p+1} = 25$ **36.** $\dfrac{1}{27} = b^{-3}$ **37.** $\dfrac{1}{81} = k^{-4}$

38. $4 = r^{2/3}$ **39.** $z^{5/2} = 32$ **40.** $27^{4z} = 9^{z+1}$ **41.** $32^t = 16^{1-t}$

42. $\left(\dfrac{1}{2}\right)^{-x} = \left(\dfrac{1}{4}\right)^{x+1}$ **43.** $\left(\dfrac{2}{3}\right)^{k-1} = \left(\dfrac{81}{16}\right)^{k+1}$

📷 *Use a graphing calculator to graph f in Exercises 44–47.*

44. $f(x) = \dfrac{e^x - e^{-x}}{2}$ **45.** $f(x) = \dfrac{e^x + e^{-x}}{2}$ **46.** $f(x) = x \cdot 2^x$ **47.** $f(x) = x^2 \cdot 2^{-x}$

Use the formula for compound interest to find each future value.

48. $8906.54 at 5% compounded semiannually for 9 years

49. $56,780 at 5.3% compounded quarterly for 23 quarters

Find the present value for each future value. See Example 7(a).

50. $25,000, if interest is 6% compounded quarterly for 11 quarters

51. $45,678.93, if interest is 9.6% compounded monthly for 11 months

Find the required annual interest rate to the nearest tenth. See Example 7(b).

52. $65,000 compounded monthly for 6 months to yield $65,325

53. $1200 compounded quarterly for 5 years to yield $1780

📷 **54.** Suppose you can borrow money at 8% compounded daily or 8.3% compounded annually. Which will cost you more? (*Hint:* Calculate a loan of $1 for 1 year both ways.) Use the TABLE feature of a graphing calculator to find the difference after 1 year, after 2 years, and after 5 years.

Solve each applied problem.

55. The exponential growth of the deer population in Massachusetts can be calculated using the equation $T = 50{,}000(1 + .06)^n$, where 50,000 is the initial deer population and .06 is the rate of growth. T is the total population after n years have passed.
 (a) Predict the total population after 4 years.
 (b) If the initial population was 30,000 and the growth rate was .12, approximately how many deer would be present after 3 years?
 (c) How many additional deer can we expect in 5 years if the initial population is 45,000 and the current growth rate is .08?

56. Since 1950, the growth in the world population in millions closely fits the exponential function defined by
$$A(t) = 2600e^{.018t},$$
where t is the number of years since 1950.
 (a) The world population was about 3700 million in 1970. How closely does the function approximate this value?

(b) Use the function to approximate the population in 1990. (The actual 1990 population was about 5320 million.)

(c) Estimate the population in the year 2000.

57. A sample of 500 g of lead 210 decays to polonium 210 according to the function given by
$$A(t) = 500e^{-.032t},$$
where t is time in years. Find the amount of the sample after each of the following times.
 (a) 4 years   **(b)** 8 years
 (c) 20 years   **(d)** Graph $y = A(t)$.

58. Vehicle theft in the United States has been rising exponentially since 1972. The number of stolen vehicles, in millions, is given by
$$f(x) = .88(1.03)^x,$$
where $x = 0$ represents the year 1972. Find the number of vehicles stolen in the following years.
 (a) 1975   **(b)** 1980
 (c) 1985   **(d)** 1990

59. Refer to Example 8. Carbon dioxide in the atmosphere traps heat from the sun. Presently, the net incoming solar radiation reaching the Earth's surface is 240 watts per square meter (w/m²). The relationship between additional watts per square meter of heat trapped by the increased carbon dioxide R and the average rise in global temperature T (in °F) is shown in the graph. This additional solar radiation trapped by carbon dioxide is called **radiative forcing.** It is measured in watts per square meter.

(a) Is T a linear or exponential function of R?
(b) Let T represent the temperature increase resulting from an additional radiative forcing of R w/m². Use the graph to write T as a function of R.
(c) Find the global temperature increase when $R = 5$ w/m².

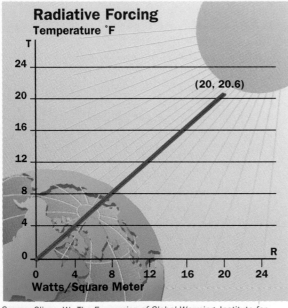

Radiative Forcing

Source: Clime, W. *The Economics of Global Warming,* Institute for International Economics, Washington, D.C., 1992.

60. The atmospheric pressure (in millibars) at a given altitude (in meters) is shown in the table. (*Source:* Miller, A. and J. Thompson, *Elements of Meteorology,* Charles E. Merrill Publishing Company, Columbus, Ohio, 1975.)

Altitude	Pressure
0	1013
1000	899
2000	795
3000	701
4000	617
5000	541
6000	472
7000	411
8000	357
9000	308
10,000	265

(a) Use a graphing calculator to plot the data for atmospheric pressure P at altitude x.
(b) Would a linear or exponential function fit the data better?
(c) The function defined by $P(x) = 1013e^{-.0001341x}$ approximates the data. Use a graphing calculator to graph P and the data on the same coordinate axes.
(d) Use P to predict the pressure at 1500 m and 11,000 m and compare it to the actual values of 846 millibars and 227 millibars, respectively.

Any points where the graphs of functions f and g intersect give solutions of the equation $f(x) = g(x)$. Use a graphing calculator and this idea to estimate the solution(s) of the equation.

61. $x = 2^x$ **62.** $5e^{3x} = 75$ **63.** $6^{-x} = 1 - x$ **64.** $3x + 2 = 4^x$

65. Graph the function $f(x) = (1 + (1/x))^x$ and the horizontal line $y = 2.71828$ with $1 \le x \le 25$ and $0 \le y \le 3$. What happens to $f(x)$ as x gets large?

66. The function e^x grows faster than any power function. See Exercise 23. Graph the function x^2/e^x for $0 \le x \le 10$ and the function x^{10}/e^x for $0 \le x \le 50$. What happens to the values of the quotient in each case as x gets large? Experiment with x^n/e^x for other values of n. Do you think x^n/e^x approaches 0 as x gets large for any n?

9.2 Logarithmic Functions ▼▼▼

The previous section dealt with exponential functions of the form $y = a^x$ for all positive values of a, where $a \neq 1$. As mentioned there, the horizontal line test shows that exponential functions are one-to-one, and thus have inverse functions. In this section we discuss the inverses of exponential functions. The equation defining the inverse of a function is found by exchanging x and y in the equation that defines the function. Doing so with $y = a^x$ gives

$$x = a^y$$

as the equation of the inverse function of the exponential function defined by $y = a^x$. This equation can be solved for y by using the following definition.

Logarithm

For all real numbers y, and all positive numbers a and x, where $a \neq 1$:

$$y = \log_a x \qquad \text{if and only if} \qquad x = a^y.$$

The "log" in the definition above is an abbreviation for *logarithm*. Read $\log_a x$ as "the logarithm to the base a of x."

Consider the following simple fill-in-the-box problems.

$$4^3 = \boxed{} \qquad 5^{\boxed{}} = 25$$

The answers, of course, are

$$4^3 = \boxed{64} \qquad 5^{\boxed{2}} = 25.$$

When we solve the problem on the left, we are "doing" exponents. When we solve the right problem, we are "doing" logarithms. That is, we are finding the power to which 5 must be raised in order to get 25. Therefore, $2 = \log_5 25$. In a certain sense, logarithms are just exponents.

By the definition of logarithm, if $s = \log_a r$, then the power to which a must be raised to get r is s, or $r = a^s$.

$$s = \log_a r \quad \text{if and only if} \quad r = a^s.$$

This key statement should be memorized. It is important to remember the location of the base and exponent in each part.

$$\text{Exponent}$$
$$\downarrow$$
$$\text{Logarithmic form: } s = \log_a r$$
$$\uparrow$$
$$\text{Base}$$

$$\text{Exponent}$$
$$\downarrow$$
$$\text{Exponential form: } a^s = r$$
$$\uparrow$$
$$\text{Base}$$

CAUTION The "log" in $y = \log_a x$ is the notation for a particular function and there must be a replacement for x following it, as in $\log_a 3$, $\log_a(2x - 1)$, or $\log_a x^2$. Avoid writing meaningless notation such as $y = \log$ or $y = \log_a$.

EXAMPLE 1
Converting between exponential and logarithmic statements

The chart below shows several pairs of equivalent statements. The same statement is written in both exponential and logarithmic forms.

Exponential Form	Logarithmic Form
$2^3 = 8$	$\log_2 8 = 3$
$\left(\dfrac{1}{2}\right)^{-4} = 16$	$\log_{1/2} 16 = -4$
$10^5 = 100{,}000$	$\log_{10} 100{,}000 = 5$
$3^{-4} = \dfrac{1}{81}$	$\log_3\left(\dfrac{1}{81}\right) = -4$
$5^1 = 5$	$\log_5 5 = 1$
$\left(\dfrac{3}{4}\right)^0 = 1$	$\log_{3/4} 1 = 0$

▶

LOGARITHMIC EQUATIONS The definition of logarithm can be used to solve logarithmic equations, as shown in the next example.

EXAMPLE 2
Solving logarithmic equations

Solve each equation.

(a) $\log_x \dfrac{8}{27} = 3$

First, write the expression in exponential form.

$$x^3 = \frac{8}{27}$$

$$x^3 = \left(\frac{2}{3}\right)^3 \qquad \scriptstyle \frac{8}{27} = \left(\frac{2}{3}\right)^3$$

$$x = \frac{2}{3} \qquad \text{Property (b) of exponents}$$

The solution is $2/3$.

(b) $\log_4 x = 5/2$

In exponential form, the given statement becomes

$$4^{5/2} = x$$
$$(4^{1/2})^5 = x$$
$$2^5 = x$$
$$32 = x.$$

The solution is 32. ▶

LOGARITHMIC FUNCTIONS   The logarithmic function with base a is defined as follows.

Logarithmic Function

If $a > 0$, $a \neq 1$, and $x > 0$, then

$$f(x) = \log_a x$$

defines the **logarithmic function** with base a.

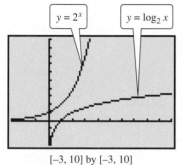

The graph of $y = \log_2 x$ can be obtained by drawing the inverse of $y = 2^x$. (It can also be graphed using the change-of-base theorem, introduced in Section 9.3.)

Exponential and logarithmic functions are inverses of each other. Since the domain of an exponential function is the set of all real numbers, the range of a logarithmic function also will be the set of all real numbers. In the same way, both the range of an exponential function and the domain of a logarithmic function are the set of all positive real numbers, so logarithms can be found for positive numbers only.

The graph of $y = 2^x$ is shown in red in Figure 10. The graph of its inverse is found by reflecting the graph of $y = 2^x$ about the line $y = x$. The graph of the inverse function, defined by $y = \log_2 x$, shown in blue, has the y-axis as a vertical asymptote.

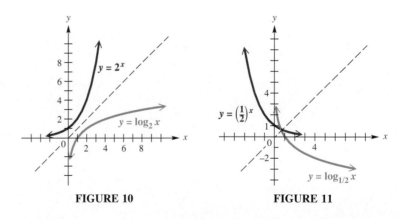

FIGURE 10   FIGURE 11

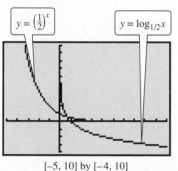

[−5, 10] by [−4, 10]

Compare to Figure 11.

The graph of $y = (1/2)^x$ is shown in red in Figure 11. The graph of its inverse, defined by $y = \log_{1/2} x$, in blue, is found by reflecting the graph of $y = (1/2)^x$ about the line $y = x$. As Figure 11 suggests, the graph of $y = \log_{1/2} x$ also has the y-axis for a vertical asymptote.

The graphs of $y = \log_2 x$ in Figure 10 and $y = \log_{1/2} x$ in Figure 11 suggest the following generalizations about the graphs of logarithmic functions of the form $f(x) = \log_a x$.

> **Graph of $f(x) = \log_a x$**
> 1. The points $(1, 0)$ and $(a, 1)$ are on the graph.
> 2. If $a > 1$, f is an increasing function; if $0 < a < 1$, f is a decreasing function.
> 3. The y-axis is a vertical asymptote.
> 4. The domain is $(0, \infty)$ and the range is $(-\infty, \infty)$.

Compare these generalizations to those for exponential functions discussed in Section 9.1.

> Calculator-generated graphs of logarithmic functions do not, in general, give an accurate picture of the behavior of the graphs near the vertical asymptotes. While it may seem as if the graph has an endpoint, this is not the case. The resolution of the calculator screen is not precise enough to indicate that the graph approaches the vertical asymptote as the value of x gets closer to it. Do not draw incorrect conclusions just because the calculator does not show this behavior.

More general logarithmic functions can be obtained by forming the composition of $h(x) = \log_a x$ with a function $g(x)$ to get

$$f(x) = h[g(x)] = \log_a[g(x)].$$

In writing composite logarithmic functions, it is important to use parentheses to make the intent clear. Just as we put parentheses around $x - 2$ in $f(x - 2)$, we put parentheses around $x - 2$ in $\log_a(x - 2)$. Similarly, we write $\log_a(xy)$ to avoid the misinterpretation $(\log_a x)y$. Also, we write $\log_a(x^2)$ to distinguish it from $(\log_a x)^2$. However, we will continue to write $\log_a x$ without parentheses, because the meaning is clear in that case.

The next examples illustrate some composite logarithmic functions.

EXAMPLE 3
Graphing a translated logarithmic function

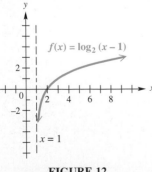

FIGURE 12

Graph each function.

(a) $f(x) = \log_2(x - 1)$

The graph of $f(x) = \log_2(x - 1)$ will be the graph of $f(x) = \log_2 x$, translated one unit to the right. The vertical asymptote is $x = 1$. The domain of the function defined by $f(x) = \log_2(x - 1)$ is $(1, \infty)$, since logarithms can be found only for positive numbers. To find some ordered pairs to plot, use the equivalent equations

$$x - 1 = 2^y \quad \text{or} \quad x = 2^y + 1,$$

choosing values for y and then calculating each of the corresponding x-values. See Figure 12.

(b) $f(x) = (\log_3 x) - 1$

This function has the same graph as $g(x) = \log_3 x$ translated down one unit. Ordered pairs to plot can be found by writing $y = (\log_3 x) - 1$ as follows.

$$y = (\log_3 x) - 1$$
$$y + 1 = \log_3 x$$
$$x = 3^{y+1}$$

Again, it is easier to choose y-values and calculate the corresponding x-values. The graph is shown in Figure 13. ◗

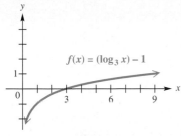

$f(x) = (\log_3 x) - 1$

FIGURE 13

CAUTION If you write a logarithmic function in exponential form, choosing y-values to calculate x-values as we did in Example 3, be careful to get the ordered pairs in the correct order.

Since a logarithmic statement can be written as an exponential statement, it is not surprising that there are properties of logarithms based on the properties of exponents. The properties of logarithms allow us to change the form of logarithmic statements so that products can be converted to sums, quotients can be converted to differences, and powers can be converted to products. These properties will be used to solve logarithmic and exponential equations later in this chapter.

> **Properties of Logarithms**
>
> If x and y are any positive real numbers, r is any real number, and a is any positive real number, $a \neq 1$, then the following properties are true.
>
> **(a)** $\log_a xy = \log_a x + \log_a y$ **(b)** $\log_a \dfrac{x}{y} = \log_a x - \log_a y$
>
> **(c)** $\log_a x^r = r \log_a x$ **(d)** $\log_a a = 1$
>
> **(e)** $\log_a 1 = 0$

Proof   To prove Property (a), let

$$m = \log_a x \quad \text{and} \quad n = \log_a y.$$

Change to exponential form.

$$a^m = x \quad \text{and} \quad a^n = y$$

Multiplication gives

$$a^m \cdot a^n = xy.$$

By a property of exponents,

$$a^{m+n} = xy.$$

Now use the definition of logarithm to write

$$\log_a xy = m + n.$$

Since $m = \log_a x$ and $n = \log_a y$,

$$\log_a xy = \log_a x + \log_a y. \quad \blacktriangleright$$

Properties (b) and (c) are proven in a similar way. Properties (d) and (e) follow directly from the definition of logarithm since $a^1 = a$ and $a^0 = 1$.

The properties of logarithms are useful for rewriting expressions with logarithms in different forms, as shown in the next examples.

EXAMPLE 4
Using the properties of logarithms

Assuming that all variables represent positive real numbers, use the properties of logarithms to rewrite each of the following expressions.

(a) $\log_6(7 \cdot 9)$
$$\log_6(7 \cdot 9) = \log_6 7 + \log_6 9$$

(b) $\log_9\left(\dfrac{15}{7}\right)$
$$\log_9\left(\dfrac{15}{7}\right) = \log_9 15 - \log_9 7$$

(c) $\log_5 \sqrt{8}$
$$\log_5 \sqrt{8} = \log_5(8^{1/2}) = \frac{1}{2} \log_5 8$$

(d) $\log_a\left(\dfrac{mnq}{p^2}\right) = \log_a m + \log_a n + \log_a q - 2 \log_a p$

(e) $\log_a \sqrt[3]{m^2} = \dfrac{2}{3} \log_a m$

(f) $\log_b \sqrt[n]{\dfrac{x^3 y^5}{z^m}} = \dfrac{1}{n} \log_b\left(\dfrac{x^3 y^5}{z^m}\right)$

$$= \frac{1}{n}(\log_b(x^3) + \log_b(y^5) - \log_b(z^m))$$

$$= \frac{1}{n}(3 \log_b x + 5 \log_b y - m \log_b z)$$

$$= \frac{3}{n} \log_b x + \frac{5}{n} \log_b y - \frac{m}{n} \log_b z$$

Notice the use of parentheses in the second step. The factor $1/n$ applies to each term. $\quad \blacktriangleright$

EXAMPLE 5
Using the properties of logarithms

Use the properties of logarithms to write each of the following as a single logarithm with a coefficient of 1. Assume that all variables represent positive real numbers.

(a) $\log_3(x + 2) + \log_3 x - \log_3 2$
Using Properties (a) and (b),

$$\log_3(x + 2) + \log_3 x - \log_3 2 = \log_3\left[\frac{(x + 2)x}{2}\right].$$

(b) $2 \log_a m - 3 \log_a n = \log_a(m^2) - \log_a(n^3) = \log_a\left(\frac{m^2}{n^3}\right)$

Here we used Property (c), then Property (b).

(c) $\frac{1}{2} \log_b m + \frac{3}{2} \log_b(2n) - \log_b(m^2 n)$

$$= \log_b(m^{1/2}) + \log_b[(2n)^{3/2}] - \log_b(m^2 n) \qquad \text{Property (c)}$$

$$= \log_b\left(\frac{m^{1/2}(2n)^{3/2}}{m^2 n}\right) \qquad \text{Properties (a) and (b)}$$

$$= \log_b\left(\frac{2^{3/2} n^{1/2}}{m^{3/2}}\right) \qquad \text{Rules for exponents}$$

$$= \log_b\left[\left(\frac{2^3 n}{m^3}\right)^{1/2}\right] \qquad \text{Rules for exponents}$$

$$= \log_b \sqrt{\frac{8n}{m^3}} \qquad \text{Definition of } a^{1/n} \quad \blacktriangleright$$

CAUTION There is no property of logarithms to rewrite a logarithm of a *sum* or *difference*. That is why, in Example 5(a), $\log_3(x + 2)$ was not written as $\log_3 x + \log_3 2$. Remember, $\log_3 x + \log_3 2 = \log_3(x \cdot 2)$.

The distributive property does not apply here, because $\log(x + y)$ is one term; "log" is not a factor.

EXAMPLE 6
Using the properties of logarithms with numerical values

Assume that $\log_{10} 2 = .3010$. Find the base 10 logarithms of 4 and 5.
By the properties of logarithms,

$$\log_{10} 4 = \log_{10}(2^2) = 2 \log_{10} 2 = 2(.3010) = .6020$$

$$\log_{10} 5 = \log_{10}\left(\frac{10}{2}\right) = \log_{10} 10 - \log_{10} 2 = 1 - .3010 = .6990.$$

We used Property (d) to replace $\log_{10} 10$ with 1. $\quad \blacktriangleright$

Compositions of the exponential and logarithmic functions can be used to get two more useful properties. If $f(x) = a^x$ and $g(x) = \log_a x$, then

$$f[g(x)] = a^{\log_a x}$$

and
$$g[f(x)] = \log_a(a^x).$$

Theorem on Inverses

For $a > 0$, $a \neq 1$:

$$a^{\log_a x} = x \qquad \text{and} \qquad \log_a(a^x) = x.$$

Proof Exponential and logarithmic functions are inverses of each other, so $f[g(x)] = x$ and $g[f(x)] = x$. Letting $f(x) = a^x$ and $g(x) = \log_a x$ gives both results. ▶

By the results of the last theorem,

$$\log_5 5^3 = 3, \qquad 7^{\log_7 10} = 10, \qquad \text{and} \quad \log_r r^{k+1} = k + 1.$$

The second statement in the theorem will be useful later when solving logarithmic or exponential equations.

9.2 Exercises ▼▼

Complete the statement or answer the question in Exercises 1–4.

1. $y = \log_a x$ if and only if _____ .

2. What is wrong with the expression $y = \log_b$?

3. The statement $\log_5 125 = 3$ tells us that _____ is the power of _____ that equals _____.

4. Let $f(x) = \log_a x$. If $a > 1$, f is a(n) _____ function; if $0 < a < 1$, f is a(n) _____ function.

For each statement, write an equivalent statement in logarithmic form. See Example 1.

5. $3^4 = 81$ 6. $2^5 = 32$ 7. $(2/3)^{-3} = 27/8$ 8. $10^{-4} = .0001$

For each statement, write an equivalent statement in exponential form. See Example 1.

9. $\log_6 36 = 2$ 10. $\log_5 5 = 1$ 11. $\log_{\sqrt{3}} 81 = 8$ 12. $\log_4\left(\dfrac{1}{64}\right) = -3$

13. Explain why logarithms of negative numbers are not defined.

14. Why does $\log_a 1$ always equal 0 for any valid base a?

Find the value of each expression. (Hint: In Exercises 15–20, let the expression equal y, and write in exponential form.)

15. $\log_5 25$

16. $\log_3 81$

17. $\log_{10} .001$

18. $\log_6\left(\dfrac{1}{216}\right)$

19. $\log_4\left(\dfrac{\sqrt[3]{4}}{2}\right)$

20. $\log_9\left(\dfrac{\sqrt[4]{27}}{3}\right)$

21. $2^{\log_2 9}$

22. $8^{\log_8 11}$

Solve the equation. See Example 2.

23. $x = \log_2 32$

24. $x = \log_2 128$

25. $\log_x 25 = -2$

26. $\log_x\left(\dfrac{1}{16}\right) = -2$

27. Compare the summary of facts about the graph of $f(x) = \log_a x$ with the similar summary about the graph of $f(x) = a^x$ in Section 9.1. Make a list of the facts that reinforce the idea that these are inverse functions.

28. Graph each function. Compare the graphs to that of $f(x) = \log_2 x$. See Example 3.
   (a) $f(x) = (\log_2 x) + 3$ **(b)** $f(x) = \log_2(x + 3)$ **(c)** $f(x) = |\log_2(x + 3)|$

29. Graph each function. Compare the graphs to that of $f(x) = \log_{1/2} x$. See Example 3.
   (a) $f(x) = (\log_{1/2} x) - 2$ **(b)** $f(x) = \log_{1/2}(x - 2)$ **(c)** $f(x) = |\log_{1/2}(x - 2)|$

30. A calculator-generated graph of $y = \log_2 x$ is shown with the values of the ordered pair with $x = 5$. What does the value of y represent?

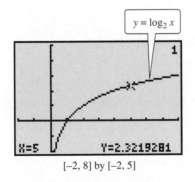

[−2, 8] by [−2, 5]

Graph each function. See Example 3.

31. $f(x) = \log_3 x$

32. $f(x) = \log_{10} x$

33. $f(x) = \log_{1/2}(1 - x)$

34. $f(x) = \log_{1/3}(3 - x)$

35. $f(x) = \log_3(x - 1)$

36. $f(x) = \log_2(x^2)$

37. Graph $y = \log_{10} x^2$ and $y = 2\log_{10} x$ on separate viewing screens. (Use the log key on your calculator; base ten is understood.) It would seem at first glance that by applying the power rule for logarithms, these graphs should be the same. Are they? If not, why not? (*Hint:* Consider the domain in each case.)

For each function, identify the corresponding graph below.

38. $f(x) = \log_2 x$

39. $f(x) = \log_2(2x)$

40. $f(x) = \log_2\left(\dfrac{1}{x}\right)$

41. $f(x) = \log_2\left(\dfrac{x}{2}\right)$

42. $f(x) = \log_2(x - 1)$

43. $f(x) = \log_2(-x)$

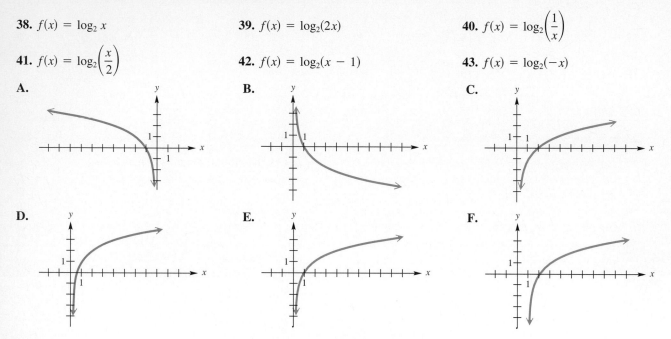

A.

B.

C.

D.

E.

F.

▦ *Use the* log *key on your graphing calculator (for* $\log_{10} x$ *) to graph each function.*

44. $f(x) = x \log_{10} x$

45. $f(x) = x^2 \log_{10} x$

▦ *Use a graphing calculator to estimate the solution(s) of each equation to the nearest hundredth.*

46. $\log_{10} x = x - 2$

47. $2^{-x} = \log_{10} x$

▼▼▼▼▼▼▼▼▼▼▼▼▼▼ **DISCOVERING CONNECTIONS** (Exercises 48–51) ▼▼▼▼▼▼▼▼▼▼▼▼▼▼

Do Exercises 48–51 in order.

48. Complete the following statement of the quotient rule for logarithms: If x and y are positive numbers, $\log_a \dfrac{x}{y} = $ _____ .

49. Use the quotient rule to explain how the graph of $f(x) = \log_2\left(\dfrac{x}{4}\right)$ can be obtained from the graph of $g(x) = \log_2 x$ by a vertical shift.

50. Graph f and g on the same axes and explain how these graphs support your answer in Exercise 49.

51. If $x = 4$, $\log_2\left(\dfrac{x}{4}\right) = $ _____ ; since $\log_2 x = $ _____ and $\log_2 4 = $ _____ , $\log_2 x - \log_2 4 = $ _____ . How does this support the quotient rule stated in Exercise 48?

Write each expression as a sum, difference, or product of logarithms. Simplify the result if possible. Assume that all variables represent positive real numbers. See Example 4.

52. $\log_2\left(\dfrac{6x}{y}\right)$ **53.** $\log_3\left(\dfrac{4p}{q}\right)$ **54.** $\log_5\left(\dfrac{5\sqrt{7}}{3}\right)$ **55.** $\log_2\left(\dfrac{2\sqrt{3}}{5}\right)$

56. $\log_4(2x + 5y)$ **57.** $\log_6(7m + 3q)$ **58.** $\log_m\sqrt{\dfrac{5r^3}{z^5}}$ **59.** $\log_p\sqrt[3]{\dfrac{m^5n^4}{t^2}}$

Write each expression as a single logarithm with a coefficient of 1. *Assume that all variables represent positive real numbers. See Example 5.*

60. $\log_a x + \log_a y - \log_a m$ **61.** $(\log_b k - \log_b m) - \log_b a$

62. $2\log_m a - 3\log_m(b^2)$ **63.** $\dfrac{1}{2}\log_y(p^3q^4) - \dfrac{2}{3}\log_y(p^4q^3)$

64. $2\log_a(z - 1) + \log_a(3z + 2), \quad z > 1$ **65.** $\log_b(2y + 5) - \dfrac{1}{2}\log_b(y + 3)$

Given $\log_{10} 2 = .3010$ *and* $\log_{10} 3 = .4771,$ *find each logarithm without using a calculator. See Example 6.*

66. $\log_{10} 6$ **67.** $\log_{10} 12$ **68.** $\log_{10}(9/4)$ **69.** $\log_{10}(20/27)$

Suppose f is a logarithmic function and $f(3) = 2$. *Determine the function values in Exercises 70–71.*

70. $f(1/9)$ **71.** $f(27)$

72. The following table lists the interest rates for various U.S. Treasury Securities in January 1996. (*Source*: Reuters.)

Time	Yield
3-month	5.71%
6-month	6.37%
1-year	6.87%
2-year	7.34%
3-year	7.52%
5-year	7.63%
10-year	7.68%
30-year	7.79%

(a) Plot the data.
(b) Discuss which type of function will model these data best: linear, exponential, or logarithmic.

9.3 Evaluating Logarithms; Change of Base ▼▼▼

COMMON LOGARITHMS The bases 10 and e are so important for logarithms that scientific and graphing calculators have keys for these bases. Base 10 logarithms are called **common logarithms.** The common logarithm of the number x, or $\log_{10} x$, is often abbreviated as just $\log x$, and we will use that convention from now on. A calculator with a log key can be used to find base 10 logarithms of any positive number. (If your calculator has an ln key, but not a log key, you will need to use the *change-of-base theorem* discussed later in this section.)

▌**EXAMPLE 1**
Evaluating common logarithms

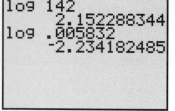

Compare these results to those in Example 1.

Use a calculator to evaluate the following logarithms.

(a) log 142

Enter 142 and press the log key. This may be a second function key on some calculators. With other calculators, these steps may be reversed. Consult your owner's manual if you have any problem using this key. The result should be 2.152 to the nearest thousandth. (This means that $10^{2.152} \approx 142$.)

(b) log .005832

A calculator gives
$$\log .005832 \approx -2.234.$$

(Thus, $10^{-2.234} \approx .005832$.) ▶

NOTE Base a, $a > 1$, logarithms of numbers less than 1 are always negative, as suggested by the graphs in Section 9.2.

In chemistry, the pH of a solution is defined as
$$pH = -\log[H_3O^+],$$
where $[H_3O^+]$ is the hydronium ion concentration in moles* per liter. The pH value is a measure of the acidity or alkalinity of solutions. Pure water has a pH of 7.0, substances with pH values greater than 7.0 are alkaline, and substances with pH values less than 7.0 are acidic.

▌**EXAMPLE 2**
Finding pH

(a) Find the pH of a solution with $[H_3O^+] = 2.5 \times 10^{-4}$.
$$
\begin{aligned}
pH &= -\log[\mathbf{H_3O^+}] \\
pH &= -\log(\mathbf{2.5 \times 10^{-4}}) &&\text{Substitute.} \\
&= -(\log 2.5 + \log 10^{-4}) &&\text{Property (a) of logarithms} \\
&= -(.3979 - 4) \\
&= -.3979 + 4 \\
&\approx 3.6
\end{aligned}
$$

It is customary to round pH values to the nearest tenth.

---

*A *mole* is the amount of a substance that contains the same number of molecules as the number of atoms in exactly 12 grams of carbon 12.

(b) Find the hydronium ion concentration of a solution with pH = 7.1.

$$\text{pH} = -\log[H_3O^+]$$
$$7.1 = -\log[H_3O^+] \qquad \text{Substitute.}$$
$$-7.1 = \log[H_3O^+] \qquad \text{Multiply by } -1.$$
$$[H_3O^+] = 10^{-7.1} \qquad \text{Write in exponential form.}$$

Evaluate $10^{-7.1}$ with a calculator to get

$$[H_3O^+] \approx 7.9 \times 10^{-8}. \quad \blacktriangleright$$

◖EXAMPLE 3
Measuring the loudness of sound

The loudness of sounds is measured in a unit called a *decibel*. To measure with this unit, we first assign an intensity of I_0 to a very faint sound, called the *threshold sound*. If a particular sound has intensity I, then the decibel rating of this louder sound is

$$d = 10 \log \frac{I}{I_0}.$$

Find the decibel rating of a sound with intensity $10{,}000I_0$.
   Let $I = 10{,}000I_0$ and find d.

$$d = 10 \log \frac{10{,}000I_0}{I_0}$$
$$= 10 \log 10{,}000$$
$$= 10(4) \qquad \log 10{,}000 = 4$$
$$= 40$$

The sound has a decibel rating of 40. $\quad \blacktriangleright$

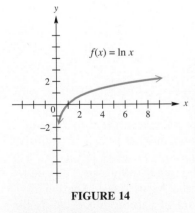

$f(x) = \ln x$

FIGURE 14

NATURAL LOGARITHMS   In most practical applications of logarithms, the number $e \approx 2.718281828$ is used as base. The number e is irrational, like π. Logarithms to base e are called **natural logarithms,** since they occur in the life sciences and economics in natural situations that involve growth and decay. The base e logarithm of x is written $\ln x$ (read "el-en x"). A graph of the natural logarithm function defined by $f(x) = \ln x$ is given in Figure 14.
   Natural logarithms can be found with a calculator that has an ln key.

◖EXAMPLE 4
Evaluating natural logarithms

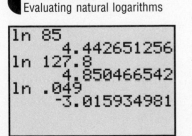

Compare these results to those in Example 4.

Use a calculator to find the following logarithms.

(a) ln 85
   With a calculator, enter 85, press the ln key, and read the result, 4.4427. The steps may be reversed with some calculators. If your calculator has an e^x key, but not a key labeled ln x, natural logarithms can be found by entering the number, pressing the INV key and then the e^x key. This works because $y = e^x$ is the inverse function of $y = \ln x$ (or $y = \log_e x$). Because $\ln 85 \approx 4.4427$, $e^{4.4427} \approx 85$.

(b) $\ln 127.8 \approx 4.850$

(c) $\ln .049 \approx -3.02$

As with common logarithms, natural logarithms of numbers between 0 and 1 are negative. ▶

EXAMPLE 5
Measuring the age of rocks

Geologists sometimes measure the age of rocks by using "atomic clocks." By measuring the amounts of potassium 40 and argon 40 in a rock, the age t of the specimen in years is found with the formula

$$t = (1.26 \times 10^9) \frac{\ln[1 + 8.33 \, (A/K)]}{\ln 2}.$$

A and K are respectively the numbers of atoms of argon 40 and potassium 40 in the specimen.

(a) How old is a rock in which $A = 0$ and $K > 0$?

If $A = 0$, $A/K = 0$ and the equation becomes

$$t = (1.26 \times 10^9) \frac{\ln 1}{\ln 2} = (1.26 \times 10^9)(0) = 0.$$

The rock is 0 years old or new.

(b) The ratio A/K for a sample of granite from New Hampshire is .212. How old is the sample?

Since A/K is .212, we have

$$t = (1.26 \times 10^9) \frac{\ln[1 + 8.33(.212)]}{\ln 2} \approx 1.85 \times 10^9.$$

The granite is about 1.85 billion years old. ▶

EXAMPLE 6
Analyzing global
temperature increase

 Carbon dioxide in the atmosphere traps heat from the sun. The additional solar radiation trapped by carbon dioxide is called *radiative forcing*. It is measured in watts per square meter. In 1896 the Swedish scientist Svante Arrhenius estimated the radiative forcing R caused by additional atmospheric carbon dioxide using the logarithmic equation $R = k \ln(C/C_0)$, where C_0 is the preindustrial amount of carbon dioxide, C is the current carbon dioxide level, and k is a constant. Arrhenius determined that $10 \leq k \leq 16$ when $C = 2C_0$.*

(a) Let $C = 2C_0$. Is the relationship between R and k linear or logarithmic?

If $C = 2C_0$, $C/C_0 = 2$, so $R = k \ln 2$ is a linear relation, because $\ln 2$ is a constant.

---

*Source: Clime, W., *The Economics of Global Warming*, Institute for International Economics, Washington, D.C., 1992.

(b) The average global temperature increase T (in °F) is given by $T(R) = 1.03R$. See Section 9.1 Exercise 59. Write T as a function of k.

Use the expression for R given in the introduction on the previous page.

$$T(R) = 1.03R$$
$$T(k) = 1.03k \ln(C/C_0) \quad \blacktriangleright$$

LOGARITHMS TO OTHER BASES   A calculator can be used to find the values of either natural logarithms (base e) or common logarithms (base 10). However, sometimes it is convenient to use logarithms to other bases. For example, base 2 logarithms are important in computer science. The following theorem can be used to convert logarithms from one base to another.

Change-of-Base Theorem

For any positive real numbers x, a, and b, where $a \neq 1$ and $b \neq 1$:

$$\log_a x = \frac{\log_b x}{\log_b a}.$$

NOTE   As an aid is remembering the change-of-base theorem, notice that x is above a on both sides of the equation.

This theorem is proved by using the definition of logarithm to write $y = \log_a x$ in exponential form.

Proof

Let

$$y = \log_a x.$$

$$a^y = x \qquad \text{Change to exponential form.}$$

$$\log_b a^y = \log_b x \qquad \text{Take logarithms on both sides.}$$

$$y \log_b a = \log_b x \qquad \text{Property (c) of logarithms}$$

$$y = \frac{\log_b x}{\log_b a} \qquad \text{Divide both sides by } \log_b a.$$

$$\log_a x = \frac{\log_b x}{\log_b a} \qquad \text{Substitute } \log_a x \text{ for } y. \quad \blacktriangleright$$

Any positive number other than 1 can be used for base b in the change of base theorem, but usually the only practical bases are e and 10, since calculators give logarithms only for these two bases. The change-of-base theorem is used to find logarithms for other bases.

> 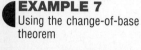 The change-of-base theorem is needed to graph logarithmic functions with bases other than 10 and e (and sometimes with one of those bases). For instance,
>
> to graph $y = \log_3(x - 1)$, graph $y = \dfrac{\log(x - 1)}{\log 3}$ or $y = \dfrac{\ln(x - 1)}{\ln 3}$.

The next example shows how the change-of-base theorem is used to find logarithms to bases other than 10 or e with a calculator.

EXAMPLE 7
Using the change-of-base theorem

The result of Example 7(a) is valid for *either* natural or common logarithms.

Use natural logarithms to find each of the following. Round to the nearest hundredth.

(a) $\log_5 17$

Use natural logarithms and the change-of-base theorem.

$$
\begin{aligned}
\log_5 17 &= \frac{\log_e 17}{\log_e 5} \\
&= \frac{\ln 17}{\ln 5} \\
&\approx \frac{2.8332}{1.6094} \\
&\approx 1.76
\end{aligned}
$$

To check, use a calculator along with the definition of logarithm, to verify that $5^{1.76} \approx 17$.

(b) $\log_2 .1$

$$
\log_2 .1 = \frac{\ln .1}{\ln 2} \approx \frac{-2.3026}{.6931} \approx -3.32 \quad \blacktriangleright
$$

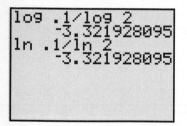

The result of Example 7(b) is valid for *either* common or natural logarithms.

NOTE In Example 7, logarithms evaluated in the intermediate steps, such as ln 17 and ln 5, were shown to four decimal places. However, the final answers were obtained *without* rounding off these intermediate values, using all the digits obtained with the calculator. In general, it is best to wait until the final step to round off the answer; otherwise, a build-up of round-off error may cause the final answer to have an incorrect final decimal place digit.

EXAMPLE 8
Solving an application with base 2 logarithms

One measure of the diversity of the species in an ecological community is given by the formula

$$
H = -[P_1 \log_2 P_1 + P_2 \log_2 P_2 + \cdots + P_n \log_2 P_n],
$$

where $P_1, P_2, \ldots, P_n$ are the proportions of a sample belonging to each of n species found in the sample. For example, in a community with two species, where there are 90 of one species and 10 of the other, $P_1 = 90/100 = .9$ and $P_2 = 10/100 = .1$. Thus,

$$H = -[.9 \log_2 .9 + .1 \log_2 .1].$$

In Example 7(b), $\log_2 .1$ was found to be -3.32. Now find $\log_2 .9$.

$$\log_2 .9 = \frac{\ln .9}{\ln 2}$$
$$\approx \frac{-.1054}{.6931}$$
$$\approx -.152$$

Therefore,

$$H \approx -[(.9)(-.152) + (.1)(-3.32)] \approx .469.$$

If the number in each species is the same, the measure of diversity is 1, representing "perfect" diversity. In a community with little diversity, H is close to 0. In this example, since $H \approx .5$, there is neither great nor little diversity. ▶

9.3 Exercises ▼▼▼▼▼▼▼▼▼▼▼▼▼▼▼▼▼▼▼▼▼▼▼▼▼▼▼▼▼▼▼▼▼▼▼▼

Use a calculator to evaluate each logarithm to four decimal places. See Examples 1 and 4.

1. log 43

2. log 1247

3. log .014

4. log .0069

5. ln 580

6. ln .08

7. ln .7

8. ln 81,000

9. The graph of $y = \log x$ is shown with the coordinates of a point displayed at the bottom of the screen. Write the logarithmic equation associated with the display.

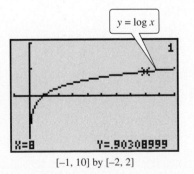

$y = \log x$

X=8   Y=.90308999

[−1, 10] by [−2, 2]

10. The graph of $y = \ln x$ is shown with the coordinates of a point displayed at the bottom of the screen. Write the logarithmic equation associated with the display.

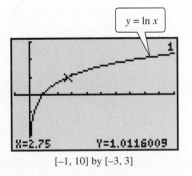

$y = \ln x$

X=2.75   Y=1.0116009

[−1, 10] by [−3, 3]

11. Is the logarithm to the base 3 of 4 written as $\log_4 3$ or $\log_3 4$?

▼▼▼▼▼▼▼▼▼▼▼▼▼ **DISCOVERING CONNECTIONS** (Exercises 12–17) ▼▼▼▼▼▼▼▼▼▼▼▼▼

Work Exercises 12–17 in order.

12. What is the exact value of $\log_3 9$?

13. What is the exact value of $\log_3 27$?

14. Between what two consecutive integers must $\log_3 16$ lie? Explain your answer.

15. Use the change-of-base theorem to support your answer for Exercise 14.

16. Repeat Exercises 12 and 13 for $\log_5(1/5)$ and $\log_5 1$.

17. Repeat Exercises 14 and 15 for $\log_5 .68$.

Use the change-of-base theorem to find each logarithm to the nearest hundredth. See Example 7.

18. $\log_5 10$ **19.** $\log_9 12$ **20.** $\log_{15} 5$ **21.** $\log_{1/2} 3$

22. $\log_{100} 83$ **23.** $\log_{200} 175$ **24.** $\log_{2.9} 7.5$ **25.** $\log_{5.8} 12.7$

26. Consider the function defined by $f(x) = \log_3 |x|$.
 (a) What is the domain of this function?
 (b) Use a graphing calculator to graph $f(x) = \log_3 |x|$ in the window $[-4, 4]$ by $[-4, 4]$.
 (c) How might one easily misinterpret the domain of the function simply by observing the calculator-generated graph?

27. The table is for $Y_1 = \log_3(4 - x)$. Why do the values of Y_1 show ERROR for $X \geq 4$?

X	Y1	
1	1	
2	.63093	
3	0	
4	ERROR	
5	ERROR	
6	ERROR	
7	ERROR	
X=1		

Graph each function.

28. $f(x) = \log_5 x$

29. $f(x) = \log_x 5$ in *connected* mode. What does the vertical line in the graph simulate?

30. Explain the error in the following "proof" that $2 < 1$.

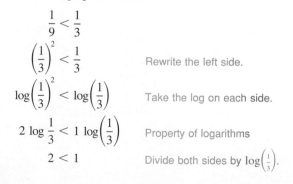

$$\frac{1}{9} < \frac{1}{3}$$

$$\left(\frac{1}{3}\right)^2 < \frac{1}{3} \qquad \text{Rewrite the left side.}$$

$$\log\left(\frac{1}{3}\right)^2 < \log\left(\frac{1}{3}\right) \qquad \text{Take the log on each side.}$$

$$2 \log \frac{1}{3} < 1 \log\left(\frac{1}{3}\right) \qquad \text{Property of logarithms}$$

$$2 < 1 \qquad \text{Divide both sides by } \log\left(\frac{1}{3}\right).$$

For each substance, find the pH from the given hydronium ion concentration. See Example 2(a).

31. Grapefruit, 6.3×10^{-4}

32. Crackers, 3.9×10^{-9}

33. Limes, 1.6×10^{-2}

34. Sodium hydroxide (lye), 3.2×10^{-14}

Find the $[H_3O^+]$ for each substance with the given pH. See Example 2(b).

35. Soda pop, 2.7   **36.** Wine, 3.4   **37.** Beer, 4.8   **38.** Drinking water, 6.5

Solve each problem. See Example 3.

39. Find the decibel ratings of sounds having the following intensities:
  (a) $100I_0$ (b) $1000I_0$
  (c) $100,000I_0$ (d) $1,000,000I_0$.

40. Find the decibel ratings of the following sounds, having intensities as given. Round each answer to the nearest whole number.
  (a) whisper, $115I_0$
  (b) busy street, $9,500,000I_0$
  (c) heavy truck, 20 m away, $1,200,000,000I_0$
  (d) rock music, $895,000,000,000I_0$
  (e) jetliner at takeoff, $109,000,000,000,000I_0$

41. The magnitude of an earthquake, measured on the Richter scale, is $\log_{10}(I/I_0)$, where I_0 is the magnitude of an earthquake of a certain (small) size. Find the Richter scale ratings for earthquakes having the following magnitudes.
  (a) $1000I_0$ (b) $1,000,000I_0$ (c) $100,000,000I_0$

42. On July 14, 1991, Peshawar, Pakistan, was shaken by an earthquake that measured 6.6 on the Richter scale.
  (a) Express this reading in terms of I_0. See Exercise 41.
  (b) In February of the same year a quake measuring 6.5 on the Richter scale killed about 900 people in the mountains of Pakistan and Afghanistan. Express the magnitude of a 6.5 reading in terms of I_0.
  (c) How much greater was the force of the earthquake with a measure of 6.6?

43. (a) The San Francisco earthquake of 1906 had a Richter scale rating of 8.3. Express the magnitude of this earthquake as a multiple of I_0.
  (b) In 1989, the San Francisco region experienced an earthquake with a Richter scale rating of 7.1. Express the magnitude of this earthquake as a multiple of I_0.

(c) Compare the magnitudes of the two San Francisco earthquakes.

Solve each problem. See Example 5.

44. The number of years, n, since two independently evolving languages split off from a common ancestral language is approximated by $n \approx -7600 \log r$, where r is the proportion of words from the ancestral language common to both languages.
  (a) Find n if $r = .9$.
  (b) Find n if $r = .3$.
  (c) How many years have elapsed since the split if half of the words of the ancestral language are common to both languages?

45. The number of species in a sample is given by

$$S(n) = a \ln\left(1 + \frac{n}{a}\right).$$

Here n is the number of individuals in the sample and a is a constant that indicates the diversity of species in the community. If $a = .36$, find $S(n)$ for the following values of n. (*Hint: n must be a whole number.*)
  (a) 100   (b) 200   (c) 150   (d) 10

46. In Exercise 45, find $S(n)$ if a changes to .88. Use the following values of n. (*Hint: $S(n)$ must be a whole number.*)
  (a) 50   (b) 100   (c) 250

In Exercises 47 and 48, refer to Example 8.

47. Suppose a sample of a small community shows two species with 50 individuals each. Find the index of diversity H.

48. A virgin forest in northwestern Pennsylvania has 4 species of large trees with the following proportions of each: hemlock, .521; beech, .324; birch, .081; maple, .074. Find the index of diversity H.

49. Given $g(x) = e^x$, evaluate the following.

   (a) $g(\ln 3)$ **(b)** $g[\ln(5^2)]$ **(c)** $g\left[\ln\left(\dfrac{1}{e}\right)\right]$

51. Given $f(x) = \ln x$, evaluate the following.

   (a) $f(e^5)$ **(b)** $f(e^{\ln 3})$ **(c)** $f(e^{2\ln 3})$

50. Given $f(x) = 3^x$, evaluate the following.

   (a) $f(\log_3 7)$ **(b)** $f[\log_3(\ln 3)]$

   (c) $f[\log_3(2 \ln 3)]$

52. Given $f(x) = \log_2 x$, evaluate the following.

   (a) $f(2^3)$ **(b)** $f(2^{\log_2 2})$ **(c)** $f(2^{2\log_2 2})$

---

53. The heights in the bar graph represent the number of visitors (in millions) to U.S. National Parks from 1950 to 1994. Suppose x represents the number of years since 1900—thus, 1950 is represented by 50, 1960 is represented by 60, and so on. The logarithmic function defined by $f(x) = -266 + 72 \ln x$ closely approximates the data. Use this function to estimate the number of visitors in the year 2000. What assumption must we make to estimate the number of visitors in years beyond 1993?

Visitors to National Parks

In millions

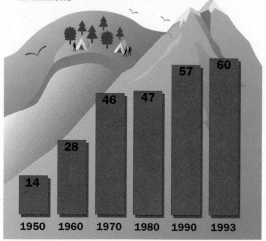

Source: Statistical Abstract of the United States 1995

54. The growth of outpatient surgery as a percent of total surgeries at hospitals is shown in the accompanying graph. Connecting the tops of the bars with a continuous curve would give a graph that indicates logarithmic growth. The function with $f(x) = -1317 + 304 \ln x$, where x represents the number of years since

1900 and $f(x)$ is the percent, approximates the curve reasonably well. What does this function predict for the percent of outpatient surgeries in 1998?

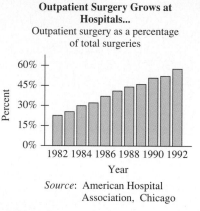

Outpatient Surgery Grows at Hospitals...

Outpatient surgery as a percentage of total surgeries

Source: American Hospital Association, Chicago

55. In Example 6, we expressed the average global temperature increase T (in °F) as $T(k) = 1.03k \ln(C/C_0)$ where C_0 is the preindustrial amount of carbon dioxide, C is the current carbon dioxide level, and k is a constant. Arrhenius determined that $10 \le k \le 16$ when C was double the value C_0. Use $T(k)$ to find the range of the rise in global temperature T (rounded to the nearest degree) that Arrhenius predicted. (*Source:* Clime, W., *The Economics of Global Warming,* Institute for International Economics, Washington, D.C., 1992.)

56. Refer to Exercise 55. According to the IPCC if present trends continue, future increases in average global temperatures (in °F) can be modeled by $T = 6.489 \cdot \ln(C/280)$ where C is the concentration of atmospheric carbon dioxide (in ppm). C can be modeled by the function with $C(x) = 353(1.006)^{x-1990}$ where x is the year. (*Source:* International Panel on Climate Change (IPCC), 1990.)

(a) Write T as a function of x.

(b) With a graphing calculator graph $C(x)$ and $T(x)$ on the interval $[1990, 2275]$ using different coordinate axes. Describe each function's graph. How are C and T related?

(c) Approximate the slope of the graph of T. What does this slope represent?

(d) Use graphing to estimate x and C when $T = 10°F$.

57. The following table contains the planets' average distances D from the sun and their periods P of revolution around the sun in years. The distances have been normalized so that Earth is one unit away from the sun. For example, since Jupiter's distance is 5.2, its distance from the sun is 5.2 times farther than Earth's. (*Source:* Ronan, C., *The Natural History of the Universe,* MacMillan Publishing Co., New York, 1991.)

(a) Plot the point $(\ln D, \ln P)$ for each planet on the xy-coordinate axes using a graphing calculator. Do the data points appear to be linear?

Planet	D	P
Mercury	.39	.24
Venus	.72	.62
Earth	1	1
Mars	1.52	1.89
Jupiter	5.2	11.9
Saturn	9.54	29.5
Uranus	19.2	84.0
Neptune	30.1	164.8

(b) Determine a linear equation that approximates the data points. Graph your line and the data on the same coordinate axes.

(c) Use this linear equation to predict the period of the planet Pluto if its distance is 39.5. Compare your answer to the true value of 248.5 years.

9.4 Exponential and Logarithmic Equations ▼▼▼

Exponential and logarithmic functions are important in many useful applications of mathematics. Using these functions in applications often requires solving exponential and logarithmic equations. Some simple equations were solved in earlier sections of this chapter. More general methods for solving these equations depend on the property below. This property follows from the fact that logarithmic functions are one-to-one.

Property of Logarithms

(f) If $x > 0$, $y > 0$, $a > 0$, and $a \neq 1$, then

$$x = y \quad \text{if and only if} \quad \log_a x = \log_a y.$$

EXPONENTIAL EQUATIONS   The first examples illustrate a general method, using the new property, for solving exponential equations.

◀**EXAMPLE 1**
Solving an exponential
equation

Solve the equation $7^x = 12$.

The properties given in Section 9.1 cannot be used to solve this equation, so we apply Property (f). While any appropriate base b can be used, the best practical base is base 10 or base e. Taking base e (natural) logarithms of both sides gives

$$7^x = 12$$
$$\ln 7^x = \ln 12$$
$$x \ln 7 = \ln 12 \qquad \text{Property (c) of logarithms}$$
$$x = \frac{\ln 12}{\ln 7} \qquad \text{Divide by } \ln 7.$$
$$\approx \frac{2.4849}{1.9459} \approx 1.277.$$

A calculator can be used to check this answer. Evaluate $7^{1.277}$; the result should be approximately 12. This step verifies that, to the nearest thousandth, the solution is 1.277. ▶

CAUTION   Be careful when evaluating a quotient like $\dfrac{\ln 12}{\ln 7}$ in Example 1. Do not confuse this quotient with $\ln\left(\dfrac{12}{7}\right)$ which can be written as $\ln 12 - \ln 7$. You *cannot* change the quotient of two *logarithms* to a difference of logarithms.

$$\frac{\ln 12}{\ln 7} \neq \ln\left(\frac{12}{7}\right)$$

◀**EXAMPLE 2**
Solving an exponential
equation

Solve $3^{2x-1} = 4^{x+2}$.

Taking natural logarithms on both sides gives

$$\ln 3^{2x-1} = \ln 4^{x+2}.$$

Now use a property of logarithms.

$$(2x - 1) \ln 3 = (x + 2) \ln 4$$
$$2x \ln 3 - \ln 3 = x \ln 4 + 2 \ln 4 \qquad \text{Distributive property}$$
$$2x \ln 3 - x \ln 4 = 2 \ln 4 + \ln 3$$
$$x(2 \ln 3 - \ln 4) = 2 \ln 4 + \ln 3 \qquad \text{Factor out } x.$$
$$x = \frac{2 \ln 4 + \ln 3}{2 \ln 3 - \ln 4}$$
$$x = \frac{\ln 16 + \ln 3}{\ln 9 - \ln 4} \qquad \text{Properties of logarithms}$$
$$x = \frac{\ln 48}{\ln\left(\dfrac{9}{4}\right)}$$

This quotient could be approximated by a decimal if desired.

$$x = \frac{\ln 48}{\ln 2.25} \approx \frac{3.8712}{.8109} \approx 4.774$$

To the nearest thousandth, the solution is 4.774. ▶

EXAMPLE 3
Solving a base *e* exponential equation

Solve $e^{x^2} = 200$.

Take natural logarithms on both sides; then use properties of logarithms.

$$e^{x^2} = 200$$

$$\ln e^{x^2} = \ln 200$$

$$x^2 = \ln 200 \qquad \text{In } e^{x^2} = x^2$$

$$x = \pm\sqrt{\ln 200}$$

$$x \approx \pm 2.302 \qquad \text{To the nearest thousandth} \ \blacktriangleright$$

CONNECTIONS Sometimes it is necessary to solve more complicated exponential equations. (A graphing calculator is very handy for these!) A typical equation that arises in some situations is

$$\frac{e^x - 1}{e^{-x} - 1} = -3.$$

To solve this equation, begin by multiplying both sides by the denominator.

$$e^x - 1 = -3(e^{-x} - 1)$$

$$e^x - 1 = -3e^{-x} + 3 \qquad \text{Distributive property}$$

$$e^x - 4 + 3e^{-x} = 0 \qquad \text{Get 0 alone on one side.}$$

$$e^{2x} - 4e^x + 3 = 0 \qquad \text{Multiply both sides by } e^x.$$

Rewrite this equation as

$$(e^x)^2 - 4e^x + 3 = 0,$$

a quadratic equation in e^x. Let $u = e^x$ and solve first for u getting the solutions 1 and 3. Then solve for x.

$$u = 1 \quad \text{or} \quad u = 3$$

$$e^x = 1 \qquad\quad e^x = 3 \qquad \text{Replace } u \text{ with } e^x.$$

$$x = 0 \qquad\quad\ x = \ln 3$$

FOR DISCUSSION OR WRITING

1. Carry out the steps of the solution of the quadratic equation to find u.
2. Check the solutions for x in the original equation. One is extraneous. Explain why one solution does not satisfy the equation.
3. In solving the equation here, we used a two-step process. What are the two steps?

LOGARITHMIC EQUATIONS   The next examples show some ways to solve logarithmic equations. The properties of logarithms given in Section 9.2 are useful here, as is Property (f).

EXAMPLE 4
Solving a logarithmic equation

Solve $\log_a(x + 6) - \log_a(x + 2) = \log_a x$.

Using a property of logarithms, rewrite the equation as

$$\log_a \frac{x + 6}{x + 2} = \log_a x. \qquad \text{Property (b) of logarithms}$$

Now the equation is in the proper form to use Property (f).

$$\frac{x + 6}{x + 2} = x \qquad\qquad \text{Property (f)}$$

$$x + 6 = x(x + 2) \qquad \text{Multiply by } x + 2.$$

$$x + 6 = x^2 + 2x \qquad \text{Distributive property}$$

$$x^2 + x - 6 = 0 \qquad\quad \text{Get 0 on one side.}$$

$$(x + 3)(x - 2) = 0 \qquad \text{Use the zero-factor property.}$$

$$x = -3 \quad \text{or} \quad x = 2$$

The negative solution ($x = -3$) cannot be used since it is not in the domain of $\log_a x$ in the original equation. For this reason, the only valid solution is the positive number 2.   ▶

CAUTION   Recall that the domain of $y = \log_b x$ is $(0, \infty)$. For this reason, it is always necessary to check that the apparent solution of a logarithmic equation results in the logarithms of positive numbers in the original equation.

EXAMPLE 5
Solving a logarithmic equation

Solve $\log(3x + 2) + \log(x - 1) = 1$.

Since $\log x$ is an abbreviation for $\log_{10} x$, and $1 = \log_{10} 10$, the properties of logarithms give

$$\log(3x + 2)(x - 1) = \log 10 \qquad \text{Property (a) of logarithms}$$

$$(3x + 2)(x - 1) = 10 \qquad\quad \text{Property (f)}$$

$$3x^2 - x - 2 = 10$$

$$3x^2 - x - 12 = 0.$$

Now use the quadratic formula to get

$$x = \frac{1 \pm \sqrt{1 + 144}}{6}.$$

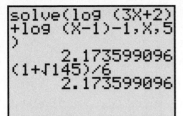

The equation-solving feature of a graphing calculator supports the result of Example 5.

The number $(1 - \sqrt{145})/6$ is negative, so $x - 1$ is negative. Therefore, $\log(x - 1)$ is not defined and this proposed solution must be discarded. Since $(1 + \sqrt{145})/6 > 1$, both $3x + 2$ and $x - 1$ are positive and the solution set is

$$\frac{1 + \sqrt{145}}{6}.$$ ▶

The definition of logarithm could have been used in Example 5 by first writing

$$\log(3x + 2) + \log(x - 1) = 1$$

$$\log_{10}(3x + 2)(x - 1) = 1 \qquad \text{Property (a)}$$

$$(3x + 2)(x - 1) = 10^1, \qquad \text{Definition of logarithm}$$

then continuing as shown above.

EXAMPLE 6
Solving a logarithmic equation

Solve $\ln e^{\ln x} - \ln(x - 3) = \ln 2$.

On the left, $\ln e^{\ln x}$ can be written as $\ln x$ using the theorem on inverses at the end of Section 9.2. The equation becomes

$$\ln x - \ln(x - 3) = \ln 2$$

$$\ln \frac{x}{x - 3} = \ln 2 \qquad \text{Property (b)}$$

$$\frac{x}{x - 3} = 2 \qquad \text{Property (f)}$$

$$x = 2x - 6 \qquad \text{Multiply by } x - 3.$$

$$6 = x.$$ ▶

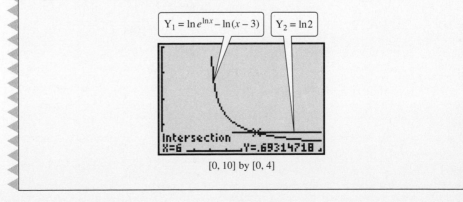

A graphing calculator is helpful for determining the number of solutions and approximating the solutions. For instance, to examine the equation in Example 6, we could graph $Y_1 = \ln e^{\ln x} - \ln(x - 3)$ and $Y_2 = \ln 2$, and find the x-values of any intersection points. The result, in the figure, supports our analytic solution.

$$Y_1 = \ln e^{\ln x} - \ln(x - 3) \qquad Y_2 = \ln 2$$

Intersection
X=6   Y=.69314718

[0, 10] by [0, 4]

A summary of the methods used for solving equations in this section follows.

Solving Exponential or Logarithmic Equations

In summary, to solve an exponential or logarithmic equation, first use the properties of algebra to change the given equation into one of the following forms, where a and b are real numbers with appropriate restrictions.

1. $a^{f(x)} = b$
   To solve, take logarithms on both sides.
2. $\log_a f(x) = b$
   Solve by changing to the exponential form $a^b = f(x)$.
3. $\log_a f(x) = \log_a g(x)$
   From the given equation, obtain the equation $f(x) = g(x)$, then solve algebraically.
4. In a more complicated equation, such as the one in the Connections box, it may be necessary to first solve for $e^{f(x)}$ or $\log_a f(x)$ and then solve the resulting equation using one of the methods given above.

The next examples show applications of exponential and logarithmic equations.

EXAMPLE 7
Solving a composite exponential equation

The strength of a habit is a function of the number of times the habit is repeated. If N is the number of repetitions and H is the strength of the habit, then, according to psychologist C. L. Hull,

$$H = 1000(1 - e^{-kN}),$$

where k is a constant. Solve this equation for k.

We must first solve the equation for e^{-kN}.

$$\frac{H}{1000} = 1 - e^{-kN} \qquad \text{Divide by 1000.}$$

$$\frac{H}{1000} - 1 = -e^{-kN} \qquad \text{Subtract 1.}$$

$$e^{-kN} = 1 - \frac{H}{1000} \qquad \text{Multiply by } -1.$$

Now solve for k. As shown earlier, we take logarithms on each side of the equation and use the fact that $\ln e^x = x$.

$$\ln e^{-kN} = \ln\left(1 - \frac{H}{1000}\right)$$

$$-kN = \ln\left(1 - \frac{H}{1000}\right) \qquad \ln e^x = x$$

$$k = -\frac{1}{N}\ln\left(1 - \frac{H}{1000}\right) \qquad \text{Multiply by } -\tfrac{1}{N}.$$

With the last equation, if one pair of values for H and N is known, k can be found, and the equation can then be used to find either H or N, for given values of the other variable. ◗

EXAMPLE 8
Solving a composite logarithmic equation

In the exercises for Section 9.3, we saw that the number of species in a sample is given by $S(n)$ or S, where

$$S = a\ln\left(1 + \frac{n}{a}\right),$$

n is the number of individuals in the sample, and a is a constant. Solve this equation for n.

We begin by solving for $\ln\left(1 + \frac{n}{a}\right)$. Then we can change to exponential form and solve the resulting equation for n.

$$\frac{S}{a} = \ln\left(1 + \frac{n}{a}\right) \qquad \text{Divide by } a.$$

$$e^{S/a} = 1 + \frac{n}{a} \qquad \text{Write in exponential form.}$$

$$e^{S/a} - 1 = \frac{n}{a} \qquad \text{Subtract 1.}$$

$$n = a(e^{S/a} - 1) \qquad \text{Multiply by } a.$$

Using this equation and given values of S and a, the number of individuals in a sample can be found. ◗

9.4 Exercises ▼▼▼▼▼▼▼▼▼▼▼▼▼▼▼▼▼▼▼▼▼▼▼▼▼▼▼▼▼▼▼▼▼▼▼▼▼▼▼

1. Between what two consecutive integers must x be if $5^x = 28$? Why is this so?

2. Is the statement $\dfrac{\log 16}{\log 3} = \dfrac{\ln 16}{\ln 3}$ true? If so, how can you support the statement?

3. Explain why an equation of the form $2^x = a$ does not always have a solution.

4. Without solving, explain why an equation of the form $\log_2 x = a$ must always have a solution for x.

Solve the equation. When necessary, give answers as decimals rounded to the nearest thousandth. See Examples 1–6.

5. $3^x = 6$

6. $4^x = 12$

7. $6^{1-2k} = 8$

8. $3^{2m-5} = 13$

9. $e^{k-1} = 4$

10. $e^{2-y} = 12$

11. $2e^{5a+2} = 8$

12. $10e^{3z-7} = 5$

13. $2^x = -3$

14. $\left(\dfrac{1}{4}\right)^p = -4$

15. $e^{2x} \cdot e^{5x} = e^{14}$

16. $e^3 \cdot e^{\ln x} = \dfrac{1}{8}$

17. $100(1 + .02)^{3+n} = 150$

18. $500(1 + .05)^{p/4} = 200$

19. $\log(t - 1) = 1$

20. $\log q^2 = 1$

21. $\ln(y + 2) = \ln(y - 7) + \ln 4$

22. $\ln p - \ln(p + 1) = \ln 5$

23. $\ln(5 + 4y) - \ln(3 + y) = \ln 3$

24. $\ln m + \ln(2m + 5) = \ln 7$

25. $2 \ln(x - 3) = \ln(x + 5) + \ln 4$

26. $\ln(k + 5) + \ln(k + 2) = \ln(14k)$

27. $\log_3(a - 3) = 1 + \log_3(a + 1)$

28. $\log w + \log(3w - 13) = 1$

29. $\ln e^x - \ln e^3 = \ln e^5$

30. $\ln e^x - 2 \ln e = \ln e^4$

31. $\log_2 \sqrt{2y^2} - 1 = \dfrac{1}{2}$

32. $\log_2(\log_2 x) = 1$

33. $\log z = \sqrt{\log z}$

34. $\log x^2 = (\log x)^2$

35. Suppose you overhear the following statement: "I must reject any negative answer when I solve an equation involving logarithms." Is this correct? Write an explanation of why it is or is not correct.

36. What values of x could not possibly be solutions of the following equation?

$$\log_a(4x - 7) + \log_a(x^2 + 4) = 0$$

Solve each equation for the indicated variable. Use logarithms to the appropriate bases. See Examples 7 and 8.

37. $I = \dfrac{E}{R}(1 - e^{-Rt/2})$ for t

38. $r = p - k \ln t$ for t

39. $p = a + \dfrac{k}{\ln x}$ for x

40. $T = T_0 + (T_1 - T_0)10^{-kt}$ for t

▼▼▼▼▼▼▼▼▼▼▼▼▼▼ **DISCOVERING CONNECTIONS** (Exercises 41–46) ▼▼▼▼▼▼▼▼▼▼▼▼▼▼

Methods of solving quadratic equations can be applied to equations that are not actually quadratic, but are quadratic in form. Consider the equation

$$e^{2x} - 4e^x + 3 = 0,$$

and work Exercises 41–46 in order.

41. The expression e^{2x} is equivalent to $(e^x)^2$. Explain why this is so.

42. The given equation is equivalent to $(e^x)^2 - 4e^x + 3 = 0$. Factor the left side of this equation.

43. Solve the equation in Exercise 42 by using the zero-factor property. Give exact values.

44. Support your solution(s) in Exercise 43 using a calculator-generated graph of $y = e^{2x} - 4e^x + 3$.

45. Use the graph from Exercise 44 to identify the x-intervals where $y > 0$. These intervals give the solutions of $e^{2x} - 4e^x + 3 > 0$.

46. Use the graph from Exercise 44 and your answer to Exercise 45 to give the intervals where $e^{2x} - 4e^x + 3 < 0$.

🖩 *Solve each equation with a graphing calculator by graphing the two functions on either side of the equals sign and finding the point(s) of intersection of the graphs. Give answers to the nearest hundredth.*

47. $e^x + \ln x = 5$

48. $e^x - \ln(x + 1) = 3$

49. $2e^x + 1 = 3e^{-x}$

50. $e^x + 6e^{-x} = 5$

51. $\log x = x^2 - 8x + 14$

52. $\ln x = -\sqrt[3]{x} + 3$

In Exercises 53 and 54 find the equation, domain, and range of $f^{-1}(x)$.

53. $f(x) = e^{3x+1}$

54. $f(x) = 3 \cdot 10^x$

🖩 *Solve each inequality with a graphing calculator by rearranging terms so that one side is zero, then graphing the expression Y on the other side and observing from the graph where Y is positive or negative as applicable. These inequalities are studied in calculus to determine where certain functions are increasing.*

55. $\log_3 x > 3$

56. $\log_x .2 < -1$

---

57. Recall from Section 9.3 that the formula for the decibel rating of the loudness of a sound is

$$d = 10 \log \frac{I}{I_0}.$$

A few years ago, there was a controversy about a proposed government limit on factory noise. One group wanted a maximum of 89 decibels, while another group wanted 86. This difference seemed very small to many people. Find the percent by which the 89-decibel intensity exceeds that for 86 decibels.

For Exercises 58–61, refer to the formula for compound interest,

$$A = P\left(1 + \frac{r}{m}\right)^{tm},$$

given in Section 9.1.

58. George Tom wants to buy a $30,000 car. He has saved $27,000. Find the number of years (to the nearest tenth) it will take for his $27,000 to grow to $30,000 at 6% interest compounded quarterly.

59. Find t to the nearest hundredth if $1786 becomes $2063.40 at 11.6%, with interest compounded monthly.

60. Find the interest rate that will produce $2500 if $2000 is left at interest compounded semiannually for 3.5 years.

61. At what interest rate will $16,000 grow to $20,000 if invested for 5.25 years, and interest is compounded quarterly?

62. The population of an animal species introduced into a certain area may grow rapidly at first but then increase more slowly as time goes on. A logarithmic function can provide an excellent description of such growth. Suppose the population of foxes in an area t months after the foxes were first introduced there is

$$F = 500 \log(2t + 3).$$

Solve the equation for t. Then find t to the nearest tenth for the following values of F.
(a) 600   **(b)** 1000

63. India has become an important exporter of software to the United States. The chart shows India's software exports (in millions of U.S. dollars) in selected years since 1985. *(Sources: NIIT, NASSCOM. From Scientific American, September, 1994, page 95.)*

Year (x)	1985	1987	1989	1991	1993	1995	1997
Million $ (y)	6	39	67	128	225	483	1000

The figure for 1997 is an estimate. Letting y represent software (in millions of dollars) and x represent the number of years since 1900, we find that the function with

$$f(x) = 6.2(10)^{-12}(1.4)^x$$

approximates the data reasonably well. According to this function, when will software exports double their 1997 value?

64. The graph shows that the percent y of U.S. children growing up without a father has increased rapidly since 1950. If x represents the number of years since 1900, the function defined by

$$f(x) = \frac{25}{1 + 1364.3e^{-x/9.316}}$$

models the data fairly well.

(a) From the graph, in what year were 20% of U.S. children living without a father?

(b) If the percent continues to increase in the same way, according to $f(x)$, in what year will 30% of U.S. children live in a home without a father?

U.S. Children in Fatherless Households

Sources: National Longitudinal Survey of Youth; U.S. Department of Commerce; Bureau of the Census

65. One action that government could take to reduce carbon emissions into the atmosphere is to place a tax on fossil fuel. This tax would be based on the amount of carbon dioxide emitted into the air when the fuel is burned. The **cost-benefit** equation $\ln(1 - P) = -.0034 - .0053T$ describes the approximate relationship between a tax of T dollars per ton of carbon and the corresponding percent reduction P (in decimal form) of emissions of carbon dioxide. (*Source:* Nordhause, W., "To Slow or Not to Slow: The Economics of the Greenhouse Effect." Yale University, New Haven, Connecticut.)

(a) Write P as a function of T.

(b) Graph P for $0 \le T \le 1000$. Discuss the benefit of continuing to raise taxes on carbon.

(c) Determine P when $T = \$60$, and interpret this result.

(d) What value of T will give a 50% reduction in carbon emissions?

66. Refer to Example 6 in Section 9.3 and Exercise 59 in Section 9.1. Using computer models the International Panel on Climate Change (IPCC) in 1990 estimated k to be 6.3 in the radiative forcing equation $R = k \ln(C/C_0)$, where C_0 is the preindustrial amount of carbon dioxide and C is the current level. (*Source:* Clime, W., *The Economics of Global Warming,* Institute for International Economics, Washington, D.C., 1992.)

(a) What radiative forcing R (in w/m^2) is expected by the IPCC if the carbon dioxide level in the atmosphere doubles from its preindustrial level?

(b) Determine the global temperature increase predicted by the IPCC if the carbon dioxide levels were to double.

Chapter 9 Summary ▼▼▼▼▼▼▼▼▼▼▼▼▼▼▼▼▼▼▼▼▼▼▼▼▼▼▼▼▼▼▼▼▼▼

SECTION	KEY IDEAS
9.1 Exponential Functions	
	Exponential Function If $a > 0$ and $a \ne 1$, then $f(x) = a^x$ defines the exponential function with base a. **The Graph of $f(x) = a^x$** **1.** The points $(0, 1)$ and $(1, a)$ are on the graph. **2.** If $a > 1$, f is an increasing function; if $0 < a < 1$, f is a decreasing function. **3.** The x-axis is a horizontal asymptote. **4.** The domain is $(-\infty, \infty)$ and the range is $(0, \infty)$.

SECTION	KEY IDEAS
9.2 Logarithmic Functions	

For all real numbers y and all positive numbers a and x, where $a \neq 1$:

$$y = \log_a x \quad \text{if and only if} \quad x = a^y.$$

Logarithmic Function
If $a > 0$, $a \neq 1$, and $x > 0$, then $f(x) = \log_a x$ defines the logarithmic function with base a.

The Graph of $f(x) = \log_a x$
1. The points $(1, 0)$ and $(a, 1)$ are on the graph.
2. If $a > 1$, f is an increasing function; if $0 < a < 1$, f is a decreasing function.
3. The y-axis is a vertical asymptote.
4. The domain is $(0, \infty)$ and the range is $(-\infty, \infty)$.

Properties of Logarithms
If x and y are positive real numbers, r is any real number, and a is any positive real number, with $a \neq 1$, then

(a) $\log_a xy = \log_a x + \log_a y$ **(b)** $\log_a \dfrac{x}{y} = \log_a x - \log_a y$

(c) $\log_a x^r = r\log_a x$ **(d)** $\log_a a = 1$ **(e)** $\log_a 1 = 0.$

SECTION	KEY IDEAS
9.3 Evaluating Logarithms; Change of Base	

Change of Base Theorem
For any positive numbers x, a, and b, where $a \neq 1$ and $b \neq 1$:

$$\log_a x = \frac{\log_b x}{\log_b a}.$$

SECTION	KEY IDEAS
9.4 Exponential and Logarithmic Equations	

Property of Logarithms
(f) If $x > 0$, $y > 0$, $b > 0$, and $b \neq 1$, then

$$\log_b x = \log_b y \quad \text{if and only if} \quad x = y.$$

Solving Exponential or Logarithmic Equations
Change the equation into one of the following forms, then solve as indicated.
1. $a^{f(x)} = b$
   To solve, take logarithms on both sides.
2. $\log_a f(x) = b$
   Solve by changing to the exponential form $a^b = f(x)$.
3. $\log_a f(x) = \log_a g(x)$

   From the given equation, obtain the equation $f(x) = g(x)$, then solve algebraically.

Chapter 9   Review Exercises ▼▼▼▼▼▼▼▼▼▼▼▼▼▼▼▼▼▼▼▼▼▼▼▼▼▼▼▼▼▼

1. $f(x) = \left(\dfrac{5}{4}\right)^x$ defines a(n) _____ function.

increasing/decreasing

2. $f(x) = \log_{2/3} x$ defines a(n) _____ function.

increasing/decreasing

Match each equation with one of the graphs below.

3. $y = \log_{.3} x$ **4.** $y = e^x$ **5.** $y = \ln x$ **6.** $y = (.3)^x$

A.   **B.**   **C.**   **D.**

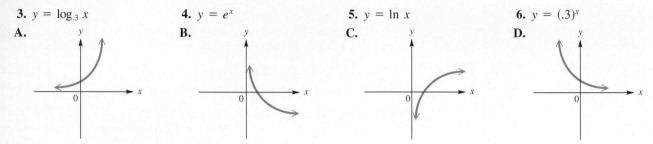

Write each equation in logarithmic form.

7. $2^5 = 32$ **8.** $100^{1/2} = 10$ **9.** $(1/16)^{1/4} = 1/2$ **10.** $(3/4)^{-1} = 4/3$ **11.** $10^{.4771} = 3$ **12.** $e^{2.4849} = 12$

13. Explain how the graph of $f(x) = 8 - 2^{x-1}$ can be obtained from the graph of $y = 2^x$ using translations and reflection.

Consider the exponential function $y = f(x) = a^x$ graphed here.

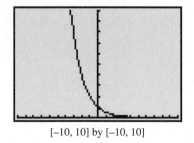

[−10, 10] by [−10, 10]

14. What is true about the value of a in comparison to 1?

15. What is the domain of f?

16. What is the range of f?

17. What is the value of $f(0)$?

18. Sketch the graph of $y = f^{-1}(x)$ by hand.

19. What is the expression that defines $f^{-1}(x)$?

Write each equation in exponential form.

20. $\log_{10} .001 = -3$ **21.** $\log_2 \sqrt{32} = \dfrac{5}{2}$ **22.** $\log 3.45 = .537819$

23. $\ln 45 = 3.806662$ **24.** $\log_9 27 = \dfrac{3}{2}$

25. One of your friends is taking a mathematics course and tells you, "I have no idea what an expression like $\log_5 27$ really means." Write a clear explanation of what it means and how you can find an approximation for it using a calculator.

26. What is the base of the logarithmic function whose graph contains the point $(81, 4)$?

27. What is the base of the exponential function whose graph contains the point $(-4, 1/16)$?

Use properties of logarithms to write each of the following logarithms as a sum, difference, or product of logarithms.

28. $\log_3\left(\dfrac{mn}{5r}\right)$

29. $\log_2\left(\dfrac{\sqrt{7}}{15}\right)$

30. $\log_5(x^2y^4\sqrt[5]{m^3p})$

31. $\log_7(7k + 5r^2)$

32. Correct the mistakes in the following equation.

$$\log_5 125 - \log_5 25 = \frac{\log_5 125}{\log_5 25} = \log_5\left(\frac{125}{25}\right) = \log_5 5 = 1$$

Find each logarithm. Round to the nearest thousandth.

33. $\log 45.6$

34. $\log .0411$

35. $\ln 470$

36. $\ln 144{,}000$

37. $\log_3 769$

38. $\log_{2/3}(5/8)$

39. A population is increasing according to the growth law $y = 2e^{.02t}$, where y is in millions and t is in years. Match each question with one of the solutions (A), (B), (C), or (D).
  (a) How long will it take for the population to triple?   **(A)** Evaluate $2e^{.02(1/3)}$.
  (b) When will the population reach 3 million?   **(B)** Solve $2e^{.02t} = 3 \cdot 2$ for t.
  (c) How large will the population be in 3 years?   **(C)** Evaluate $2e^{.02(3)}$.
  (d) How large will the population be in 4 months?   **(D)** Solve $2e^{.02t} = 3$ for t.

40. The population of the world is expected to double in the next 44 years. Without solving for the growth constant k, determine how much the population will increase in 22 years.

41. If the world population continues to grow at the current rate, by what factor will the population grow in the next 220 years? See Exercise 40. (*Hint:* $220 = 5 \cdot 44$.)

Solve each equation. Round to the nearest thousandth if necessary.

42. $8^k = 32$

43. $\dfrac{8}{27} = b^{-3}$

44. $10^{2r-3} = 17$

45. $e^{p+1} = 10$

46. $\log_{64} y = \dfrac{1}{3}$

47. $\ln(6x) - \ln(x + 1) = \ln 4$

48. $\log_{16}\sqrt{x + 1} = \dfrac{1}{4}$

49. $\ln x + 3\ln 2 = \ln\dfrac{2}{x}$

50. $\ln[\ln(e^{-x})] = \ln 3$

51. What annual interest rate, to the nearest tenth, will produce $8780 if $3500 is left at interest for 10 years?

52. Find the number of years (to the nearest tenth) needed for $48,000 to become $58,344 at 5% interest compounded semiannually.

53. Manuel deposits $10,000 for 12 years in an account paying 12% compounded annually. He then puts this total amount on deposit in another account paying 10% compounded semiannually for another 9 years. Find the total amount on deposit after the entire 21-year period.

54. Anne Kelly deposits $12,000 for 8 years in an account paying 5% compounded annually. She then leaves the money alone with no further deposits at 6% compounded annually for an additional 6 years. Find the total amount on deposit after the entire 14-year period.

55. The function defined by

$$A(t) = (5 \times 10^{12})e^{-.04t}$$

gives the known coal reserves in the world in year t (in tons), where $t = 0$ corresponds to 1970, and $-.04$ indicates the rate of consumption.
  (a) Find the amount of coal available in 1990.
  (b) When were the coal reserves half of what they were in 1970?

56. Give the property that justifies each step of the following derivation. Let a be any number.
  (a) (a, e^a) is on the graph of $f(x) = e^x$.
  (b) (e^a, a) is on the graph of $g(x) = \ln x$.
  (c) $\ln e^a = a$.

57. The graphs of $y = x^2$ and $y = 2^x$ have the points $(2, 4)$ and $(4, 16)$ in common. There is a third point in common to the graphs whose coordinates can be approximated by using a graphing calculator. Find the coordinates, giving as many decimal places as your calculator displays.

58. Let $Y_1 = 3^{x+4}$ and $Y_2 = 27^{x+1}$. Graph Y_1 and Y_2 in the same window with a graphing calculator and find the coordinates of the point of intersection. (Use the window $[0, 3]$ by $[0, 300]$.)

59. Use the same functions for Y_1 and Y_2 as in Exercise 58. Graph $Y_3 = Y_1 - Y_2$, and explain how the graph of Y_3 supports your answer in Exercise 58.

60. Solve $\log_2 x + \log_2(x + 2) = 3$.

61. To support the solution in Exercise 60, we could graph $Y_1 = \log_2 x + \log_2(x + 2) - 3$ and find the x-intercept. Write an expression for Y_1 using the change-of-base theorem, with base 10.

62. Consider $f(x) = \log_4(2x^2 - x)$.
  (a) Use the change-of-base theorem with base e to write $\log_4(2x^2 - x)$ in a suitable form to graph with a calculator.
  (b) Graph the function using a graphing calculator. Use the window $[-2.5, 2.5]$ by $[-5, 2.5]$.
  (c) What are the x-intercepts?
  (d) Give the equations of the vertical asymptotes.
  (e) Explain why there is no y-intercept.

63. After a medical drug is injected directly into the bloodstream it is gradually eliminated from the body.

Graph each of the following functions on the interval $[0, 10]$. Use $[0, 500]$ for the range of $A(t)$. Determine the function that best describes the amount $A(t)$ (in milligrams) of a drug remaining in the body after t hours if 350 milligrams were initially injected.
  (a) $A(t) = t^2 - t + 350$
  (b) $A(t) = 350 \log(t + 1)$
  (c) $A(t) = 350(.75)^t$
  (d) $A(t) = 100(.95)^t$

64. Computing power of personal computers has increased dramatically as a result of the ability to place an increasing number of transistors on a single processor chip. The table lists the number of transistors on some popular computer chips made by Intel. (*Source:* Intel.)

Year	Chip	Transistors
1971	4004	2300
1986	386DX	275,000
1989	486DX	1,200,000
1993	Pentium	3,300,000
1995	P6	5,500,000

  (a) Plot the data. Let the x-axis represent the year, where $x = 0$ corresponds to 1971, and let the y-axis represent the number of transistors.
  (b) The function f defined by $f(x) = 2300e^{.3241x}$ approximates these data fairly well. Plot f and the data on the same coordinate system.
  (c) Assuming that the present trend continues, use f to predict the number of transistors on a chip in the year 2000.

Answers to Selected Exercises

TO THE STUDENT

If you need further help with trigonometry, you may want to obtain a copy of the *Student's Solution Manual* that goes with this book. It contains solutions to most of the odd-numbered exercises plus extra review exercises with solutions. Your college bookstore either has the *Manual* or can order it for you.

In this section we provide the answers that we think most students will obtain when they work the exercises using the methods explained in the text. If your answer does not look exactly like the one given here, it is not necessarily wrong. In many cases there are equivalent forms of the answer. For example, if the answer section shows 3/4 and your answer is .75, you have obtained the correct answer but written it in a different (yet equivalent) form. Unless the directions specify otherwise, .75 is just as valid an answer as 3/4. In general, if your answer does not agree with the one given in the text, see whether it can be transformed into the other form. If it can, then it is the correct answer. If you still have doubts, talk with your instructor.

CHAPTER 1   The Trigonometric Functions ▼▼▼

CONNECTIONS (page 5)   **1.** $(-1/2, 1)$ **2.** $(-4, 2)$

1.1 EXERCISES (page 11)   **1–7.**   **9.** II   **11.** III   **15.** I and III   **17.** $\sqrt{34}$ **19.** $\sqrt{34}$

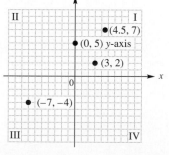

21. $\sqrt{133}$ **23.** 6   **25.** $(-2, 1)$ and $(3, -\pi)$ **29.** yes   **31.** no   **33.** It is true. For example, let $a = 3$, $b = 4$, and $c = 5$. Then $a^2 + b^2 = c^2$, or $3^2 + 4^2 = 5^2$ is true, but $3 + 4 \neq 5$.   **35.** yes   **37.** no   **39.** $5, -1$ **41.** $9 + \sqrt{119}, 9 - \sqrt{119}$

43. $x^2 + y^2 = 25$ y

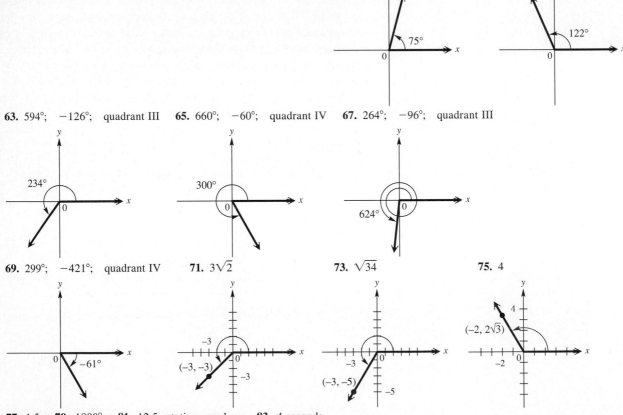

45. 31.6 ft   **47.** 231.3 m   **51.** (1.5, −.5)   **53.** (2, −1)   **55.** (14, 11)   **57.** (6, ∞)
59. [10, ∞)   **61.** [8, 13]   **63.** (−∞, 2)   **67.** 6   **69.** 0   **71.** −2a² + 8
73. (a) 1.2   (b) 5   (c) (0, 3.6)   (d) (3, 0)   **75.** (−∞, ∞); (−∞, ∞); function
77. (−∞, ∞); [4, ∞); function   **79.** (−∞, ∞); (−∞, 4]; function
81. [0, ∞); (−∞, ∞); not a function   **83.** (−∞, ∞); [1, ∞); function
85. [−5, 4]; [−2, 6]; function   **87.** (−∞, −3] ∪ [3, ∞); (−∞, ∞); not a function
89. (−∞, 0) ∪ (0, ∞)   **91.** (−∞, ∞)   **93.** $a + b$; $a + b$; $a^2 + 2ab + b^2$
94. $\left(\dfrac{1}{2}\right)ab$; $2ab$; c^2 **95.** $2ab + c^2$ **96.** $a^2 + 2ab + b^2$; $2ab + c^2$
97. $a^2 + b^2$; c^2

1.2 EXERCISES (page 21)   **3.** 45°   **5.** counterclockwise   **7.** 70°; 110°   **9.** 55°; 35°   **11.** 80°; 100°
13. 90 − x degrees   **15.** x − 360 degrees   **17.** 83° 59′   **19.** 119° 27′   **21.** 38° 32′   **23.** 17° 1′ 49″   **25.** 20.900°
27. 91.598°   **29.** 274.316°   **31.** 31° 25′ 47″   **33.** 89° 54′ 1″   **35.** 178° 35′ 58″   **39.** 320°   **41.** 235°   **43.** 179°
45. 130°   **47.** 30° + n · 360°   **49.** 60° + n · 360°   **51.** 135° + n · 360°   **53.** −90° + n · 360°   **57.** 320°
Angles other than those given are possible in Exercises 59–69.   **59.** 435°; −285°; quadrant I   **61.** 482°; −238°; quadrant II

63. 594°; −126°; quadrant III   **65.** 660°; −60°; quadrant IV   **67.** 264°; −96°; quadrant III

69. 299°; −421°; quadrant IV   **71.** $3\sqrt{2}$ **73.** $\sqrt{34}$ **75.** 4

77. 1.5   **79.** 1800°   **81.** 12.5 rotations per hour   **83.** 4 seconds

1.3 EXERCISES (page 29)   **1.** vertical angles   **3.** 51°; 51°   **5.** 50°; 60°; 70°   **7.** 60°; 60°; 60°   **9.** 65°; 115°
11. 49°; 49°   **13.** 48°; 132°   **15.** 91°   **17.** 2° 29′   **19.** 25.4°   **23.** Answers are given in numerical order: 55°, 65°, 60°, 65°,
60°, 120°, 60°, 60°, 55°, 55°   **25.** right; scalene   **27.** acute; equilateral   **29.** right; scalene   **31.** right; isosceles
33. obtuse; scalene   **35.** acute; isosceles   **41.** A and P; B and Q; C and R; AC and PR; BC and QR; AB and PQ
43. A and C; E and D; ABE and CBD; EB and DB; AB and CB; AE and CD   **45.** P = 90°; Q = 42°; B = R = 48°

47. $B = 106°$; $A = M = 44°$ **49.** $X = M = 52°$ **51.** $a = 20$; $b = 15$ **53.** $a = 6$; $b = 7\frac{1}{2}$ **55.** $x = 6$ **57.** 30 m

59. 500 m, 700 m **61.** 112.5 ft **63.** $x = 110$ **65.** $c = 111.1$ **67.** The unknown side in the first quadrilateral is 40 cm; the unknown sides in the second quadrilateral are 27 cm and 36 cm. **69.** $x = 10$; $y = 2$ **71. (a)** ≈ 2856 miles **(b)** no

CONNECTIONS (page 38) **1.** $\sin \theta = \frac{y}{1} = y = PQ$; $\cos \theta = \frac{x}{1} = x = OQ$; $\tan \theta = \frac{y}{x} = \frac{PQ}{OQ} = \frac{BA}{1}$, so $BA = \tan \theta$

2.

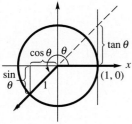

θ in Quadrant III

θ in Quadrant IV

1.4 EXERCISES (page 41) **1.**

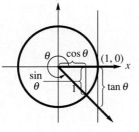

3.

In Exercises 5–11 and 25–29 we give, in order, sine, cosine, tangent, cotangent, secant, and cosecant.

5. $\frac{4}{5}$; $-\frac{3}{5}$; $-\frac{4}{3}$; $-\frac{3}{4}$; $-\frac{5}{3}$; $\frac{5}{4}$ **7.** 1; 0; undefined; 0; undefined; 1 **9.** $\frac{\sqrt{3}}{2}$; $\frac{1}{2}$; $\sqrt{3}$; $\frac{\sqrt{3}}{3}$; 2; $\frac{2\sqrt{3}}{3}$

11. $-.34727$; $.93777$; $-.37031$; -2.7004; 1.0664; -2.8796 **17.** positive **19.** negative **21.** positive **23.** negative

25. $-\frac{2\sqrt{5}}{5}$; $\frac{\sqrt{5}}{5}$; -2; $-\frac{1}{2}$; $\sqrt{5}$; $-\frac{\sqrt{5}}{2}$ **27.** $-\frac{4\sqrt{65}}{65}$; $-\frac{7\sqrt{65}}{65}$; $\frac{4}{7}$; $\frac{7}{4}$; $-\frac{\sqrt{65}}{7}$; $-\frac{\sqrt{65}}{4}$

29. $\frac{5\sqrt{34}}{34}$; $-\frac{3\sqrt{34}}{34}$; $-\frac{5}{3}$; $-\frac{3}{5}$; $-\frac{\sqrt{34}}{3}$; $\frac{\sqrt{34}}{5}$

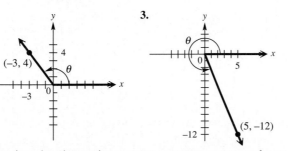

33. -3 **35.** -3 **37.** 5 **39.** 1 **41.** -1 **43.** 0 **45.** undefined **47.** They are equal. **49.** They are negatives of each other. **53.** about .940; about .342 **55.** $35°$ **57.** decrease; increase **59. (a)** $\tan\theta = \dfrac{y}{x}$ **(b)** $x = \dfrac{y}{\tan\theta}$

1.5 EXERCISES (page 51) **1.** 1; $\theta = 90°$ **3.** $\dfrac{1}{3}$ **5.** -5 **7.** $2\sqrt{2}$ **9.** $-\dfrac{3\sqrt{5}}{5}$ **11.** .70069071

15. $\cot\theta = \dfrac{1}{\tan\theta}$; $\cot\theta\tan\theta = 1$ **17.** $\dfrac{1}{2}$ **19.** $\sqrt{3}$ **21.** -100 **23.** $2°$ **25.** $1°$ **27.** $2°$ **29.** II

31. I or III **33.** II or IV **35.** II or III
In Exercises 37–45, we give, in order, sine and cosecant, cosine and secant, and tangent and cotangent.
37. +; +; + **39.** −; −; + **41.** −; +; − **43.** +; +; + **45.** −; +; − **47.** $\tan 30°$
49. $\sec 33°$ **51.** impossible **53.** possible **55.** impossible **57.** possible **59.** impossible **61.** possible **63.** $-2\sqrt{2}$
65. $\dfrac{\sqrt{15}}{4}$ **67.** $-2\sqrt{2}$ **69.** $-\dfrac{\sqrt{15}}{4}$ **71.** .38443820 **73.** -1.4269034 **75.** .36 **77.** yes
In Exercises 79–87 we give, in order, sine, cosine, tangent, cotangent, secant, and cosecant.
79. $\dfrac{15}{17}$; $-\dfrac{8}{17}$; $-\dfrac{15}{8}$; $-\dfrac{8}{15}$; $-\dfrac{17}{8}$; $\dfrac{17}{15}$ **81.** $-\dfrac{4}{5}$; $-\dfrac{3}{5}$; $\dfrac{4}{3}$; $\dfrac{3}{4}$; $-\dfrac{5}{3}$; $-\dfrac{5}{4}$
83. $-\dfrac{\sqrt{3}}{2}$; $-\dfrac{1}{2}$; $\sqrt{3}$; $\dfrac{\sqrt{3}}{3}$; -2; $-\dfrac{2\sqrt{3}}{3}$ **85.** $\dfrac{\sqrt{5}}{7}$; $\dfrac{2\sqrt{11}}{7}$; $\dfrac{\sqrt{55}}{22}$; $\dfrac{2\sqrt{55}}{5}$; $\dfrac{7\sqrt{11}}{22}$; $\dfrac{7\sqrt{5}}{5}$
87. $-.555762$; .831342; $-.668512$; -1.49586; 1.20287; -1.79933
91. False; for example, $\sin 30° + \cos 30° \approx .5 + .8660 = 1.3660 \neq 1$.

CHAPTER 1 REVIEW EXERCISES (page 56) **3.** $(-\infty, -4]$ **5.** 5 **7.** yes **9.** 2 **11.** $-x^2 + x + 4$
In Exercises 13–19, we give the domain, then the range, and then tell whether it is a function.
13. $(-\infty, \infty)$; $(-\infty, \infty)$; function **15.** $(-\infty, \infty)$; $[0, \infty)$; function **17.** $[-1, \infty)$; $(-\infty, \infty)$; not a function
19. $[-3, 3]$; $[-5, 5]$; not a function **21.** $309°$ **23.** $72°$ **25.** $1280°$ **27.** $47.420°$ **29.** $74°\,17'\,54''$ **31.** $183°\,5'\,50''$
33. $58°$; $58°$ **35.** $V = 41°$; $Z = 32°$; $Y = U = 107°$ **37.** $m = 45$; $n = 60$ **39.** $r = 108/7$ **41.** proportional; equal
In Exercises 43–53 and 67–71, we give, in order, sine, cosine, tangent, cotangent, secant, cosecant.
43. $-\dfrac{\sqrt{2}}{2}$; $-\dfrac{\sqrt{2}}{2}$; 1; 1; $-\sqrt{2}$; $-\sqrt{2}$ **45.** 0; -1; 0; undefined; -1; undefined
47. $\dfrac{15}{17}$; $-\dfrac{8}{17}$; $-\dfrac{15}{8}$; $-\dfrac{8}{15}$; $-\dfrac{17}{8}$; $\dfrac{17}{15}$ **49.** $-\dfrac{5\sqrt{26}}{26}$; $\dfrac{\sqrt{26}}{26}$; -5; $-\dfrac{1}{5}$; $\sqrt{26}$; $-\dfrac{\sqrt{26}}{5}$
51. $-\dfrac{1}{2}$; $\dfrac{\sqrt{3}}{2}$; $-\dfrac{\sqrt{3}}{3}$; $-\sqrt{3}$; $\dfrac{2\sqrt{3}}{3}$; -2 **53.** $\dfrac{5\sqrt{34}}{34}$; $\dfrac{3\sqrt{34}}{34}$; $\dfrac{5}{3}$; $\dfrac{3}{5}$; $\dfrac{\sqrt{34}}{3}$; $\dfrac{\sqrt{34}}{5}$
55. **57.** -5 **59.** -4 **61.** 99 **63.** impossible **65.** impossible
67. $-\dfrac{\sqrt{39}}{8}$; $-\dfrac{5}{8}$; $\dfrac{\sqrt{39}}{5}$; $\dfrac{5\sqrt{39}}{39}$; $-\dfrac{8}{5}$; $-\dfrac{8\sqrt{39}}{39}$

graph showing $y = -5x$, $x \le 0$ with point $(-1, 5)$ and angle θ

69. $\dfrac{2\sqrt{5}}{5}$; $-\dfrac{\sqrt{5}}{5}$; -2; $-\dfrac{1}{2}$; $-\sqrt{5}$; $\dfrac{\sqrt{5}}{2}$
71. $-\dfrac{3}{5}$; $\dfrac{4}{5}$; $-\dfrac{3}{4}$; $-\dfrac{4}{3}$; $\dfrac{5}{4}$; $-\dfrac{5}{3}$ **75.** ≈ 9500 ft

CHAPTER 2 Acute Angles and Right Triangles ▼▼▼

2.1 EXERCISES (page 69)
In Exercises 1–5, we give, in order, sine, cosine, tangent, cotangent, secant, and cosecant.
1. $\dfrac{3}{5}$; $\dfrac{4}{5}$; $\dfrac{3}{4}$; $\dfrac{4}{3}$; $\dfrac{5}{4}$; $\dfrac{5}{3}$ **3.** $\dfrac{21}{29}$; $\dfrac{20}{29}$; $\dfrac{21}{20}$; $\dfrac{20}{21}$; $\dfrac{29}{20}$; $\dfrac{29}{21}$ **5.** $\dfrac{n}{p}$; $\dfrac{m}{p}$; $\dfrac{n}{m}$; $\dfrac{m}{n}$; $\dfrac{p}{m}$; $\dfrac{p}{n}$
In Exercises 7–9, we give, in order, the unknown side, sine, cosine, tangent, cotangent, secant, and cosecant.
7. $c = 13$; $\dfrac{12}{13}$; $\dfrac{5}{13}$; $\dfrac{12}{5}$; $\dfrac{5}{12}$; $\dfrac{13}{5}$; $\dfrac{13}{12}$ **9.** $b = \sqrt{13}$; $\dfrac{\sqrt{13}}{7}$; $\dfrac{6}{7}$; $\dfrac{\sqrt{13}}{6}$; $\dfrac{6\sqrt{13}}{13}$; $\dfrac{7}{6}$; $\dfrac{7\sqrt{13}}{13}$ **13.** $\tan 17°$
15. $\csc 51°$ **17.** $\tan(100° - \beta)$ **19.** $\cos 51.3°$ **21.** $40°$ **23.** $20°$ **25.** $12°$ **27.** true **29.** false **31.** true

33. $0° \leq A < 45°$ **35.** $\dfrac{\sqrt{3}}{3}$ **37.** $\dfrac{1}{2}$ **39.** $\sqrt{2}$ **41.** $\dfrac{\sqrt{2}}{2}$ **43.** $\dfrac{\sqrt{3}}{2}$ **45.** $\sqrt{3}$ **47.** $\sin x,\ \tan x$ **49.** $60°$

51. $\left(\dfrac{\sqrt{2}}{2}, \dfrac{\sqrt{2}}{2}\right);\ 45°$ **53.** $y = \dfrac{\sqrt{3}}{3}x$ **55.** $60°$ **57. (a)** $60°$ **(b)** k **(c)** $k\sqrt{3}$ **(d)** $2;\ \sqrt{3};\ 30°;\ 60°$

59. $a = 12;\ b = 12\sqrt{3};\ d = 12\sqrt{3};\ c = 12\sqrt{6}$ **61.** $m = \dfrac{7\sqrt{3}}{3};\ a = \dfrac{14\sqrt{3}}{3};\ n = \dfrac{14\sqrt{3}}{3};\ q = \dfrac{14\sqrt{6}}{3}$ **63.** $\dfrac{s^2\sqrt{3}}{4}$

65. yes   **67.**

θ	$\cos \theta$
$0°$	$\sqrt{4}/2 = 1$
$30°$	$\sqrt{3}/2$
$45°$	$\sqrt{2}/2$
$60°$	$\sqrt{1}/2 = 1/2$
$90°$	$\sqrt{0}/2 = 0$

69. ≈ 78 mph

2.2 EXERCISES (page 78)   **5.** $82°$ **7.** $45°$ **9.** $30°$

In Exercises 11–29, we give, in order, sine, cosine, tangent, cotangent, secant, and cosecant.

11. $\dfrac{\sqrt{3}}{2};\ -\dfrac{1}{2};\ -\sqrt{3};\ -\dfrac{\sqrt{3}}{3};\ -2;\ \dfrac{2\sqrt{3}}{3}$ **13.** $\dfrac{1}{2};\ -\dfrac{\sqrt{3}}{2};\ -\dfrac{\sqrt{3}}{3};\ -\sqrt{3};\ -\dfrac{2\sqrt{3}}{3};\ 2$

15. $-\dfrac{\sqrt{3}}{2};\ -\dfrac{1}{2};\ \sqrt{3};\ \dfrac{\sqrt{3}}{3};\ -2;\ -\dfrac{2\sqrt{3}}{3}$ **17.** $-\dfrac{\sqrt{2}}{2};\ \dfrac{\sqrt{2}}{2};\ -1;\ -1;\ \sqrt{2};\ -\sqrt{2}$

19. $\dfrac{\sqrt{3}}{2};\ \dfrac{1}{2};\ \sqrt{3};\ \dfrac{\sqrt{3}}{3};\ 2;\ \dfrac{2\sqrt{3}}{3}$ **21.** $\dfrac{\sqrt{2}}{2};\ -\dfrac{\sqrt{2}}{2};\ -1;\ -1;\ -\sqrt{2};\ \sqrt{2}$ **23.** $\dfrac{1}{2};\ \dfrac{\sqrt{3}}{2};\ \dfrac{\sqrt{3}}{3};\ \sqrt{3};\ \dfrac{2\sqrt{3}}{3};\ 2$

25. $\dfrac{\sqrt{3}}{2};\ \dfrac{1}{2};\ \sqrt{3};\ \dfrac{\sqrt{3}}{3};\ 2;\ \dfrac{2\sqrt{3}}{3}$ **27.** $-\dfrac{1}{2};\ \dfrac{\sqrt{3}}{2};\ -\dfrac{\sqrt{3}}{3};\ -\sqrt{3};\ \dfrac{2\sqrt{3}}{3};\ -2$

29. $\dfrac{\sqrt{3}}{2};\ \dfrac{1}{2};\ \sqrt{3};\ \dfrac{\sqrt{3}}{3};\ 2;\ \dfrac{2\sqrt{3}}{3}$ **31.** $\dfrac{\sqrt{3}}{3};\ \sqrt{3}$ **33.** $\dfrac{\sqrt{3}}{2};\ \dfrac{\sqrt{3}}{3};\ \dfrac{2\sqrt{3}}{3}$ **35.** $-1;\ -1$ **37.** $-\dfrac{\sqrt{3}}{2};\ -\dfrac{2\sqrt{3}}{3}$

39. yes   **41.** positive   **43.** negative   **45.** negative   **49.** $-.4$ **51.** $135°, 315°$ **53.** false; $\dfrac{1 + \sqrt{3}}{2} \neq 1$ **55.** true

57. false; $\dfrac{\sqrt{3}}{2} \neq 0$ **59.** true   **61. (a)** ≈ 550 ft **(b)** ≈ 369 ft

2.3 EXERCISES (page 82)   **1.** $.5657728$ **3.** 1.1342773 **5.** 1.0273488 **7.** $.6383201$ **9.** 1.7768146 **11.** $.4771588$
13. -5.7297416 **15.** 1.9074147 **17.** $.9668234$ **19.** $.4327386$ **21.** $.2308682$ **23.** $.0825664$ **25.** 1.2162701
29. $57.997172°$ **31.** $30.502748°$ **33.** $46.173581°$ **35.** $81.168073°$ **37.** $56°$ **39.** 1 **41.** 0 **43.** 2×10^8 m per sec
45. $19°$ **47.** $48.7°$ **49.** false   **51.** true   **53.** false   **55.** false   **57.** ≈ -100.5 lb   **59.** $\approx 2.866°$ **61.** ≈ 2771 lb
63. the 2200-pound car   **65. (a)** ≈ 703 ft **(b)** ≈ 1701 ft **(c)** R would decrease.

2.4 EXERCISES (page 90)   **3.** $.05$ **5.** $28,999.5$ to $29,000.5$ **7.** 1649.5 to $1650.5;\ 159.5$ to 160.5 **9.** Both 2 and 65 are
exact measurements.   **11.** $B = 53°\ 40';\ a = 571$ m; $b = 777$ m   **13.** $M = 38.8°;\ n = 154$ m; $p = 198$ m
15. $A = 47.9108°;\ c = 84.816$ cm; $a = 62.942$ cm   **23.** $B = 62.00°;\ a = 8.17$ ft; $b = 15.4$ ft
25. $A = 17.00°;\ a = 39.1$ in; $c = 134$ in   **27.** $c = 85.9$ yd; $A = 62°\ 50';\ B = 27°\ 10'$ **29.** $b = 42.3$ cm;
$A = 24°\ 10';\ B = 65°\ 50'$ **31.** $B = 36°\ 36';\ a = 310.8$ ft; $b = 230.8$ ft   **33.** $A = 50°\ 51';\ a = .4832$ m; $b = .3934$ m
35. The angle of elevation from X to Y is $90°$ whenever Y is directly above X.   **39.** 9.35 m   **41.** 88.3 m   **43.** 33.4 m
45. 134.7 cm   **47.** 28.0 m   **49.** 469 m   **51.** 146 m   **53.** $37°\ 35'$ **55.** 3.342 mm   **57. (a)** $\approx 29,008$ ft **(b)** shorter
59. (a) ≈ 23.4 ft **(b)** ≈ 48.3 ft **(c)** The faster the speed, the more land needs to be cleared on the inside of the curve.

CONNECTIONS (page 98)   **1.** $y = (\tan 36.7°)(x - 50)$ **2.** $(110.49675, 45.092889);\ $ yes

2.5 EXERCISES (page 99)   **1.** $270°;\ $ N $90°$ W   **3.** $315°;\ $ N $45°$ W   **5.** $y = \dfrac{\sqrt{3}}{3}x,\ x \leq 0$ **7.** 220 mi

9. 47 nautical mi   **11.** 148 mi   **13.** 120 mi   **15.** $x = \dfrac{b}{a - c}$ **17.** $y = (\tan 35°)(x - 25)$ **19.** 433 ft   **21.** 114 ft

23. 5.18 m   **25. (a)** $d = \dfrac{b}{2}\left(\cot \dfrac{\alpha}{2} + \cot \dfrac{\beta}{2}\right)$ **(b)** 345.3951 m   **27.** 10.8 ft   **29.** 84.7 m

CHAPTER 2 REVIEW EXERCISES (page 103)
In Exercises 1 and 13–17, we give answers in the order sine, cosine, tangent, cotangent, secant, and cosecant.

1. $\frac{60}{61}$; $\frac{11}{61}$; $\frac{60}{11}$; $\frac{11}{60}$; $\frac{61}{11}$; $\frac{61}{60}$ **3.** 10° **5.** 7° **7.** true **9.** true **13.** $-\frac{\sqrt{2}}{2}$; $-\frac{\sqrt{2}}{2}$; 1; 1; $-\sqrt{2}$; $-\sqrt{2}$

15. $\frac{1}{2}$; $\frac{\sqrt{3}}{2}$; $\frac{\sqrt{3}}{3}$; $\sqrt{3}$; $\frac{2\sqrt{3}}{3}$; 2 **17.** $-\frac{1}{2}$; $\frac{\sqrt{3}}{2}$; $-\frac{\sqrt{3}}{3}$; $-\sqrt{3}$; $\frac{2\sqrt{3}}{3}$; -2 **19.** 120°; 240° **21.** 150°; 210°

23. $3 - \frac{2\sqrt{3}}{3}$ **25.** $\frac{7}{2}$ **27.** .95371695 **29.** $-.71592968$ **31.** 1.9362132 **33.** .99984900 **37.** 55.673870°

39. 12.733938° **41.** 63.008286° **43.** 47.1°; 132.9° **45.** false; 1.4088321 ≠ 1 **47.** true **49.** no **51.** III
53. $B = 31° 30'$; $a = 638$; $b = 391$ **55.** $B = 50.28°$; $a = 32.38$ m; $c = 50.66$ m; $C = 90°$ **57.** 73.7 ft
59. 18.75 cm **61.** 1200 m **63.** 140 mi **67.** AB **69.** OB **71.** (a) $x_Q = x_P + d \cos \theta$, $y_Q = y_P + d \sin \theta$
(b) (308.69, 395.67)

CHAPTER 3   Radian Measure and the Circular Functions ▼▼▼

3.1 EXERCISES (page 113)
1. 1 **3.** 3 **5.** $\frac{\pi}{3}$ **7.** $\frac{\pi}{2}$ **9.** $\frac{5\pi}{6}$ **11.** $\frac{5\pi}{3}$ **13.** $\frac{5\pi}{2}$ **21.** 60° **23.** 315° **25.** 330°

27. $-30°$ **29.** 126° **31.** 48° **33.** 153° **35.** .68 **37.** .742 **39.** 2.43 **41.** 1.122 **43.** .9847 **45.** .832391

47. 114°35′ **49.** 99°42′ **51.** 19°35′ **53.** 287°6′ **57.** $\frac{\sqrt{3}}{2}$ **59.** 1 **61.** $\frac{2\sqrt{3}}{3}$ **63.** 1 **65.** $-\sqrt{3}$ **67.** $\frac{1}{2}$ **69.** -1

71. $-\frac{\sqrt{3}}{2}$ **73.** $\frac{1}{2}$ **75.** We begin the answers with the blank next to 30°, and then proceed counterclockwise from there:

$\frac{\pi}{6}$; 45; $\frac{\pi}{3}$; 120; 135; $\frac{5\pi}{6}$; π; $\frac{7\pi}{6}$; $\frac{5\pi}{4}$; 240; 300; $\frac{7\pi}{4}$; $\frac{11\pi}{6}$. **77.** radian **79. (a)** 4π **(b)** $\frac{2\pi}{3}$
81. (a) ≈ 1.998 w/m² **(b)** ≈ -46.461 w/m² **(c)** 46.478 w/m² **(d)** $N = 82.5$ or 165 (Since N represents a day number, which should be a natural number, we might interpret day 82.5 as noon on the 82nd day.)

3.2 EXERCISES (page 119)
1. 2π **3.** 8 **5.** 1 **7.** 25.8 cm **9.** 318 m **11.** 5.05 m **13.** 2.5 **15.** $\frac{20}{\pi}$ in

17. The length is doubled. **19.** 3500 km **21.** 5900 km **23.** 44° N **25. (a)** 11.6 in **(b)** 37°5′ **27.** 38.5° **29.** 146 in
31. .20 km **33.** 2100 mi **35.** 6π **37.** 1.5 **39.** 1120 m² **41.** 1300 cm² **43.** 114 cm² **45.** 3.6 **47.** 16 m

49. The area of a circle of radius r is πr^2. **51.** $\frac{\pi}{3}$ **53. (a)** 13.85° **(b)** 76 m² **55.** $V = \frac{r^2 \theta h}{2}$ (θ in radians)

57. (a) ≈ 550 m **(b)** ≈ 1800 m

3.3 EXERCISES (page 129)
1. 17 **3.** $-\frac{1}{2}$ **5.** -1 **7.** $-\frac{\sqrt{3}}{2}$ **9.** -2 **11.** $-\sqrt{3}$ **13.** $-\frac{1}{2}$ **15.** $\frac{2\sqrt{3}}{3}$ **17.** $-\frac{\sqrt{2}}{2}$

19. .73135046 **21.** .80036052 **23.** .99813420 **25.** 1.0170372 **27.** .96364232 **29.** .42442278 **31.** 1.2131367
33. $-.44357977$ **35.** $-.75469733$ **37.** $-.99668945$ **39.** .20952066 **41.** 1.4429646 **43.** 1.0151896 **45.** .95409991

47. 1.4747226 **49.** 1.0181269 **51.** $\frac{\pi}{3}$ **53.** $\frac{2\pi}{3}$ **55.** $\frac{7\pi}{6}$ **57.** $\frac{11\pi}{6}$ **59.** ≈ 2.782 **61.** $(-.96679819, -.25554110)$

63. $(-.72593230, .68776616)$ **65.** IV **67.** III **69. (a)** 1° **(b)** 19° **(c)** 53° **(d)** 58° **(e)** 48° **(f)** 12°
71. $2\pi - .5$, $\pi - .5$ **73.** 8.6 hr; 15.4 hr

3.4 EXERCISES (page 134)
1. 2π sec **5.** 2π radians **7.** $\frac{3\pi}{32}$ radians per sec **9.** $\frac{6}{5}$ min **11.** .180311 radians per sec

13. 8π m per sec **15.** $\frac{9}{5}$ radians per sec **17.** 1.83333 radians per sec **19.** 18π cm **21.** 12 sec **23.** $\frac{3\pi}{32}$ radians per sec

27. $\frac{\pi}{6}$ radians per hr **29.** $\frac{\pi}{30}$ radians per sec **31.** $\frac{7\pi}{30}$ cm per min **33.** 168π m per min **35.** 1500π m per min

37. $\frac{16\pi}{3}$ m per sec; $\frac{16\pi}{3}$ radians per sec **39.** .24 radian per sec **41.** .303 m **43. (a)** 2π radians per day; $\frac{\pi}{12}$ radians per hr

(b) 0 **(c)** 12,800π km per day or about 533π km per hr **(d)** about 9050π km per day or about 377π km per hr
45. ≈ 523.6 radians per sec

CHAPTER 3 REVIEW EXERCISES (page 137)   **3.** Three of the many possible answers are $2 - 2\pi$, $2 - 4\pi$, and $2 - 6\pi$.

5. $\dfrac{\pi}{4}$ **7.** $\dfrac{4\pi}{9}$ **9.** $\dfrac{11\pi}{6}$ **11.** $\dfrac{17\pi}{3}$ **13.** 225°   **15.** 480°   **17.** $-110°$ **19.** 168°   **21.** $\sqrt{3}$ **23.** $-\dfrac{1}{2}$ **25.** $-\sqrt{3}$ **27.** 2

29. 35.8 cm   **31.** 7.683 cm   **33.** 273 m^2 **35.** 4500 km   **37.** $\dfrac{3}{4}$;   1.5 square units   **39. (a)** $\dfrac{\pi}{3}$ radians   **(b)** 2π in

41. .86602663   **43.** .97030688   **45.** 1.6755332   **47.** 1.1311943   **49.** .38974894   **51.** .51489440   **53.** 1.1053762   **55.** $\dfrac{\pi}{4}$

57. $\dfrac{7\pi}{6}$ **59.** $-.4$ **61.** 35°; $\dfrac{7\pi}{36}$ radians   **63.** $\dfrac{15}{32}$ sec   **65.** $\dfrac{\pi}{20}$ radians per sec   **67.** 285.3 cm   **69. (b)** $\dfrac{\pi}{6}$

(c) less ultraviolet light when $\theta = \dfrac{\pi}{3}$

CHAPTER 4   Graphs of the Circular Functions ▼▼▼

4.1 EXERCISES (page 151)   **5.** .92387953; .38268343   **7.** (0, 1)   **9.** 2

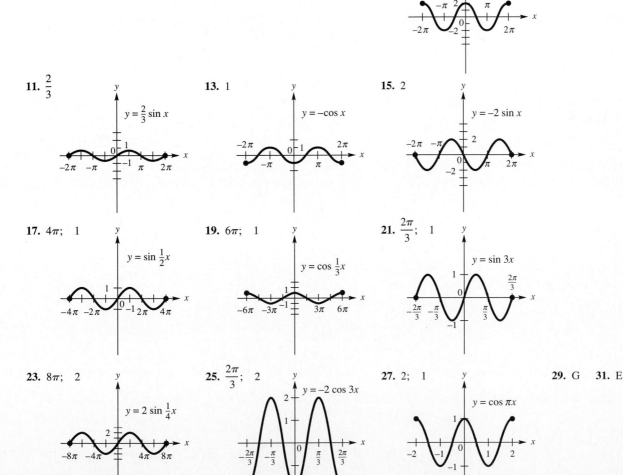

11. $\dfrac{2}{3}$

13. 1

15. 2

17. 4π; 1

19. 6π; 1

21. $\dfrac{2\pi}{3}$; 1

23. 8π; 2

25. $\dfrac{2\pi}{3}$; 2

27. 2; 1

29. G   **31.** E

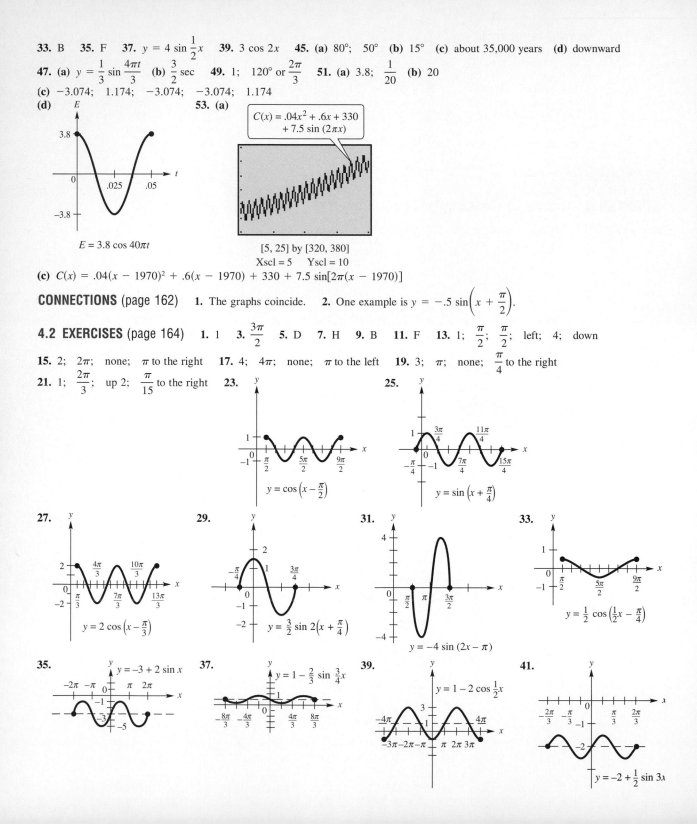

33. B   **35.** F   **37.** $y = 4 \sin \frac{1}{2}x$ **39.** $3 \cos 2x$ **45. (a)** 80°; 50° **(b)** 15° **(c)** about 35,000 years   **(d)** downward

47. (a) $y = \frac{1}{3} \sin \frac{4\pi t}{3}$ **(b)** $\frac{3}{2} \sec$ **49.** 1; 120° or $\frac{2\pi}{3}$ **51. (a)** 3.8; $\frac{1}{20}$ **(b)** 20

(c) −3.074; 1.174; −3.074; −3.074; 1.174

(d)

$E = 3.8 \cos 40\pi t$

53. (a)

$C(x) = .04x^2 + .6x + 330 + 7.5 \sin(2\pi x)$

[5, 25] by [320, 380]

Xscl = 5   Yscl = 10

(c) $C(x) = .04(x − 1970)^2 + .6(x − 1970) + 330 + 7.5 \sin[2\pi(x − 1970)]$

CONNECTIONS (page 162)   **1.** The graphs coincide.   **2.** One example is $y = -.5 \sin\left(x + \frac{\pi}{2}\right)$.

4.2 EXERCISES (page 164)   **1.** 1   **3.** $\frac{3\pi}{2}$ **5.** D   **7.** H   **9.** B   **11.** F   **13.** 1; $\frac{\pi}{2}$; $\frac{\pi}{2}$; left; 4; down

15. 2; 2π; none; π to the right   **17.** 4; 4π; none; π to the left   **19.** 3; π; none; $\frac{\pi}{4}$ to the right

21. 1; $\frac{2\pi}{3}$; up 2; $\frac{\pi}{15}$ to the right   **23.**   **25.**

$y = \cos\left(x − \frac{\pi}{2}\right)$ $y = \sin\left(x + \frac{\pi}{4}\right)$

27.   **29.**   **31.**   **33.**

$y = 2 \cos\left(x − \frac{\pi}{3}\right)$ $y = \frac{3}{2} \sin 2\left(x + \frac{\pi}{4}\right)$ $y = -4 \sin(2x − \pi)$ $y = \frac{1}{2} \cos\left(\frac{1}{2}x − \frac{\pi}{4}\right)$

35.   **37.**   **39.**   **41.**

$y = -3 + 2 \sin x$ $y = 1 − \frac{2}{3} \sin \frac{3}{4}x$ $y = 1 − 2 \cos \frac{1}{2}x$ $y = -2 + \frac{1}{2} \sin 3x$

43.

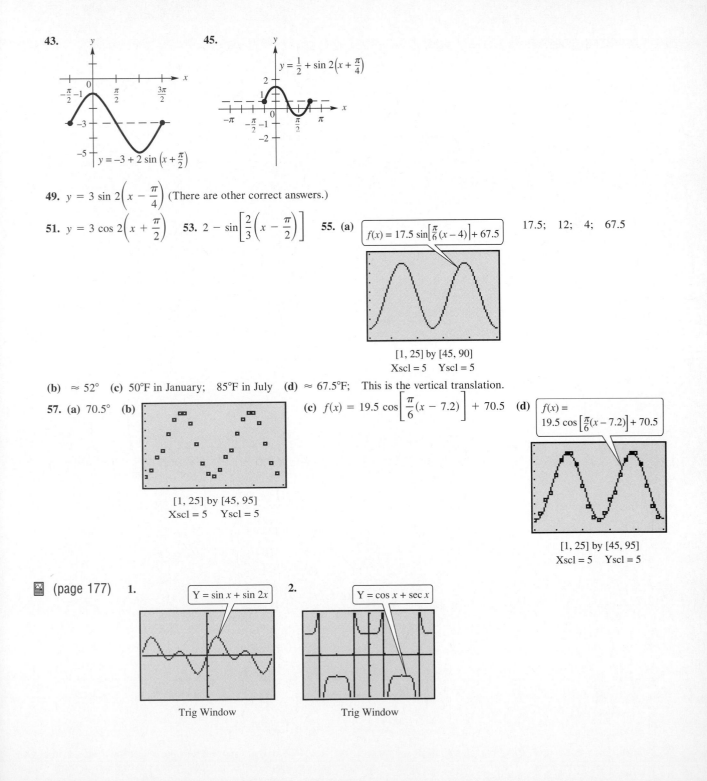

$y = -3 + 2 \sin\left(x + \frac{\pi}{2}\right)$

45.

$y = \frac{1}{2} + \sin 2\left(x + \frac{\pi}{4}\right)$

49. $y = 3 \sin 2\left(x - \frac{\pi}{4}\right)$ (There are other correct answers.)

51. $y = 3 \cos 2\left(x + \frac{\pi}{2}\right)$ **53.** $2 - \sin\left[\frac{2}{3}\left(x - \frac{\pi}{2}\right)\right]$ **55. (a)** $f(x) = 17.5 \sin\left[\frac{\pi}{6}(x-4)\right] + 67.5$ 17.5;   12;   4;   67.5

[1, 25] by [45, 90]
Xscl = 5   Yscl = 5

(b) $\approx 52°$ **(c)** 50°F in January;   85°F in July   **(d)** $\approx 67.5°$F;   This is the vertical translation.

57. (a) 70.5°   **(b)**

[1, 25] by [45, 95]
Xscl = 5   Yscl = 5

(c) $f(x) = 19.5 \cos\left[\frac{\pi}{6}(x - 7.2)\right] + 70.5$ **(d)** $f(x) = 19.5 \cos\left[\frac{\pi}{6}(x - 7.2)\right] + 70.5$

[1, 25] by [45, 95]
Xscl = 5   Yscl = 5

(page 177)   **1.**

Y = sin x + sin 2x

Trig Window

2.

Y = cos x + sec x

Trig Window

4.3 EXERCISES (page 178)   3. false; $3 \csc x = \dfrac{3}{\sin x}$ 5. true   7. B   9. E   11. D

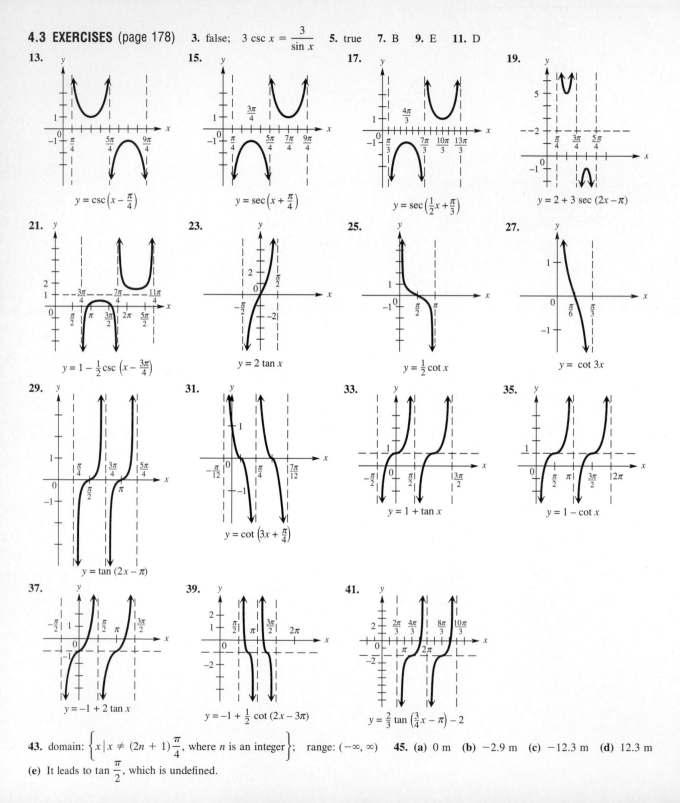

13. $y = \csc\left(x - \dfrac{\pi}{4}\right)$

15. $y = \sec\left(x + \dfrac{\pi}{4}\right)$

17. $y = \sec\left(\dfrac{1}{2}x + \dfrac{\pi}{3}\right)$

19. $y = 2 + 3 \sec(2x - \pi)$

21. $y = 1 - \dfrac{1}{2}\csc\left(x - \dfrac{3\pi}{4}\right)$

23. $y = 2 \tan x$

25. $y = \dfrac{1}{2}\cot x$

27. $y = \cot 3x$

29. $y = \tan(2x - \pi)$

31. $y = \cot\left(3x + \dfrac{\pi}{4}\right)$

33. $y = 1 + \tan x$

35. $y = 1 - \cot x$

37. $y = -1 + 2 \tan x$

39. $y = -1 + \dfrac{1}{2}\cot(2x - 3\pi)$

41. $y = \dfrac{2}{3}\tan\left(\dfrac{3}{4}x - \pi\right) - 2$

43. domain: $\left\{x \mid x \neq (2n + 1)\dfrac{\pi}{4},\text{ where } n \text{ is an integer}\right\}$;   range: $(-\infty, \infty)$ **45. (a)** 0 m   **(b)** -2.9 m   **(c)** -12.3 m   **(d)** 12.3 m
(e) It leads to $\tan \dfrac{\pi}{2}$, which is undefined.

CHAPTER 4 REVIEW EXERCISES (page 181)   **1.** sin x, csc x, cos x, sec x **3.** sin x, cos x, tan x, cot x

5. 2; 2π; none; none   **7.** $\dfrac{1}{2}$; $\dfrac{2\pi}{3}$; none; none   **9.** 2; 8π; 1; none   **11.** 3; 2π; none; $\dfrac{\pi}{2}$ units to the left

13. not applicable; π; none; $\dfrac{\pi}{8}$ units to the right   **15.** not applicable; $\dfrac{\pi}{3}$; none; $\dfrac{\pi}{9}$ units to the right

17. tangent   **19.** cosine   **21.** cotangent

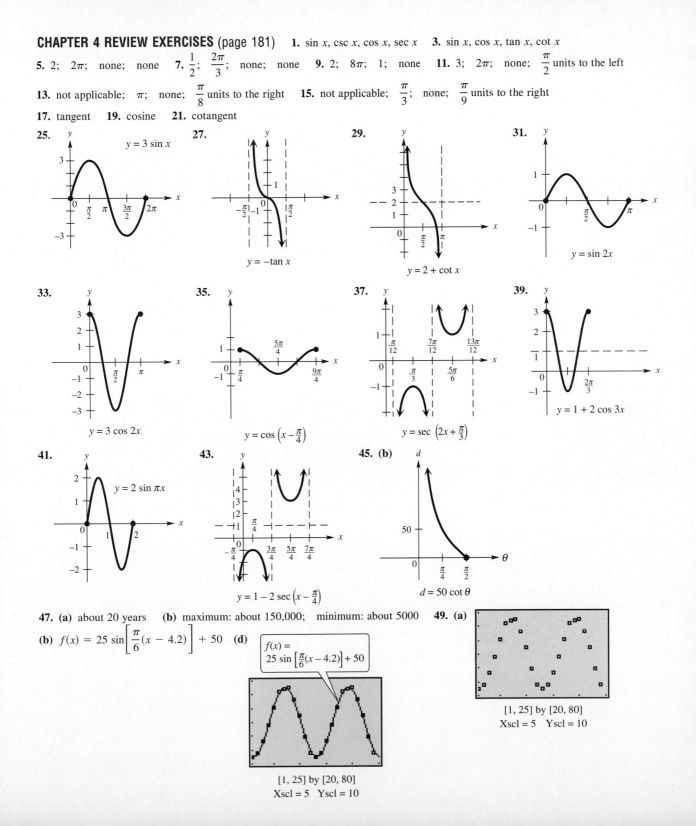

25. $y = 3 \sin x$

27. $y = -\tan x$

29. $y = 2 + \cot x$

31. $y = \sin 2x$

33. $y = 3 \cos 2x$

35. $y = \cos\left(x - \dfrac{\pi}{4}\right)$

37. $y = \sec\left(2x + \dfrac{\pi}{3}\right)$

39. $y = 1 + 2 \cos 3x$

41. $y = 2 \sin \pi x$

43. $y = 1 - 2 \sec\left(x - \dfrac{\pi}{4}\right)$

45. (b) $d = 50 \cot \theta$

47. (a) about 20 years   **(b)** maximum: about 150,000;   minimum: about 5000   **49. (a)**

(b) $f(x) = 25 \sin\left[\dfrac{\pi}{6}(x - 4.2)\right] + 50$ **(d)**

$f(x) = 25 \sin\left[\dfrac{\pi}{6}(x - 4.2)\right] + 50$

[1, 25] by [20, 80]
Xscl = 5   Yscl = 10

[1, 25] by [20, 80]
Xscl = 5   Yscl = 10

CHAPTER 5 Trigonometric Identities ▼▼▼

CONNECTIONS (page 187) The tangent function is odd. Yes, $\tan(-x) = -\tan x$, because tangent is an odd function.

5.1 EXERCISES (page 191) **1.** -2.6 **3.** 2.5 **5.** $\dfrac{\sqrt{7}}{4}$ **7.** $-\dfrac{2\sqrt{5}}{5}$ **9.** $-\dfrac{\sqrt{105}}{11}$ **13.** $\dfrac{\sqrt{21}}{2}$

15. $\cos\theta = -\dfrac{\sqrt{5}}{3}$; $\tan\theta = -\dfrac{2\sqrt{5}}{5}$; $\cot\theta = -\dfrac{\sqrt{5}}{2}$; $\sec\theta = -\dfrac{3\sqrt{5}}{5}$; $\csc\theta = \dfrac{3}{2}$ **17.** $\sin\theta = -\dfrac{\sqrt{17}}{17}$; $\cos\theta = \dfrac{4\sqrt{17}}{17}$; $\cot\theta = -4$; $\sec\theta = \dfrac{\sqrt{17}}{4}$; $\csc\theta = -\sqrt{17}$ **19.** $\sin\theta = \dfrac{3}{5}$; $\cos\theta = \dfrac{4}{5}$; $\tan\theta = \dfrac{3}{4}$; $\sec\theta = \dfrac{5}{4}$; $\csc\theta = \dfrac{5}{3}$

21. $\sin\theta = -\dfrac{\sqrt{7}}{4}$; $\cos\theta = \dfrac{3}{4}$; $\tan\theta = -\dfrac{\sqrt{7}}{3}$; $\cot\theta = -\dfrac{3\sqrt{7}}{7}$; $\csc\theta = -\dfrac{4\sqrt{7}}{7}$ **23.** (b) **25.** (e) **27.** (a) **29.** (a)

31. (d) **35.** $\sin\theta = \dfrac{\pm\sqrt{2x+1}}{x+1}$ **37.** $\cos\theta$ **39.** $\cot\theta$ **41.** $\cos^2\theta$ **43.** $\sec\theta - \cos\theta$ **45.** $\cot\theta - \tan\theta$ **47.** $\sec\theta\csc\theta$

49. $\cos^2\theta$ **51.** $\sec^2\theta$ **53.** $\dfrac{\pm\sqrt{1+\cot^2\theta}}{1+\cot^2\theta}$; $\dfrac{\pm\sqrt{\sec^2\theta-1}}{\sec\theta}$ **55.** $\dfrac{\pm\sin\theta\sqrt{1-\sin^2\theta}}{1-\sin^2\theta}$; $\dfrac{\pm\sqrt{1-\cos^2\theta}}{\cos\theta}$; $\pm\sqrt{\sec^2\theta-1}$;

$\dfrac{\pm\sqrt{\csc^2\theta-1}}{\csc^2\theta-1}$ **57.** $\dfrac{\pm\sqrt{1-\sin^2\theta}}{1-\sin^2\theta}$; $\pm\sqrt{\tan^2\theta+1}$; $\dfrac{\pm\sqrt{1+\cot^2\theta}}{\cot\theta}$; $\dfrac{\pm\csc\theta\sqrt{\csc^2\theta-1}}{\csc^2\theta-1}$ **59.** $\dfrac{25\sqrt{6}-60}{12}$; $\dfrac{-25\sqrt{6}-60}{12}$

61. $\cot x$ and $\csc x$ are odd, $\sec x$ is even. **63.** $-\sin(2x)$ **64.** It is the negative of $\sin(2x)$. **65.** $\cos(4x)$ **66.** It is the same function. **67.** (a) $-\sin(4x)$ (b) $\cos(2x)$ (c) $5\sin(3x)$ **69.** The graph of $y = \csc(-x)$ is a reflection across the x-axis as compared to the graph of $y = \csc x$. **71.** The graph of $y = \cot(-x)$ is a reflection across the x-axis as compared to the graph of $y = \cot x$.

CONNECTIONS (page 200) $\sqrt{(1-x^2)^3} = \sin^3\theta$

5.2 EXERCISES (page 201) **1.** $\dfrac{1}{\sin\theta\cos\theta}$ or $\csc\theta\sec\theta$ **3.** $1 + \cos s$ **5.** 1 **7.** 1 **9.** $2 + 2\sin t$

11. $\dfrac{-2\cos x}{\sin^2 x}$ or $-2\cot x\csc x$ **13.** $(\sin\gamma + 1)(\sin\gamma - 1)$ **15.** $4\sin x$ **17.** $(2\sin x + 1)(\sin x + 1)$ **19.** $(\cos^2 x + 1)^2$

21. $(\sin x - \cos x)(1 + \sin x\cos x)$ **23.** $\sin\theta$ **25.** 1 **27.** $\tan^2\beta$ **29.** $\tan^2 x$ **31.** $\sec^2 x$ **71.** $\cos\theta$ **73.** $\cot\theta$

75. identity **77.** not an identity **79.** not an identity **81.** not an identity **83.** identity

5.3 EXERCISES (page 209) **3.** $\dfrac{\sqrt{6}-\sqrt{2}}{4}$ **5.** $\dfrac{\sqrt{6}-\sqrt{2}}{4}$ **7.** $\dfrac{\sqrt{2}-\sqrt{6}}{4}$ **9.** $\dfrac{\sqrt{6}+\sqrt{2}}{4}$ **11.** 0 **13.** 0 **15.** $\cot 3°$

17. $\sin\dfrac{5\pi}{12}$ **19.** $\sec 104° 24'$ **21.** $\cos\left(-\dfrac{\pi}{8}\right)$ **23.** $\csc(-56° 42')$ **25.** $\tan(-86.9814°)$ **27.** $\tan$ **29.** $\cos$ **31.** $\csc$

33. $15°$ **35.** $\dfrac{140°}{3}$ **37.** $20°$ **39.** $\cos\theta$ **41.** $-\cos\theta$ **43.** $\cos\theta$ **45.** $-\cos\theta$ **47.** $\dfrac{4-6\sqrt{6}}{25}$; $\dfrac{4+6\sqrt{6}}{25}$

49. $\dfrac{16}{65}$; $-\dfrac{56}{65}$ **51.** $\dfrac{2\sqrt{638}-\sqrt{30}}{56}$; $\dfrac{2\sqrt{638}+\sqrt{30}}{56}$ **53.** true **55.** false **57.** true **59.** true **61.** false

74. $-\dfrac{\sqrt{6}+\sqrt{2}}{4}$ **75.** $-\dfrac{\sqrt{6}+\sqrt{2}}{4}$ **76.** (a) $\dfrac{\sqrt{2}-\sqrt{6}}{4}$ (b) $-\dfrac{\sqrt{6}+\sqrt{2}}{4}$ **81.** 3 **83.** (a) The pressure P is oscillating.

(b) The pressure oscillates and amplitude decreases as r increases.

For $x = r$,
$P(r) = \dfrac{3}{r}\cos\left[\dfrac{2\pi r}{4.9} - 10{,}260\right]$

(c) $P = \dfrac{a}{n\lambda}\cos(ct)$

For $x = t$,
$P(t) = \dfrac{4}{10}\cos\left[\dfrac{20\pi}{4.9} - 1026t\right]$

$[0, 20]$ by $[-2, 2]$
Xscl $= 1$ Yscl $= 1$

$[0, .05]$ by $[-.05, .05]$
Xscl $= .01$ Yscl $= .01$

5.4 EXERCISES (page 215) **3.** $\dfrac{\sqrt{6} - \sqrt{2}}{4}$ **5.** $2 - \sqrt{3}$ **7.** $\dfrac{-\sqrt{6} - \sqrt{2}}{4}$ **9.** $\dfrac{\sqrt{6} + \sqrt{2}}{4}$ **11.** $2 - \sqrt{3}$

13. $\dfrac{-\sqrt{6} - \sqrt{2}}{4}$ **15.** $\dfrac{\sqrt{2}}{2}$ **17.** -1 **19.** 0 **21.** 1 **23.** $\dfrac{\sqrt{3}\cos\theta - \sin\theta}{2}$ **25.** $\dfrac{\cos\theta - \sqrt{3}\sin\theta}{2}$

27. $\dfrac{\sqrt{2}(\sin x - \cos x)}{2}$ **29.** $\dfrac{\sqrt{3}\tan\theta + 1}{\sqrt{3} - \tan\theta}$ **31.** $\dfrac{\sqrt{2}(\cos x + \sin x)}{2}$ **33.** $-\cos\theta$ **35.** $-\tan\theta$ **37.** $-\tan\theta$

41. $\dfrac{63}{65}$; $\dfrac{33}{65}$; $\dfrac{63}{16}$; $\dfrac{33}{56}$; I; I **43.** $\dfrac{4\sqrt{2} + \sqrt{5}}{9}$; $\dfrac{4\sqrt{2} - \sqrt{5}}{9}$; $\dfrac{-8\sqrt{5} - 5\sqrt{2}}{20 - 2\sqrt{10}}$ or $\dfrac{4\sqrt{2} + \sqrt{5}}{2 - 2\sqrt{10}}$; $\dfrac{-8\sqrt{5} + 5\sqrt{2}}{20 + 2\sqrt{10}}$ or

$\dfrac{-4\sqrt{2} + \sqrt{5}}{2\sqrt{10} + 2}$; II; II **45.** $\dfrac{77}{85}$; $\dfrac{13}{85}$; $-\dfrac{77}{36}$; $\dfrac{13}{84}$; II; I **47.** $-\dfrac{33}{65}$; $-\dfrac{63}{65}$; $\dfrac{33}{56}$; $\dfrac{63}{16}$; III; III

49. $\dfrac{-(3\sqrt{22} + \sqrt{21})}{20}$; $\dfrac{-3\sqrt{22} + \sqrt{21}}{20}$; $\dfrac{-(66\sqrt{7} + 7\sqrt{66})}{154 - 3\sqrt{462}}$; $\dfrac{-66\sqrt{7} + 7\sqrt{66}}{154 + 3\sqrt{462}}$; IV; IV

51. $\sin\left(\dfrac{\pi}{2} + x\right) = \cos x$ **53.** $\tan\left(\dfrac{\pi}{2} + x\right) = -\cot x$ **65.** $\dfrac{\sqrt{6} - \sqrt{2}}{4}$ **67.** $2 + \sqrt{3}$ **69.** $\dfrac{-\sqrt{6} - \sqrt{2}}{4}$

71. $\dfrac{\sqrt{6} - \sqrt{2}}{4}$ **73.** $-2 + \sqrt{3}$ **77.** $\sin(A + B + C) = \sin A \cos B \cos C + \cos A \sin B \cos C + \cos A \cos B \sin C - \sin A \cdot$

$\sin B \sin C$ **80.** $180° - \beta$ **81.** $\theta = \beta - \alpha$ **82.** $\tan\theta = \dfrac{\tan\beta - \tan\alpha}{1 + \tan\alpha\tan\beta}$ **84.** $18.4°$ **85.** $80.8°$ **87. (a)** 425 lb **(c)** $0°$

5.5 EXERCISES (page 223) **1.** $.2$ **3.** $\cos\theta = \dfrac{2\sqrt{5}}{5}$; $\sin\theta = \dfrac{\sqrt{5}}{5}$; $\tan\theta = \dfrac{1}{2}$; $\sec\theta = \dfrac{\sqrt{5}}{2}$; $\csc\theta = \sqrt{5}$; $\cot\theta = 2$

5. $\cos x = -\dfrac{\sqrt{42}}{12}$; $\sin x = \dfrac{\sqrt{102}}{12}$; $\tan x = -\dfrac{\sqrt{119}}{7}$; $\sec x = -\dfrac{2\sqrt{42}}{7}$; $\csc x = \dfrac{2\sqrt{102}}{17}$; $\cot x = -\dfrac{\sqrt{119}}{17}$

7. $\cos 2\theta = \dfrac{17}{25}$; $\sin 2\theta = -\dfrac{4\sqrt{21}}{25}$; $\tan 2\theta = -\dfrac{4\sqrt{21}}{17}$; $\sec 2\theta = \dfrac{25}{17}$; $\csc 2\theta = -\dfrac{25\sqrt{21}}{84}$; $\cot 2\theta = -\dfrac{17\sqrt{21}}{84}$

9. $\tan 2x = -\dfrac{4}{3}$; $\sec 2x = -\dfrac{5}{3}$; $\cos 2x = -\dfrac{3}{5}$; $\cot 2x = -\dfrac{3}{4}$; $\sin 2x = \dfrac{4}{5}$; $\csc 2x = \dfrac{5}{4}$

11. $\sin 2\alpha = -\dfrac{4\sqrt{55}}{49}$; $\cos 2\alpha = \dfrac{39}{49}$; $\tan 2\alpha = -\dfrac{4\sqrt{55}}{39}$; $\cot 2\alpha = -\dfrac{39\sqrt{55}}{220}$; $\sec 2\alpha = \dfrac{49}{39}$; $\csc 2\alpha = -\dfrac{49\sqrt{55}}{220}$

13. $\dfrac{\sqrt{3}}{2}$ **15.** $\dfrac{\sqrt{3}}{3}$ **17.** $\dfrac{\sqrt{3}}{2}$ **19.** $\dfrac{\sqrt{2}}{2}$ **21.** $-\dfrac{\sqrt{2}}{2}$ **23.** $\dfrac{\sqrt{2}}{4}$ **25.** $\dfrac{1}{2}\tan 102°$ **27.** $\dfrac{1}{4}\cos 94.2°$ **29.** $-\cos\dfrac{4\pi}{5}$

31. $\sin 10x$ **35.** 1 **37.** $-\dfrac{1}{2}$ **39.** $-\dfrac{1}{2}$ **41.** $\sqrt{3}$ **43.** $\sqrt{3}$ **45.** 0 **47.** $\cos^4 x - \sin^4 x = \cos 2x$

49. $\dfrac{\cot^2 x - 1}{2\cot x} = \cot 2x$ **69.** $\tan^2 2x = \dfrac{4\tan^2 x}{1 - 2\tan^2 x + \tan^4 x}$ **71.** $\cos 3x = 4\cos^3 x - 3\cos x$

73. $\tan 3x = \dfrac{3\tan x - \tan^3 x}{1 - 3\tan^2 x}$ **75.** $\tan 4x = \dfrac{4(\tan x - \tan^3 x)}{1 - 6\tan^2 x + \tan^4 x}$ **77.** 1 **79.** $\csc^2 3r$ **81.** 980.799 cm per sec^2

83. $\dfrac{1}{2}(\sin 70° - \sin 20°)$ **85.** $\dfrac{3}{2}(\cos 8x + \cos 2x)$ **87.** $\dfrac{1}{2}[\cos 2\theta - \cos(-4\theta)] = \dfrac{1}{2}(\cos 2\theta - \cos 4\theta)$

89. $-4[\cos 9y + \cos(-y)] = -4(\cos 9y + \cos y)$ **91.** $a = -885.6$, $c = 885.6$, $\omega = 240\pi$

93. (a) $\dfrac{1}{\omega}$

For $x = t$,
$W = -885.6\cos 240\pi t + 885.6$

$[0, .05]$ by $[-500, 2000]$
Xscl = .01 Yscl = 500

5.6 EXERCISES (page 230)   **1.** .1270166538   **3.** −   **5.** +   **7.** $\dfrac{\sqrt{2-\sqrt{3}}}{2}$ **9.** $\dfrac{\sqrt{2+\sqrt{2}}}{2}$ **11.** $1+\sqrt{2}$

13. $\dfrac{\sqrt{2+\sqrt{2}}}{2}$ **15.** $-\dfrac{\sqrt{2+\sqrt{3}}}{2}$ **17.** $-\dfrac{\sqrt{2+\sqrt{3}}}{2}$ **21.** $\dfrac{\sqrt{10}}{4}$ **23.** 3   **25.** $\dfrac{\sqrt{50-10\sqrt{5}}}{10}$ **27.** $-\sqrt{7}$ **29.** $\dfrac{\sqrt{5}}{5}$

31. $-\dfrac{\sqrt{42}}{12}$ **33.** $\sin 20°$ **35.** $\tan 73.5°$ **37.** $\tan 29.87°$ **39.** $\cos 9x$ **41.** $\tan 4\theta$ **43.** $\cos\dfrac{x}{8}$ **45.** $\dfrac{\sin x}{1+\cos x}=\tan\dfrac{x}{2}$

47. $\dfrac{\tan\dfrac{x}{2}+\cot\dfrac{x}{2}}{\cot\dfrac{x}{2}-\tan\dfrac{x}{2}}=\sec x$ **59.** 106°   **61.** 47°   **63.** 2   **64.** They are both radii of the circle.   **65.** It is the supplement of a

30° angle.   **66.** Their sum is $180-150=30$ degrees and they are equal.   **67.** $2+\sqrt{3}$ **69.** $\dfrac{\sqrt{6}+\sqrt{2}}{4}$ **70.** $\dfrac{\sqrt{6}-\sqrt{2}}{4}$

71. $2-\sqrt{3}$

CHAPTER 5 REVIEW EXERCISES (page 234)   **1.** sine, tangent, cotangent, cosecant   **3.** $-.96$

5. $\sin x=-\dfrac{4}{5};\quad \tan x=-\dfrac{4}{3};\quad \sec x=\dfrac{5}{3};\quad \csc x=-\dfrac{5}{4};\quad \cot x=-\dfrac{3}{4}$

7. (a) $\sin\dfrac{\pi}{12}=\dfrac{\sqrt{6}-\sqrt{2}}{4};\quad \cos\dfrac{\pi}{12}=\dfrac{\sqrt{6}+\sqrt{2}}{4};\quad \tan\dfrac{\pi}{12}=2-\sqrt{3}$ **(b)** $\sin\dfrac{\pi}{12}=\dfrac{\sqrt{2-\sqrt{3}}}{2};\quad \cos\dfrac{\pi}{12}=\dfrac{\sqrt{2+\sqrt{3}}}{2};$

$\tan\dfrac{\pi}{12}=2-\sqrt{3}$ **9.** (e)   **11.** (j)   **13.** (i)   **15.** (h)   **17.** (g)   **19.** (a)   **21.** (f)   **23.** (e)

25. 1   **27.** $\dfrac{1}{\cos^2\theta}$ **29.** $\dfrac{1}{\sin^2\theta\cos^2\theta}$ **31.** $\dfrac{4+3\sqrt{15}}{20};\quad \dfrac{4\sqrt{15}+3}{20};\quad \dfrac{192+25\sqrt{15}}{231};$ I

33. $\dfrac{4-9\sqrt{11}}{50};\quad \dfrac{12\sqrt{11}-3}{50};\quad \dfrac{\sqrt{11}-16}{21};$ IV   **35.** $\sin\theta=\dfrac{\sqrt{14}}{4};\quad \cos\theta=\dfrac{\sqrt{2}}{4}$ **37.** $\sin 2x=\dfrac{3}{5};\quad \cos 2x=-\dfrac{4}{5}$

39. $\dfrac{1}{2}$ **41.** $\dfrac{\sqrt{5}-1}{2}$ **43.** $-\dfrac{\sin 2x+\sin x}{\cos 2x-\cos x}=\cot\dfrac{x}{2}$ **45.** $\dfrac{\sin x}{1-\cos x}=\cot\dfrac{x}{2}$ **47.** $\dfrac{2(\sin x-\sin^3 x)}{\cos x}=\sin 2x$

CHAPTER 6   Inverse Trigonometric Functions and Trigonometric Equations ▼▼▼

6.1 EXERCISES (page 248)   **1.** $-\dfrac{\pi}{6}$ **3.** $\cos^{-1}\!\left(\dfrac{1}{a}\right)$ **5.** $-\dfrac{\pi}{6}$ **7.** $\dfrac{\pi}{4}$ **9.** π **11.** $-\dfrac{\pi}{3}$ **13.** 0   **15.** $\dfrac{\pi}{2}$ **17.** $\dfrac{\pi}{4}$

19. $\dfrac{5\pi}{6}$ **21.** $\dfrac{3\pi}{4}$ **23.** $-\dfrac{\pi}{6}$ **25.** $\dfrac{\pi}{6}$ **27.** $\dfrac{\pi}{3}$ **29.** $-45°$ **31.** $-60°$ **33.** 120°   **35.** $-30°$ **37.** .83798122

39. 2.3154725   **41.** 1.1900238   **43.** $-7.6713835°$ **45.** 113.500970°   **47.** 30.987961°

49. $(-\infty,\infty);(0,\pi)$ **51.** $(-\infty,-1]\cup[1,\infty);\left[0,\dfrac{\pi}{2}\right)\cup\left(\dfrac{\pi}{2},\pi\right]$ **57.** $\dfrac{\sqrt{7}}{3}$ **59.** $\dfrac{\sqrt{5}}{5}$ **61.** $-\dfrac{\sqrt{5}}{2}$ **63.** 2

$y=\cot^{-1}x$

$y=\text{arcsec } x$

65. $\dfrac{\pi}{4}$ **67.** $\dfrac{\pi}{3}$ **69.** $\dfrac{120}{169}$ **71.** $-\dfrac{7}{25}$ **73.** $\dfrac{4\sqrt{6}}{25}$ **75.** $-\dfrac{24}{7}$ **77.** $\dfrac{\sqrt{10}-3\sqrt{30}}{20}$ **79.** $-\dfrac{16}{65}$ **81.** .89442719

83. .12343998   **85.** $\sqrt{1-u^2}$ **87.** $\dfrac{\sqrt{1-u^2}}{u}$ **89.** $\dfrac{\sqrt{u^2-4}}{|u|}$ **91.** $\dfrac{u\sqrt{2}}{2}$ **95. (a)** 113°   **(b)** 84°   **(c)** 60°   **(d)** 47°

6.2 EXERCISES (page 257)

1. $-1 < b < 1$; $b = \pm 1$; $b > 1$ or $b < -1$ **3.** $a + n\pi$, where n is an integer **5.** $\dfrac{\pi}{3}, \dfrac{5\pi}{3}$

7. $\dfrac{3\pi}{4}, \dfrac{7\pi}{4}$ **9.** $\dfrac{\pi}{6}, \dfrac{5\pi}{6}$ **11.** no solution **13.** $\dfrac{\pi}{4}, \dfrac{2\pi}{3}, \dfrac{5\pi}{4}, \dfrac{5\pi}{3}$ **15.** π **17.** $\dfrac{7\pi}{6}, \dfrac{3\pi}{2}, \dfrac{11\pi}{6}$ **19.** $\dfrac{\pi}{4}, \dfrac{\pi}{2}, \dfrac{3\pi}{4}, \dfrac{5\pi}{4}, \dfrac{3\pi}{2}, \dfrac{7\pi}{4}$

21. $\dfrac{\pi}{2} + 2n\pi, \dfrac{7\pi}{6} + 2n\pi, \dfrac{11\pi}{6} + 2n\pi$, where n is an integer **23.** $\dfrac{\pi}{3} + 2n\pi, \dfrac{2\pi}{3} + 2n\pi, \dfrac{4\pi}{3} + 2n\pi, \dfrac{5\pi}{3} + 2n\pi$, where n is an

integer **25.** no solution **27.** $30°, 210°, 240°, 300°$ **29.** $90°, 210°, 330°$ **31.** $45°, 135°, 225°, 315°$ **33.** $45°, 225°$
35. $0°, 30°, 150°, 180°$ **37.** $0°, 45°, 135°, 180°, 225°, 315°$ **39.** $0°, 90°$ **41.** $90°, 221.8°, 318.2°$ **43.** $135°, 315°, 71.6°, 251.6°$
45. $71.6°, 90°, 251.6°, 270°$ **47.** $53.6°, 126.4°, 187.9°, 352.1°$ **49.** $149.6°, 329.6°, 106.3°, 286.3°$ **51.** no solution
53. $57.7°, 159.2°$ **55.** $360° \cdot n, 120° + 360° \cdot n, 240° + 360° \cdot n$, where n is an integer **59. (a)** $\dfrac{1}{4}$ sec **(b)** $\dfrac{1}{6}$ sec **(c)** .21 sec

61. $14°$ **63.** $f = \dfrac{1}{T}$ or $T = \dfrac{1}{f}$ **65.** One such value is $t = \dfrac{\pi}{3}$.

6.3 EXERCISES (page 263)

1. $.5, 3.1, 5.8$ **3.** $0, \dfrac{2\pi}{3}$ **5.** $\dfrac{\pi}{12}, \dfrac{11\pi}{12}, \dfrac{13\pi}{12}, \dfrac{23\pi}{12}$ **7.** $\dfrac{\pi}{2}, \dfrac{7\pi}{6}, \dfrac{11\pi}{6}$

9. $\dfrac{\pi}{18}, \dfrac{7\pi}{18}, \dfrac{13\pi}{18}, \dfrac{19\pi}{18}, \dfrac{25\pi}{18}, \dfrac{31\pi}{18}$ **11.** $\dfrac{3\pi}{8}, \dfrac{5\pi}{8}, \dfrac{11\pi}{8}, \dfrac{13\pi}{8}$ **13.** $\dfrac{\pi}{2}, \dfrac{3\pi}{2}$ **15.** $0, \dfrac{\pi}{4}, \dfrac{\pi}{2}, \dfrac{3\pi}{4}, \pi, \dfrac{5\pi}{4}, \dfrac{3\pi}{2}, \dfrac{7\pi}{4}$

17. no solution **19.** $\dfrac{\pi}{2}$ **21.** $\dfrac{\pi}{3}, \pi, \dfrac{5\pi}{3}$ **23.** $15°, 45°, 135°, 165°, 255°, 285°$ **25.** $0°$ **27.** $120°, 240°$ **29.** $30°, 150°, 270°$

31. $0°, 30°, 150°, 180°$ **33.** $60°, 300°$ **35.** $11.8°, 78.2°, 191.8°, 258.2°$ **37.** $30°, 90°, 150°, 210°, 270°, 330°$
41. (a) 91.3 days after March 21, on June 20 **(b)** 273.8 days after March 21, on December 19 **(c)** 228.7 days after March 21, on
November 4, and again after 318.8 days, on February 2 **43.** .001 sec **45.** .004 sec
47. (a) The final graph is as follows. **(b)** The graph approximates a sawtooth shape. **(c)** The maximum value of P is approximately
.00317 and occurs when $x \approx .000188, .00246, .00474, .00701,$ and $.00928$.

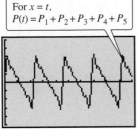

For $x = t$,
$P(t) = P_1 + P_2 + P_3 + P_4 + P_5$

[0, .01] by [−.005, .005]
Xscl = .001 Yscl = .001

6.4 EXERCISES (page 270)

1. $\left(\dfrac{\sqrt{2}}{2}, \dfrac{\pi}{4} \right)$; $\dfrac{\sqrt{2}}{2}$ **3.** 0 **5.** $x = \arccos \dfrac{y}{5}$ **7.** $x = \dfrac{1}{3}\operatorname{arccot} 2y$ **9.** $x = \dfrac{1}{2}\arctan \dfrac{y}{3}$

11. $x = 4 \arccos \dfrac{y}{6}$ **13.** $x = \dfrac{1}{5}\arccos\left(-\dfrac{y}{2} \right)$ **15.** $x = -3 + \arccos y$ **17.** $x = \arcsin(y + 2)$ **19.** $x = \arcsin\left(\dfrac{y + 4}{2} \right)$

23. $-2\sqrt{2}$ **25.** $\pi - 3$ **27.** $\dfrac{3}{5}$ **29.** $\dfrac{4}{5}$ **31.** 0 **33.** $\dfrac{1}{2}$ **35.** $-\dfrac{1}{2}$ **37.** 0 **39.** $t = \dfrac{50}{\pi}\arccos\left(\dfrac{d - 550}{450} \right)$

41. (a) $t = \dfrac{1}{2\pi f}\arcsin \dfrac{e}{E_{\max}}$ **(b)** .00068 sec **43. (a)** $x = \sin u, -\dfrac{\pi}{2} \le u \le \dfrac{\pi}{2}$ **(b)**

(c) $\tan u = \dfrac{x\sqrt{1 - x^2}}{1 - x^2}$ **(d)** $u = \arctan \dfrac{x\sqrt{1 - x^2}}{1 - x^2}$

45. (a) $A \approx .0035$, $\phi \approx .470$, $P = .0035 \sin(600\pi t + .47)$ **(b)** The two graphs are the same.

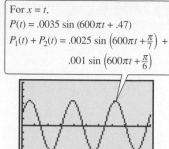

For $x = t$,
$P(t) = .0035 \sin (600\pi t + .47)$
$P_1(t) + P_2(t) = .0025 \sin \left(600\pi t + \frac{\pi}{7}\right) +$
$.001 \sin \left(600\pi t + \frac{\pi}{6}\right)$

[0, .01] by [−.006, .006]
Xscl = .001   Yscl = .001

47. (a) $\approx .94$ and 4.26 **(b)** $\approx .60$ and 6.64

CHAPTER 6 REVIEW EXERCISES (page 274)   **5.** $\frac{\pi}{4}$ **7.** $-\frac{\pi}{3}$ **9.** $\frac{3\pi}{4}$ **11.** $\frac{2\pi}{3}$ **13.** $\frac{3\pi}{4}$ **15.** $-60°$ **17.** $60.679245°$

19. $36.489508°$ **21.** $73.262206°$ **25.** $(-\infty, \infty)$ **27.** $\frac{1}{2}$ **29.** -1 **31.** $\frac{3\pi}{4}$ **33.** $\frac{\pi}{4}$ **35.** $\frac{\sqrt{7}}{4}$ **37.** $\frac{\sqrt{3}}{2}$ **39.** $\frac{294 + 125\sqrt{6}}{92}$

41. $\sqrt{1 - u^2}$ **43.** $[-1, 1];$ $\left[-\frac{\pi}{2}, \frac{\pi}{2}\right]$ **45.** $(-\infty, \infty);$ $(0, \pi)$ **47.** $.46364761, 3.6052403$

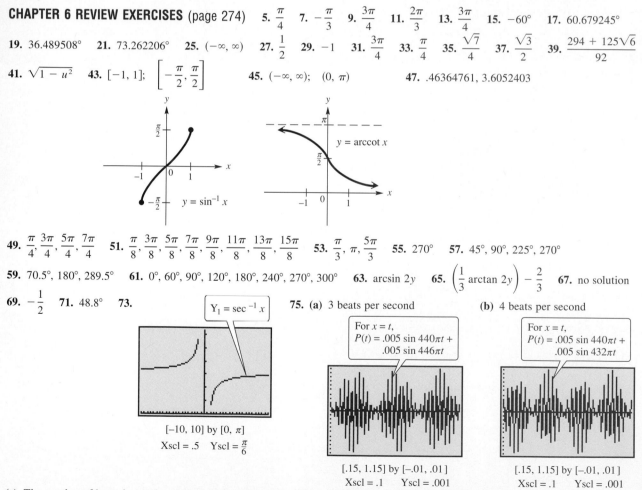

49. $\frac{\pi}{4}, \frac{3\pi}{4}, \frac{5\pi}{4}, \frac{7\pi}{4}$ **51.** $\frac{\pi}{8}, \frac{3\pi}{8}, \frac{5\pi}{8}, \frac{7\pi}{8}, \frac{9\pi}{8}, \frac{11\pi}{8}, \frac{13\pi}{8}, \frac{15\pi}{8}$ **53.** $\frac{\pi}{3}, \pi, \frac{5\pi}{3}$ **55.** $270°$ **57.** $45°, 90°, 225°, 270°$

59. $70.5°, 180°, 289.5°$ **61.** $0°, 60°, 90°, 120°, 180°, 240°, 270°, 300°$ **63.** $\arcsin 2y$ **65.** $\left(\frac{1}{3}\arctan 2y\right) - \frac{2}{3}$ **67.** no solution

69. $-\frac{1}{2}$ **71.** $48.8°$ **73.**

$Y_1 = \sec^{-1} x$

[−10, 10] by [0, π]
Xscl = .5   Yscl = $\frac{\pi}{6}$

75. (a) 3 beats per second   **(b)** 4 beats per second

For $x = t$,
$P(t) = .005 \sin 440\pi t +$
$.005 \sin 446\pi t$

[.15, 1.15] by [−.01, .01]
Xscl = .1   Yscl = .001

For $x = t$,
$P(t) = .005 \sin 440\pi t +$
$.005 \sin 432\pi t$

[.15, 1.15] by [−.01, .01]
Xscl = .1   Yscl = .001

(c) The number of beats is equal to the absolute value of the difference of the frequencies in the two tones.

CHAPTER 7 Applications of Trigonometry and Vectors ▼▼▼

CONNECTIONS (page 285) **1.** House: $X_H = 1131.8$ ft, $Y_H = 4390.2$ ft; Fire: $X_F = 2277.5$ ft, $Y_F = -2596.2$ ft **2.** 7079.7 ft

7.1 EXERCISES (page 286) **1.** $6\sqrt{2}$ **3.** $\sqrt{3}$ **5.** $C = 95°$, $b = 13$ m, $a = 11$ m **7.** $B = 37.3°$, $a = 38.5$ ft, $b = 51.0$ ft
9. $C = 57.36°$, $b = 11.13$ ft, $c = 11.55$ ft **11.** $B = 18.5°$, $a = 239$ yd, $c = 230$ yd **13.** $A = 56°\ 00'$, $AB = 361$ ft, $BC = 308$ ft
15. $B = 110.0°$, $a = 27.01$ m, $c = 21.36$ m **17.** $A = 34.72°$, $a = 3326$ ft, $c = 5704$ ft **19.** $C = 97°\ 34'$, $b = 283.2$ m,
$c = 415.2$ m **25.** yes **27.** 118 m **29.** 1.93 mi **31.** 10.4 in **33.** 111° **35.** first location: 5.1 mi; second location: 7.2 mi
37. $\dfrac{\sqrt{3}}{2}$ **39.** $\dfrac{\sqrt{2}}{2}$ **41.** 46.4 m² **43.** 356 cm² **45.** 722.9 in² **47.** 100 m² **49.** increasing **51.** $b = \dfrac{a \sin B}{\sin A}$
52. $b = \dfrac{a \sin B}{\sin A} = a \cdot \dfrac{\sin B}{\sin A}$. Since $\dfrac{\sin B}{\sin A} < 1$, $b = a \cdot \dfrac{\sin B}{\sin A} < a \cdot 1 = a$, so $b < a$. **55.** ≈ 1.95 mi **57.** 5126 ft

7.2 EXERCISES (page 294) **1.** 45, 135 **3. (a)** $4 < h < 5$ **(b)** $h = 4$ and $h > 5$ **(c)** $h < 4$ **5.** 1 **7.** 2 **9.** 0
11. 45° **13.** $B_1 = 49.1°$, $C_1 = 101.2°$, $B_2 = 130.9°$, $C_2 = 19.4°$ **15.** $B = 26°\ 30'$, $A = 112°\ 10'$ **17.** no such triangle
19. $B = 27.19°$, $C = 10.68°$ **21.** $B = 20.6°$, $C = 116.9°$, $c = 20.6$ ft **23.** no such triangle **25.** $B_1 = 49°\ 20'$, $C_1 = 92°\ 00'$,
$c_1 = 15.5$ km, $B_2 = 130°\ 40'$, $C_2 = 10°\ 40'$, $c_2 = 2.88$ km **27.** $B = 37.77°$, $C = 45.43°$, $c = 4.174$ ft **29.** $A_1 = 53.23°$,
$C_1 = 87.09°$, $c_1 = 37.16$ m, $A_2 = 126.77°$, $C_2 = 13.55°$, $c_2 = 8.719$ m **31.** 1; 90°; a right triangle **35.** does not exist

CONNECTIONS (page 302) All three formulas give the area as 9.5 square units.

7.3 EXERCISES (page 302) **3.** 7 **5.** 30° **7.** $c = 2.83$ in, $A = 44.9°$, $B = 106.8°$ **9.** $c = 6.46$ m, $A = 53.1°$, $B = 81.3°$
11. $a = 156$ cm, $B = 64°\ 50'$, $C = 34°\ 30'$ **13.** $b = 9.529$ in, $A = 64.59°$, $C = 40.61°$ **15.** $a = 15.7$ m, $B = 21.6°$, $C = 45.6°$
17. $A = 30°$, $B = 56°$, $C = 94°$ **19.** $A = 82°$, $B = 37°$, $C = 61°$ **21.** $A = 42.0°$, $B = 35.9°$, $C = 102.1°$
23. $A = 47.7°$, $B = 44.9°$, $C = 87.4°$ **25.** 16.26° **27.** $24\sqrt{3}$ or ≈ 41.57 **29.** 78 m² **31.** 12,600 cm² **33.** 3650 ft²
35. 33 cans **37.** 392,000 mi² **39.** 257 m **41.** 281 km **43.** 22 ft **45.** 18 ft **47.** 2000 km **49.** 163.5°
51. 25.24983 mi **63.** Each side is equal to $-2 + \sqrt{3}$. **65.** Since A is obtuse, $90° < A < 180°$. The cosine of a quadrant II angle
is negative. **66.** In $a^2 = b^2 + c^2 - 2bc \cos A$, $\cos A$ is negative, so $a^2 = b^2 + c^2$ plus a positive quantity. Thus $a^2 > b^2 + c^2$.
67. $b^2 + c^2 > b^2$ and $b^2 + c^2 > c^2$. If $a^2 > b^2 + c^2$, then $a^2 > b^2$ and $a^2 > c^2$ from which $a > b$ and $a > c$ because a, b, and c
are nonnegative. **68.** Because A is obtuse it is the largest angle, so the longest side should be a, not c.

CONNECTIONS (page 309) **1.** $|\mathbf{u}| = \sqrt{13}$, $\theta = 326.3°$ **2.** $\mathbf{u} = \left\langle \dfrac{5}{2}, \dfrac{5\sqrt{3}}{2} \right\rangle$ **3. (a)** $\langle 2, 4 \rangle$ **(b)** $\langle 4, -2 \rangle$ **(c)** $\langle -20, -5 \rangle$

7.4 EXERCISES (page 310) **3.** $\mathbf{m}$ and $\mathbf{p}$; $\mathbf{n}$ and $\mathbf{r}$ **5.** $\mathbf{m}$ and $\mathbf{p}$ equal $2\mathbf{t}$, or $\mathbf{t}$ is one half $\mathbf{m}$ or $\mathbf{p}$; also, $\mathbf{m} = 1\mathbf{p}$ and $\mathbf{n} = 1\mathbf{r}$

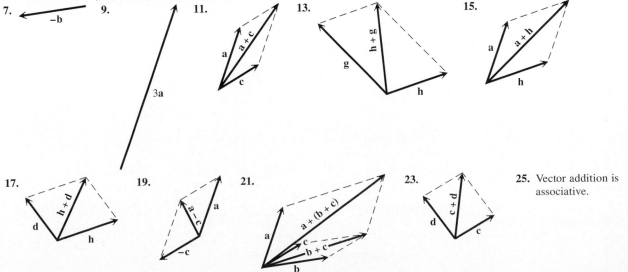

25. Vector addition is associative.

27.    **29.**   **31.**   **33.** 9.5, 7.4

35. 17, 20   **37.** 13.7, 7.11   **39.** 198, 132   **41. a** and **b** have the same direction.   **43.** The angle between **a** and **b** must be less than 90°.   **45.** 530 newtons   **47.** 27.2 lb   **49.** 88.2 lb   **51.** $|\mathbf{u}| = 13$, $\theta = 67.4°$ **53.** $|\mathbf{u}| = 5$, $\theta = 126.9°$ **55.** $\langle 3\sqrt{3}, 3 \rangle$ **57.** $\langle -2, 2\sqrt{3} \rangle$ **65.** -22 **67.** -50 **71.** 151°

CONNECTIONS (page 316)   **1.** $y - b\sin\theta = (-\cot\theta)(x - b\cos\theta)$ (Other answers are possible.)

2. $(a, -a\cot\theta + b\cos\theta\cot\theta + b\sin\theta)$ **3.** $|\mathbf{v}| = \dfrac{\sqrt{a^2 + b^2 - 2ab\cos\theta}}{\sin\theta}$ **4.** the line joining the endpoints of **a** and **b**

5. From the right triangle with base a and hypotenuse $|\mathbf{v}|$, $\cos\alpha = a/|\mathbf{v}|$.

7.5 EXERCISES (page 316)   **1.** 93.9°   **3.** 18°   **5.** 2.4 tons   **7.** 2640 lb at an angle of 167.2° with the 1480-lb force   **9.** weight: 64.8 lb;   tension: 61.9 lb   **11.** 190 lb, 283 lb, respectively   **13.** 173.1°   **15.** 39.2 km   **17.** 237°;   470 mph   **19.** 358°;   170 mph   **21.** 230 km per hr;   167°   **23.** 3:21 P.M.   **25. (a)** ≈ 56 mi/sec   **(b)** ≈ 87

CHAPTER 7 REVIEW EXERCISES (page 320)   **1.** 63.7 m   **3.** 41.7°   **5.** 54° 20′ or 125° 40′   **9. (a)** $b = 5$, $b \geq 10$ **(b)** $5 < b < 10$ **(c)** $b < 5$ **11.** 19.87° or 19°52′   **13.** 55.5 m   **15.** 148 cm   **17.** $B = 17.3°$, $C = 137.5°$, $c = 11.0$ yd   **19.** $c = 18.7$ cm, $A = 91° 40′$, $B = 45° 50′$ **21.** 153,600 m²   **23.** .234 km²   **25.** Each expression is equal to $\dfrac{1 + \sqrt{3}}{2}$.   **27.** He needs about 2.5 cans, so he must buy 3 cans.   **29.** 13 m   **31.** 10.8 mi   **33.** 115 km   **35.** 5500 m   **37.** 438.14 ft

39. 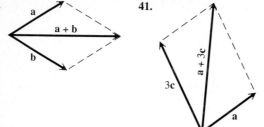   **41.**   **43.** false   **45.** 17.9, 66.8   **47.** 28 lb   **49.** 135 newtons

51. 29, 316.4°   **53.** $\langle 3, 3\sqrt{3} \rangle$ **55.** 280 newtons, 30.4°   **57.** 3° 50′   **59.** speed: 21 km per hr;   bearing: 118°   **61.** $AB = 1978.28$ ft; $BC = 975.05$ ft

CHAPTER 8   Complex Numbers and Polar Equations ▼▼▼

8.1 EXERCISES (page 331)   **3.** $5i, -5i$ **5.** $2i$ **7.** $\dfrac{5}{3}i$ **9.** $5i\sqrt{6}$ **11.** $4i\sqrt{5}$ **13.** -3 **15.** $-\sqrt{30}$ **17.** $\dfrac{\sqrt{6}}{2}$

19. $\dfrac{\sqrt{3}}{3}i$ **21.** $4i, -4i$ **23.** $2i\sqrt{3}, -2i\sqrt{3}$ **25.** $-\dfrac{2}{3} + \dfrac{\sqrt{2}}{3}i, -\dfrac{2}{3} - \dfrac{\sqrt{2}}{3}i$ **27.** $3 + i\sqrt{5}, 3 - i\sqrt{5}$ **29.** $\dfrac{1}{2} + \dfrac{\sqrt{6}}{2}i, \dfrac{1}{2} - \dfrac{\sqrt{6}}{2}i$

31. $-\dfrac{1}{2} + \dfrac{\sqrt{3}}{2}i, -\dfrac{1}{2} - \dfrac{\sqrt{3}}{2}i$ **37.** $5 - 3i$ **39.** $-5 + 2i$ **41.** $-4 + i$ **43.** $8 - i$ **45.** $-14 + 2i$ **47.** 5   **49.** $-5i$

51. $\dfrac{7}{25} - \dfrac{24}{25}i$ **53.** $\dfrac{13}{20} - \dfrac{1}{20}i$ **57.** 1   **59.** -1 **61.** i **63.** -1 **65.** 0   **67.** $x = 2, y = -3$ **69.** $x = \dfrac{1}{2}, y = 15$

71. $x = 14, y = 8$ **79. (a)** $110 + 32i$ **(b)** ≈ 16.22°   **81.** $E = 50 + 98i$ **83.** $I = \dfrac{215}{26} + \dfrac{95}{26}i$

CONNECTIONS (page 334)   **1.** $10 + i$;

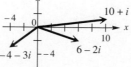

8.2 EXERCISES (page 339)

1. magnitude (length)

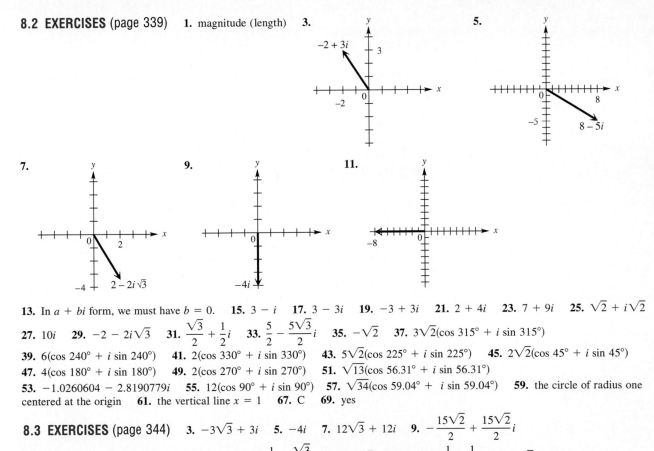

3. $-2 + 3i$

5. $8 - 5i$

7. $2 - 2i\sqrt{3}$

9. $-4i$

11. -8

13. In $a + bi$ form, we must have $b = 0$.   **15.** $3 - i$ **17.** $3 - 3i$ **19.** $-3 + 3i$ **21.** $2 + 4i$ **23.** $7 + 9i$ **25.** $\sqrt{2} + i\sqrt{2}$

27. $10i$ **29.** $-2 - 2i\sqrt{3}$ **31.** $\dfrac{\sqrt{3}}{2} + \dfrac{1}{2}i$ **33.** $\dfrac{5}{2} - \dfrac{5\sqrt{3}}{2}i$ **35.** $-\sqrt{2}$ **37.** $3\sqrt{2}(\cos 315° + i \sin 315°)$

39. $6(\cos 240° + i \sin 240°)$ **41.** $2(\cos 330° + i \sin 330°)$ **43.** $5\sqrt{2}(\cos 225° + i \sin 225°)$ **45.** $2\sqrt{2}(\cos 45° + i \sin 45°)$

47. $4(\cos 180° + i \sin 180°)$ **49.** $2(\cos 270° + i \sin 270°)$ **51.** $\sqrt{13}(\cos 56.31° + i \sin 56.31°)$

53. $-1.0260604 - 2.8190779i$ **55.** $12(\cos 90° + i \sin 90°)$ **57.** $\sqrt{34}(\cos 59.04° + i \sin 59.04°)$ **59.** the circle of radius one

centered at the origin   **61.** the vertical line $x = 1$ **67.** C   **69.** yes

8.3 EXERCISES (page 344)

3. $-3\sqrt{3} + 3i$ **5.** $-4i$ **7.** $12\sqrt{3} + 12i$ **9.** $-\dfrac{15\sqrt{2}}{2} + \dfrac{15\sqrt{2}}{2}i$

11. $-3i$ **13.** $\sqrt{3} - i$ **15.** $-1 - i\sqrt{3}$ **17.** $-\dfrac{1}{6} - \dfrac{\sqrt{3}}{6}i$ **19.** $2\sqrt{3} - 2i$ **21.** $-\dfrac{1}{2} - \dfrac{1}{2}i$ **23.** $\sqrt{3} + i$

25. $.65366807 + 7.4714602i$ **27.** $30.858023 + 18.541371i$ **29.** $.20905693 + 1.9890438i$ **31.** $-3.7587705 - 1.3680806i$

33. 2   **34.** $w = \sqrt{2} \text{ cis } 135°; \quad z = \sqrt{2} \text{ cis } 225°$ **35.** $2 \text{ cis } 0°$ **36.** 2; it is the same   **37.** $-i$ **38.** $\text{cis}(-90°)$

39. $-i$; it is the same   **43.** $1.2 - .14i$ **45.** **(a)** $\approx 27.43 + 11.5i$ **(b)** $\approx 22.75°$

8.4 EXERCISES (page 350)

1. $27i$ **3.** 1   **5.** $\dfrac{27}{2} - \dfrac{27\sqrt{3}}{2}i$ **7.** $-16\sqrt{3} + 16i$ **9.** $-128 + 128i\sqrt{3}$

11. $128 + 128i$ **13.** $(\cos 0° + i \sin 0°)$,
$(\cos 120° + i \sin 120°)$,
$(\cos 240° + i \sin 240°)$

15. $2 \text{ cis } 20°$, $2 \text{ cis } 140°$,
$2 \text{ cis } 260°$

17. $2(\cos 90° + i \sin 90°)$,
$2(\cos 210° + i \sin 210°)$,
$2(\cos 330° + i \sin 330°)$

19. 4(cos 60° + i sin 60°),
4(cos 180° + i sin 180°),
4(cos 300° + i sin 300°)

21. $\sqrt[3]{2}$(cos 20° + i sin 20°),
$\sqrt[3]{2}$(cos 140° + i sin 140°),
$\sqrt[3]{2}$(cos 260° + i sin 260°)

23. $\sqrt[3]{4}$(cos 50° + i sin 50°),
$\sqrt[3]{4}$(cos 170° + i sin 170°),
$\sqrt[3]{4}$(cos 290° + i sin 290°)

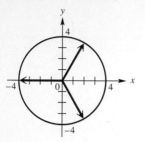
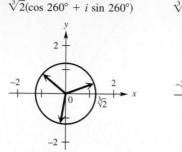
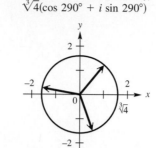

25. (cos 0° + i sin 0°), (cos 180° + i sin 180°)

27. (cos 0° + i sin 0°), (cos 60° + i sin 60°),
(cos 120° + i sin 120°), (cos 180° + i sin 180°),
(cos 240° + i sin 240°), (cos 300° + i sin 300°)

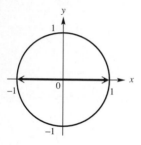
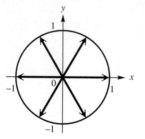

29. (cos 45° + i sin 45°), (cos 225° + i sin 225°)   **33.** false

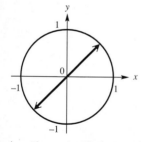

39. (cos 0° + i sin 0°), (cos 120° + i sin 120°), (cos 240° + i sin 240°)
41. (cos 90° + i sin 90°), (cos 210° + i sin 210°), (cos 330° + i sin 330°)
43. 2(cos 0° + i sin 0°), 2(cos 120° + i sin 120°), 2(cos 240° + i sin 240°)
45. (cos 45° + i sin 45°), (cos 135° + i sin 135°), (cos 225° + i sin 225°), (cos 315° + i sin 315°)
47. (cos 22.5° + i sin 22.5°), (cos 112.5° + i sin 112.5°), (cos 202.5° + i sin 202.5°), (cos 292.5° + i sin 292.5°)
49. 2(cos 20° + i sin 20°), 2(cos 140° + i sin 140°), 2(cos 260° + i sin 260°)
51. 1.3606 + 1.2637i, −1.7747 + .5464i, .4141 − 1.8102i
53. 1, $-\dfrac{1}{2} + \dfrac{\sqrt{3}}{2}i$, $-\dfrac{1}{2} - \dfrac{\sqrt{3}}{2}i$ **55.** −4, 2 − 2$i\sqrt{3}$ **56.** cos 2θ + i sin 2θ   **57.** (cos² θ − sin² θ) + i(2 cos θ sin θ) =
cos 2θ + i sin 2θ   **58.** cos 2θ = cos² θ − sin² θ;   sin 2θ = 2 cos θ sin θ   **59.** 1, .30901699 + .95105652i,
−.809017 + .58778525i, −.809017 − .5877853i, .30901699 − .9510565i **61. (a)** yes **(b)** no **(c)** yes

8.5 EXERCISES (page 359)   Answers may vary in Exercises 3–9.

3. $(1, 405°), (-1, 225°)$ **5.** $(-2, 495°), (2, 315°)$
7. $(5, 300°), (-5, 120°)$
9. $(-3, 150°), (3, -30°)$
11. $(3, 660°), (-3, 120°)$ **13.** quadrantal

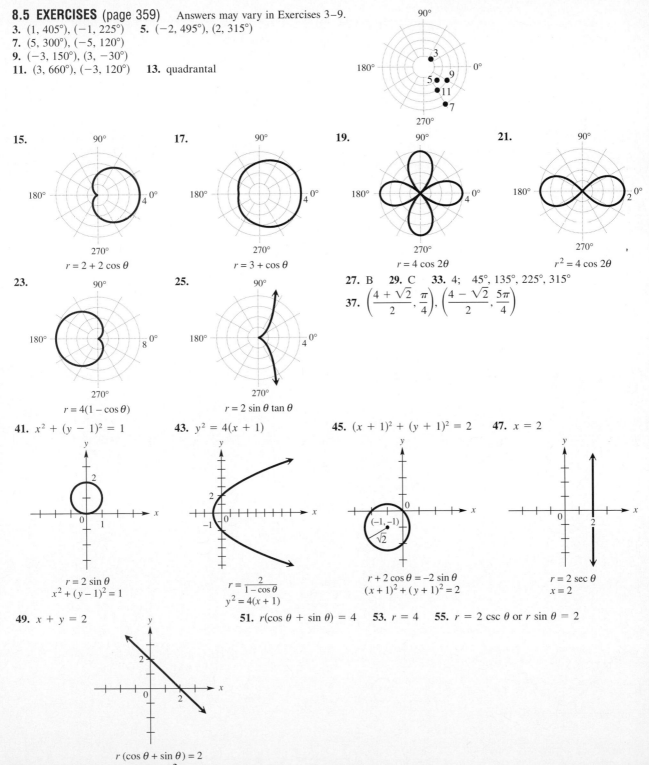

15.
$r = 2 + 2\cos\theta$

17.
$r = 3 + \cos\theta$

19.
$r = 4\cos 2\theta$

21.
$r^2 = 4\cos 2\theta$

23.
$r = 4(1 - \cos\theta)$

25.
$r = 2\sin\theta\tan\theta$

27. B   **29.** C   **33.** 4; 45°, 135°, 225°, 315°

37. $\left(\dfrac{4 + \sqrt{2}}{2}, \dfrac{\pi}{4}\right), \left(\dfrac{4 - \sqrt{2}}{2}, \dfrac{5\pi}{4}\right)$

41. $x^2 + (y - 1)^2 = 1$

$r = 2\sin\theta$
$x^2 + (y - 1)^2 = 1$

43. $y^2 = 4(x + 1)$

$r = \dfrac{2}{1 - \cos\theta}$
$y^2 = 4(x + 1)$

45. $(x + 1)^2 + (y + 1)^2 = 2$

$r + 2\cos\theta = -2\sin\theta$
$(x + 1)^2 + (y + 1)^2 = 2$

47. $x = 2$

$r = 2\sec\theta$
$x = 2$

49. $x + y = 2$

$r(\cos\theta + \sin\theta) = 2$
$x + y = 2$

51. $r(\cos\theta + \sin\theta) = 4$ **53.** $r = 4$ **55.** $r = 2\csc\theta$ or $r\sin\theta = 2$

57.

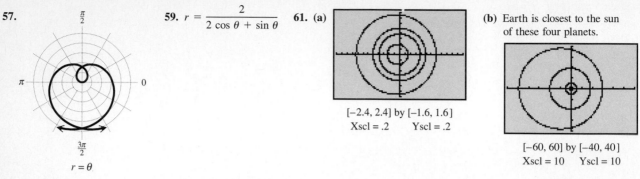

59. $r = \dfrac{2}{2\cos\theta + \sin\theta}$ **61. (a)**

$[-2.4, 2.4]$ by $[-1.6, 1.6]$
Xscl $= .2$ Yscl $= .2$

$r = \theta$

(b) Earth is closest to the sun of these four planets.

$[-60, 60]$ by $[-40, 40]$
Xscl $= 10$ Yscl $= 10$

(c) not always

8.6 EXERCISES (page 366)   **1.** $(-2, 1)$; $(4, 3)$; $(1.6, 2.2)$ **3.** the second set of equations
5. $y = (1/2)x + 1$, for x in $[-4, 6]$ **7.** $y = 3x^2 - 4$, for x in $[0, 2]$ **9.** $y = x - 2$, for x in $(-\infty, \infty)$

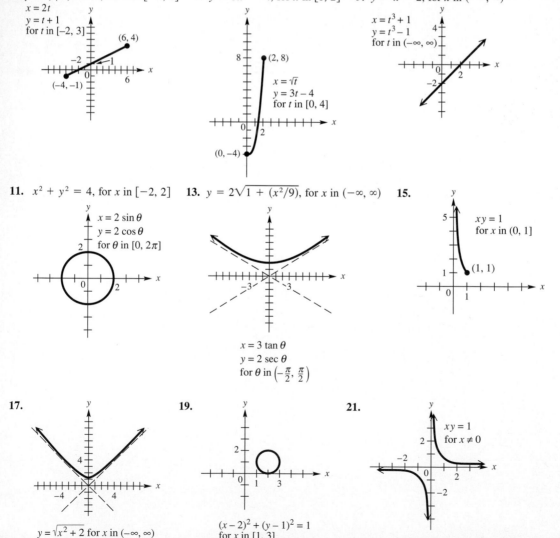

11. $x^2 + y^2 = 4$, for x in $[-2, 2]$ **13.** $y = 2\sqrt{1 + (x^2/9)}$, for x in $(-\infty, \infty)$ **15.**

23. (a)   **(b)**   **25.**   **27. (a)** 17.7 sec
                                             (b) 5000 ft
                                             (c) 1250 ft

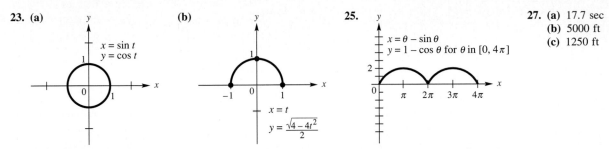

31. Many answers are possible, two of which are $x = t$, $y = m(t - x_1) + y_1$ and $x = t^2$, $y = m(t^2 - x_1) + y_1$.

33. Many answers are possible; for example, $x = a \sec \theta$, $y = b \tan \theta$ and $x = t$, $y^2 = \dfrac{b^2}{a^2}(t^2 - a^2)$.

CHAPTER 8 REVIEW EXERCISES (page 369)   **1.** $3i$ **3.** $-9i, 9i$ **5.** $-2 - 3i$ **7.** $5 + 4i$ **9.** $29 + 37i$

11. $-32 + 24i$ **13.** $-2 - 2i$ **15.** $\dfrac{8}{5} + \dfrac{6}{5}i$ **17.** $-\dfrac{3}{26} + \dfrac{11}{26}i$ **19.** i **21.** -1 **23.** $-30i$ **25.** $-\dfrac{1}{8} + \dfrac{\sqrt{3}}{8}i$ **27.** $8i$

29. $-\dfrac{1}{2} - \dfrac{\sqrt{3}}{2}i$ **31.** x **33.**   **35.**   **37.** $5 + 4i$

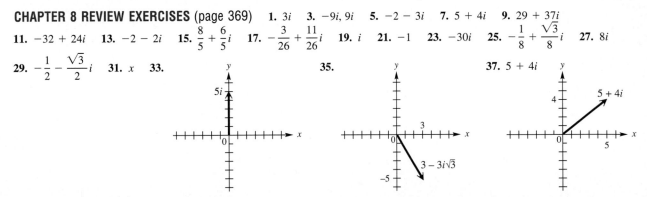

39. $2\sqrt{2}(\cos 135° + i \sin 135°)$ **41.** $-\sqrt{2} - i\sqrt{2}$ **43.** $\sqrt{2}(\cos 315° + i \sin 315°)$ **45.** $4(\cos 270° + i \sin 270°)$
47. a circle of radius 2 with the origin as center   **49.** The vector (a, b) lies on the y-axis.   **51.** $\sqrt[10]{8}(\cos 27° + i \sin 27°)$,
$\sqrt[10]{8}(\cos 99° + i \sin 99°)$, $\sqrt[10]{8}(\cos 171° + i \sin 171°)$, $\sqrt[10]{8}(\cos 243° + i \sin 243°)$, $\sqrt[10]{8}(\cos 315° + i \sin 315°)$ **53.** one
55. $5(\cos 60° + i \sin 60°)$, $5(\cos 180° + i \sin 180°)$, $5(\cos 300° + i \sin 300°)$ **57.** $(\cos 135° + i \sin 135°)$, $(\cos 315° + i \sin 315°)$

59.   **61.**   **63.** $y^2 = -6\left(x - \dfrac{3}{2}\right)$ or $y^2 + 6x - 9 = 0$

65. $\left(x - \dfrac{1}{2}\right)^2 + \left(y - \dfrac{1}{2}\right)^2 = \dfrac{1}{2}$ or $x^2 + y^2 - x - y = 0$

67. $\sin \theta = \cos \theta$ or $\tan \theta = 1$ **69.** $r = \dfrac{\cos \theta}{\sin^2 \theta}$ or $r = \cos \theta \csc^2 \theta$

$r = -1 + \cos \theta$ $r = 2 \sin 4\theta$

71. $r = 2 \sec \theta$ **73.**   **75.** $x - 3y = 5$, for x in $[-13, 17]$ **77.** $y = \sqrt{x^2 + 1}$, for x in $[0, \infty)$
79. $y = 3\sqrt{1 + (x^2/25)}$, for x in $(-\infty, \infty)$

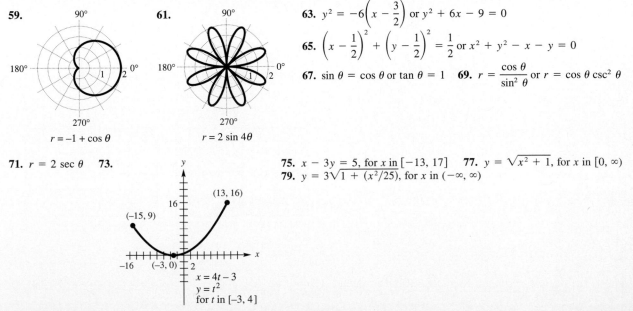

CHAPTER 9   Exponential and Logarithmic Functions ▼▼▼

CONNECTIONS (page 377)   **1.** One possibility is $h(x) = 2^x$ and $g(x) = x + 3$.   **2.** $h(g(x)) = 2^{-x^2}$ **3.** One possibility is $f(x) = kx^{nt}$, $g(x) = 1 + x$.

CONNECTIONS (page 383)   **1.** 2.717   **2.** .9512   **3.** $\dfrac{x^6}{6 \cdot 5 \cdot 4 \cdot 3 \cdot 2 \cdot 1}$

9.1 EXERCISES (page 385)   **1.** true   **3.** false   **5.** true   **7. (a)**

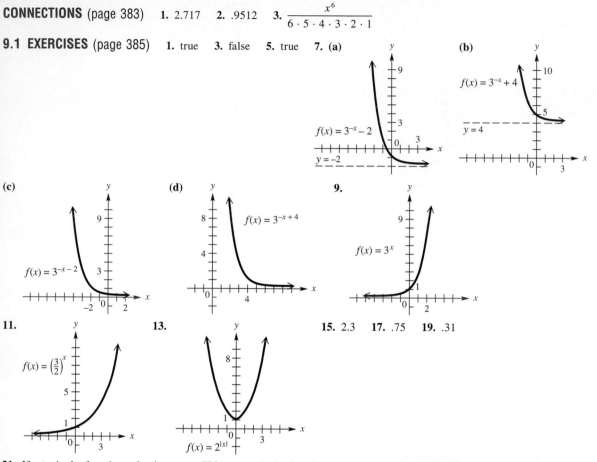

15. 2.3   **17.** .75   **19.** .31

21. If $a > 1$, the function value increases. If $0 < a < 1$, the function value decreases.   **25.** $f(x) = 2^x$ **27.** $f(t) = 27 \cdot 9^t$

31. $\dfrac{1}{3}$ **33.** -2 **35.** $\dfrac{1}{2}$ **37.** $-3, 3$ **39.** 4   **41.** $\dfrac{4}{9}$ **43.** $-\dfrac{3}{5}$ **45.**

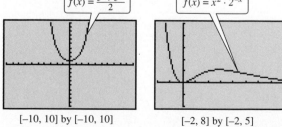

$$f(x) = \frac{e^x + e^{-x}}{2}$$

$[-10, 10]$ by $[-10, 10]$

47. $f(x) = x^2 \cdot 2^{-x}$

$[-2, 8]$ by $[-2, 5]$

49. $76,855.95   **51.** $41,845.63   **53.** 8.0%   **55. (a)** about 63,000   **(b)** about 42,000   **(c)** about 21,000
57. (a) 440 g   **(b)** 387 g   **(c)** 264 g   **(d)**   **59. (a)** linear   **(b)** $T(R) = 1.03R$ **(c)** 5.15°F

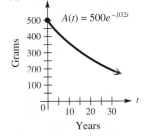

61. no solution   **63.** 0, .73   **65.** $f(x)$ approaches the line $y = 2.71828$.

9.2 EXERCISES (page 396)   **1.** $x = a^y$ **3.** 3; 5; 125   **5.** $\log_3 81 = 4$ **7.** $\log_{2/3}\left(\dfrac{27}{8}\right) = -3$ **9.** $6^2 = 36$

11. $(\sqrt{3})^8 = 81$ **15.** 2   **17.** -3 **19.** $-\dfrac{1}{6}$ **21.** 9   **23.** 5   **25.** $\dfrac{1}{5}$

29. (a)   **(b)**   **(c)**   **31.**

33.   **35.**   **37.** They are not the same because the domains are different.

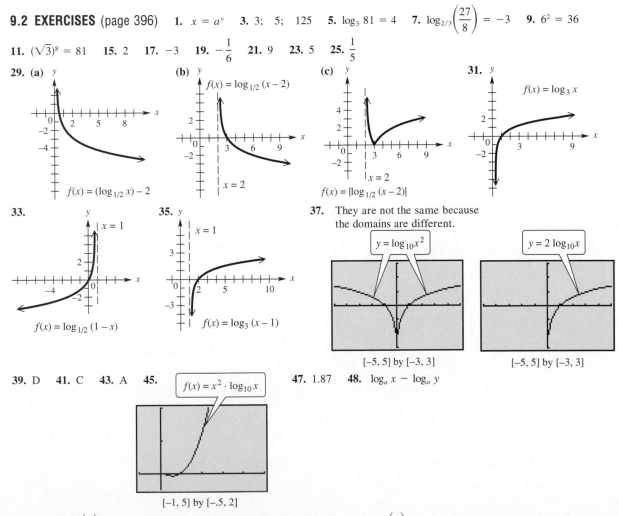

39. D   **41.** C   **43.** A   **45.**   **47.** 1.87   **48.** $\log_a x - \log_a y$

49. Since $\log_2\left(\dfrac{x}{4}\right) = \log_2 x - \log_2 4$ by the quotient rule, the graph of $y = \log_2\left(\dfrac{x}{4}\right)$ can be obtained by shifting the graph of $y = \log_2 x$ down $\log_2 4 = 2$ units.

50.

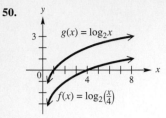

51. 0; 2; 2; 0; By the quotient rule, $\log_2\left(\dfrac{x}{4}\right) = \log_2 x - \log_2 4$. Both sides should equal 0. Since $2 - 2 = 0$, they do.

53. $\log_3 4 + \log_3 p - \log_3 q$ **55.** $1 + \left(\dfrac{1}{2}\right)\log_2 3 - \log_2 5$ **57.** cannot be simplified **59.** $\left(\dfrac{1}{3}\right)(5 \log_p m + 4 \log_p n - 2 \log_p t)$

61. $\log_b\left(\dfrac{k}{ma}\right)$ **63.** $\log_y(p^{-7/6})$ **65.** $\log_b\left(\dfrac{2y + 5}{\sqrt{y + 3}}\right)$ **67.** 1.0791 **69.** $-.1303$ **71.** 6

9.3 EXERCISES (page 405) **1.** 1.6335 **3.** -1.8539 **5.** 6.3630 **7.** $-.3567$ **9.** $\log 8 \approx .90308999$ **11.** $\log_3 4$
12. 2 **13.** 3 **14.** It lies between 2 and 3. Because the function defined by $y = \log_3 x$ is increasing and $9 < 16 < 27$, we have
$\log_3 9 < \log_3 16 < \log_3 27$. **15.** By the change-of-base-theorem, $\log_3 16 = \dfrac{\log 16}{\log 3} = \dfrac{\ln 16}{\ln 3} \approx 2.523719014$. **16.** -1; 0

17. It lies between -1 and 0. $\dfrac{1}{5} = .2 < .68 < 1$, so $\log_5 .2 < \log_5 .68 < \log_5 1$; $\log_5 .68 = \dfrac{\log .68}{\log 5} = \dfrac{\ln .68}{\ln 5} \approx -.2396255723$
19. 1.13 **21.** -1.58 **23.** .97 **25.** 1.45 **27.** The function is undefined for $X \geq 4$ because the domain of $Y = \log_a X$ is
$X > 0$. This means $4 - X > 0$ here, so $X < 4$ is the domain. **29.** The vertical line simulates an asymptote at $x = 1$. The base

$f(x) = \log_x 5$

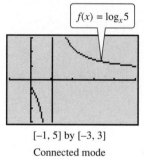

$[-1, 5]$ by $[-3, 3]$
Connected mode

must be greater than 0 and not equal to 1. **31.** 3.2 **33.** 1.8 **35.** 2.0×10^{-3} **37.** 1.6×10^{-5} **39. (a)** 20 **(b)** 30
(c) 50 **(d)** 60 **41. (a)** 3 **(b)** 6 **(c)** 8 **43. (a)** about $200,000,000I_0$ **(b)** about $13,000,000I_0$ **(c)** The 1906 earthquake had
a magnitude more than 15 times greater than the 1989 earthquake. **45. (a)** 2 **(b)** 2 **(c)** 2 **(d)** 1 **47.** 1 **49. (a)** 3
(b) $5^2 = 25$ **(c)** $1/e$ **51. (a)** 5 **(b)** $\ln 3$ **(c)** $2 \ln 3$ or $\ln 9$ **53.** about 66 million; We must assume that the rate of
increase continues to be logarithmic. **55.** between $7°F$ and $11°F$

57. (a) Let $x = \ln D$ and $y = \ln P$ for each planet. From the graph, the data appear to be linear.

Planet	ln D	ln P
Mercury	−.94	−1.43
Venus	−.33	−.48
Earth	0	0
Mars	.42	.64
Jupiter	1.65	2.48
Saturn	2.26	3.38
Uranus	2.95	4.43
Neptune	3.40	5.10

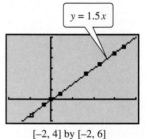

[−2, 4] by [−2, 6]

(b) The points (0, 0) and (3.40, 5.10) determine the line $y = 1.5x$ or $\ln P = 1.5 \ln D$. (Answers will vary.) **(c)** $P \approx 248.3$ years

$y = 1.5x$

[−2, 4] by [−2, 6]

CONNECTIONS (page 411)   **2.** 0 is extraneous because $e^0 = 1$, so the denominator of the original equation becomes 0.
3. First, solve for e^x;   second, solve for x.

9.4 EXERCISES (page 415)   **5.** 1.631   **7.** −.080   **9.** 2.386   **11.** −.123   **13.** no solution   **15.** 2   **17.** 17.475
19. 11   **21.** 10   **23.** 4   **25.** 11   **27.** no solution   **29.** 8   **31.** −2, 2   **33.** 1, 10   **37.** $t = -\dfrac{2}{R} \ln\left(1 - \dfrac{RI}{E}\right)$
39. $x = e^{k/(p-a)}$ **42.** $(e^x - 1)(e^x - 3) = 0$ **43.** 0, ln 3   **44.** The graph intersects the x-axis at 0 and $1.099 \approx \ln 3$.

$y = e^{2x} - 4e^x + 3$

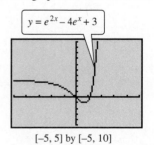

[−5, 5] by [−5, 10]

45. $(-\infty, 0) \cup (\ln 3, \infty)$ **46.** (0, ln 3)   **47.** 1.52   **49.** 0   **51.** 2.45, 5.66   **53.** $f^{-1}(x) = \dfrac{1}{3}(\ln(x) - 1)$; $(0, \infty)$; $(-\infty, \infty)$
55. $(27, \infty)$ **57.** 89 decibels is about twice as loud as 86 decibels, for a 100% increase.   **59.** 1.25 yr   **61.** 4.27%   **63.** 1999

65. (a) $P(T) = 1 - e^{-.0034 - .0053T}$ **(b)**

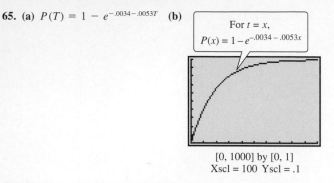

For $t = x$,
$P(x) = 1 - e^{-.0034 - .0053x}$

[0, 1000] by [0, 1]
Xscl = 100 Yscl = .1

(c) $P(60) \approx .275$ or 27.5%. The reduction in carbon emissions from a tax of $60 per ton of carbon is 27.5%.
(d) $T = \$130.14$

CHAPTER 9 REVIEW EXERCISES (page 420)

1. increasing **3.** B **5.** C **7.** $\log_2 32 = 5$ **9.** $\log_{1/16}\left(\frac{1}{2}\right) = \frac{1}{4}$

11. $\log_{10} 3 = .4771$ or $\log 3 = .4771$ **15.** $(-\infty, \infty)$ **17.** 1 **19.** $f^{-1}(x) = \log_a x$ **21.** $2^{5/2} = \sqrt{32}$ **23.** $e^{3.806662} = 45$

27. 2 **29.** $\left(\frac{1}{2}\right)\log_2 7 - \log_2 15$ **31.** cannot be simplified **33.** 1.659 **35.** 6.153 **37.** 6.049 **39. (a)** B **(b)** D **(c)** C

(d) A **41.** by a factor of $2^5 = 32$ **43.** $\frac{3}{2}$ **45.** 1.303 **47.** 2 **49.** $\frac{1}{2}$ **51.** 9.6% **53.** \$93,761.31 **55. (a)** 2.2×10^{12}

tons **(b)** 1987 **57.** $(-.7666647, .58777476)$ **59.** The x-intercept is .5, supporting the solution found in Exercise 58.

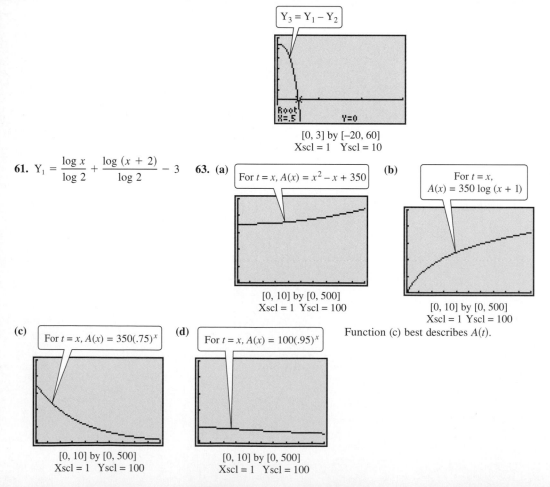

$Y_3 = Y_1 - Y_2$

Root
X=.5   Y=0

[0, 3] by [−20, 60]
Xscl = 1   Yscl = 10

61. $Y_1 = \dfrac{\log x}{\log 2} + \dfrac{\log (x + 2)}{\log 2} - 3$ **63. (a)**

For $t = x$, $A(x) = x^2 - x + 350$

[0, 10] by [0, 500]
Xscl = 1 Yscl = 100

(b)

For $t = x$,
$A(x) = 350 \log (x + 1)$

[0, 10] by [0, 500]
Xscl = 1 Yscl = 100

Function (c) best describes $A(t)$.

(c)

For $t = x$, $A(x) = 350(.75)^x$

[0, 10] by [0, 500]
Xscl = 1   Yscl = 100

(d)

For $t = x$, $A(x) = 100(.95)^x$

[0, 10] by [0, 500]
Xscl = 1   Yscl = 100

Index